Nonlinear Systems and Complexity

Series Editor
Albert C. J. Luo
Southern Illinois University
Edwardsville, IL, USA

For further volumes:
http://www.springer.com/series/11433

J. A. Tenreiro Machado • Dumitru Baleanu
Albert C. J. Luo
Editors

Discontinuity and Complexity in Nonlinear Physical Systems

Springer

Editors
J. A. Tenreiro Machado
Department of Electrical Engineering
Polytechnic of Porto
Institute of Engineering
Porto, Portugal

Albert C. J. Luo
Department of Mechanical Engineering
Southern Illinois University Edwardsville
Edwardsville, IL, USA

Dumitru Baleanu
Faculty of Art and Sciences
Mathematics and Computer Sciences
Cankaya University
Ankara, Turkey

Institute of Space Sciences
Magurele-Bucharest, Romania

ISSN 2195-9994 ISSN 2196-0003 (electronic)
ISBN 978-3-319-37534-2 ISBN 978-3-319-01411-1 (eBook)
DOI 10.1007/978-3-319-01411-1
Springer Cham Heidelberg New York Dordrecht London

Printed on acid-free paper

Springer is part of Springer Science+Business Media (www.springer.com)

Preface

This edited book is selected from *International Conference on Nonlinear Science and Complexity*, held at Budapest, Hungary, during August 6–11, 2012. The aims of this edited book are to present the new results in the fundamental and frontier theories and techniques in science and technology, and to stimulate more research interest in the community of nonlinear science and complexity. This is the fourth of a series of events held during the last years reflecting the progress in this challenging area. The first conference on *Nonlinear Science and Complexity* was held in 2006 at Beijing, China. The second conference was held in 2008 at Porto, Portugal. The third conference was held in 2010 at Ankara, Turkey. The edited book included 24 chapters selected and extended from 60 accepted papers in NSC 2012 after peer-review. Presented are the following four issues:

1. Fractional dynamics and nonlinearity
2. Chaos and complexity
3. Discontinuous dynamics
4. Engineering and financial nonlinearity

In the first topic, eight chapters present Lie group analysis, fractional dynamical systems and control. The second topic includes six papers on stability, bifurcation, and chaos in nonlinear dynamics. Discontinuous dynamics constitutes the third topic and includes four chapters presenting impact vibro-dynamical systems and chaos in piecewise linear systems. The fourth topic presents six chapters in engineering and financial nonlinearity.

Herein, editors would like to thank authors and reviewers to support the projects. The results presented in this edited book will constitute an important contribution for the progress in scientific arena of nonlinear science and complexity.

Porto, Portugal — J. A. Tenreiro Machado
Ankara, Turkey — Dumitru Baleanu
Edwardsville, IL, USA — Albert C. J. Luo

Contents

Part I Fractional Dynamics and Nonlinearity

1 **Nonlinear Self-Adjointness for some Generalized KdV Equations** ... 3
M.L. Gandarias and M. Rosa

2 **Weak Self-Adjointness and Conservation Laws for a Family of Benjamin-Bona-Mahony-Burgers Equations** ... 23
M.S. Bruzón

3 **Some Analytical Techniques in Fractional Calculus: Realities and Challenges** ... 35
Dumitru Baleanu, Guo-Cheng Wu, and Jun-Sheng Duan

4 **Application of the Local Fractional Fourier Series to Fractal Signals** ... 63
Xiao-Jun Yang, Dumitru Baleanu, and J. A. Tenreiro Machado

5 **Parameter Optimization of Fractional Order $PI^{\lambda}D^{\mu}$ Controller Using Response Surface Methodology** ... 91
Beyza Billur İskender, Necati Özdemir, and Aslan Deniz Karaoglan

6 **Dynamical Response of a Van der Pol System with an External Harmonic Excitation and Fractional Derivative** ... 107
Arkadiusz Syta and Grzegorz Litak

7 **Fractional Calculus: From Simple Control Solutions to Complex Implementation Issues** ... 113
Cristina I. Muresan

8 **Emerging Tools for Quantifying Unconscious Analgesia: Fractional-Order Impedance Models** ... 135
Amélie Chevalier, Dana Copot, Clara M. Ionescu, J. A. Tenreiro Machado, and Robin De Keyser

Part II Chaos and Complexity

9 1D Cahn–Hilliard Dynamics: Coarsening and Interrupted Coarsening 153
Simon Villain-Guillot

10 Nonlinear Analysis of Phase-locked Loop-Based Circuits 169
R.E. Best, N.V. Kuznetsov, G.A. Leonov, M.V. Yuldashev, and R.V. Yuldashev

11 Approaches to Defining and Measuring Assembly Supply Chain Complexity 193
V. Modrak and D. Marton

12 Non-commutative Tomography: Applications to Data Analysis 215
Françoise Briolle and Xavier Leoncini

13 Projective Synchronization of Two Gyroscope Systems with Different Motions 255
Fuhong Min and Albert C. J. Luo

14 Measuring and Analysing Nonlinearities in the Lung Tissue 273
Clara M. Ionescu

Part III Discontinuous Dynamics

15 Drilling Systems: Stability and Hidden Oscillations 287
M.A. Kiseleva, N.V. Kuznetsov, G.A. Leonov, and P. Neittaanmäki

16 Chaos in a Piecewise Linear System with Periodic Oscillations 305
Chunqing Lu

17 Basins of Attraction in a Simple Harvesting System with a Stopper 315
Marek Borowiec, Grzegorz Litak, and Stefano Lenci

18 Analytical Dynamics of a Mass –Damper –Spring Constrained System 323
Albert C. J. Luo and Richard George

Part IV Engineering and Financial Nonlinearity

19 Formations of Transitional Zones in Shock Wave with Saddle-Node Bifurcations 347
Jia-Zhong Zhang, Yan Liu, Pei-Hua Feng, and Jia-Hui Chen

20 Dynamics of Composite Milling: Application of Recurrence Plots to Huang Experimental Modes 359
G. Litak, R. Rusinek, K. Kecik, A. Rysak, and A. Syta

21 The Dynamics of Shear-Type Frames Equipped with Chain-Based Nonlinear Braces ... 369
Enrico Babilio

22 In-Plane Free Vibration and Stability of High Speed Rotating Annular Disks and Rings ... 389
Hamid R. Hamidzadeh and Ehsan Sarfaraz

23 Patent Licensing: Stackelberg Versus Cournot Models ... 409
Oana Bode and Flávio Ferreira

24 Privatization and Government Preferences in a Mixed Duopoly: Stackelberg Versus Cournot ... 421
Fernanda A. Ferreira and Flávio Ferreira

Index ... 431

Contributors

Enrico Babilio Department of Structures for Engineering and Architecture (DiSt), University of Naples, Naples, Italy

Dumitru Baleanu Department of Mathematics and Computer Sciences, Cankaya University, Ankara, Turkey

Institute of Space Sciences, Magurele-Bucharest, Romania

Department of Chemical and Materials Engineering, King Abdulaziz University, Jeddah, Saudi Arabia

R.E. Best Best Engineering, Oberwil, Switzerland

Oana Bode Babeş-Bolyai University, Cluj-Napoca, Romania

Marek Borowiec Lublin University of Technology, Lublin, Poland

Françoise Briolle Centre de Physique Théorique, Campus de Luminy, CReA, BA 701, 13300 Salon de Provence, France

Aix Marseille Université, CNRS, CPT, UMR 7332, 13288 Marseille, France

M.S. Bruzón Departamento de Matemáticas, Universidad de Cádiz, Cádiz, Spain

Jia-Hui Chen School of Energy and Power Engineering, Xi'an Jiaotong University, Xi'an, P.R. China

Amélie Chevalier Ghent University, Ghent, Belgium

Dana Copot Ghent University, Ghent, Belgium

Robin De Keyser Ghent University, Ghent, Belgium

Jun-Sheng Duan School of Mathematics and Information Sciences, Zhaoqing University, Zhaoqing, Guang Dong, P. R. China

Pei-Hua Feng School of Energy and Power Engineering, Xi'an Jiaotong University, Xi'an, P. R. China

Fernanda A. Ferreira ESEIG - Polytechnic Institute of Porto, Rua D. Sancho I, 981, 4480-876 Vila do Conde, Portugal

Flávio Ferreira ESEIG - Polytechnic Institute of Porto, Rua D. Sancho I, 981, 4480-876 Vila do Conde, Portugal

M.L. Gandarias Departamento de Matemáticas, Universidad de Cádiz, Cádiz, Spain

Richard George Southern Illinois University Edwardsville, Edwardsville, IL, USA

Hamid R. Hamidzadeh Department of Mechanical and Manufacturing Engineering, Tennessee State University, Nashville, TN, USA

Clara M. Ionescu Ghent University, Ghent, Belgium

Beyza Billur Iskender Department of Mathematics, Balikesir University, Balikesir, Turkey

Aslan Deniz Karaoglan Department of Industrial Engineering, Balikesir University, Balikesir, Turkey

K. Kecik Lublin University of Technology, Lublin, Poland

M.A. Kiseleva University of Jyväskylä, Finland, Saint Petersburg State University, Russia

N. V. Kuznetsov University of Jyväskylä, Finland, Saint Petersburg State University, Russia

University of Jyväskylä, Jyväskylä, Finland

Stefano Lenci Department of Civil and Building Engineering, and Architecture, Polytechnic University of Marche, Ancona, Italy

Xavier Leoncini Centre de Physique Théorique, Campus de Luminy, Aix-Marseille Universitisé, Marseille, France

G.A. Leonov Saint Petersburg State University, Saint Petersburg, Russia

University of Jyväskylä, Jyväskylä, Finland

G. Litak Lublin University of Technology, Lublin, Poland

Yan Liu School of Mechanical Engineering, Northwestern Polytechnical University, Xi'an, P.R. China

Chunqing Lu Southern Illinois University Edwardsville, Edwardsville, IL, USA

Albert C. J. Luo Southern Illinois University Edwardsville, Edwardsville, IL, USA

J. A. Tenreiro Machado Department of Electrical Engineering, ISEP-Institute of Engineering, Polytechnic of Porto, Porto, Portugal

D. Marton Department of Manufacturing Management, Technical University of Kosice, Presov, Slovakia

Fuhong Min Nanjing Normal University, Nanjing, Jiangsu, China

V. Modrak Department of Manufacturing Management, Technical University of Kosice, Presov, Slovakia

Cristina I. Muresan Department of Automatic Control, Technical University of Cluj-Napoca, Cluj-Napoca, Romania

P. Neittaanmäki University of Jyväskylä, Finland, Saint Petersburg State University, Russia

Necati Ozdemir Department of Mathematics, Balikesir University, Balikesir, Turkey

M. Rosa Departamento de Matemáticas, Universidad de Cádiz, Cádiz, Spain

R. Rusinek Lublin University of Technology, Lublin, Poland

A. Rysak Lublin University of Technology, Lublin, Poland

Ehsan Sarfaraz Department of Mechanical and Manufacturing Engineering, Tennessee State University, Nashville, TN, USA

Arkadiusz Syta Lublin University of Technology, Lublin, Poland

Simon Villain-Guillot Laboratoire Onde et Matière d'Aquitaine, Université Bordeaux I, Talence Cedex, France

Guo-Cheng Wu College of Mathematics and Information Science, Neijiang Normal University, Neijiang, P.R. China

Xiao-Jun Yang Department of Mathematics and Mechanics, China University of Mining and Technology, Xuzhou, Jiangsu, China

Institute of Software Science, Zhengzhou Normal University, Zhengzhou, China

Institute of Applied mathematics, Qujing Normal University, Qujing, China

M.V. Yuldashev Saint Petersburg State University, Saint Petersburg, Russia

University of Jyväskylä, Jyväskylä, Finland

R.V. Yuldashev Saint Petersburg State University, Saint Petersburg, Russia

University of Jyväskylä, Jyväskylä, Finland

Jia-Zhong Zhang School of Energy and Power Engineering, Xi'an Jiaotong University, Xi'an, P. R. China

Part I
Fractional Dynamics and Nonlinearity

Chapter 1
Nonlinear Self-Adjointness for some Generalized KdV Equations

M.L. Gandarias and M. Rosa

Abstract The new concepts of self-adjoint equations formulated in Gandarias (J Phys A: Math Theor 44:262001, 2011) and Ibragimov (J Phys A: Math Theor 44:432002, 2011) are applied to some classes of third order equations. Then, from Ibragimov's theorem on conservation laws, conservation laws for two generalized equations of KdV type and a potential Burgers equation are established.

Keywords Self-adjointness • Conservation laws • Lie symmetries

1.1 Introduction

The classical KdV equation arises in various physical contexts and it models weakly nonlinear unidirectional long waves. A more complicated equation is obtained if one allows the appearance of higher-order terms. This equation is non-integrable but still admits some special wave solutions [16]. This equation,

$$u_t + ku_x + \alpha uu_x + \beta u_{xxx} + \alpha^2\rho_1 u^2 u_x + \alpha\beta(\rho_2 uu_{xxx} + \rho_3 u_x u_{xx}) = 0 \quad (1.1)$$

which will be referred to as a generalized KdV equation, was studied in [3] by Fokas, who presented a local transformation connecting it with an integrable partial differential equation (PDE). The higher-order wave equations of KdV type model strongly nonlinear long wavelength and short amplitude waves. It is for the reason that the strongly nonlinear character and integrability of these equations attract many researchers to study them. In [19], for some special sets of parameters, the authors derived some analytical expressions for solitary wave solutions and they carried

M.L. Gandarias (✉) • M. Rosa
Departamento de Matemáticas, Universidad de Cádiz, 11510 Puerto Real, Cádiz, Spain
e-mail: mariluz.gandarias@uca.es; maria.rosa@uca.es

J.A.T. Machado et al. (eds.), *Discontinuity and Complexity in Nonlinear Physical Systems*, Nonlinear Systems and Complexity 6, DOI 10.1007/978-3-319-01411-1_1,

out a detailed numerical study of these solutions using a Fourier pseudospectral method combined with a finite-difference scheme. The integral bifurcation method was used in [15] to study (1.1) and some new travelling wave solutions with singular or nonsingular character were obtained for some special sets of parameters. In [16], Marinakis considered as well the third order approximation

$$\begin{aligned} &u_t + ku_x + \alpha uu_x + \beta u_{xxx} + \alpha^2\rho_1 u^2 u_x + \alpha\beta(\rho_2 uu_{xxx} + \rho_3 u_x u_{xx}) \\ &\quad + \alpha^3\rho_4 u^3 u_x + \alpha^2\beta(\rho_5 u^2 u_{xxx} + \rho_6 uu_x u_{xx} + \rho_7 u_x^3) = 0. \end{aligned} \tag{1.2}$$

Equation (1.2) is equivalent to an integrable equation recently studied in [17] and the study in [16] reveals two integrable cases for (1.2). After some changes of variables for particular values of the parameters, (1.2) is transformed into

$$u_t + u^2 u_x + \frac{4}{9}u_x^3 - uu_x u_{xx} + u^2 u_{xxx} = 0 \tag{1.3}$$

Recently Marinakis proved that (1.3) is integrable.
In [6] (see also [5]), a general theorem on conservation laws for arbitrary differential equations which do not require the existence of Lagrangians has been proved. This new theorem is based on the concept of adjoint equations for nonlinear equations. There are many equations with physical significance that are not self-adjoint. Therefore, one cannot eliminate the nonlocal variables from conservation laws of these equations by setting $v = u$. In [7], Ibragimov generalized the concept of self-adjoint equations by introducing the definition of quasi-self-adjoint equations. Recently, some works have been done in this direction to get conservation laws for nonlinear wave equations [9]. In [21], Yasar and Özer have derived conservation laws for one-layer shallow water wave systems and, by using these conserved systems, they have found potential symmetries for the plane flow case. In [11], the authors have proved that the Camassa–Holm equation is self-adjoint and they have constructed conservation laws for the generalized Camassa–Holm equation using its symmetries. In [12], the conservation laws for a $(1 + n)$-dimensional heat equation on curved surfaces have been constructed by using a partial Noether's approach associated with partial Lagrangian [14]. In [18], conservation laws were derived for a nonlocal shallow water wave equation. In [20], by using the nonlocal conservation theorem method [5] and the partial Lagrangian approach [14], conservation laws for the modified KdV equation were presented. It was observed that only the nonlocal conservation theorem method leads to the nontrivial and infinite conservation laws. It happens that many equations having remarkable symmetry properties, such as the forced KdV equation, are neither self-adjoint nor quasi-self-adjoint. In [4], Gandarias has generalized the concept of quasi-self-adjoint equations by introducing the concept of weak self-adjoint equations. Thus, substitution $v = h(u)$ can be replaced with a more general substitution where h involves not only the variable u but also the independent variables $h = h(x, t, u)$. In [8], the concept of quasi-self-adjoint equations has been generalized by introducing the definition of nonlinear self-adjoint equations. Thus, substitution $v = h(u)$ can be replaced by a more general substitution where h involves not only the variable u but also its

derivatives as well as the independent variables $v = h(x, t, u, u_t, u_x, ..)$. This will be a differential substitution. By using these two recent developments in [2], Freire and Sampaio have determined the nonlinear self-adjoint class of a generalized fifth order equation, and by using Ibragimov's theorem [5], the authors have established some local conservation laws. In [13], Johnpillai and Khalique have studied the conservation laws of some special forms of the nonlinear scalar evolution equation, the modified Korteweg-de-Vries (mKdV) equation with time-dependent variable coefficients of damping and dispersion

$$u_t + u^2 u_x + a(t)u + b(t)u_{xxx} = 0.$$

The authors use the new conservation theorem [5] and the partial Lagrangian approach in [14].

In this work we will consider equations (1.2), (1.3) as well as the third order potential Burgers equation

$$u_t = u_{xxx} + 3u_x u_{xx} + u_x^3. \tag{1.4}$$

The aim of this work is to determine the subclasses of equations which are weak and nonlinear self-adjoint. And to determine, by using the Lie generators of equations (1.2), (1.3), and (1.4) and the notation and techniques of [6], some nontrivial conservation laws for equations (1.2), (1.3), and (1.4).

1.2 The Class of Nonlinear Self-Adjoint Equations

Recently, the definitions of adjoint equations and self-adjoint equations have been extended, and the definitions of weak self-adjointness and nonlinear self adjointness have been introduced.

Consider an sth-order partial differential equation

$$F(x, u, u_{(1)}, \dots, u_{(s)}) = 0 \tag{1.5}$$

with independent variables $x = (x^1, \dots, x^n)$ and a dependent variable u, where $u_{(1)} = \{u_i\}$, $u_{(2)} = \{u_{ij}\}, \dots$ denote the sets of the partial derivatives of the first, second, etc. orders, $u_i = \partial u/\partial x^i$, $u_{ij} = \partial^2 u/\partial x^i \partial x^j$.

The adjoint equation to (1.5) is

$$F^*(x, u, v, u_{(1)}, v_{(1)}, \dots, u_{(s)}, v_{(s)}) = 0, \tag{1.6}$$

with

$$F^*(x, u, v, u_{(1)}, v_{(1)}, \dots, u_{(s)}, v_{(s)}) = \frac{\delta(v F)}{\delta u}, \tag{1.7}$$

where

$$\frac{\delta}{\delta u} = \frac{\partial}{\partial u} + \sum_{s=1}^{\infty}(-1)^s D_{i_1} \cdots D_{i_s} \frac{\partial}{\partial u_{i_1 \cdots i_s}} \tag{1.8}$$

denotes the variational derivatives (the Euler-Lagrange operator), and v is a new dependent variable. Here

$$D_i = \frac{\partial}{\partial x^i} + u_i \frac{\partial}{\partial u} + u_{ij} \frac{\partial}{\partial u_j} + \cdots$$

are the total differentiations.

Definition. Equation (1.5) is said to be *nonlinear self-adjoint* if the equation obtained from the adjoint equation (1.6) by the substitution $v = h(x, u, u_{(1)}, \ldots)$, with a certain function $h(x, u, u_{(1)}, \ldots)$ such that $h(x, u, u_{(1)}, \ldots) \neq constant$,

$$F^*(x, u, u, u_{(1)}, u_{(1)}, \ldots, u_{(s)}, u_{(s)}) = 0,$$

is identical with the original equation (1.5).
In other words, if

$$F^*(x, u, u_{(1)}, u_{(1)}, \ldots, u_{(s)}, u_{(s)}) = \lambda(x, u, u_{(1)}, \ldots)\, F(x, u, u_{(1)}, \ldots, u_{(s)}). \tag{1.9}$$

In particular:

Definition. Equation (1.5) is said to be self-adjoint if the adjoint equation (1.6) is equivalent to the original equation (1.5) upon the substitution $v = u$.

Definition. Equation (1.5) is said to be quasi-self-adjoint if the adjoint equation (1.6) is equivalent to the original equation (1.5) upon the substitution $v = h(u)$ with a certain function $h(u)$ such that $h'(u) \neq 0$.

Definition. Equation (1.5) is said to be weak self-adjoint if the adjoint equation (1.6) is equivalent to the original equation (1.5) upon the substitution $v = h(x, t, u)$ with a certain function $h(x, t, u)$ such that $h_u \neq 0$ and $h_x \neq 0$ or $h_t \neq 0$.

1.2.1 The Subclass of Nonlinear Self-Adjoint Equations

Let us single out some nonlinear self-adjoint equations from the equations of the form (1.2)

$$\begin{aligned} &u_t + ku_x + \alpha uu_x + \beta u_{xxx} + \alpha^2 \rho_1 u^2 u_x + \alpha\beta(\rho_2 uu_{xxx} + \rho_3 u_x u_{xx}) \\ &\quad + \alpha^3 \rho_4 u^3 u_x + \alpha^2 \beta(\rho_5 u^2 u_{xxx} + \rho_6 uu_x u_{xx} + \rho_7 u_x^3) = 0. \end{aligned}$$

Theorem. *Equation* (1.2) *is nonlinear self-adjoint for any arbitrary parameters* ρ_i, $i = 1, \ldots, 7$.

Proof. Equation (1.7) yields

$$\begin{aligned}
F^* &= \frac{\delta}{\delta u}[v(u_t + \rho u_x + \alpha u u_x + \beta u_{xxx} + \alpha^2 \rho_1 u^2 u_x + \alpha^3 \rho_4 u^3 u_x \\
&\quad + \alpha\beta(\rho_2 u u_{xxx} + \rho_3 u_x u_{xx}) + \alpha^2\beta(\rho_5 u^2 u_{xxx} + \rho_6 u u_x u_{xx} + \rho_7 u_x^3))] \\
&= -\alpha^2 \beta \rho_5 u^2 v_{xxx} - \alpha \beta \rho_2 u v_{xxx} - \beta v_{xxx} + \alpha^2 \beta \rho_6 u u_x v_{xx} \\
&\quad - 6\alpha^2 \beta \rho_5 u u_x v_{xx} + \alpha \beta \rho_3 u_x v_{xx} - 3\alpha \beta \rho_2 u_x v_{xx} + \alpha^2 \beta \rho_6 u u_{xx} v_x \\
&\quad - 6\alpha^2 \beta \rho_5 u u_{xx} v_x + \alpha \beta \rho_3 u_{xx} v_x - 3\alpha \beta \rho_2 u_{xx} v_x - 3\alpha^2 \beta \rho_7 (u_x)^2 v_x \\
&\quad + 2\alpha^2 \beta \rho_6 (u_x)^2 v_x - 6\alpha^2 \beta \rho_5 (u_x)^2 v_x - \alpha^3 \rho_4 u^3 v_x - \alpha^2 \rho_1 u^2 v_x - \alpha u v_x \\
&\quad - \rho v_x - v_t - 6\alpha^2 \beta \rho_7 u_x u_{xx} v + 3\alpha^2 \beta \rho_6 u_x u_{xx} v - 6\alpha^2 \beta \rho_5 u_x u_{xx} v
\end{aligned} \tag{1.10}$$

By substituting $v = h(x, t, u)$ and its derivatives

$$\begin{aligned}
v &= h(x,t,u), \\
v_t &= h_u u_t + h_t, \\
v_x &= h_u u_x + h_x, \\
v_{xx} &= h_u u_{xx} + u_x (h_{uu} u_x + h_{ux}) + h_{ux} u_x + h_{xx}, \\
v_{xxx} &= h_u u_{xxx} + u_x (h_{uu} u_{xx} + u_x (h_{uuu} u_x + h_{uux}) + h_{uux} u_x + h_{uxx}) \\
&\quad + 2(h_{uu} u_x + h_{ux}) u_{xx} + h_{ux} u_{xx} + u_x (h_{uux} u_x + h_{uxx}) \\
&\quad + h_{uxx} u_x + h_{xxx}
\end{aligned}$$

into the adjoint equation (1.16) we obtain:

$$\begin{aligned}
&-\alpha^2 \beta h_u \rho_5 u^2 u_{xxx} - \alpha \beta h_u \rho_2 u u_{xxx} - \beta h_u u_{xxx} - 3\alpha^2 \beta h_{uu} \rho_5 u^2 u_x u_{xx} \\
&\quad + 2\alpha^2 \beta h_u \rho_6 u u_x u_{xx} - 12\alpha^2 \beta h_u \rho_5 u u_x u_{xx} - 3\alpha \beta h_{uu} \rho_2 u u_x u_{xx} \\
&\quad - 6\alpha^2 \beta h \rho_7 u_x u_{xx} + 3\alpha^2 \beta h \rho_6 u_x u_{xx} - 6\alpha^2 \beta h \rho_5 u_x u_{xx} \\
&\quad + 2\alpha \beta h_u \rho_3 u_x u_{xx} - 6\alpha \beta h_u \rho_2 u_x u_{xx} - 3\beta h_{uu} u_x u_{xx} \\
&\quad - 3\alpha^2 \beta h_{ux} \rho_5 u^2 u_{xx} + \alpha^2 \beta h_x \rho_6 u u_{xx} - 6\alpha^2 \beta h_x \rho_5 u u_{xx} \\
&\quad - 3\alpha \beta h_{ux} \rho_2 u u_{xx} + \alpha \beta h_x \rho_3 u_{xx} - 3\alpha \beta h_x \rho_2 u_{xx} - 3\beta h_{ux} u_{xx} \\
&\quad - \alpha^2 \beta h_{uuu} \rho_5 u^2 (u_x)^3 + \alpha^2 \beta h_{uu} \rho_6 u (u_x)^3 - 6\alpha^2 \beta h_{uu} \rho_5 u (u_x)^3 \\
&\quad - \alpha \beta h_{uuu} \rho_2 u (u_x)^3 - 3\alpha^2 \beta h_u \rho_7 (u_x)^3 + 2\alpha^2 \beta h_u \rho_6 (u_x)^3 \\
&\quad - 6\alpha^2 \beta h_u \rho_5 (u_x)^3 + \alpha \beta h_{uu} \rho_3 (u_x)^3 - 3\alpha \beta h_{uu} \rho_2 (u_x)^3
\end{aligned}$$

$$-\beta\, h_{u\,u\,u}\, (u_x)^3 - 3\,\alpha^2\,\beta\, h_{u\,u\,x}\,\rho_5\, u^2\, (u_x)^2 + 2\,\alpha^2\,\beta\, h_{u\,x}\,\rho_6\, u\, (u_x)^2$$
$$-12\,\alpha^2\,\beta\, h_{u\,x}\,\rho_5\, u\, (u_x)^2 - 3\,\alpha\,\beta\, h_{u\,u\,x}\,\rho_2\, u\, (u_x)^2 - 3\,\alpha^2\,\beta\, h_x\,\rho_7\, (u_x)^2$$
$$+2\,\alpha^2\,\beta\, h_x\,\rho_6\, (u_x)^2 - 6\,\alpha^2\,\beta\, h_x\,\rho_5\, (u_x)^2 + 2\,\alpha\,\beta\, h_{u\,x}\,\rho_3\, (u_x)^2$$
$$-6\,\alpha\,\beta\, h_{u\,x}\,\rho_2\, (u_x)^2 - 3\,\beta\, h_{u\,u\,x}\, (u_x)^2 - \alpha^3\, h_u\,\rho_4\, u^3\, u_x - 3\,\alpha^2\,\beta\, h_{u\,x\,x}\,\rho_5\, u^2\, u_x$$
$$-\alpha^2\, h_u\,\rho_1\, u^2\, u_x + \alpha^2\,\beta\, h_{x\,x}\,\rho_6\, u\, u_x - 6\,\alpha^2\,\beta\, h_{x\,x}\,\rho_5\, u\, u_x - 3\,\alpha\,\beta\, h_{u\,x\,x}\,\rho_2\, u\, u_x$$
$$-\alpha\, h_u\, u\, u_x + \alpha\,\beta\, h_{x\,x}\,\rho_3\, u_x - 3\,\alpha\,\beta\, h_{x\,x}\,\rho_2\, u_x - h_u\,\rho\, u_x - 3\,\beta\, h_{u\,x\,x}\, u_x - h_u\, u_t$$
$$-\alpha^3\, h_x\,\rho_4\, u^3 - \alpha^2\,\beta\, h_{x\,x\,x}\,\rho_5\, u^2 - \alpha^2\, h_x\,\rho_1\, u^2 - \alpha\,\beta\, h_{x\,x\,x}\,\rho_2\, u - \alpha\, h_x\, u$$
$$-h_x\,\rho - \beta\, h_{x\,x\,x} - h_t = 0$$

Hence the condition of nonlinear self-adjointness is written as follows:

$$F^* - \lambda[v(u_t + \rho u_x + \alpha u u_x + \beta u_{xxx} + \alpha^2\rho_1 u^2 u_x + \alpha^3\rho_4 u^3 u_x$$
$$+\alpha\beta(\rho_2 u u_{xxx} + \rho_3 u_x u_{xx}) + \alpha^2\beta(\rho_5 u^2 u_{xxx} + \rho_6 u u_x u_{xx} + \rho_7 u_x^3))]$$
$$= -\alpha^2\,\beta\,\rho_5\, u^2\, u_{x\,x\,x}\,\lambda - \alpha\,\beta\,\rho_2\, u\, u_{x\,x\,x}\,\lambda - \beta\, u_{x\,x\,x}\,\lambda - \alpha^2\,\beta\,\rho_6\, u\, u_x\, u_{x\,x}\,\lambda$$
$$-\alpha\,\beta\,\rho_3\, u_x\, u_{x\,x}\,\lambda - \alpha^2\,\beta\,\rho_7\, (u_x)^3\,\lambda - \alpha^3\,\rho_4\, u^3\, u_x\,\lambda - \alpha^2\,\rho_1\, u^2\, u_x\,\lambda - \alpha\, u\, u_x\,\lambda$$
$$-\rho\, u_x\,\lambda - u_t\,\lambda - \alpha^2\,\beta\, h_u\,\rho_5\, u^2\, u_{x\,x\,x} - \alpha\,\beta\, h_u\,\rho_2\, u\, u_{x\,x\,x} - \beta\, h_u\, u_{x\,x\,x}$$
$$-3\,\alpha^2\,\beta\, h_{u\,u}\,\rho_5\, u^2\, u_x\, u_{x\,x} + 2\,\alpha^2\,\beta\, h_u\,\rho_6\, u\, u_x\, u_{x\,x} - 12\,\alpha^2\,\beta\, h_u\,\rho_5\, u\, u_x\, u_{x\,x}$$
$$-3\,\alpha\,\beta\, h_{u\,u}\,\rho_2\, u\, u_x\, u_{x\,x} - 6\,\alpha^2\,\beta\, h\,\rho_7\, u_x\, u_{x\,x} + 3\,\alpha^2\,\beta\, h\,\rho_6\, u_x\, u_{x\,x}$$
$$-6\,\alpha^2\,\beta\, h\,\rho_5\, u_x\, u_{x\,x} + 2\,\alpha\,\beta\, h_u\,\rho_3\, u_x\, u_{x\,x} - 6\,\alpha\,\beta\, h_u\,\rho_2\, u_x\, u_{x\,x} - 3\,\beta\, h_{u\,u}\, u_x\, u_{x\,x}$$
$$-3\,\alpha^2\,\beta\, h_{u\,x}\,\rho_5\, u^2\, u_{x\,x} + \alpha^2\,\beta\, h_x\,\rho_6\, u\, u_{x\,x} - 6\,\alpha^2\,\beta\, h_x\,\rho_5\, u\, u_{x\,x}$$
$$-3\,\alpha\,\beta\, h_{u\,x}\,\rho_2\, u\, u_{x\,x} + \alpha\,\beta\, h_x\,\rho_3\, u_{x\,x} - 3\,\alpha\,\beta\, h_x\,\rho_2\, u_{x\,x} - 3\,\beta\, h_{u\,x}\, u_{x\,x}$$
$$-\alpha^2\,\beta\, h_{u\,u\,u}\,\rho_5\, u^2\, (u_x)^3 + \alpha^2\,\beta\, h_{u\,u}\,\rho_6\, u\, (u_x)^3 - 6\,\alpha^2\,\beta\, h_{u\,u}\,\rho_5\, u\, (u_x)^3$$
$$-\alpha\,\beta\, h_{u\,u\,u}\,\rho_2\, u\, (u_x)^3 - 3\,\alpha^2\,\beta\, h_u\,\rho_7\, (u_x)^3 + 2\,\alpha^2\,\beta\, h_u\,\rho_6\, (u_x)^3$$
$$-6\,\alpha^2\,\beta\, h_u\,\rho_5\, (u_x)^3 + \alpha\,\beta\, h_{u\,u}\,\rho_3\, (u_x)^3 - 3\,\alpha\,\beta\, h_{u\,u}\,\rho_2\, (u_x)^3$$
$$-\beta\, h_{u\,u\,u}\, (u_x)^3 - 3\,\alpha^2\,\beta\, h_{u\,u\,x}\,\rho_5\, u^2\, (u_x)^2 + 2\,\alpha^2\,\beta\, h_{u\,x}\,\rho_6\, u\, (u_x)^2$$
$$-12\,\alpha^2\,\beta\, h_{u\,x}\,\rho_5\, u\, (u_x)^2 - 3\,\alpha\,\beta\, h_{u\,u\,x}\,\rho_2\, u\, (u_x)^2 - 3\,\alpha^2\,\beta\, h_x\,\rho_7\, (u_x)^2$$
$$+2\,\alpha^2\,\beta\, h_x\,\rho_6\, (u_x)^2 - 6\,\alpha^2\,\beta\, h_x\,\rho_5\, (u_x)^2 + 2\,\alpha\,\beta\, h_{u\,x}\,\rho_3\, (u_x)^2$$
$$-6\,\alpha\,\beta\, h_{u\,x}\,\rho_2\, (u_x)^2 - 3\,\beta\, h_{u\,u\,x}\, (u_x)^2 - \alpha^3\, h_u\,\rho_4\, u^3\, u_x - 3\,\alpha^2\,\beta\, h_{u\,x\,x}\,\rho_5\, u^2\, u_x$$
$$-\alpha^2\, h_u\,\rho_1\, u^2\, u_x + \alpha^2\,\beta\, h_{x\,x}\,\rho_6\, u\, u_x - 6\,\alpha^2\,\beta\, h_{x\,x}\,\rho_5\, u\, u_x - 3\,\alpha\,\beta\, h_{u\,x\,x}\,\rho_2\, u\, u_x$$
$$-\alpha\, h_u\, u\, u_x + \alpha\,\beta\, h_{x\,x}\,\rho_3\, u_x - 3\,\alpha\,\beta\, h_{x\,x}\,\rho_2\, u_x - h_u\,\rho\, u_x$$
$$-3\,\beta\, h_{u\,x\,x}\, u_x - h_u\, u_t - \alpha^3\, h_x\,\rho_4\, u^3 - \alpha^2\,\beta\, h_{x\,x\,x}\,\rho_5\, u^2 - \alpha^2\, h_x\,\rho_1\, u^2$$
$$-\alpha\,\beta\, h_{x\,x\,x}\,\rho_2\, u - \alpha\, h_x\, u - h_x\,\rho - \beta\, h_{x\,x\,x} - h_t = 0, \tag{1.11}$$

where λ is an undetermined coefficient. Hence comparing the coefficients for the different derivatives of u we obtain that $\lambda = -h_u$ and the following conditions must be satisfied:

$$
\begin{aligned}
-3\,\beta\,h_{u\,u}\left(\alpha^2\,\rho_5\,u^2+\alpha\,\rho_2\,u+1\right)-3\,\alpha^2\,\beta\,h\,\left(2\,\rho_7-\rho_6+2\,\rho_5\right)&\\
+3\,\alpha\,\beta\,h_u\,\left(\alpha\,\rho_6\,u-4\,\alpha\,\rho_5\,u+\rho_3-2\,\rho_2\right)&=0,\\
\alpha\,\beta\,h_x\,\left(\alpha\,\rho_6\,u-6\,\alpha\,\rho_5\,u+\rho_3-3\,\rho_2\right)-3\,\beta\,h_{u\,x}\left(\alpha^2\,\rho_5\,u^2+\alpha\,\rho_2\,u+1\right)&=0,\\
-\beta\,h_{u\,u\,u}\left(\alpha^2\,\rho_5\,u^2+\alpha\,\rho_2\,u+1\right)-2\,\alpha^2\,\beta\,h_u\,\left(\rho_7-\rho_6+3\,\rho_5\right)&\\
+\alpha\,\beta\,h_{u\,u}\,\left(\alpha\,\rho_6\,u-6\,\alpha\,\rho_5\,u+\rho_3-3\,\rho_2\right)&=0,\\
-3\,\beta\,h_{u\,u\,x}\left(\alpha^2\,\rho_5\,u^2+\alpha\,\rho_2\,u+1\right)-\alpha^2\,\beta\,h_x\,\left(3\,\rho_7-2\,\rho_6+6\,\rho_5\right)&\\
+2\,\alpha\,\beta\,h_{u\,x}\,\left(\alpha\,\rho_6\,u-6\,\alpha\,\rho_5\,u+\rho_3-3\,\rho_2\right)&=0,\\
\alpha\,\beta\,h_{x\,x}\,\left(\alpha\,\rho_6\,u-6\,\alpha\,\rho_5\,u+\rho_3-3\,\rho_2\right)&\\
-3\,\beta\,h_{u\,x\,x}\left(\alpha^2\,\rho_5\,u^2+\alpha\,\rho_2\,u+1\right)&=0,\\
-\alpha^3\,h_x\,\rho_4\,u^3+\left(-\alpha^2\,\beta\,h_{x\,x\,x}\,\rho_5-\alpha^2\,h_x\,\rho_1\right)u^2&\\
+\left(-\alpha\,\beta\,h_{x\,x\,x}\,\rho_2-\alpha\,h_x\right)u-h_x\,k-\beta\,h_{x\,x\,x}-h_t&=0.
\end{aligned}
$$

Equations third and fifth are differential consequences of the first and second equations, respectively.
Consequently, if (1.2) is weak self-adjoint, $h = h(x,t,u)$ must satisfy the following conditions:

$$
\begin{aligned}
-3\,\beta\,h_{u\,u}\left(\alpha^2\,\rho_5\,u^2+\alpha\,\rho_2\,u+1\right)-3\,\alpha^2\,\beta\,h\,\left(2\,\rho_7-\rho_6+2\,\rho_5\right)&\\
+3\,\alpha\,\beta\,h_u\,\left(\alpha\,\rho_6\,u-4\,\alpha\,\rho_5\,u+\rho_3-2\,\rho_2\right)&=0,\\
\alpha\,\beta\,h_x\,\left(\alpha\,\rho_6\,u-6\,\alpha\,\rho_5\,u+\rho_3-3\,\rho_2\right)-3\,\beta\,h_{u\,x}\left(\alpha^2\,\rho_5\,u^2+\alpha\,\rho_2\,u+1\right)&=0,\\
h_{u\,u\,x}\left(\alpha^2\,\rho_5\,u^2+\alpha\,\rho_2\,u+1\right)-\alpha^2\,\beta\,h_x\,\left(3\,\rho_7-2\,\rho_6+6\,\rho_5\right)&\\
+2\alpha,\beta\,h_{u\,x}\,\left(\alpha\,\rho_6\,u-6\,\alpha\,\rho_5\,u+\rho_3-3\,\rho_2\right)&=0,\\
-\alpha^3\,h_x\,\rho_4\,u^3+\left(-\alpha^2\,\beta\,h_{x\,x\,x}\,\rho_5-\alpha^2\,h_x\,\rho_1\right)u^2&\\
+\left(-\alpha\,\beta\,h_{x\,x\,x}\,\rho_2-\alpha\,h_x\right)u-h_x\,k-\beta\,h_{x\,x\,x}-h_t&=0.
\end{aligned}
$$

However, setting $h = h(u)$ we get that equation (1.2) is nonlinear self-adjoint for any arbitrary parameter ρ_i setting $h(u)$ so that it satisfies the following condition:

$$
\begin{aligned}
&-3\,\beta\,h_{u\,u}\left(\alpha^2\,\rho_5\,u^2+\alpha\,\rho_2\,u+1\right)-3\,a^2\,\beta\,h\,\left(2\,\rho_7-\rho_6+2\,\rho_5\right)\\
&+3\,\alpha\,\beta\,h_u\,\left(\alpha\,\rho_6\,u-4\,\alpha\,\rho_5\,u+\rho_3-2\,\rho_2\right)=0.
\end{aligned}
\tag{1.12}
$$

1.2.2 Nonlinear and Weak Self-Adjointness

Equation (1.3) is not self-adjoint; however, we prove that:

Theorem. *Equation* (1.3) *is weak self-adjoint and nonlinear self-adjoint.*

We substitute

$$\begin{aligned} v &= h(x,t,u),\\ v_t &= h_u u_t + h_t,\\ v_x &= h_u u_x + h_x,\\ v_{xx} &= h_u\, u_{xx} + u_x\, (h_{uu}\, u_x + h_{ux}) + h_{ux}\, u_x + h_{xx},\\ v_{xxx} &= h_u\, u_{xxx} + u_x\, (h_{uu}\, u_{xx} + u_x\, (h_{uuu}\, u_x + h_{uux}) + h_{uux}\, u_x + h_{uxx})\\ &\quad + 2\,(h_{uu}\, u_x + h_{ux})\, u_{xx} + h_{ux}\, u_{xx} + u_x\,(h_{uux}\, u_x + h_{uxx})\\ &\quad + h_{uxx}\, u_x + h_{xxx}, \end{aligned}$$

into the adjoint equation

$$-u^2\, v_{xxx} - 7\,u\,u_x\, v_{xx} - 7\,u\,u_{xx}\, v_x - \frac{28\,(u_x)^2\, v_x}{3} - u^2\, v_x - v_t - \frac{35\,u_x\, u_{xx}\, v}{3} = 0$$

obtaining

$$\begin{aligned} &-h_u\, u^2\, u_{xxx} - 3\,h_{uu}\, u^2\, u_x\, u_{xx} - 14\,h_u\, u\, u_x\, u_{xx} - \frac{35\,h\, u_x\, u_{xx}}{3} - 3\,h_{ux}\, u^2\, u_{xx}\\ &\quad - 7\,h_x\, u\, u_{xx} - h_{uuu}\, u^2\,(u_x)^3 - 7\,h_{uu}\, u\,(u_x)^3 - \frac{28\,h_u\,(u_x)^3}{3} - 3\,h_{uux}\, u^2\,(u_x)^2\\ &\quad - 14\,h_{ux}\, u\,(u_x)^2 - \frac{28\,h_x\,(u_x)^2}{3} - 3\,h_{uxx}\, u^2\, u_x - h_u\, u^2\, u_x - 7\,h_{xx}\, u\, u_x\\ &\quad - h_u\, u_t - h_{xxx}\, u^2 - h_x\, u^2 - h_t = 0. \end{aligned}$$

Hence, the condition of nonlinear self-adjointness is written as follows:

$$\begin{aligned} &-u^2\, u_{xxx}\,\lambda + u\, u_x\, u_{xx}\,\lambda - \frac{4\,(u_x)^3\,\lambda}{9} - u^2\, u_x\,\lambda - u_t\,\lambda - h_u\, u^2\, u_{xxx}\\ &\quad - 3\,h_{uu}\, u^2\, u_x\, u_{xx} - 14\,h_u\, u\, u_x\, u_{xx} - \frac{35\,h\, u_x\, u_{xx}}{3} - 3\,h_{ux}\, u^2\, u_{xx} - 7\,h_x\, u\, u_{xx} \end{aligned}$$

$$-h_{uuu}\,u^2\,(u_x)^3-7\,h_{uu}\,u\,(u_x)^3-\frac{28\,h_u\,(u_x)^3}{3}-3\,h_{uux}\,u^2\,(u_x)^2$$
$$-14\,h_{ux}\,u\,(u_x)^2-\frac{28\,h_x\,(u_x)^2}{3}-3\,h_{uxx}\,u^2\,u_x-h_u\,u^2\,u_x$$
$$-7\,h_{xx}\,u\,u_x-h_u\,u_t-h_{xxx}\,u^2-h_x\,u^2-h_t=0.$$

Here λ is an undetermined coefficient such that

$$\lambda+h_u=0$$

and the following conditions must be satisfied:

$$\begin{aligned} 3\,h_{uu}\,u^2+15\,h_u\,u+\frac{35\,h}{3}&=0,\\ 3\,h_{ux}\,u^2+7\,h_x\,u&=0,\\ -h_{uuu}\,u^2-7\,h_{uu}\,u-\frac{80\,h_u}{9}&=0,\\ -3\,h_{uux}\,u^2-14\,h_{ux}\,u-\frac{28\,h_x}{3}&=0,\\ -3\,h_{uxx}\,u^2-7\,h_{xx}\,u&=0,\\ -h_{xxx}\,u^2-h_x\,u^2-h_t&=0. \end{aligned}$$

The solution is

$$h=\frac{a}{u^{\frac{5}{3}}}+\frac{b(x,t)}{u^{\frac{7}{3}}},$$

where $a=constant$ and $b=b(x)$ satisfies

$$b_{xxx}+b_x=0. \tag{1.13}$$

Namely the adjoint equation becomes equivalent to the original equation upon the substitution

$$v=\frac{a}{u^{\frac{5}{3}}}+\frac{b}{u^{\frac{7}{3}}},$$

with $a=constant$ and being $b=b(x)$ any solution of (1.13).

1.2.3 The Condition of Quasi-Self-Adjointness

Let us see if (1.4) is quasi-self-adjoint.
Equation (1.7) yields

$$F^* = \frac{\delta}{\delta u}[v(-u_{xxx} - 3\,u_x\,u_{xx} - (u_x)^3 + u_t)]$$
$$= v_{xxx} - 3\,u_x\,v_{xx} - 3\,u_{xx}\,v_x + 3\,(u_x)^2\,v_x - v_t + 6\,u_x\,u_{xx}\,v \quad (1.14)$$

Setting $v = h(u)$ in (1.14) we have

$$F^* = h_u\,u_{xxx} + 3\,h_{uu}\,u_x\,u_{xx} - 6\,h_u\,u_x\,u_{xx} + 6\,h\,u_x\,u_{xx}$$
$$+ h_{uuu}\,(u_x)^3 - 3\,h_{uu}\,(u_x)^3 + 3\,h_u\,(u_x)^3 - h_u\,u_t.$$

Using (1.9) yields:

$$F^* - \lambda(u_t - u_{xxx} - 3u_xu_{xx} - u_x^3) = u_{xxx}\,\lambda + 3\,u_x\,u_{xx}\,\lambda + (u_x)^3\,\lambda - u_t\,\lambda$$
$$+ h_u\,u_{xxx} + 3\,h_{uu}\,u_x\,u_{xx} - 6\,h_u\,u_x\,u_{xx} + 6\,h\,u_x\,u_{xx} + h_{uuu}\,(u_x)^3$$
$$- 3\,h_{uu}\,(u_x)^3 + 3\,h_u\,(u_x)^3 - h_u\,u_t = 0.$$

Comparing the coefficients for u_t, we obtain $\lambda + h_u = 0$ and the following conditions must be satisfied:

$$3\,h_{uu} - 9\,h_u + 6\,h = 0,$$
$$h_{uuu} - 3\,h_{uu} + 2\,h_u = 0. \quad (1.15)$$

From (1.15) we get that

$$h(u) = ae^u + be^{2u},$$

where $a = constant$ and $b = constant$. We can state the following:

Theorem. *Equation* (1.3) *is not self-adjoint and it is quasi-self-adjoint, upon the substitution*

$$h(u) = ae^u + be^{2u},$$

where $a = constant$ *and* $b = constant$.

1.2.4 The Condition of Weak Self-Adjointness

Let us see if (1.4) is weak self-adjoint.
Equation (1.7) yields

$$
\begin{aligned}
F^* &= \frac{\delta}{\delta u}[v(-u_{xxx} - 3\,u_x\,u_{xx} - (u_x)^3 + u_t)] \\
&= v_{xxx} - 3\,u_x\,v_{xx} - 3\,u_{xx}\,v_x + 3\,(u_x)^2\,v_x - v_t + 6\,u_x\,u_{xx}\,v. \qquad (1.16)
\end{aligned}
$$

Setting $v = h(x,t,u)$ in (1.16) we have

$$
\begin{aligned}
F^* &= u\,u_{xxx} + 3\,h_{uu}\,u_x\,u_{xx} - 6\,h_u\,u_x\,u_{xx} + 6\,h\,u_x\,u_{xx} - 3\,h_x\,u_{xx} + 3\,h_{ux}\,u_{xx} \\
&\quad + h_{uuu}\,(u_x)^3 - 3\,h_{uu}\,(u_x)^3 + 3\,h_u\,(u_x)^3 + 3\,h_x\,(u_x)^2 + 3\,h_{uux}\,(u_x)^2 \\
&\quad - 6\,h_{ux}\,(u_x)^2 - 3\,h_{xx}\,u_x + 3\,h_{uxx}\,u_x - h_u\,u_t + h_{xxx} - h_t.
\end{aligned}
$$

Equation (1.9) yields:

$$
\begin{aligned}
&F^* - \lambda(u_t - u_{xxx} - 3u_x u_{xx} - u_x^3) = u_{xxx}\,\lambda + 3\,u_x\,u_{xx}\,\lambda + (u_x)^3\,\lambda - u_t\,\lambda \\
&+ h_u\,u_{xxx} + 3\,h_{uu}\,u_x\,u_{xx} - 6\,h_u\,u_x\,u_{xx} + 6\,h\,u_x\,u_{xx} - 3\,h_x\,u_{xx} + 3\,h_{ux}\,u_{xx} \\
&+ h_{uuu}\,(u_x)^3 - 3\,h_{uu}\,(u_x)^3 + 3\,h_u\,(u_x)^3 + 3\,h_x\,(u_x)^2 + 3\,h_{uux}\,(u_x)^2 \\
&- 6\,h_{ux}\,(u_x)^2 - 3\,h_{xx}\,u_x + 3\,h_{uxx}\,u_x - h_u\,u_t + h_{xxx} - h_t = 0.
\end{aligned}
$$

Comparing the coefficients of the u derivatives we obtain that $\lambda = -h_u$ and the following conditions must be satisfied

$$
\begin{aligned}
3\,h_{uu} - 9\,h_u + 6\,h &= 0, \\
3\,h_{ux} + 3\,h_x &= 0, \\
h_{uuu} - 3\,h_{uu} + 2\,h_u &= 0, \\
3\,h_x + 3\,h_{uux} - 6\,h_{ux} &= 0, \\
3\,h_{uu} - 9\,h_u + 3\,h_{xx} + 3\,h_{uxx} &= 0, \\
h_{xxx} - h_t &= 0. \qquad (1.17)
\end{aligned}
$$

From (1.17) we get that

$$h(x,t,u) = a(x,t)e^u,$$

where $a = a(x,t)$ must satisfy the linear equation

$$a_t - a_{xxx} = 0.$$

We can state the following:

Theorem. *Equation* (1.3) *is weak self-adjoint, upon the substitution*

$$h(u,x,t) = a(x,t)e^{u},$$

where $a = a(x,t)$ *satisfies*

$$a_t - a_{xxx} = 0.$$

1.2.5 The Condition of Nonlinear Self-Adjointness

Let us see if (1.4) is nonlinear self-adjoint.
Equation (1.7) yields

$$\begin{aligned} F^* &= \frac{\delta}{\delta u}[v(-u_{xxx} - 3\,u_x\,u_{xx} - (u_x)^3 + u_t)] \\ &= v_{xxx} - 3\,u_x\,v_{xx} - 3\,u_{xx}\,v_x + 3\,(u_x)^2\,v_x - v_t + 6\,u_x\,u_{xx}\,v. \end{aligned} \tag{1.18}$$

Setting $v = h(u,u_x)$ in (1.18) and denoting $u_x = w$ we have

$$\begin{aligned} F^* = {} & h_w w_{xxx} + 3h_{ww} w_x w_{xx} - 3h_w w w_{xx} + 3h_{uw} w w_{xx} + h_u w_{xx} + h_{www}\,(w_x)^3 \\ & - 3h_{ww} w\,(w_x)^2 + 3h_{uww} w\,(w_x)^2 - 3h_w\,(w_x)^2 + 3h_{uw}\,(w_x)^2 + 3h_w w^2 w_x \\ & + 3h_{uuw} w^2 w_x - 6h_{uw} w^2 w_x + 3h_{uu} w w_x - 6h_u w w_x + 6h w w_x - h_w w_t \\ & + h_{uuu} w^3 - 3h_{uu} w^3 + 3h_u w^3 - h_u u_t. \end{aligned}$$

Using (1.9) and the derivative with respect to x of (1.4)

$$-u_{xxxx} - 3\,u_x\,u_{xxx} - 3\,(u_{xx})^2 - 3\,(u_x)^2\,u_{xx} + u_{tx} = 0$$

which in terms of w can be written as

$$-w_{xxx} - 3\,w\,w_{xx} - 3\,(w_x)^2 - 3\,w^2\,w_x + w_t = 0$$

yields:

$$\begin{aligned} & F^* - \mu(-u_{xxxx} - 3u_x u_{xxx} - 3\,(u_{xx})^2 - 3\,(u_x)^2\,u_{xx} + u_{tx}) \\ & - \lambda(u_t - u_{xxx} - 3u_x u_{xx} - u_x^3) = -\left(-u_{xxx} - 3u_x u_{xx} - (u_x)^3 + u_t\right)\lambda \\ & + h_{u_x} u_{xxx} + 3h_{u_x u_x} u_{xx} u_{xxx} - 3h_{u_x} u_x u_{xxx} + 3h_{uu_x} u_x u_{xxx} + h_u u_{xxx} \\ & + h_{u_x u_x u_x}\,(u_{xx})^3 - 3h_{u_x u_x} u_x\,(u_{xx})^2 + 3h_{uu_x u_x} u_x\,(u_{xx})^2 - 3h_{u_x}\,(u_{xx})^2 \end{aligned}$$

$$+ 3h_{uu_x}(u_{xx})^2 + 3h_{u_x}u_x^2u_{xx} + 3h_{uuu_x}u_x^2u_{xx} - 6h_{uu_x}u_x^2u_{xx} + 3h_{uu}u_xu_{xx}$$
$$- 6h_uu_xu_{xx} + 6hu_xu_{xx} - h_{u_x}u_{xt} + h_{uuu}(u_x)^3 - 3h_{uu}(u_x)^3 + 3h_u(u_x)^3 - h_uu_t$$
$$- \mu\left(-u_{xxxx} - 3u_xu_{xxx} - 3(u_{xx})^2 - 3(u_x)^2u_{xx} + u_{tx}\right) = 0.$$

Setting $u_x = w$

$$-\left(-w_{xx} - 3ww_x - w^3 + u_t\right)\lambda - \mu\left(-w_{xxx} - 3ww_{xx} - 3(w_x)^2 - 3w^2w_x + w_t\right)$$
$$+ 3h_{ww}w_xw_{xx} - 3h_www_{xx} + 3h_{uw}ww_{xx} + h_uw_{xx} + h_{www}(w_x)^3 - 3h_{ww}w(w_x)^2$$
$$+ 3h_{uww}w(w_x)^2 - 3h_w(w_x)^2 + 3h_{uw}(w_x)^2 + 3h_ww^2w_x + 3h_{uuw}w^2w_x$$
$$- 6h_{uw}w^2w_x + 3h_{uu}ww_x + h_ww_{xxx} - 6h_uww_x + 6hww_x - h_ww_t + h_{uuu}w^3$$
$$- 3h_{uu}w^3 + 3h_uw^3 - h_uu_t = 0.$$

Comparing the coefficients for u_t, we obtain $\lambda + h_u = 0$.
Comparing the coefficients for w_t, we obtain $\mu + h_w = 0$ and the following conditions must be satisfied:

$$h_{ww} = 0, \tag{1.19}$$

$$h_u - 2h = 0. \tag{1.20}$$

From (1.19) we get that

$$h(u, w) = ce^{2u}w.$$

We can state the following:

Theorem. *Equation* (1.4) *is nonlinear self-adjoint, upon the substitution*

$$h(u, u_x) = ce^{2u}u_x.$$

1.2.6 General Theorem on Conservation Laws

We use the following theorem on conservation laws proved in [6].

Theorem. *Any Lie point, Lie-Bäcklund or nonlocal symmetry*

$$X = \xi^i(x, u, u_{(1)}, \ldots)\frac{\partial}{\partial x^i} + \eta(x, u, u_{(1)}, \ldots)\frac{\partial}{\partial u} \tag{1.21}$$

of equation (1.5) *provides a conservation law* $D_i(C^i) = 0$ *for the system of differential equations* (1.5) *and* (1.6). *The conserved vector is given by*

$$C^i = \xi^i \mathscr{L} + W\left[\frac{\partial \mathscr{L}}{\partial u_i} - D_j\left(\frac{\partial \mathscr{L}}{\partial u_{ij}}\right) D_j D_k\left(\frac{\partial \mathscr{L}}{\partial u_{ijk}}\right) - \cdots\right]$$
$$+ D_j(W)\left[\frac{\partial \mathscr{L}}{\partial u_{ij}} - D_k\left(\frac{\partial \mathscr{L}}{\partial u_{ijk}}\right) + \cdots\right] + D_j D_k(W)\left[\frac{\partial \mathscr{L}}{\partial u_{ijk}} - \cdots\right] + \cdots, \tag{1.22}$$

where W *and* $\mathscr{L}$ *are defined as follows:*

$$W = \eta - \xi^j u_j, \quad \mathscr{L} = v\, F\left(x, u, u_{(1)}, \ldots, u_{(s)}\right). \tag{1.23}$$

We will write generators of point transformation group in the form

$$X = \xi^1 \frac{\partial}{\partial t} + \xi^2 \frac{\partial}{\partial x} + \eta \frac{\partial}{\partial u}$$

by setting $t = x^1$ and $x = x^2$. The conservation law will be written

$$D_t(C^1) + D_x(C^2) = 0. \tag{1.24}$$

1.2.7 Conservation Laws for a Subclass of Self-Adjoint Equations

Let us apply the general theorem on conservation laws to the self-adjoint equation (1.2) with

$$\begin{aligned}
&\rho_3 = 2\rho_2, \quad \rho_7 = \rho_6 - 3\rho_5, \ \rho_5 = \rho_2^2/4,\\
&\rho_6 = \rho_2^2, \quad \rho_2 = 1/\rho, \qquad \rho_4 = 0,\\
&\rho_1 = 1/4\rho\,.
\end{aligned}$$

Let us find the conservation law provided by the following symmetry of (1.2):

$$X = t\frac{\partial}{\partial t} - \left(\frac{u}{2} + \frac{1}{a\rho_2}\right)\frac{\partial}{\partial u}. \tag{1.25}$$

In this case, we have that $W = -\frac{u}{2} - \frac{1}{a\rho_2} - t u_t$ and (1.22) yield the conservation law (1.24) with

$$C^1 = -\frac{u\,(\alpha\, u + 2\, k)}{2\,\alpha} + D_x(B^1),$$

$$C^2 = -\frac{\alpha^2 \beta u^3 u_{xx}}{4k^2} - \frac{5\alpha \beta u^2 u_{xx}}{4k} - 2\beta u u_{xx} - \frac{\beta k u_{xx}}{\alpha} - \frac{\alpha^2 \beta u^2 (u_x)^2}{8k^2} - \frac{\alpha b u (u_x)^2}{4\rho} - \frac{\alpha^2 u^4}{16k} - \frac{5\alpha u^3}{12} - \rho u^2 - \frac{\rho^2 u}{a} - D_t(B^1),$$

where

$$B^1 = \left(\frac{\alpha^2 \beta t u^3}{4\rho^2} + \frac{\alpha \beta t u^2}{\rho} + \beta t u\right) u_{xx} + \frac{\alpha^2 \beta t u^2 (u_x)^2}{8\rho^2} - \frac{\beta t (u_x)^2}{2} - \frac{\beta t (u_x)^2}{2} + \frac{\alpha^2 t u^4}{16\rho} + \frac{\alpha t u^3}{3} + \frac{\rho t u^2}{2}.$$

We simplify the conserved vector by transferring the terms of the form $D_x(\ldots)$ from C^1 to C^2 and obtain:

$$C^1 = -\frac{u(\alpha u + 2\rho)}{2\alpha},$$

$$C^2 = -\frac{\alpha^2 \beta u^3 u_{xx}}{4\rho^2} - \frac{5\alpha \beta u^2 u_{xx}}{4k} - 2\beta u u_{xx} - \frac{\beta k u_{xx}}{\alpha} - \frac{\alpha^2 \beta u^2 (u_x)^2}{8\rho^2} - \frac{\alpha b u (u_x)^2}{4\rho} - \frac{\alpha^2 u^4}{16\rho} - \frac{5\alpha u^3}{12} - \rho u^2 - \frac{\rho^2 u}{\alpha}.$$

1.2.8 Conservation Laws for a Subclass of Self-Adjoint Third Order Equations

Let us apply the general theorem on conservation laws to the quasi-self-adjoint equation (1.3).
In this case, we have

$$\mathscr{L} = \left(u_t + u^2 u_x + \frac{4}{9} u_x^3 - u u_x u_{xx} + u^2 u_{xxx}\right) v. \tag{1.26}$$

Let us find the conservation law provided by the following obvious scaling symmetry of (1.3):

$$X = t\frac{\partial}{\partial t} - \frac{u}{2}\frac{\partial}{\partial u}. \tag{1.27}$$

In this case, we have that $W = -\dfrac{u}{2} - t u_t$ and (1.22) yield the conservation law (1.24) with

$$C^1 = -\frac{k}{2\,u^{\frac{2}{3}}} + D_x\left(k\,t\,u^{\frac{1}{3}}\,u_{x\,x} - \frac{2\,k\,t\,(u_x)^2}{3\,u^{\frac{2}{3}}} + \frac{3\,k\,t\,u^{\frac{4}{3}}}{4}\right),$$

$$C^2 = \frac{k\,u^{\frac{1}{3}}\,u_{x\,x}}{3} - \frac{2\,k\,(u_x)^2}{9\,u^{\frac{2}{3}}} + \frac{k\,u^{\frac{4}{3}}}{4} - D_t\left(k\,t\,u^{\frac{1}{3}}\,u_{x\,x} - \frac{2\,k\,t\,(u_x)^2}{3\,u^{\frac{2}{3}}} + \frac{3\,k\,t\,u^{\frac{4}{3}}}{4}\right).$$

We simplify the conserved vector by transferring the terms of the form $D_x(\ldots)$ from C^1 to C^2 and obtain

$$C^1 = -\frac{k}{2\,u^{\frac{2}{3}}},$$

$$C^2 = \frac{k\,u^{\frac{1}{3}}\,u_{x\,x}}{3} - \frac{2\,k\,(u_x)^2}{9\,u^{\frac{2}{3}}} + \frac{k\,u^{\frac{4}{3}}}{4}.$$

1.2.9 Conservation Laws

Let us apply general theorem on conservation laws to the weak self-adjoint and nonlinear self-adjoint equation

$$u_t - u_{xxx} - 3u_x u_{xx} - u_x^3 = 0, \tag{1.28}$$

with $h(x,t,u) = a(x,t)e^u$ where $a = a(x,t)$ satisfies

$$a_t - a_{xxx} = 0 \tag{1.29}$$

and $h(u,u_x) = e^{2u}u_x$, $u_x = w$. In this case we have

$$\mathscr{L} = \left(u_t - u_{xxx} - 3u_x u_{xx} - u_x^3\right)v. \tag{1.30}$$

1. Let us find the conservation law provided by the following symmetry of (1.4):

$$\mathbf{v} = x\frac{\partial}{\partial x} + 3t\frac{\partial}{\partial t} \tag{1.31}$$

and $h(u,x,t) = a(x,t)e^u$, where $a = a(x,t)$ satisfies (1.29).
In this case, we find that $W = -xu_x - 3tu_t$ and (1.22) yield the conservation law (1.24) where after simplifying the conserved vector by transferring the terms of the form $D_x(\ldots)$ from C^1 to C^2 we obtain:

$$C^1 = e^u\,(a_x\,x + 3\,a_{x\,x\,x}\,t + a),$$

$$
\begin{aligned}
C^2 =& -a_x\, e^u\, u_{xx}\, x - a_x\, e^u\, (u_x)^2\, x + a_{xx}\, e^u\, u_x\, x - a_t\, e^u\, x - 3\, a_t\, t\, e^u\, u_{xx}\\
&- a\, e^u\, u_{xx} - 3\, a_t\, t\, e^u\, (u_x)^2 - a\, e^u\, (u_x)^2 + 3\, a_{t\,x}\, t\, e^u\, u_x\\
&+ 2\, a_x\, e^u\, u_x - 3\, a_{t\,x\,x}\, t\, e^u - 3\, a_{xx}\, e^u,
\end{aligned}
$$

and $a = a(x,t)$ satisfies (1.29).

2. Let us find the conservation law provided by the following symmetry of (1.4):

$$
\mathbf{v} = k_1 \frac{\partial}{\partial x} + k_2 \frac{\partial}{\partial t} \tag{1.32}
$$

and $h(u,x,t) = a(x,t)e^u$, where $a = a(x,t)$ satisfies (1.29).
In this case, we find that $W = -k_1 u_x - k_2 u_t$ and (1.22) yield the conservation law (1.24), where after simplifying the conserved vector by transferring the terms of the form $D_x(\ldots)$ from C^1 to C^2 we obtain:

$$
\begin{aligned}
C^1 =& -(a_{xx}\, k_2 + a\, k_1)\; e^u\, u_x,\\
C^2 =& -a_t\, k_2\, e^u\, u_{xx} - a_x\, k_1\, e^u\, u_{xx} - a_t\, k_2\, e^u\, (u_x)^2 - a_x\, k_1\, e^u\, (u_x)^2\\
&+ a_{t\,x}\, k_2\, e^u\, u_x + a_{xx}\, k_1\, e^u\, u_x + a_{xx}\, k_2\, e^u\, u_t + a\, k_1\, e^u\, u_t,
\end{aligned}
$$

where $a = a(x,t)$ satisfies (1.29).

3. Let us find the conservation law provided by the following symmetry of (1.4):

$$
\mathbf{v} = k_1 \frac{\partial}{\partial x} + k_2 \frac{\partial}{\partial t} \tag{1.33}
$$

and $h(u,u_x) = u_x e^{2u}$.
In this case, we find $W = -k_1 u_x - k_2 u_t$ and (1.22) yield the conservation law (1.24) where after simplifying the conserved vector by transferring the terms of the form $D_x(\ldots)$ from C^1 to C^2 we obtain:

$$
\begin{aligned}
C^1 =& -k_2\, e^{2u}\, (u_{xx})^2 + \frac{k_2\, e^{2u}\, (u_x)^4}{3} - k_1\, e^{2u}\, (u_x)^2,\\
C^2 =&\; 2k_2 e^{2u} u_{xx} u_{xxxx} - k_2 e^{2u} (u_{xxx})^2 + 2k_2 e^{2u} u_x u_{xx} u_{xxx} - \frac{4k_2 e^{2u} (u_x)^3\, u_{xxx}}{3}\\
&+ 2k_1 e^{2u} u_x u_{xxx} + 6k_2 e^{2u} (u_{xx})^3 + 3k_2 e^{2u} (u_x)^2 (u_{xx})^2 - k_1 e^{2u} (u_{xx})^2\\
&- 2k_2 e^{2u} (u_x)^4\, u_{xx} + 4k_1 e^{2u} (u_x)^2\, u_{xx} - \frac{k_2 e^{2u} (u_x)^6}{3} + k_1 e^{2u} (u_x)^4.
\end{aligned}
$$

4. Let us find the conservation law provided by the following symmetry of (1.4):

$$
\mathbf{v} = x \frac{\partial}{\partial x} + 3t \frac{\partial}{\partial t} \tag{1.34}
$$

and $h(u,u_x) = u_x e^{2u}$.

In this case, we find that $W = -xu_x - 3tu_t$ and (1.22) yield the conservation law (1.24), where, after simplifying the conserved vector by transferring the terms of the form $D_x(\ldots)$ from C^1 to C^2 we obtain:

$$C^1 = -e^{2u}(u_x)^2 x + 3te^{2u}(u_{xx})^2 - te^{2u}(u_x)^4,$$

$$\begin{aligned}C^2 = {} & 2e^{2u}u_x u_{xxx}x - e^{2u}(u_{xx})^2 x + 4e^{2u}(u_x)^2 u_{xx}x + e^{2u}(u_x)^4 x \\ & - 6te^{2u}u_{xx}u_{xxxx} + 3te^{2u}(u_{xxx})^2 - 6te^{2u}u_x u_{xx}u_{xxx} \\ & + 4te^{2u}(u_x)^3 u_{xxx} - 18te^{2u}(u_{xx})^3 - 9te^{2u}(u_x)^2(u_{xx})^2 \\ & + 6te^{2u}(u_x)^4 u_{xx} - 2e^{2u}u_x u_{xx} + te^{2u}(u_x)^6 .\end{aligned}$$

1.3 Conclusions

In this work we have considered three third order equations: a potential Burgers equation and two third order wave equations of the KdV type. We have determined the subclasses of these equations which are weak and nonlinear self-adjoint. By using the general theorem on conservation laws proved by Nail Ibragimov, we found some conservation laws for some of these partial differential equations without classical Lagrangians.

Acknowledgements The support of DGICYT project MTM2009-11875 and Junta de Andalucía group FQM-201 is gratefully acknowledged.

References

1. Adem KR, Khalique CM (2012) Exact solutions and conservation laws of Zakharov-Kuznetsov modified equal width equation with power law nonlinearity. Nonlin Anal: Real World Appl 13:1692–1702
2. Freire IL, Sampaio JCS (2012) Nonlinear self-adjointness of a generalized fifth-order KdV equation. J Phys A: Math Theoret 45:032001
3. Fokas AS (1995) On a class of physically important integrable equations. Physica D 87: 1451–1550
4. Gandarias ML (2011) Weak self-adjoint differential equations. J Phys A: Math Theor 44:262001
5. Ibragimov NH (2006) The answer to the question put to me by LV Ovsiannikov 33 years ago. Arch ALGA 3:53–80
6. Ibragimov NH (2007) A new conservation theorem. J Math Anal Appl 333:311–328
7. Ibragimov NH (2007) Quasi-self-adjoint differential equations Arch. ALGA 4:55–60
8. Ibragimov NH (2011) Nonlinear self-adjointness and conservation laws. J Phys A: Math Theor 44:432002
9. Ibragimov NH, Torrisi M, Tracina R (2010) Quasi self-adjoint nonlinear wave equations. J Phys A: Math Theor 43:442001

10. Ibragimov NH, Torrisi M, Tracina R (2011) Self-adjointness and conservation laws of a generalized Burgers equation. J Phys A: Math Theor 44:145201
11. Ibragimov NH, Khamitova RS, Valenti A (2011) Self-adjointness of a generalized Camassa-Holm equation. Appl Math Comp 218:2579–2583
12. Jhangeer A, Naeem I, Qureshi MN (2012) Conservation laws for heat equation on curved surfaces. Nonlinear Anal Real World Appl 13:340–347
13. Johnpillai AG, Khalique CM (2011) Variational approaches to conservation laws for a nonlinear evolution equation with time dependent coefficients. Quaestiones Mathematicae 34:235–245
14. Kara AH, Mahomed FM (2006) Noether-type symmetries and conservation laws via partial Lagrangians. Nonlin Dyn 45:367–383
15. Li J, Rui W, Long Y, He B (2006) Travelling wave solutions for higher-order wave equations of KdV type III. Math Biosci Eng 3:125135
16. Marinakis V, Bountis TC (2000) Special solutions of a new class of water wave equations. Comm Appl Anal 4:43345
17. Qiao ZJ (2009) A new integrable equation with no smooth solitons. Chaos Solitons Fractals 41:587–593
18. Rezvan F, Yasar E, Özer MN (2011) Group properties and conservation laws for nonlocal shallow water wave equation. J Appl Math Comput 218:974–979
19. Tzirtzilakis E, Marinakis V, Apokis C, Bountis T (2002) Soliton-like solutions of higher order wave equations of the Korteweg-de-Vries type. J Math Phys 43:6151–6165
20. Yasar E (2002) On the conservation laws and invariants solutions of the mKdV equation. J Math Anal Appl 363:174–181
21. Yasar E, Özer T (2010) Conservation laws for one layer shallow water wave systems. Nonlin Anal: Real World Appl 11:838–848

Chapter 2
Weak Self-Adjointness and Conservation Laws for a Family of Benjamin-Bona-Mahony-Burgers Equations

M.S. Bruzón

Abstract Ibragimov introduced the concepts of self-adjoint and quasi-self-adjoint equations. Gandarias generalized these concepts and defined the concept of weak self-adjoint equations. In this paper we consider a family of Benjamin-Bona-Mahony-Burgers equations and we determine the subclass of equations which are self-adjoint, quasi-self-adjoint and weak self-adjoint. By using a general theorem on conservation laws proved by Ibragimov we obtain conservation laws for these equations.

Keywords Weak self-adjointness • Conservation laws

2.1 Introduction

Nonlinear PDEs that admit conservation laws arise in many disciplines of the applied sciences including physical chemistry, fluid mechanics, particle and quantum physics, plasma physics, elasticity, gas dynamics, electromagnetism, magneto-hydro-dynamics, nonlinear optics, and the bio-sciences. Conservation laws are fundamental laws of physics. They maintain that a certain quantity, e.g. momentum, mass, or energy, will not change with time during physical processes.

In [16] (see also [15]) Ibragimov proved a general theorem on conservation laws for arbitrary differential equations which do not require the existence of Lagrangians. This new theorem is based on the concept of adjoint equations for nonlinear equations. There are many equations with physical significance which are not self-adjoint. Therefore one cannot eliminate the nonlocal variables from the conservation laws of these equations. Ibragimov in [15]) extended the

M.S. Bruzón (✉)
Departamento de Matemáticas, Universidad de Cádiz, Puerto Real, Cádiz, 11510, Spain
e-mail: m.bruzon@uca.es

J.A.T. Machado et al. (eds.), *Discontinuity and Complexity in Nonlinear Physical Systems*, Nonlinear Systems and Complexity 6, DOI 10.1007/978-3-319-01411-1_2,

concept of self-adjointness to quasi-self-adjointness. In [9] extended the concept of quasi-self-adjointness to weak-self-adjointness. Next, in [17] Ibragimov introduced a new concept: nonlinear self-adjointness.

Symmetry groups have several different applications in the context of nonlinear differential equations [3–5]. For example, they are used to obtain exact solutions and conservation laws of partial differential equations (PDEs) [8, 10]. The classical method for finding symmetry reductions of partial differential equations is the Lie group method [13, 18, 19]. The fundamental basis of this method is that, when a differential equation is invariant under a Lie group of transformations, a reduction transformation exists. For PDEs with two independent variables a single group reduction transforms the PDE into an ordinary differential equation (ODE), which in general is easier to solve.

The Benjamin-Bona-Mahony-Burgers (BBMB) equation

$$\Delta \equiv u_t - u_{xxt} - \alpha u_{xx} + \beta u_x + (g(u))_x = 0, \tag{2.1}$$

where $u(x,t)$ represents the fluid velocity in the horizontal direction x, α is a positive constant, $\beta \in \mathrm{R}$ and $g(u)$ is a C^2-smooth nonlinear function appears in [11]. Equation (2.1) is the alternative regularized long-wave equation proposed by Peregrine [20] and Benjamin [2]. In [5, 6] we studied similarity reductions of the BBMB equation (2.1) and we derived a set of new solitons, kinks, antikinks, compactons, and Wadati solitons.

Wang et al. [21] introduced a method which is called the $\frac{G'}{G}$-expansion method to look for travelling wave solutions of nonlinear evolution equations. In [7] we found the functions $g(u) = u^m$ for which we can apply the $\frac{G'}{G}$-expansion method to (2.1). We obtained new travelling wave solutions which did not appear in [5, 6]. In [1] the $\frac{G'}{G}$-expansion method is used to establish travelling wave solutions for special form of the generalized (2.1) with $\alpha = 0$, $\beta = 1$, and $g(u) = \frac{u^2}{2}$. The solutions given in [1] were obtained by Bruzón and Gandarias in [7] and Kudryashov in [12].

The aim of this work is to determine, for (2.1), the subclasses of equations which are self-adjoint, quasi-self-adjoint, and weak self-adjoint. We also determine, by using the notation and techniques of the work [15, 16], some nontrivial conservation laws for (2.1). The paper is organized as follows. In Sect. 2.2 we determine the subclasses of equations of (2.1) which are self-adjoint, quasi-self-adjoint, and weak self-adjoint. In Sect. 2.3 we give the Lie symmetries of (2.1) equation obtained by Bruzón and Gandarias in [5–7]. In Sect. 2.4 we obtain some nontrivial conservation laws for (2.1). Finally, in Sect. 2.5 we give conclusions.

2.2 Determination of Self-Adjoint Equations

In [16] Ibragimov introduced a new theorem on conservation laws. The theorem is valid for any system of differential equations where the number of equations is equal to the number of dependent variables. The new theorem does not require existence of a Lagrangian and this theorem is based on a concept of an adjoint equation for nonlinear equations.

Definition 1. Consider an sth-order partial differential equation

$$F(x, u, u_{(1)}, \ldots, u_{(s)}) = 0 \tag{2.2}$$

with independent variables $x = (x^1, \ldots, x^n)$ and a dependent variable u, where $u_{(1)} = \{u_i\}$, $u_{(2)} = \{u_{ij}\}, \ldots$ denote the sets of the partial derivatives of the first, second, etc. orders, $u_i = \partial u/\partial x^i$, $u_{ij} = \partial^2 u/\partial x^i \partial x^j$. The adjoint equation to (2.2) is

$$F^*(x, u, v, u_{(1)}, v_{(1)}, \ldots, u_{(s)}, v_{(s)}) = 0, \tag{2.3}$$

with

$$F^*(x, u, v, u_{(1)}, v_{(1)}, \ldots, u_{(s)}, v_{(s)}) = \frac{\delta(vF)}{\delta u}, \tag{2.4}$$

where

$$\frac{\delta}{\delta u} = \frac{\partial}{\partial u} + \sum_{s=1}^{\infty}(-1)^s D_{i_1} \cdots D_{i_s} \frac{\partial}{\partial u_{i_1 \cdots i_s}} \tag{2.5}$$

denotes the variational derivative (the Euler-Lagrange operator), and v is a new dependent variable. Here

$$D_i = \frac{\partial}{\partial x^i} + u_i \frac{\partial}{\partial u} + u_{ij} \frac{\partial}{\partial u_j} + \cdots$$

are the total differentiations.

Proposition 1. *Given the generalized BBMB equation (2.1), by applying definition (1), the adjoint equation to (2.1) is defined by*

$$F^* \equiv -\alpha u_{xx} - g_u u_x - \beta u_x + u_{txx} - u_t. \tag{2.6}$$

2.2.1 *Weak Self-Adjoint Equations*

We use the following definitions given in [15, 16].

Definition 2. Equation (2.2) is said to be **self-adjoint** if the equation obtained from the adjoint equation (2.3) by the substitution

$$v = u, \tag{2.7}$$

$$F^*(x, u, v, u_{(1)}, v_{(1)}, \ldots, u_{(s)}, v_{(s)})$$

is identical to the original equation (2.2).

Definition 3. Equation (2.2) is said to be **quasi-self-adjoint** if the equation obtained from the adjoint equation (2.3) is equivalent to the original equation (2.2) upon the substitution

$$v = h(u), \tag{2.8}$$

with a certain function $h(u)$ such that $h'(u) \neq 0$.

And the following definition given in [9].

Definition 4. Equation (2.2) is said to be **weak self-adjoint** if the equation obtained from the adjoint equation (2.3) by the substitution

$$v = h(x, t, u), \tag{2.9}$$

such that $h_x(x, t, u) \neq 0$, $h_u(x, t, u) \neq 0$, is identical to the original equation, i.e.

$$F^* |_{v=h} = \lambda F. \tag{2.10}$$

Given the generalized BBMB equation (2.1) we apply definition (4). Taking into account the expression (2.6) and using (2.9) and its derivatives we rewrite (2.10)

$$\begin{aligned} &u_t h_{uu} u_{xx} - \alpha h_u u_{xx} + h_{tu} u_{xx} - \alpha h_{xx} + u_t h_{uuu} u_x^2 - \alpha h_{uu} u_x^2 + h_{tuu} u_x^2 \\ &\quad -2\alpha h_{ux} u_x + 2u_t h_{uux} u_x + 2u_{tx} h_{uu} u_x - g_u h_u u_x - \beta h_u u_x + 2h_{tux} u_x \\ &\quad -g_u h_x - \beta h_x + u_t h_{uxx} + 2u_{tx} h_{ux} + u_{txx} h_u - u_t h_u + h_{txx} - h_t \\ &= \lambda(-\alpha\, u_{xx} + g_u\, u_x + \beta\, u_x - u_{txx} + u_t). \end{aligned} \tag{2.11}$$

Comparing the coefficients for u_{txx}, we obtain $\lambda + h_u = 0$ and the following conditions must be satisfied:

$$\begin{aligned} h_{uxx} &= 0, \\ h_{tu} - 2\alpha\, h_u &= 0, \\ 2\, h_{tux} - 2\alpha\, h_{ux} &= 0, \\ h_{tuu} - \alpha\, h_{uu} &= 0, \\ h_{uuu} &= 0, \\ h_{ux} &= 0, \\ h_{uu} &= 0, \\ h_{uux} &= 0, \\ \alpha\, h_{xx} + g_u\, h_x + \beta\, h_x - h_{txx} + h_t &= 0. \end{aligned} \tag{2.12}$$

Table 2.1 Weak self-adjoint equations (2.1)

Case_i	α	β	$g(u)$	h
1.	Arbitrary	Arbitrary	$k_3 - \dfrac{(k_2+\beta\,k_1)\,u}{k_1}$	$k_1\,x+k\,u+k_2\,t+k_3$
2.	0	Arbitrary	$k_3 - \dfrac{(k_2+\beta\,k_1)\,u}{k_1}$	$k_1\exp(-k\,t)$
3.	Arbitrary	Arbitrary	Arbitrary	C
4.	0	Arbitrary	Arbitrary	$k_1\,u+k_2$

Solving the system (2.12) we obtain that $h = k\,e^{2\alpha t}\,u + a(x,t)$ and α, β, $g(u)$ and $a(x,t)$ must satisfy the equation

$$2\,\alpha\,k\,e^{2\alpha\,t}\,u + a_x\,g_u + a_x\,\beta + \alpha\,a_{x\,x} - a_{t\,x\,x} + a_t = 0. \tag{2.13}$$

From (2.13) we obtain

- For $g(u) = k_3 - \dfrac{(k_2+\beta\,k_1)\,u}{k_1}$, with $k_1 \neq 0$ and α arbitrary constant

$$h = k_1\,x + k_2\,t + k_3.$$

- For $g(u) = k_3 - \dfrac{(k_2+\beta\,k_1)\,u}{k_1}$, with $k_1 \neq 0$ and $\alpha = 0$

$$h = k_1\,x + k\,u + k_2\,t + k_3.$$

- For α and β arbitrary constants and g arbitrary function

$$h = C, \qquad \text{with C constant.}$$

- For $\alpha = 0$, β arbitrary constants and g arbitrary function

$$h = k_1\,u + k_2.$$

Consequently, we deduce that

Proposition 2. *Equation (2.1) is weak self-adjoint in cases given in Table 2.1. We remark that for $\alpha = 0$, β arbitrary constants and g arbitrary function equation (2.1) is self-adjoint. For α and β arbitrary constants and g arbitrary function (2.1) is quasi-self-adjoint with $h = C$.*

2.3 Classical Symmetries

To apply the Lie classical method to (2.1) we consider the one-parameter Lie group of infinitesimal transformations in (x,t,u) given by

$$x^* = x + \epsilon\xi(x,t,u) + O(\epsilon^2), \tag{2.14}$$

$$t^* = t + \epsilon\tau(x,t,u) + O(\epsilon^2), \tag{2.15}$$

$$u^* = u + \epsilon\eta(x,t,u) + O(\epsilon^2), \tag{2.16}$$

where ϵ is the group parameter. We require that this transformation leaves invariant the set of solutions of (2.1). This yields to an overdetermined, linear system of equations for the infinitesimals $\xi(x,t,u)$, $\tau(x,t,u)$, and $\eta(x,t,u)$. The associated Lie algebra of infinitesimal symmetries is the set of vector fields of the form

$$\mathbf{v} = \xi(x,t,u)\frac{\partial}{\partial x} + \tau(x,t,u)\frac{\partial}{\partial t} + \eta(x,t,u)\frac{\partial}{\partial u}. \tag{2.17}$$

Having determined the infinitesimals, the symmetry variables are found by solving the characteristic equation which is equivalent to solving the invariant surface condition

$$\eta(x,t,u) - \xi(x,t,u)\frac{\partial u}{\partial x} - \tau(x,t,u)\frac{\partial u}{\partial t} = 0. \tag{2.18}$$

The set of solutions of (2.1) is invariant under the transformation (2.14)-(2.16) provided that

$$\mathrm{pr}^{(3)}\mathbf{v}(\Delta) = 0 \quad \text{when} \quad \Delta = 0,$$

where $\mathrm{pr}^{(3)}\mathbf{v}$ is the third prolongation of the vector field (2.17) given by

$$pr^{(3)}\mathbf{v} = \mathbf{v} + \sum_J \eta^J(x,t,u^{(3)})\frac{\partial}{\partial u_J}$$

where

$$\eta^J(x,t,u^{(3)}) = D_J(\eta - \xi u_x - \tau u_t) + \xi u_{Jx} + \eta u_{Jt},$$

with $J = (j_1,\ldots,j_k)$, $1 \le j_k \le 2$ y $1 \le k \le 3$. Hence we obtain the following ten determining equations for the infinitesimals:

$$\begin{aligned}
\tau_u &= 0,\\
\tau_x &= 0,\\
\xi_u &= 0,\\
\xi_t &= 0,\\
\eta_{uu} &= 0,\\
\alpha\tau_t + \eta_{tu} &= 0,\\
2\eta_{ux} - \xi_{xx} &= 0,\\
\eta_{uxx} - 2\xi_x &= 0,\\
\eta_x g_u - \alpha\eta_{xx} + \beta\eta_x - \eta_{txx} + \eta_t &= 0,\\
-\alpha\xi_{xx} - g_u\xi_x - \beta\xi_x - g_u\tau_t - \beta\tau_t - \eta g_{uu} + 2\alpha\eta_{ux} + 2\eta_{tux} &= 0.
\end{aligned} \tag{2.19}$$

From system (2.19) $\xi = \xi(x)$, $\tau = \tau(t)$ and $\eta = \gamma(x,t)u + \delta(x,t)$ where α, β, ξ, τ, γ, δ, and g satisfy

$$\begin{aligned}\gamma_t + \alpha\,\tau_t &= 0,\\ 2\,\gamma_x - \xi_{xx} &= 0,\\ \gamma_{xx} - 2\,\xi_x &= 0,\\ 2\,\alpha\,\gamma_x + 2\,\gamma_{tx} - g_{uu}\,u\,\gamma - \alpha\xi_{xx} - g_u\,\xi_x - \beta\xi_x - g_u\,\tau_t - \beta\,\tau_t - \delta\,g_{uu} &= 0,\\ -\alpha u\,\gamma_{xx} + g_u\,u\,\gamma_x + \beta\,u\,\gamma_x - u\,\gamma_{txx} + u\gamma_t + \delta_x\,g_u - \alpha\,\delta_{xx} + \beta\delta_x - \delta_{txx} + \delta_t &= 0.\end{aligned} \tag{2.20}$$

From (2.20) we obtain

$$\gamma = \frac{e^{-2x}}{8}\left((k_4 + 2\,k_3)\,e^{4\,x} + (4k_1 - 8\alpha\tau)\,e^{2x} - k_4 + 2k_3\right),$$

$$\xi - \frac{(k_4 + 2k_3)\;e^{2x}}{8} + \frac{(k_4 - 2k_3)\;e^{-2x}}{8} - \frac{k_4 - 4k_2}{4},$$

and α, β, τ, δ, and g are related by the following conditions:

$$\begin{aligned}&((g_u + \beta - 2\alpha)\,k_4 + (2g_u + 2\,\beta - 4\alpha)\,k_3)\,ue^{4x}\\ &+ \left(-4\alpha\tau_t u + \delta_x\,(4g_u + 4\beta) - 4\alpha\delta_{xx} - 4\delta_{txx} + 4\,\delta_t\right)\,e^{2\,x}\\ &+ ((g_u + \beta + 2\,\alpha)\;k_4 + (-2\,g_u - 2\,\beta - 4\,\alpha)\;k_3)\;u = 0,\end{aligned} \tag{2.21}$$

$$\begin{aligned}&((g_{uu}\,k_4 + 2\,g_{uu}\,k_3)\;u + (2\,g_u + 2\,\beta)\;k_4 + (4g_u + 4\beta)\,k_3)\,e^{4\,x}\\ &+ ((4g_{uu}k_1 - 8\alpha g_{uu}\tau)\,u + 8g_u\tau_t + 8\,\beta\,\tau_t + 8\,\delta g_{uu})\;e^{2\,x} + (2\,g_{uu}k_3 - g_{uu}k_4)\;u\\ &+ (-2g_u - 2\,\beta)\;k_4 + (4g_u + 4\,\beta)\;k_3 = 0.\end{aligned} \tag{2.22}$$

Solving system (2.21)-(2.22) we obtain that if g is an arbitrary function the only symmetries admitted by (2.1) are

$$\xi = k_1, \quad \tau = k_2, \quad \eta = 0. \tag{2.23}$$

The generators of this are $\mathbf{v}_1 = \dfrac{\partial}{\partial x}$ (corresponding to space translational invariance) and $\mathbf{v}_2 = \dfrac{\partial}{\partial t}$ (time translational invariance). In the following cases (2.1) has extra symmetries:

(i) If $\alpha = 0$, $g(u) = -\beta u + \frac{k}{a(n+1)}(au + b)^{n+1}$, $a \neq 0$,

$$\xi = k_1, \quad \tau = k_2 t + k_3, \quad \eta = -\frac{k_2}{an}(au + b).$$

Besides $\mathbf{v}_1$ and $\mathbf{v}_2$, we obtain the infinitesimal generator

$$\mathbf{v}_3 = t\partial_t - \frac{au+b}{an}\partial_u.$$

(ii) If $\alpha \neq 0$, $\beta \neq 0$ and $g(u) = au + b$,

$$\xi = k_1, \quad \tau = k_2, \quad \eta = \delta(x,t),$$

where δ satisfy

$$\alpha\delta_{xx} - g_u\delta_x - \beta\delta_x + \delta_{txx} - \delta_t = 0.$$

2.4 General Theorem on Conservation Laws

Much of the research on conservation laws centers around applications of Noether's theorem, which requires the existence of a Lagrangian. Anco and Bluman developed a procedure. The advantage of this procedure is that, in the Lagrangian case, it bypasses the actual formulation of the Lagrangian, and more importantly, it is applicable to non-Lagrangian systems.

Given a PDE (2.2) a conservation law for (2.2) is a relation of the form

$$\nabla \cdot \mathbf{C} = D_t(C^1) + D_x(C^2) = 0 \tag{2.24}$$

where $\mathbf{C} = (C^1, C^2)$ represents the conserved flux and density, respectively, and D_x, D_t denote the total derivative operators with respect to x and t, respectively. If (2.24) is a conservation law for (2.2), then it can be shown that there exists an operator λ such that

$$\nabla \cdot \mathbf{C} = \lambda(u)F$$

The operator λ is called the characteristic of the conservation law.

The conservation laws determined via Noether's theorem need to have a Lagrangian formulation. Noether's theorem connects conservation laws with variational symmetries with infinitesimal generators

We use the following theorem on conservation laws proved in [16]. Any Lie point, Lie-Bäcklund, or non-local symmetry

$$X = \xi^i(x, u, u_{(1)}, \ldots)\frac{\partial}{\partial x^i} + \eta(x, u, u_{(1)}, \ldots)\frac{\partial}{\partial u} \tag{2.25}$$

of Eq. (2.2) provides a conservation law $D_i(C^i) = 0$ for the simultaneous system (2.2), (2.3). The conserved vector is given by

$$C^i = \xi^i \mathcal{L} + W \left[\frac{\partial \mathcal{L}}{\partial u_i} - D_j \left(\frac{\partial \mathcal{L}}{\partial u_{ij}} \right) + D_j D_k \left(\frac{\partial \mathcal{L}}{\partial u_{ijk}} \right) - \cdots \right]$$
$$+ D_j(W) \left[\frac{\partial \mathcal{L}}{\partial u_{ij}} - D_k \left(\frac{\partial \mathcal{L}}{\partial u_{ijk}} \right) + \cdots \right] + D_j D_k(W) \left[\frac{\partial \mathcal{L}}{\partial u_{ijk}} - \cdots \right] + \cdots, \tag{2.26}$$

where W and $\mathcal{L}$ are defined as follows:

$$W = \eta - \xi^j u_j, \quad \mathcal{L} = v\, F\left(x, u, u_{(1)}, \ldots, u_{(s)}\right). \tag{2.27}$$

The proof is based on the following operator identity (N.H. Ibragimov, 1979):

$$X + D_i(\xi^i) = W \frac{\delta}{\delta u} + D_i \mathcal{N}^i, \tag{2.28}$$

where X is operator (2.25) taken in the prolonged form:

$$X = \xi^i \frac{\partial}{\partial x^i} + \eta \frac{\partial}{\partial u} + \zeta_i \frac{\partial}{\partial u_i} + \zeta_{i_1 i_2} \frac{\partial}{\partial u_{i_1 i_2}} + \cdots,$$
$$\zeta_i = D_i(\eta) - u_j D_i(\xi^j), \quad \zeta_{i_1 i_2} = D_{i_2}(\zeta_{i_1}) - u_{j i_1} D_{i_2}(\xi^j), \ \ldots.$$

For the expression of operator $\mathcal{N}^i$ and a discussion of the identity (2.28) in the general case of several dependent variables, see [14] (Sect. 8.4.4).

We will write the generators of a point transformation group admitted by (2.1) in the form

$$X = \xi^1 \frac{\partial}{\partial t} + \xi^2 \frac{\partial}{\partial x} + \eta \frac{\partial}{\partial u}$$

by setting $t = x^1,\ x = x^2$. The conservation law will be written as (2.24)

Now we use the Ibragimov's Theorem on conservation laws to establish the conservation laws of (2.1). We have obtained that equation (2.1) is self-adjoint when it has the following form

$$u_t - u_{xxt} + \beta u_x + (g(u))_x = 0. \tag{2.29}$$

In this case, the formal Lagragian is

$$\mathcal{L} = v(u_t - u_{xxt} - \alpha u_{xx} + \beta u_x + (g(u))_x).$$

For α and β arbitrary constants, $g(u)$ arbitrary function and $h = C$, (2.29) admits the generator $\mathbf{v}_1 + \mathbf{v}_2$. In this case we obtain trivial conservation laws.

Equation (2.29) admits the generator

$$\mathbf{v}_3 = t\partial t - \frac{1}{an}(au+b)\partial u,$$

and the normal form for this group is

$$W = -\frac{1}{an}(au+b) - t\,u_t.$$

The vector components are

$$\begin{aligned}
C^1 &= \frac{tu_t v_{xx}}{3} + \frac{uv_{xx}}{3n} + \frac{bv_{xx}}{3an} - \frac{u_x v_x}{3n} - \frac{tu_{tx} v_x}{3} + \frac{u_{xx} v}{3n} \\
&\quad + kt\,(au+b)^n\,u_x v - \frac{2tu_{txx} v}{3} - \frac{uv}{n} - \frac{bv}{an} \\
C^2 &= -\frac{tu_{tt} v_x}{3} - \frac{u_t v_x}{3n} - \frac{u_t v_x}{3} + \frac{2tu_t v_{tx}}{3} + \frac{2uv_{tx}}{3n} \\
&\quad + \frac{2bv_{tx}}{3an} - \frac{u_x v_t}{3n} - \frac{tu_{tx} v_t}{3} + \frac{2tu_{ttx} v}{3} + \frac{2u_{tx} v}{3n} \\
&\quad + \frac{2u_{tx} v}{3} - kt\,(au+b)^n\,u_t v - \frac{ku\,(au+b)^n\,v}{n} \\
&\quad - \frac{bk\,(au+b)^n\,v}{an}
\end{aligned} \tag{2.30}$$

Setting $v = u$ in (2.30)

$$\begin{aligned}
C^1 &= \frac{tu_t u_{xx}}{3} + \frac{2uu_{xx}}{3n} + \frac{bu_{xx}}{3an} - \frac{(u_x)^2}{3n} - \frac{tu_{tx} u_x}{3} \\
&\quad + ktu\,(au+b)^n\,u_x - \frac{2tuu_{txx}}{3} - \frac{u^2}{n} - \frac{bu}{an}, \\
C^2 &= -\frac{tu_{tt} u_x}{3} - \frac{2u_t u_x}{3n} - \frac{u_t u_x}{3} + \frac{2tuu_{ttx}}{3} + \frac{tu_t u_{tx}}{3} \\
&\quad + \frac{4uu_{tx}}{3n} + \frac{2uu_{tx}}{3} + \frac{2bu_{tx}}{3an} - ktu\,(au+b)^n\,u_t \\
&\quad - \frac{ku^2\,(au+b)^n}{n} - \frac{bku\,(au+b)^n}{an}.
\end{aligned} \tag{2.31}$$

We simplify the conserved vector by transferring the terms of the form $D_x(\ldots)$ from C^1 to C^2 and obtain

$$\begin{aligned}
C^1 &= -\frac{(u_x)^2}{n} - \frac{u\,(a\,u+b)}{a\,n} \\
C^2 &= \frac{(2au+b)\,u_{tx}}{an} - \frac{k\,(au+b)^{n+1}\,(2a(n+1)u+bn)}{a^2 n\,(n+1)\,(n+2)}
\end{aligned} \tag{2.32}$$

For $g(u) = k_3 - \dfrac{(k_2 + \beta\, k_1)\, u}{k_1}$, with $k_1 \neq 0$, α arbitrary constant and $h = k_1\, x + k_2\, t + k_3$ (2.29) admits the generator $\mathbf{v}_1 + \mathbf{v}_2$. In this case, we do as before and we obtain

$$\begin{aligned} C^1 &= ({k_2}^2 + {k_1}^2)\, u \\ C^2 &= -\alpha\, ({k_2}^2 + {k_1}^2)\, u_x - ({k_2}^2 + {k_1}^2)\, u_{t\,x} - k_2 \left(\frac{{k_2}^2}{k_1} + k_1 \right) u \end{aligned} \tag{2.33}$$

For $g(u) = k_3 - \dfrac{(k_2 + \beta\, k_1)\, u}{k_1}$, with $k_1 \neq 0$ and $\alpha = 0$ and $h = k_1\, x + k\, u + k_2\, t + k_3$ (2.29) admits the generator $\mathbf{v}_1 + \mathbf{v}_2$. In this case, we proceed as before and we obtain the conservation law (2.33) with $\alpha = 0$.

2.5 Conclusions

In this work we have considered a generalized Benjamin-Bona-Mahony-Burgers equation (2.1). We have determined the subclasses of equations (2.1) which are self-adjoint, quasi-self-adjoint, and weak self-adjoint. By using a general theorem on conservation laws proved by Nail Ibragimov we found conservation laws for some of these partial differential equations without classical Lagrangians.

References

1. Abdollahzadeh M, Hosseini M, Ghanbarpour M, Kashani S (2011) Exact travelling solutions for Benjamin-Bona-Mahony-Burgers equations by $\frac{G'}{G}$-expansion method. Inter J Appl Math Comput 3(1):70–76
2. Benjamin TB, Bona JL, Mahony JJ (1972) Model equations for long waves in nonlinear dispersive systems. Phil Trans R Soc A 272:47–78
3. Bruzón MS, Gandarias ML, Camacho JC (2007) Classical and nonclassical symmetries for a Kuramoto-Sivashinsky equation with dispersive effects. Math Meth Appl Sci 30:2091–2100
4. Bruzón MS, Gandarias ML, Camacho JC (2008) Symmetry for a family of BBM equations. J Nonlin Math Phys 15:81–90
5. Bruzón MS, Gandarias ML (2008) New solutions for a generalized Benjamin-Bona-Mahony-Burgers equation. In: Proceedings of American conference on applied mathematics, Cambridge, Massachusetts, 2008, pp 159–164
6. Bruzón MS, Gandarias ML (2008) Travelling wave solutions for a generalized Benjamin-Bon-Mahony-Burgers equation. Inter J Math Models Methods Appl Sci Ed Elect 2:103–108
7. Bruzón MS, Gandarias ML (2009) Symmetry reductions and exact solutions of Benjamin-Bona-Mahony-Burgers equation. In: 4th workshop Group Analysis of Differential Equations & Integrable Systems, pp 45–61
8. Bruzón MS, Gandarias ML, Ibragimov NH (2009) Self-adjoint sub-classes of generalized thin film equations. J Math Anal Appl 357:307–313

9. Gandarias ML (2011) Weak self-adjoint differential equations. J Phys A: Math Theoret 44:262001
10. Gandarias ML, Redondo M, Bruzon MS (2011) Some weak self-adjoint Hamilton-Jacobi-Bellman equations arising in financial mathematics. Nonlin Anal: Real World Appl 13: 340–347
11. Khaled KA, Momani S, Alawneh A (2005) Approximate wave solutions for generalized Benjamin-Bona-Mahony-Burgers equations. Appl Math Comput 171:281–292
12. Kudryashov NA (2005) Simplest equation method to look for exact solutions of nonlinear differential equations. Chaos, Solitons and Fractals 24:1217–1231
13. Ibragimov NH (1985) Transformation groups applied to mathematical physics. Reidel-Dordrecht, Holland
14. Ibragimov NH (1999) Elementary lie group analysis and ordinary differential equations. Wiley, Chichester
15. Ibragimov NH (2007) Quasi self-adjoint differential equations. Arch ALGA 4:55–60
16. Ibragimov NH (2007) A new conservation theorem. J Math Anal Appl 333:311–328
17. Ibragimov NH (2011) Nonlinear self-adjointness and conservation laws. J Phys A: Math Theor 44:432002–432010
18. Olver P (1993) Applications of Lie groups to differential equations. Springer, New York
19. Ovsyannikov LV (1982) Group analysis of differential equations. Academic, New York
20. Peregrine DH (1996) Calculations of the development of an undular bore. J Fluid Mech 25:321–330
21. Wang M, Li X, Zhang J (2008) The $\left(\frac{G'}{G}\right)$-expansion method and travelling wave solutions of nonlinear evolution equations in mathematical physics. Phys Lett A 372:417–423

Chapter 3
Some Analytical Techniques in Fractional Calculus: Realities and Challenges

Dumitru Baleanu, Guo-Cheng Wu, and Jun-Sheng Duan

Abstract In the last decades, much effort has been dedicated to analytical aspects of the fractional differential equations. The Adomian decomposition method and the variational iteration method have been developed from ordinary calculus and become two frequently used analytical methods. In this article, the recent developments of the methods in the fractional calculus are reviewed. The realities and challenges are comprehensively encompassed.

Keywords Fractional differential equations • Adomian decomposition method • Variational iteration method • Riemann–Liouville derivative • Caputo derivative • Adomian polynomials • One-step numeric algorithms • Approximate solutions

---

D. Baleanu (✉)
Department of Mathematics and Computer Sciences, Cankaya University, 06530 Balgat, Ankara, Turkey

Institute of Space Sciences, Magurele-Bucharest, Romania

Department of Chemical and Materials Engineering, Faculty of Engineering, King Abdulaziz University, Jeddah, Saudi Arabia
e-mail: dumitru@cankaya.edu.tr

G.-C. Wu
College of Mathematics and Information Science, Neijiang Normal University, Neijiang 641112, P.R. China
e-mail: wuguocheng@gmail.com

J.-S. Duan
School of Mathematics and Information Sciences, Zhaoqing University, Zhaoqing, Guang Dong 526061, P.R. China
e-mail: duanjsssdu@sina.com

J.A.T. Machado et al. (eds.), *Discontinuity and Complexity in Nonlinear Physical Systems*, Nonlinear Systems and Complexity 6, DOI 10.1007/978-3-319-01411-1_3,

3.1 Introduction

We review the basic definitions of the Riemann–Liouville (R–L) and the Caputo derivatives. For additional details readers can refer to references [19, 21, 30, 58, 60, 61, 65, 70].

Definition 1. Let $f(t)$ be a function of class $\mathfrak{C}$, i.e. piecewise continuous on $(t_0, +\infty)$ and integrable on any finite subinterval of $(t_0, +\infty)$. Then for $t > t_0$, the Riemann–Liouville integral of $f(t)$ of β order is defined as

$$ {}_{t_0}I_t^{\beta} f(t) = \frac{1}{\Gamma(\beta)} \int_{t_0}^{t} (t-\tau)^{\beta-1} f(\tau) d\tau, \tag{3.1}$$

where β is a positive real number and $\Gamma(\cdot)$ is Euler's Gamma function.

The fractional integral satisfies the following equalities,

$$ {}_{t_0}I_t^{\beta}\ {}_{t_0}I_t^{\mu} f(t) = {}_{t_0}I_t^{\beta+\mu} f(t),\ \beta \geq 0,\ \mu \geq 0, \tag{3.2}$$

$$ {}_{t_0}I_t^{\nu} (t-t_0)^{\mu} = \frac{\Gamma(\mu+1)}{\Gamma(\mu+\nu+1)} (t-t_0)^{\mu+\nu},\ \nu \geq 0,\ \mu > -1. \tag{3.3}$$

Definition 2. Let $f(t)$ be a function of class $\mathfrak{C}$ and α be a positive real number satisfying $m-1 < \alpha \leq m$, $m \in \mathbb{N}^+$, where $\mathbb{N}^+$ is the set of positive integers. Then, the Riemann–Liouville derivative of $f(t)$ of order α is defined as (when it exists)

$$ {}_{t_0}\mathbf{D}_t^{\alpha} f(t) = \frac{d^m}{dt^m} \left({}_{t_0}I_t^{m-\alpha} f(t) \right),\ t > t_0. \tag{3.4}$$

Defining for complementarity ${}_{t_0}\mathbf{D}_t^0 = I$, the identity operator, then ${}_{t_0}\mathbf{D}_t^{\alpha} f(t) = f^{(\alpha)}(t)$ if $\alpha = m, m = 0, 1, 2, \ldots$.

Note that the Riemann–Liouville fractional derivative ${}_{t_0}\mathbf{D}_t^{\alpha} f(t)$ is not zero for the constant function $f(t) \equiv C$ if $\alpha > 0$ and $\alpha \notin \mathbb{N}^+$.

For the power functions, the following holds

$$ {}_{t_0}\mathbf{D}_t^{\alpha} (t-t_0)^{\mu} = \frac{\Gamma(\mu+1)}{\Gamma(\mu-\alpha+1)} (t-t_0)^{\mu-\alpha}, \tag{3.5}$$

where $\mu > -1,\ 0 \leq m-1 < \alpha \leq m,\ t > t_0$.

Definition 3. Let α be a positive real number, $m-1 < \alpha \leq m$, $m \in \mathbb{N}^+$, and $f^{(m)}(t)$ exist and be a function of class $\mathfrak{C}$. Then the Caputo fractional derivative of $f(t)$ of order α is defined as

$$ {}_{t_0}D_t^{\alpha} f(t) = {}_{t_0}I_t^{m-\alpha} f^{(m)}(t),\ t > t_0. \tag{3.6}$$

Defining for complementarity ${}_{t_0}D_t^0 = I$, the identity operator, then ${}_{t_0}D_t^{\alpha} f(t) = f^{(\alpha)}(t)$ if $\alpha = m, m = 0, 1, 2, \ldots$.

For the Caputo fractional derivative, the following equality holds

$$ {}_{t_0}D_t^{\alpha}(a_0 t^r + a_1 t^{r-1} + \cdots + a_r) = 0,\ m-1 < \alpha \le m, \tag{3.7}$$

where the degree of the polynomial about t is no more than $m-1$, i.e. $r \le m-1$. Moreover, the α-order integral of the α-order Caputo derivative requires the initial values of the function and its integer order derivatives,

$$ {}_{t_0}I_t^{\alpha}\ {}_{t_0}D_t^{\alpha} f(t) = f(t) - \sum_{k=0}^{m-1} f^{(k)}(t_0^+)\frac{(t-t_0)^k}{k!},\ m-1 < \alpha \le m. \tag{3.8}$$

Furthermore, for $\beta > \alpha > 0$, $m-1 < \alpha \le m$, we have

$$\begin{aligned} {}_{t_0}I_t^{\beta}\ {}_{t_0}D_t^{\alpha} f(t) &= {}_{t_0}I_t^{\beta-\alpha}\ {}_{t_0}I_t^{\alpha}\ {}_{t_0}D_t^{\alpha} f(t) \\ &= {}_{t_0}I_t^{\beta-\alpha} f(t) - \sum_{k=0}^{m-1} f^{(k)}(t_0^+)\frac{(t-t_0)^{k+\beta-\alpha}}{\Gamma(k+1+\beta-\alpha)}. \end{aligned} \tag{3.9}$$

For the Caputo derivative of the power function $(t-t_0)^{\mu}$, $\mu > 0$, if $0 \le m-1 < \alpha \le m < \mu+1$, then we have

$$ {}_{t_0}D_t^{\alpha}(t-t_0)^{\mu} = \frac{\Gamma(\mu+1)}{\Gamma(\mu-\alpha+1)}(t-t_0)^{\mu-\alpha},\ t > t_0 \tag{3.10}$$

and

$$ {}_{t_0}D_t{}^{\alpha} f(t) = {}_{t_0}\mathbf{D}_t{}^{\alpha}\left[f(t) - \sum_{k=0}^{m-1}\frac{(t-t_0)^k}{k!} f^{(k)}(t_0^+)\right]. \tag{3.11}$$

3.2 Adomian Decomposition Method

3.2.1 A Review of the Method

The Adomian decomposition method (ADM) [3–7, 24, 40, 40, 83, 84] is a powerful tool solving both linear and nonlinear functional equations, including ordinary differential equations (ODEs), partial differential equations (PDEs), integral equations, integro-differential equations, etc. The ADM provides efficient algorithms for analytic approximate solutions and numeric simulations for real-world applications in the applied sciences and engineering. It permits us to solve both nonlinear initial value problems (IVPs) and boundary value problems (BVPs) [6, 10, 11, 18, 26, 36, 37, 40, 62, 82, 83] without unphysical restrictive assumptions such as required by linearization, perturbation, and guessing the initial term or a set of basis functions.

Furthermore the ADM does not require the use of Green's functions, which would complicate such analytic calculations since Green's functions are not easily determined in most cases. The accuracy of the analytic approximate solutions can be verified by direct substitution. Advantages of the ADM over Picard's iterated method were demonstrated in [72]. Advantages of the ADM in computation were demonstrated in [86]. A key notion is the Adomian polynomials [7], which are tailored to the particular nonlinearity to solve nonlinear operator equations.

The ADM solves nonlinear operator equations with any analytic nonlinearity, including polynomial, exponential, trigonometric, hyperbolic, negative-power, and even decimal-power nonlinearities [40], providing us with an easily computable, readily verifiable, and rapidly convergent sequence of analytic approximate solutions.

Let first recall the basic principles of the ADM using an IVP for a nonlinear ODE in the form

$$L[u] + R[u] + N[u] = g(t), \tag{3.12}$$

where g is the system input and u is the system output, and where L is the linear operator to be inverted, which usually is just the highest order differential operator, R is the linear remainder operator, and N is the nonlinear operator, which is assumed to be analytic.

We emphasize that the choice for L and concomitantly its inverse L^{-1} are determined by the particular equation to be solved; hence, the choice is nonunique [40]. Generally, we choose $L = \frac{d^p}{dt^p}(\cdot)$ for pth-order differential equations and thus its inverse L^{-1} follows as the p-fold definite integration operator from t_0 to t. We have $L^{-1}Lu = u - \Phi$, where Φ incorporates the initial values.

Applying the inverse linear operator L^{-1} to both sides of (3.12) it gives

$$u = \gamma(t) - L^{-1}[R[u] + N[u]], \tag{3.13}$$

where $\gamma(t) = \Phi + L^{-1}g$.

The ADM decomposes the solution $u(t)$ into a series of solution components, and then decomposes the analytic nonlinearity $N[u]$ into the series of the Adomian polynomials [3, 4, 7, 39, 40, 73]

$$u(t) = \sum_{n=0}^{\infty} u_n, \; N[u] = \sum_{n=0}^{\infty} A_n, \tag{3.14}$$

where $A_n = A_n(u_0, u_1, \ldots, u_n)$ are the Adomian polynomials, which are defined by the formula [7]

$$A_n = \frac{1}{n!}\frac{d^n}{d\lambda^n} N(\sum_{k=0}^{\infty} u_k \lambda^k)\bigg|_{\lambda=0}, \; n \geq 0. \tag{3.15}$$

For convenient reference, we list the first five Adomian polynomials for the general analytic nonlinearity $N[u] = f(u)$ as follows

$$
\begin{aligned}
A_0 &= f(u_0),\\
A_1 &= f'(u_0)u_1,\\
A_2 &= f'(u_0)u_2 + f''(u_0)\frac{u_1^2}{2!},\\
A_3 &= f'(u_0)u_3 + f''(u_0)u_1u_2 + f'''(u_0)\frac{u_1^3}{3!},\\
A_4 &= f'(u_0)u_4 + f''(u_0)\left(\frac{u_2^2}{2!} + u_1u_3\right) + f'''(u_0)\frac{u_1^2u_2}{2!} + f^{(4)}(u_0)\frac{u_1^4}{4!}.
\end{aligned}
$$

Several algorithms [7, 71, 73, 81] for symbolic programming have been devised to efficiently generate the Adomian polynomials quickly and to high orders. New, efficient algorithms and subroutines in MATHEMATICA for rapid computer-generation of the Adomian polynomials to high orders have been provided by Duan in [31–33], including the single variable and multivariable cases.

For the case of the one-variable Adomian polynomials, we list Duan's Corollary 3 algorithm [33] as follows

$$
A_0 = f\left(u_0\right), \quad A_n = \sum_{k=1}^{n} C_n^k\, f^{(k)}\left(u_0\right), \text{ for } n \geq 1, \tag{3.16}
$$

where the coefficients C_n^k are defined recursively as

$$
\begin{aligned}
C_n^1 &= u_n,\ n \geq 1,\\
C_n^k &= \frac{1}{n}\sum_{j=0}^{n-k}(j+1)\,u_{j+1}C_{n-1-j}^{k-1},\ 2 \leq k \leq n.
\end{aligned} \tag{3.17}
$$

We emphasize that in this algorithm, the recursion operations for the coefficients C_n^k do not involve the differentiation, but only require the elementary operations of addition and multiplication, and are thus eminently convenient for computer algebra systems such as MATHEMATICA, MAPLE, or MATLAB.

Upon substitution of the Adomian decomposition series for the solution $u(t)$ and the series of Adomian polynomials tailored to the nonlinearity $N[u]$ from (3.14) into (3.13), we have

$$
\sum_{n=0}^{\infty} u_n = \gamma(t) - L^{-1}\left[R\sum_{n=0}^{\infty} u_n + \sum_{n=0}^{\infty} A_n\right]. \tag{3.18}
$$

The solution components $u_n(t)$ may be determined by one of the several advantageous recursion schemes, which differ from one another by the choice of the initial solution component $u_0(t)$, beginning with the classic Adomian recursion scheme

$$\begin{aligned} u_0(t) &= \gamma(t), \\ u_{n+1}(t) &= -L^{-1}[R[u_n] + A_n],\ n \geq 0, \end{aligned} \tag{3.19}$$

where Adomian has chosen the initial solution component as $u_0 = \gamma(t)$. The n-term approximation of the solution is

$$\varphi_n(t) = \sum_{k=0}^{n-1} u_k(t). \tag{3.20}$$

By various partitions of the original initial term and then delaying the contribution of its remainder by different algorithms, we can design alternate recursion schemes, such as the Adomian–Rach [10,11], Wazwaz [80], Wazwaz-El-Sayed [85], Duan [31], and Duan–Rach [37, 42] modified recursion schemes for different computational advantages.

Several researchers [2, 23, 24, 47, 73] have previously proved convergence of the Adomian decomposition series and the series of the Adomian polynomials. For example, Cherruault and Adomian [24] have proved convergence of the decomposition series without appealing to the fixed point theorem, which is too restrictive for most physical and engineering applications. Furthermore, Abdelrazec and Pelinovsky [2] have published a rigorous proof of convergence for the ADM under the aegis of the Cauchy–Kovalevskaya theorem for IVPs. A key concept is that the Adomian decomposition series is a computationally advantageous rearrangement of the Banach-space analog of the Taylor expansion series about the initial solution component function, which permits solution by recursion. A remarkable measure of success of the ADM is demonstrated by its widespread adoption and many adaptations to enhance computability for specific purposes, such as the various modified recursion schemes. The choice of decomposition is nonunique, which provides a valuable advantage to the analyst, permitting the freedom to design modified recursion schemes for ease of computation in realistic systems.

In [5], Adomian introduced the concept of the accelerated Adomian polynomials $\hat{A}_n$. In [12], Adomian and Rach presented two new kinds of modified Adomian polynomials $\overline{A}_n$ and $\overline{\overline{A}}_n$. Rach [73] gave a new definition of the Adomian polynomials, in which different classes of the Adomian polynomials were defined within the same premise. Duan [34] presented new recurrence algorithms for these nonclassic Adomian polynomials. Generalized forms of the Adomian polynomials were also proposed by Duan [35].

We remark that the domain of the convergence for the decomposition series solution, like other series solutions, may not always be sufficiently large for engineering purposes. But we can readily address this issue by means of one of the several common convergence acceleration techniques, such as the diagonal Padé approximants [6,40,67,75,83] or the iterated Shanks transform [6,41]. For example, the MATHEMATICA built-in command "PadeApproximant" can be used to easily generate the Padé approximants.

Rach and Duan [75] presented the combined solution of the near-field and far-field approximations by the Adomian and asymptotic decomposition methods, where the Padé approximant technique was used in the mid-field region as necessary. In the ADM, Duan's parametrized recursion scheme [31, 37, 42] was also proposed in order to obtain decomposition solutions with large effective regions of convergence.

The multistage ADM and its numeric schemes were considered in [13,15,36,38]. In [36], Duan and Rach considered one-step numeric algorithms for IVPs based on the ADM and the Rach–Adomian–Meyers MDM, respectively. In [38] higher-order numeric schemes based on the Wazwaz-El-Sayed modified ADM were proposed.

Duan and Rach [37] and Duan et al. [43, 44] have introduced new error analysis formulas for the approximate decomposition solutions when the exact solution is unknown in advance. When the exact solution is known in advance, we can use the usual error function, e.g. [45]. However when the exact solution is unknown in advance, we instead compute the following error remainder function

$$ER_n(t) = L[\varphi_n(t)] + R[\varphi_n(t)] + N[\varphi_n(t)] - g(t), \tag{3.21}$$

which we recommend as the best objective measure of how well the sequence of solution approximants $\varphi_n(t)$ satisfy the original nonlinear differential equation. In our error analysis, we also compute the maximal error remainder parameter

$$MER_n = \max_{a \le t \le b} |ER_n(t)|, \tag{3.22}$$

where the logarithmic plots of the maximal error remainder parameter versus the number of solution components per solution approximant characterize the rate of convergence, e.g. a linear relation signifies an exponential rate of convergence. Thus our new approach yields an analytic, readily verifiable and rapidly convergent approximation to the solution of the authentic nonlinear differential equation that represents the actual physical process under consideration.

For generalization and applications of the ADM to linear or nonlinear and ordinary or partial fractional differential equations (FDEs), see [16,25,39,40,48,67, 77, 78, 92]. A numeric scheme solving the FDEs based on the ADM was designed in [63].

Example 1. Consider the IVP for the nonlinear FDE with a composite nonlinearity

$${}_0D_t^{\alpha} u(t) + {}_0D_t^{\beta} u(t) + e^{-u^2(t)} = 1, \tag{3.23}$$

$$u(0) = 1,\ u'(0) = -1, \tag{3.24}$$

where α and β are real numbers satisfying $1 < \alpha \le 2$ and $0 < \beta \le 1$.

Applying the fractional integral operator ${}_0I_t^{\alpha}$ to both sides of (3.23) yields

$$u(t) = 1 - t + \frac{t^{\alpha}}{\Gamma(1+\alpha)} + \frac{t^{\alpha-\beta}}{\Gamma(\alpha-\beta+1)} - {}_0I_t^{\alpha-\beta}u(t) - {}_0I_t^{\alpha}(e^{-u^2(t)}), \quad (3.25)$$

where we have used the formula ${}_0I_t^{\alpha}\ {}_0D_t^{\beta}u(t) = {}_0I_t^{\alpha-\beta}u(t) - \frac{u(0)\,t^{\alpha-\beta}}{\Gamma(\alpha-\beta+1)}$. We decompose the solution as $u(t) = \sum_{n=0}^{\infty} u_n$ and the nonlinearity as $e^{-u^2(t)} = \sum_{n=0}^{\infty} A_n$, where the Adomian polynomials in terms of the solution components u_n are

$$
\begin{aligned}
A_0 &= e^{-u_0^2},\\
A_1 &= -2e^{-u_0^2}u_0u_1,\\
A_2 &= e^{-u_0^2}\left(-u_1^2 + 2u_0^2u_1^2 - 2u_0u_2\right),\\
A_3 &= e^{-u_0^2}\left(2u_0u_1^3 - \frac{4}{3}u_0^3u_1^3 - 2u_1u_2 + 4u_0^2u_1u_2 - 2u_0u_3\right),\\
A_4 &= e^{-u_0^2}\left(\frac{u_1^4}{2} - 2u_0^2u_1^4 + 6u_0u_1^2u_2 - 4u_0^3u_1^2u_2 - u_2^2 + 2u_0^2u_2^2 - 2u_1u_3\right.\\
&\quad \left. +4u_0^2u_1u_3 - 2u_0u_4 + \frac{2}{3}u_0^4u_1^4\right),\\
&\cdots .
\end{aligned}
$$

Substituting the decompositions of the solution and the nonlinearity into (3.25), and using Wazwaz's modified recursion scheme [80, 83] for the solution components in order to develop easy-to-integrate series, we calculate

$$u_0 = 1, \quad (3.26)$$

$$u_1 = -t + \frac{t^{\alpha}}{\Gamma(1+\alpha)} + \frac{t^{\alpha-\beta}}{\Gamma(\alpha-\beta+1)} - {}_0I_t^{\alpha-\beta}u_0 - {}_0I_t^{\alpha}A_0, \quad (3.27)$$

$$u_{n+1} = -{}_0I_t^{\alpha-\beta}u_n - {}_0I_t^{\alpha}A_n,\ n = 1, 2, \cdots, \quad (3.28)$$

where the recurrence operation involves fractional integrations. Further computations lead to

$$
\begin{aligned}
u_1 &= -t + \frac{(-1+e)t^{\alpha}}{e\Gamma(1+\alpha)},\\
u_2 &= -\frac{2t^{1+\alpha}}{e\Gamma(2+\alpha)} + \frac{t^{1-\beta+\alpha}}{\Gamma(2-\beta+\alpha)} - \frac{2t^{2\alpha}}{e^2\Gamma(1+2\alpha)} + \frac{2t^{2\alpha}}{e\Gamma(1+2\alpha)}\\
&\quad - \frac{t^{-\beta+2\alpha}}{\Gamma(1-\beta+2\alpha)} + \frac{t^{-\beta+2\alpha}}{e\Gamma(1-\beta+2\alpha)},\\
&\cdots .
\end{aligned}
$$

The nth-stage solution approximant is $\varphi_n(t;\alpha,\beta) = \sum_{k=0}^{n-1} u_k$. For the case of integer orders, i.e. $\alpha = 2$ and $\beta = 1$, we list the computed 5th-stage solution approximant

$$\begin{aligned}\varphi_5(t;2,1) = 1 - t &+ \left(1 - \frac{1}{2e}\right)t^2 + \left(-\frac{1}{3} - \frac{1}{6e}\right)t^3 + \left(\frac{1}{12} - \frac{1}{12e^2} + \frac{1}{8e}\right)t^4 \\ &+ \left(-\frac{1}{120} - \frac{1}{20e^2} + \frac{3}{40e}\right)t^5 + \left(-\frac{1}{72e^3} + \frac{13}{180e^2} - \frac{5}{72e}\right)t^6 \\ &+ \left(\frac{1}{280e^3} - \frac{1}{45e^2} + \frac{43}{2520e}\right)t^7 \\ &+ \left(-\frac{1}{2016e^4} - \frac{1}{1440e^3} + \frac{3}{1120e^2} - \frac{1}{672e}\right)t^8.\end{aligned}$$

For the case of $\alpha = 1.5$ and $\beta = 0.5$, we list the 4th-stage solution approximant

$$\begin{aligned}\varphi_4(t;1.5,0.5) = 1 - t &+ \frac{4(-1+e)t^{3/2}}{3e\sqrt{\pi}} + \frac{t^2}{2} - \frac{8(1+e)t^{5/2}}{15e\sqrt{\pi}} \\ &+ \left(-\frac{1}{6} - \frac{1}{3e^2} + \frac{1}{3e}\right)t^3 + \left(\frac{16}{105\sqrt{\pi}} + \frac{16}{105e\sqrt{\pi}}\right)t^{7/2} \\ &+ \left(-\frac{5}{24e^2} + \frac{1}{24e}\right)t^4 \\ &+ \left(-\frac{1024(-1+e)^2}{2835e^3\pi^{3/2}} - \frac{128}{945e^3\sqrt{\pi}} + \frac{128}{945e^2\sqrt{\pi}}\right)t^{9/2}.\end{aligned}$$

3.2.2 *The Rach–Adomian–Meyers Modified Decomposition Method*

In 1992, Rach et al. [76] proposed a modified decomposition method (MDM) based on the nonlinear transformation of series by the Adomian–Rach theorem [8, 9]:

$$\text{If } u(t) = \sum_{n=0}^{\infty} a_n (t-t_0)^n, \text{ then } f(u(t)) = \sum_{n=0}^{\infty} A_n (t-t_0)^n, \tag{3.29}$$

where $A_n = A_n(a_0, a_1, \ldots, a_n)$ are the Adomian polynomials in terms of the solution coefficients. The Rach–Adomian–Meyers MDM combines the power series solution and the Adomian–Rach theorem and has been efficiently applied to solve various nonlinear models [6]. Higher-order numerical one-step methods based on the Rach–Adomian–Meyers MDM were developed by Adomian et al. [13] and Duan and Rach [36, 38].

For the FDEs as we will be discussed in (3.33), the solution can be expressed as a generalized power series in the form of

$$u(t) = \sum_{n=0}^{\infty} a_n (t - t_0)^{\alpha n}, \tag{3.30}$$

where α is a real number. In this case, we have the following generalized Adomian–Rach theorem [39, 40]

$$f(\sum_{n=0}^{\infty} a_n (t - t_0)^{\alpha n}) = \sum_{n=0}^{\infty} A_n (t - t_0)^{\alpha n}, \tag{3.31}$$

where $A_n = A_n(a_0, a_1, \ldots, a_n)$ are the Adomian polynomials in terms of the solution coefficients.

The multivariable version of the generalized Adomian–Rach theorem is

$$f(\sum_{n=0}^{\infty} a_{1,n}(t - t_0)^{\alpha n}, \sum_{n=0}^{\infty} a_{2,n}(t - t_0)^{\alpha n}, \ldots, \sum_{n=0}^{\infty} a_{m,n}(t - t_0)^{\alpha n})$$
$$= \sum_{n=0}^{\infty} A_n (t - t_0)^{\alpha n}, \tag{3.32}$$

where f is an m-ary analytic function, then

$$A_n = A_n(a_{1,0}, a_{1,1}, \ldots, a_{1,n}; a_{2,0}, a_{2,1}, \ldots, a_{2,n}; \ldots; a_{m,0}, a_{m,1}, \ldots, a_{m,n})$$

are the m-variable Adomian polynomials [6, 8, 9, 32, 33, 40].

We consider the IVP for the nonlinear FDE

$$\sum_{k=0}^{q-1} \alpha_k \cdot {}_{t_0}D_t^{\frac{q-k}{p}} u(t) + \alpha_q u(t) + \alpha_{q+1} f(u(t)) = g(t), \; t_0 < t < T, \tag{3.33}$$

$$u(t_0) = C_0, \; u'(t_0) = C_1, \; \ldots, \; u^{(m-1)}(t_0) = C_{m-1}, \tag{3.34}$$

where p, q are positive integers, $p \geq 2$, satisfying $m - 1 < \frac{q}{p} \leq m$, $m \in \mathbb{N}^+$, $t_0, T, C_i, i = 0, 1, \ldots, m-1$, and $\alpha_n, n = 1, 2, \ldots, q+1$, are real constants and $\alpha_0 = 1$, f is an analytic nonlinear function and $g(t)$ is the system input function that can be written in the form of a generalized power series [39]

$$g(t) = \sum_{n=0}^{\infty} g_n (t - t_0)^{n/p}. \tag{3.35}$$

We decompose the solution as the form

$$u(t) = \sum_{n=0}^{\infty} a_n (t - t_0)^{n/p}. \tag{3.36}$$

Considering the initial conditions in (3.34), we take

$$a_0 = C_0,\ a_p = C_1,\ a_{2p} = \frac{C_2}{2!}, \ldots, a_{(m-1)p} = \frac{C_{m-1}}{(m-1)!}, \tag{3.37}$$

and

$$a_j = 0, \ \text{for all}\ j \leq q-1 \ \text{and}\ j \neq kp,\ k = 0, 1, 2, \ldots, m-1. \tag{3.38}$$

Thus the solution (3.36) may be written as

$$u(t) = \sum_{k=0}^{m-1} a_{kp}(t - t_0)^k + \sum_{n=q}^{\infty} a_n (t - t_0)^{n/p}. \tag{3.39}$$

Substituting (3.35), (3.36), and (3.39) into (3.33) and using the generalized Adomian–Rach theorem (3.31) we can obtain a recursion scheme for the solution coefficients a_n.

The Padé approximants can be directly applied to an analytic function, such as a polynomial approximation $\phi_n(t) = \sum_{k=0}^{n-1} a_k t^k$. The Padé approximant of $\phi_n(t)$ is a rational function in t. We denote the $[m/m]$ diagonal Padé approximant of $\phi_n(t)$ in t by $[m/m]\{\phi_n(t)\}$.

For the n-term approximation $\phi_n(t) = \sum_{k=0}^{n-1} a_k t^{\lambda k}$ of the solution for an FDE, where usually λ is not an integer, we need to indirectly apply the Padé approximant technique. We first make the replacement $t^\lambda = s$, so $\phi_n(t)$ becomes

$$\bar{\phi}_n(s) = \sum_{k=0}^{n-1} a_k s^k. \tag{3.40}$$

Calculating the diagonal Padé approximants for $\bar{\phi}_n(s)$ in s, then transforming $s = t^\lambda$, we obtain the desired result, denoted by

$$\text{Padé}_{m/m}\{\phi_n(t)\} := [m/m]\{\bar{\phi}_n(s)\}\big|_{s=t^\lambda}, \tag{3.41}$$

where usually we take $m = (n-1)/2$ if $n = 3, 5, 7, \ldots$, and $m = n/2$ if $n = 4, 6, 8, \ldots$.

We remark that in the generalized MDM the recurrence scheme for the coefficients a_n does not involve integration, while in the ADM the recurrence scheme for the solution components u_n does. This offers a computational advantage for the MDM [39].

Duan et al. [46] investigated eigenvalue problems for fractional ODEs. Jafari and Daftardar-Gejji [56, 57] considered a system of nonlinear FDEs and the nonlinear fractional BVPs using the ADM, including the fractional planar Bratu-type problem

$$D_x^\alpha u(x) + \mu e^{u(x)} = 0,\ 1 < \alpha \leq 2,\ 0 \leq x \leq 1,$$
$$u(0) = u(1) = 0.$$

Linear and nonlinear fractional PDEs have also been solved using the ADM. Here we list the coupled Burgers equations with time- and space-fractional derivatives considered by Chen and An [22]

$$D_t^{\alpha} u = L_{2x} u + 2u D_x^{\alpha} u - L_x(uv),\ 0 < \alpha \leq 1, \tag{3.42}$$

$$D_t^{\beta} u = L_{2x} v + 2v D_x^{\beta} v - L_x(uv),\ 0 < \beta \leq 1, \tag{3.43}$$

subject to the initial conditions

$$u(x,0) = f(x),\ v(x,0) = g(x),$$

where we have adopted the notation $L_{nx} = \frac{\partial^n}{\partial x^n}$.

For a comprehensive bibliography featuring many new engineering applications and a modern review of the ADM, see [40, 74].

3.3 Variational Iteration Method for Fractional Calculus

The variational iteration method (VIM) was developed in (1999) [50, 51]. The method doesn't require specific treatments as in the ADM and perturbation techniques for the nonlinear terms. It has been shown by many authors that this method provides improvements over existing analytical techniques. Several review articles have been dedicated to the topic [52,53]. Let's firstly revisit the basics of the method.

3.3.1 *Basic Principles of the Variational Iteration Method*

The basic character of the method is to construct a correction functional for the system (3.12) which reads

$$u_{n+1}(t) = u_n(t) + \int_0^t \lambda(t,\tau)(L[u_n(\tau)] + N[\tilde{u}_n(\tau)] - g(\tau))d\tau$$

where $\lambda(t,s)$ is a general Lagrange multiplier. It can be identified optimally via variational theory, u_n is the n-th approximate solution, and u_n denotes a restricted variation, i.e., $\delta \tilde{u}_n = 0$.

For example, consider the following simple linear equation

$$\frac{du}{dt} + u(t) = 0, u(0) = 1. \tag{3.44}$$

According to the VIM's rule, construct the correction functional

$$u_{n+1}(t) = u_n(t) + \int_0^t \lambda(t,\tau)(\frac{du_n}{d\tau} + u_n(\tau))d\tau \tag{3.45}$$

In order to find necessary extremum conditions, make the functional stationary with respect to $u_n(t)$. Take the variational derivative δ to both sides of (3.45)

$$\delta u_{n+1}(t) = \delta u_n(t) + \delta \int_0^t \lambda(t,\tau)(\frac{du_n}{d\tau} + \tilde{u}_n)d\tau \tag{3.46}$$

Through the integration by parts and calculus of variations, one can derive

$$\begin{aligned} \delta u_{n+1}(t) &= \delta u_n(t) + \delta[\lambda(t,\tau)u_n \mid_0^t - \int_0^t \frac{\partial\lambda(t,\tau)}{\partial\tau}u_n d\tau] + \delta \int_0^t \lambda(t,\tau)\tilde{u}_n d\tau \\ &= \delta u_n(t) + \lambda(t,\tau) \mid_{\tau=t} \delta u_n - \int_0^t \frac{\partial\lambda(t,\tau)}{\partial\tau}\delta u_n d\tau + \int_0^t \lambda(t,\tau)\delta\tilde{u}_n d\tau \end{aligned} \tag{3.47}$$

The condition $\delta u_{n+1}(t) = 0$ leads to the system

$$1 + \lambda(t,\tau) \mid_{\tau=t} = 0, \quad \frac{\partial\lambda(t,\tau)}{\partial\tau} = 0,$$

from which the Lagrange multiplier can be identified as

$$\lambda(t,\tau) = -1.$$

As a result, the variational iteration formula (3.45) reads

$$u_{n+1}(t) = u_n(t) - \int_0^t (\frac{du_n}{d\tau} + u_n(\tau))d\tau \tag{3.48}$$

Remark.

I. When constructing the variational iteration formulae, we note that the integration by parts plays a crucial role in the identification of the Lagrange multipliers;
II. There may be various choices of the Lagrange multipliers for a given equation. Generally speaking, the more explicit the Lagrange multiplier is, the higher accuracy the approximate solution. For example, if we use a Lagrange multiplier in (3.44) as $\lambda(t,\tau) = -e^{\tau-t}$ and only within one step, we can obtain the exact solution.

3.3.2 Formation of the Problems in Fractional Calculus

The VIM was first applied to the fractional system by He [50] as early as 1998

$$\frac{d^2u}{dt^2} + {}_0\mathbf{D}_t^{\frac{1}{2}} u(t) + \varepsilon u(t)^3 = 0, \ 0 < \varepsilon \ll 1. \tag{3.49}$$

$u(t)$ in the term ${}_0\mathbf{D}_t^{\alpha}u(t)$ was assumed as a restricted variation in the correction functional and the variational iteration formula was given as

$$\begin{cases} u_{n+1}(t) = u_n(t) + \int_0^t \lambda(t,\tau)(\frac{d^2u_n}{d\tau^2} + {}_0\mathbf{D}_\tau^{\frac{1}{2}}u_n(\tau) + \varepsilon u_n(\tau)^3)d\tau, \\ \lambda(t,\tau) = \tau - t. \end{cases} \tag{3.50}$$

In later years, the methodology was suggested as a sample for solving time fractional PDEs [55, 66, 68, 69, 94]. See, for example, the application to the Burgers equation [55]. The variational iteration formula was identified as

$$\begin{cases} u_{n+1}(t) = u_n(t) + \int_0^t \lambda(t,\tau)({}_0D_\tau^{\alpha}u_n + u_n u_{n,x} - u_{n,xx})d\tau, \\ \lambda(t,\tau) = -1. \end{cases} \tag{3.51}$$

But we can check the approximate solution here doesn't tend to the exact one for $n \to \infty$. The main reason is that the Lagrange multiplier there wasn't identified explicitly.

In view of this point, several modified versions have been suggested. For the nonlinear FDEs,

$${}_0\mathbf{D}_t^{\alpha}u + f(t,u) = 0, \quad {}_0D_t^{\alpha}u + f(t,u) = 0, \tag{3.52}$$

Ghorbani [49] embedded the parameter h to control the iteration scheme's convergence

$$u_{n+1}(t) = u_n(t) + h\, {}_0I^{\alpha}({}_0\mathbf{D}_\tau^{\alpha}u_n + f(\tau,u_n))d\tau \tag{3.53}$$

where h is a constant.

Yang et al. [93] proved the convergence of the variational iteration formula

$$u_{n+1}(t) = u_n(t) - \int_0^t \frac{(t-\tau)^{(m-1)}}{\Gamma(m)}(\frac{d^m u}{d\tau^m} +_0 D_\tau^{m-\alpha}f(\tau,u_n))d\tau \tag{3.54}$$

where $m - 1 < \alpha \leq m$.

But some other drawbacks arise in their applications. The parameter h cannot be chosen arbitrarily in (3.53) and the approximate solutions (3.54) should satisfy the property ${}_0D_t^{m-\alpha}\,{}_0D_t^{\alpha}u_n = {}_0D_t^{m}u_n$ which generally cannot hold for the Caputo derivative.

3.3.3 Recent Developments

The identification of the Lagrange multiplier is the most crucial step in the VIM. In the fractional calculus, it is difficult to identify the Lagrange multiplier according to the classical VIM's rule [51]. In order to overcome the drawbacks in the applications to FDEs, two accurate ways [87–91] are suggested by means of Laplace transform. In this section, let's revisit the methodology and the proof.

3.3.3.1 VIM-I Using Laplace Transform

Firstly, an iteration formula for finding the solution of an algebraic equation

$$f(u) = 0 \tag{3.55}$$

can be constructed as

$$u_{n+1} = u_n + \lambda f(u_n). \tag{3.56}$$

The optimality condition $\delta u_{n+1} = 0$ leads to

$$\lambda = -\frac{1}{f'(u_n)}. \tag{3.57}$$

For a given initial value u_0, we can find the approximate solution u_{n+1} by the iterative scheme for (3.55)

$$u_{n+1} = u_n - \frac{f(u_n)}{f'(u_n)}, \quad f'(u_0) \neq 0, \; n = 0, 1, 2, \ldots. \tag{3.58}$$

This algorithm is well known as Newton–Raphson method and has quadratic convergence. Now, we extend this idea for finding the unknown Lagrange multiplier. The main step is to first take Laplace transform to (3.12). The linear part is assumed as $L[u] = d^m u/dt^m$. Then, (3.12) is transformed into an algebraic equation as

$$s^m \bar{u}(s) - u^{(m-1)}(0) - \ldots - s^{m-1}u(0) + \mathscr{L}[N[u]] - \mathscr{L}[g(t)] = 0 \tag{3.59}$$

where $\bar{u}(s) = \mathscr{L}[u(t)] = \int\limits_0^\infty e^{-st}u(t)dt$.

The main iterative scheme involving the Lagrange multiplier can be constructed as

$$\begin{aligned}\bar{u}_{n+1}(s) = \bar{u}_n(s) + \lambda(s)[s^m u_n(s) - u^{(m-1)}(0) - \ldots - s^{m-1}u(0)\\ +\mathscr{L}[N[u_n] - g(t)]].\end{aligned} \tag{3.60}$$

One can derive a Lagrange multiplier as

$$\lambda(s) = -\frac{1}{s^m}. \tag{3.61}$$

With the inverse-Laplace transform $\mathscr{L}^{-1}$, the iteration formula (3.60) can be explicitly given as

$$u_{n+1}(t) = u_0(t) - \mathscr{L}^{-1}[\tfrac{1}{s^m}\mathscr{L}[N[u_n]]], \tag{3.62}$$

where $u_0(t)$ the initial iteration can be determined by

$$u_0(t) = u(0) + u'(0)t + \ldots + \frac{u^{(m-1)}(0)t^{m-1}}{(m-1)!} + \mathscr{L}[\frac{1}{s^m}\mathscr{L}[g(t)]]. \tag{3.63}$$

Example 2. [89] Consider the relaxation oscillator equation

$${}_0D_t^\alpha u + \omega^\alpha u = 0,\ u(0) = 1,\ u'(0) = 0,\ 0 < t,\ 0 < \alpha < 2, 0 < \omega. \tag{3.64}$$

It was found to have the exact solution $E_\alpha((-\omega t)^\alpha)$ [64] and $E_\alpha(z)$ is the Mittag-Leffler function.

With Laplace transform, we get the following iteration formula

$$\bar{u}_{n+1}(s) = \bar{u}_n(s) + \lambda(s)[s^\alpha \bar{u}_n(s) - u(0)s^{\alpha-1} - u'(0^+)s^{\alpha-2} + \omega^\alpha \mathscr{L}[u_n]]. \tag{3.65}$$

Setting $\mathscr{L}[u_n(t)]$ as a restrict variation term, the Lagrange multiplier can be identified as

$$\lambda(s) = -\frac{1}{s^\alpha}. \tag{3.66}$$

The approximate solution of (3.64) can be given as

$$u_{n+1}(t) = u_0(t) - \mathscr{L}^{-1}[\tfrac{\omega^\alpha}{s^\alpha}\mathscr{L}[u_n]] \tag{3.67}$$

which reads

$$\begin{aligned} u_0(t) &= 1, \\ u_1(t) &= 1 - \tfrac{\omega^\alpha t^\alpha}{\Gamma(1+\alpha)}, \\ u_2(t) &= 1 - \tfrac{\omega^\alpha t^\alpha}{\Gamma(1+\alpha)} + \tfrac{\omega^{2\alpha} t^{2\alpha}}{\Gamma(1+2\alpha)}, \\ &\vdots \end{aligned} \tag{3.68}$$

For $n \to \infty, u_n(t)$ rapidly tends to the exact solution.

3.3.3.2 VIM-II Using Laplace Transform

Theorem 1. *[87, 91] The fractionalized system (3.12)*

$${}_0D_t^\alpha u + R[u] + N[u] = g(t), \tag{3.69}$$

has one variational iteration formula

$$u_{n+1} = u_n - \int_0^t \frac{(t-\tau)^{\alpha-1}}{\Gamma(\alpha)}({}_0D_\tau^\alpha u_n + R[u_n] + N[u_n] - g(\tau))d\tau,\ 0 < \alpha. \tag{3.70}$$

Proof. We can construct a correction functional through the R–L integration

$$u_{n+1} = u_n +_0 I_t^\alpha \lambda(t,\tau)[{}_0D_\tau^\alpha u_n + R[u_n] + N[u_n] - g(\tau)]. \tag{3.71}$$

Take Laplace transform to both sides

$$\bar{u}_{n+1}(s) = \bar{u}_n(s) + \mathcal{L}[{}_0I_t^\alpha \lambda(t,\tau)({}_0^C D_\tau^\alpha u_n + R[u_n] + N[u_n] - g(\tau)]. \tag{3.72}$$

We consider the term

$${}_0I_t^\alpha[\lambda(t,\tau)\,{}_0D_t^\alpha u_n] = \frac{1}{\Gamma(\alpha)}\int_0^t (t-\tau)^{\alpha-1}\lambda(t,\tau)\,{}_0D_\tau^\alpha u_n(\tau)\,d\tau. \tag{3.73}$$

Setting the Lagrange multiplier $\lambda(t,\tau) = \lambda(X)/_{X=t-\tau}$, (3.73) becomes a convolution of the function $a(t) = \lambda(t)t^{\alpha-1}/\Gamma(\alpha)$ and the term ${}_0D_t^\alpha u_n(t)$. This strategy was meanwhile suggested in the two-point value problems of differential equations in [59].

Making the correction functional stationary with respect to $\bar{u}_n(s)$ in (3.72), we can get

$$\delta\bar{u}_{n+1}(s) = \delta\bar{u}_n(s) + \delta[\bar{a}(s)s^\alpha\bar{u}_n(s) - \sum_{k=0}^{m-1} u^{(k)}(0^+)s^{\alpha-1-k}] = (1 + \bar{a}(s)s^\alpha)\delta\bar{u}_n(s). \tag{3.74}$$

The extremum condition $\delta\bar{u}_{n+1}(s) = 0$ requires $1+\bar{a}(s)s^\alpha = 0$. With the inverse Laplace transform, we can have

$$a(t) = \mathscr{L}^{-1}[\bar{a}_(s)] = -\frac{t^{\alpha-1}}{\Gamma(\alpha)}, 0 < \alpha. \tag{3.75}$$

As a result, the Lagrange multiplier can be explicitly identified as

$$\lambda(t,\tau) = -1 \tag{3.76}$$

and the iteration formula is given as

$$u_{n+1} = u_n - {}_0I_t^\alpha[{}_0D_\tau u_n + R[u_n] + N[u_n] - g(\tau)],\ 0 < \alpha.$$

This completes the proof.

For $\alpha = 2$, we can check the variational iteration formula (3.50)'s validness since it is a special case of (3.70). On the other hand, we only derive the simplest Lagrange multiplier $\lambda(t,\tau) = -1$ here. In fact, more explicit Lagrange multipliers can be identified if more terms in $R[u_n]$ of (3.12) (if it exists) are used. For example, we can derive a variational iteration formula

$$u_{n+1} = u_n - \int_0^t (t-\tau)^{\alpha-1} E_{\alpha-\beta,\alpha}(-(t-\tau)^{\alpha})({}_0D_\tau^\alpha u_n + {}_0D_\tau^\beta u_n + f(\tau,u_n))d\tau,$$
$$0 < \beta < \alpha \tag{3.77}$$

for a multi-term FDE

$${}_0D_t^\alpha u + {}_0D_t^\beta u + f(t,u) = 0, 0 < t,\ 0 < \beta < \alpha. \tag{3.78}$$

For $\alpha = 1, \beta = 0$, and $\alpha = 2,\ \beta = 0$, (3.77) reduces to the formulae in [54]

$$\begin{cases} u_{n+1} = u_n + \int_0^t \lambda(t,\tau)(u_n^{(')} + u_n + f(\tau,u_n))d\tau, \\ \lambda(t,\tau) = -(t-\tau)^{\alpha-1} E_{\alpha-\beta,\alpha}(-(t-\tau)^{\alpha-\beta})\,\big|_{\alpha=1,\beta=0} = -e^{-(t-\tau)} \end{cases} \tag{3.79}$$

and

$$\begin{cases} u_{n+1} = u_n + \int_0^t \lambda(t,\tau)(u_n^{(2)} + u_n + f(\tau,u_n))d\tau, \\ \lambda = -(t-\tau)^{\alpha-1} E_{\alpha-\beta,\alpha}(-(t-\tau)^{\alpha-\beta})\,\big|_{\alpha=2,\beta=0} = \sin(\tau - t). \end{cases} \tag{3.80}$$

For the details, readers are referred to our recent work [91].

Example 3. [91] The VIM is applied to the time-fractional Burgers equation

$$\begin{cases} {}_0D_t^\alpha u + u\frac{\partial u}{\partial x} = v\frac{\partial^2 u}{\partial x^2}, 0 < \alpha \le 1, \\ u_0 = u(x,0) = g(x) = \sin(2\pi x),\ u(0,t) = u(1,t) = 0 \end{cases} \tag{3.81}$$

where u is the velocity and v is the viscosity coefficient of the flow.

We can have the following variational iteration formula from *Theorem 1*

$$u_{n+1} = u_n - \int_0^t \frac{(t-\tau)^{\alpha-1}}{\Gamma(\alpha)}({}_0D_\tau^\alpha u_n + u_n\frac{\partial u_n}{\partial x} - v\frac{\partial^2 u_n}{\partial x^2})d\tau. \tag{3.82}$$

The successive approximate solutions can be obtained

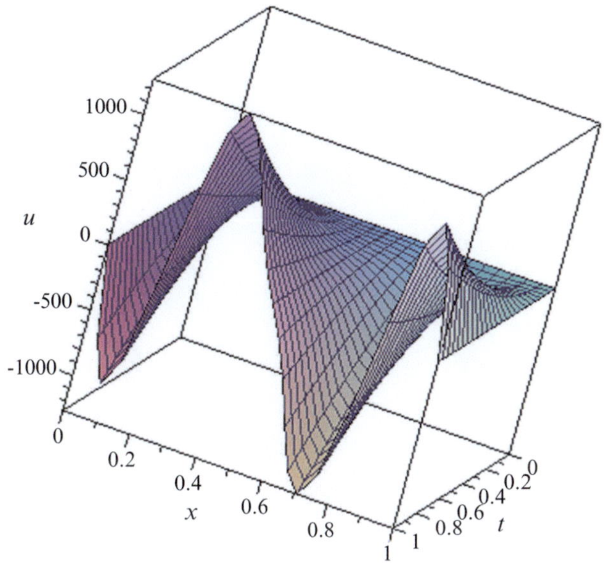

Fig. 3.1 The velocity of the flow for $\alpha = 0.5$ and $\nu = 0.5$ [91]

$$
\begin{aligned}
u_0 &= g(x) = g, \\
u_1 &= g - (gg' - \nu g^{(2)})\frac{t^{\alpha}}{\Gamma(\alpha+1)}, \\
u_2 &= g - (gg' - \nu g^{(2)})\frac{t^{\alpha}}{\Gamma(1+\alpha)} + (2gg'^2 + g^2 g^{(2)} \\
&\quad -2\nu g g^{(3)} - 4\nu g' g^{(2)} + \nu^2 g^{(4)})\frac{t^{2\alpha}}{\Gamma(1+2\alpha)} \\
&\quad -(gg' - \nu g^{(2)})(g'^2 + g g^{(2)} - \nu g^{(3)})\frac{\Gamma(1+2\alpha)}{\Gamma^2(1+\alpha)}\frac{t^{3\alpha}}{\Gamma(1+3\alpha)}, \\
&\vdots
\end{aligned}
\tag{3.83}
$$

where $g' = \frac{dg}{dx}$ and $g^{(m)} = \frac{d^m g}{dx^m}$. We get the approximate solution u_2 as the second term approximation. For $g(x) = \sin(2\pi x)$ and the fractional order $\alpha = 0.5$, Figs. 3.1 and 3.2 show the velocity of the flow at the various viscosity coefficients ν.

Remark. As we mentioned in the Sect. 3.3.1, the classical VIM should use the integration by parts in the identification of the Lagrange multipliers. *Sections 3.3.1* and *3.3.2* provide two ways to calculate the Lagrange multipliers more accurately but without using the integration by parts. Both of the two VIMs lead to the same approximate solutions. They have different advantages, respectively.

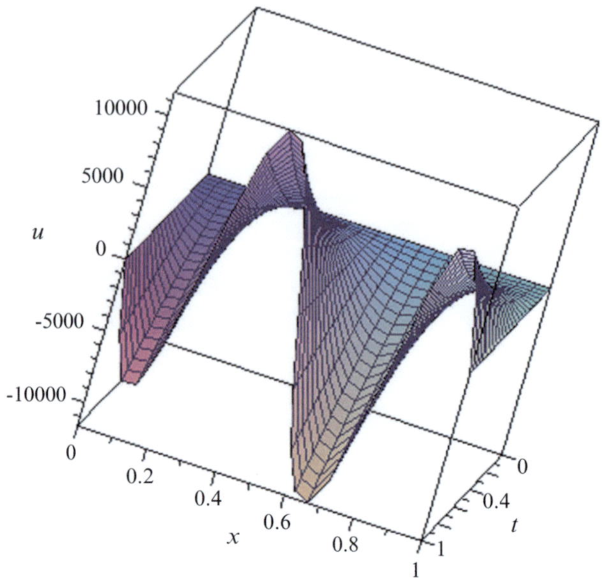

Fig. 3.2 The velocity of the flow for $\alpha = 0.5$ and $v = 1.5$

For the VIM-I, the correction functional is constructed by Laplace transform instead an integral. In this way, the Lagrange multipliers can be readily given in form of complex variable s and the initial iteration solution also can be easily determined. However, readers need to be familiar with the numerical inverse of Laplace transform.

In the VIM-II, the correction functional is constructed through the Riemann-Liouville integral which strictly follows the classical VIM's rule. In order to avoid the using of the integral by parts in the fractional case, the convolution of Laplace transform is applied to identify the Lagrange multipliers.

3.3.4 The Coupled Analytical Method Based on the VIM and the ADM

The VIM and the ADM have their own merits. Abbasbandy [1] initially modified the variational iteration method by using the Adomian series. The Adomian series treated the nonlinear terms which can improve the efficiency. The coupled methods can fully use the methods' advantages.

Example 4. More generally, we can consider the following type

$$ {}_0D_t^{\alpha} u(t) = 2u(t)^2 + t, u(0) = 0, 0 < \alpha \leq 1. \tag{3.84}$$

The correction functional can be constructed by Laplace transform

$$u_{n+1}(t) = u_n(t) + \mathscr{L}^{-1}[\lambda(s)\mathscr{L}[{}_0D_t^\alpha u_n - 2u_n(\tau)^2 - t]]. \tag{3.85}$$

With the identified Lagrange multiplier $\lambda(s) = -1/s^\alpha$, we can derive the iteration formula

$$u_{n+1}(t) = \frac{t^{1+\alpha}}{\Gamma(\alpha+2)} + 2\mathscr{L}^{-1}[\frac{1}{s^\alpha}\mathscr{L}[u_n^2]]. \tag{3.86}$$

In order to decompose the term u_n^2, assume $u = \sum_{i=0}^{\infty} v_i$ and the n-th order approximation is $u_n = \sum_{i=0}^{n} v_i$. We can give the recurrence scheme involving the Adomian series

$$\begin{aligned} v_{n+1}(t) &= 2\mathscr{L}^{-1}[\frac{1}{s^\alpha}\mathscr{L}[A_n]], \\ v_0(t) &= \frac{t^{1+\alpha}}{\Gamma(\alpha+2)}. \end{aligned} \tag{3.87}$$

As a result, we can obtain the approximate solutions successively as

$$\begin{aligned} u_0(t) &= \frac{t^{1+\alpha}}{\Gamma(\alpha+2)}, \\ u_1(t) &= \frac{t^{1+\alpha}\Gamma(\alpha+2)\Gamma(3+3\alpha) + 2\,\Gamma(2\alpha+3)t^{3\alpha+2}}{\Gamma(3+3\alpha)\Gamma(\alpha+2)^2}, \\ &\vdots \end{aligned} \tag{3.88}$$

Through the defined remainder function (3.21)

$$g(t) = {}_0D_t^\alpha u_n(t) - 2u_n(t)^2 - t, \tag{3.89}$$

we can illustrate the higher order approximation's validness in Fig. 3.3.

3.3.5 *New Numerical Schemes Using the VIM*

For the FDEs with multi-terms, the explicit Lagrange multiplier can be given in form of the Mittag-Leffler functions and has longer memory. However, the analytical calculation of the approximate solutions becomes even impossible for a linear FDE. In our recent work [79], we consider the numerical methods when the Lagrange multiplier is complicated.

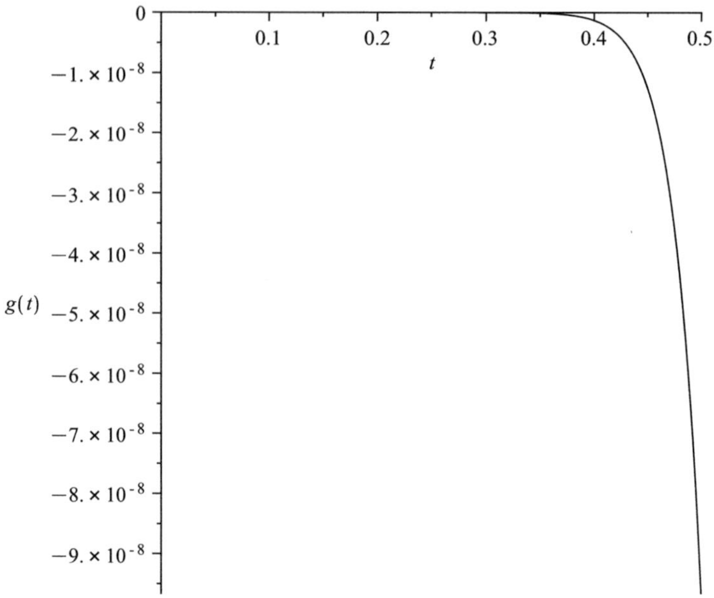

Fig. 3.3 The error remainder function $g(t)$ for $\alpha = 0.8$ and $n = 6$

Example 5. The Bagley–Torvik equation of fractional order reads

$$\frac{d^2u}{dt^2} + {}_0D_t^\alpha u + u(t) = f(t), 1 < \alpha \leq 2, u(0) = 0, u'(0) = 0. \tag{3.90}$$

It was found to have the exact solution [70]

$$\int_0^t \sum_{j=0}^{\infty} \frac{(-1)^j}{j!}(t-\tau)^{2j+1} E^{(j)}_{2-\alpha,2+\alpha j}(-(t-\tau)^{2-\alpha}) f(\tau)d\tau \tag{3.91}$$

and $E^{(j)}_{\alpha,\beta}(t)$ is defined as

$$E^{(j)}_{\alpha,\beta}(t) = \frac{d^j}{dt^j} E_{\alpha,\beta}(t) = \sum_{k=0}^{\infty} \frac{(k+j)!t^k}{k!\Gamma(\alpha k + \alpha j + \beta)}.$$

We can obtain the following variational iteration formulae, respectively

$$\begin{cases} u_{n+1}(t) = u_0(t) + \int_0^t \lambda(t,\tau)({}_0D_t^\alpha u_n + u_n - f(\tau))d\tau, \ u_0(t) = 0, \\ \lambda(t,\tau) = \tau - t \end{cases} \tag{3.92}$$

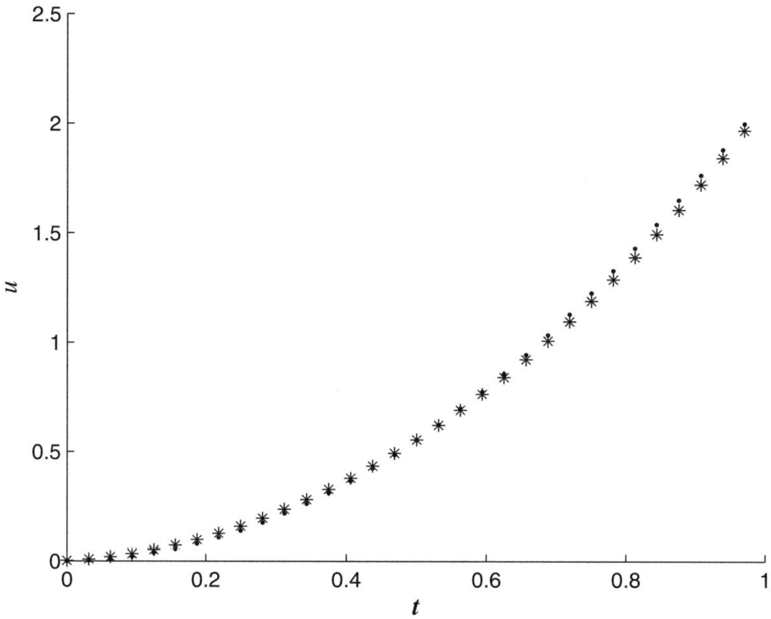

Fig. 3.4 Comparison among the numerical solutions using the predictor–corrector method and the exact solution for $\alpha = 1.9$ and the step size $h = 1/32$

and

$$\begin{cases} u_{n+1}(t) = u_0(t) + \int_0^t \lambda(t,\tau)(u_n(\tau) - f(\tau))d\tau, \ u_0(t) = 0, \\ \lambda = (\tau - t)E_{2-\alpha,2}(-(t-\tau)^{2-\alpha}). \end{cases} \tag{3.93}$$

From the convergence conditions [79], the iteration formulae lead to the integral equation accordingly

$$u(t) = \int_0^t (\tau - t)({}_0D_t^{\alpha}u(\tau) + u(\tau) - f(\tau))d\tau \tag{3.94}$$

and

$$u(t) = \int_0^t (\tau - t)E_{2-\alpha,2}(-(t-\tau)^{2-\alpha})(u(\tau) - f(\tau))d\tau. \tag{3.95}$$

For the variational iteration formula (3.92), one can easily obtain the approximate solutions. But in the variational iteration formula (3.93), the Lagrange multiplier is complicated and the analytical calculation by hand becomes tedious and even impossible with symbolic computation. As a result, we consider the numerical discretization of the equivalent integral equation. The Adams–Moulton formula and the predictor–corrector approach [27, 29] are adopted to establish a numerical scheme and the numerical results in Fig. 3.4 show its efficiency.

3.4 Conclusions

In the last decades, the ADM and the VIM have been two often used analytical methods in the area of fractional calculus. We have presented a contemporary review of the ADM and the VIM in the FDEs and discussed its utility and advantages for solving linear or nonlinear FDEs. The efficiency of the ADM is greatly increased since the new algorithms for the Adomian polynomials [31–33]. In view of the recent developments, some new applications become possible.

- Approximate solutions of other fractional models with the ADM and the VIM

Recently, some models with fractional derivatives newly appear in applied sciences and describe the nonlocal effects characterized by the nonlocal structure of the fractional derivatives. See, for example, the fractional delay equations[17], the fractional fuzzy equations[14], and the fractional sequential equations[20]. The two methods can be used to analytically investigate these modes and the authors believe that they can play the same role as that in ordinary calculus.

- New numerical schemes with the ADM and the VIM

The combined solution of the near-field and far-field approximations by the Adomian and asymptotic decomposition methods was presented in [75]. The parametrized recursion scheme [31, 37, 42] was also proposed to obtain decomposition solutions with large effective regions of convergence. The multistage ADM and its numeric schemes were considered in [13, 15, 36, 38]. For the FDEs, the ADM with the convergence acceleration techniques, such as the diagonal Padé approximants or the iterated Shanks transform was considered in [41, 67]. A numeric scheme solving the FDEs based on the ADM was designed in [63]. For a comprehensive bibliography featuring many new engineering applications and a modern review of the ADM, see [40, 74].

The VIM provides a successive iteration formulae for the FDEs in Sect. 3.3. With the convergence conditions, the formula can lead to an integral equation. Since there may be various choices of the Lagrange multipliers in the VIM, rich equivalent integral equations can be derived. Choosing the optimal one, the accuracies of the approximate solutions can be improved. Furthermore, numerical methods become more convenient since there are no terms containing fractional derivatives and this idea can be further extended to PDEs of fractional order. This chapter mainly concentrates on the analytical methods, for the numerical aspects and the analysis of the FDEs, readers are referred to the monographs [19, 28].

Acknowledgments This work was partly supported by the National Natural Science Foundation of China (No. 11171295, No. 11061028) and the key program (51134018).

References

1. Abbasbandy S (2007) An approximation solution of a nonlinear equation with Riemann–Liouville's fractional derivatives by He's variational iteration method. J Comput Appl Math 207:53–58
2. Abdelrazec A, Pelinovsky D (2011) Convergence of the Adomian decomposition method for initial-value problems. Numer Methods Partial Differ Equ 27:749–766
3. Adomian G (1983) Stochastic systems. Academic, New York
4. Adomian G (1986) Nonlinear stochastic operator equations. Academic, Orlando
5. Adomian G (1989) Nonlinear stochastic systems theory and applications to physics. Kluwer, Dordrecht
6. Adomian G (1994) Solving Frontier problems of physics: the decomposition method. Kluwer Academic, Dordrecht
7. Adomian G, Rach R (1983) Inversion of nonlinear stochastic operators. J Math Anal Appl 91:39–46
8. Adomian G, Rach R (1991) Transformation of series. Appl Math Lett 4:69–71
9. Adomian G, Rach R (1992) Nonlinear transformation of series – part II. Comput Math Appl 23:79–83
10. Adomian G, Rach R (1993) Analytic solution of nonlinear boundary-value problems in several dimensions by decomposition. J Math Anal Appl 174:118–137
11. Adomian G, Rach R (1993) A new algorithm for matching boundary conditions in decomposition solutions. Appl Math Comput 58:61–68
12. Adomian G, Rach R (1996) Modified Adomian polynomials. Math Comput Model 24:39–46
13. Adomian G, Rach R, Meyers RE (1997) Numerical integration, analytic continuation, and decomposition. Appl Math Comput 88:95–116
14. Agarwal RP, Arshad S, O'Regan D, Lupulescu V (2012) Fuzzy fractional integral equations under compactness type condition. Fract Calc Appl Anal 15:572–590
15. Al-Sawalha MM, Noorani MSM, Hashim I (2008) Numerical experiments on the hyperchaotic Chen system by the Adomian decomposition method. Int J Comput Methods 5:403–412
16. Arora HL, Abdelwahid FI (1993) Solution of non-integer order differential equations via the Adomian decomposition method. Appl Math Lett 6:21–23
17. Bhalekar S, Daftardar-Gejji V, Baleanu D, Magin R (2011) Fractional Bloch equation with delay. Comput Math Appl 61:1355–1365
18. Bigi D, Riganti R (1986) Solution of nonlinear boundary value problems by the decomposition method. Appl Math Model 10:49–52
19. Băleanu D, Diethelm K, Scalas E, Trujillo JJ (2012) Fractional calculus models and numerical methods (series on complexity, nonlinearity and chaos). World Scientific, Boston
20. Băleanu D, Mustafa OG, Agarwal RP (2010) On the solution set for a class of sequential fractional differential equations. J Phys A Math Theor 43:385209
21. Băleanu D, Mustafa OG, O'Regan D (2012) On a fractional differential equation with infinitely many solutions. Adv Differ Equ 2012:145
22. Chen Y, An HL (2008) Numerical solutions of coupled Burgers equations with time- and space-fractional derivatives. Appl Math Comput 200:87–95
23. Cherruault Y (1989) Convergence of Adomian's method. Kybernetes 18:31–38
24. Cherruault Y, Adomian G (1993) Decomposition methods: a new proof of convergence. Math Comput Model 18:103–106
25. Daftardar-Gejji V, Jafari H (2005) Adomian decomposition: a tool for solving a system of fractional differential equations. J Math Anal Appl 301:508–518
26. Dehghan M, Tatari M (2010) Finding approximate solutions for a class of third-order nonlinear boundary value problems via the decomposition method of Adomian. Int J Comput Math 87:1256–1263
27. Deng WH (2007) Numerical algorithm for the time fractional Fokker-Planck equation. J Comput Phys 227:1510–1522

28. Diethelm K (2004) The analysis of fractional differential equations. Springer, New York
29. Diethelm K, Ford NJ (2004) Multi-order fractional differential equations and their numerical solution. Appl Math Comput 154:621–640
30. Duan JS (2005) Time- and space-fractional partial differential equations. J Math Phys 46:13504–13511
31. Duan JS (2010) Recurrence triangle for Adomian polynomials. Appl Math Comput 216: 1235–1241
32. Duan JS (2010) An efficient algorithm for the multivariable Adomian polynomials. Appl Math Comput 217:2456–2467
33. Duan JS (2011) Convenient analytic recurrence algorithms for the Adomian polynomials. Appl Math Comput 217:6337–6348
34. Duan JS (2011) New recurrence algorithms for the nonclassic Adomian polynomials. Comput Math Appl 62:2961–2977
35. Duan JS (2011) New ideas for decomposing nonlinearities in differential equations. Appl Math Comput 218:1774–1784
36. Duan JS, Rach R (2011) New higher-order numerical one-step methods based on the Adomian and the modified decomposition methods. Appl Math Comput 218:2810–2828
37. Duan JS, Rach R (2011) A new modification of the Adomian decomposition method for solving boundary value problems for higher order nonlinear differential equations. Appl Math Comput 218:4090–4118
38. Duan JS, Rach R (2012) Higher-order numeric Wazwaz-El-Sayed modified Adomian decomposition algorithms. Comput Math Appl 63:1557–1568
39. Duan JS, Chaolu T, Rach R (2012) Solutions of the initial value problem for nonlinear fractional ordinary differential equations by the Rach-Adomian-Meyers modified decomposition method. Appl Math Comput 218:8370–8392
40. Duan JS, Rach R, Băleanu D, Wazwaz AM (2012) A review of the Adomian decomposition method and its applications to fractional differential equations, Commun Fract Calc 3:73–99
41. Duan JS, Chaolu T, Rach R, Lu L (2013) The Adomian decomposition method with convergence acceleration techniques for nonlinear fractional differential equations. Comput Math Appl. 66:728–736
42. Duan JS, Rach R, Wang Z (2013) On the effective region of convergence of the decomposition series solution. J Algorithm Comput Tech 7:227–247
43. Duan JS, Wang Z, Fu SZ, Chaolu T (2013) Parametrized temperature distribution and efficiency of convective straight fins with temperature-dependent thermal conductivity by a new modified decomposition method. Int J Heat Mass Transf 59:137–143
44. Duan JS, Rach R, Wazwaz AM (2013) Solution of the model of beam-type micro- and nano-scale electrostatic actuators by a new modified Adomian decomposition method for nonlinear boundary value problems. Int J Non Linear Mech 49:159–169
45. Duan JS, Rach R, Wazwaz AM, Chaolu T, Wang Z (2013) A new modified Adomian decomposition method and its multistage form for solving nonlinear boundary value problems with Robin boundary conditions. Appl Math Model. doi:10.1016/j.apm.2013.02.002
46. Duan JS, Wang Z, Liu YL, Qiu X (2013) Eigenvalue problems for fractional ordinary differential equations. Chaos Solitons Fractals 46:46–53
47. Gabet L (1994) The theoretical foundation of the Adomian method. Comput Math Appl 27: 41–52
48. George AJ, Chakrabarti A (1995) The Adomian method applied to some extraordinary differential equations. Appl Math Lett 8:391–397
49. Ghorbani A (2008) Toward a new analytical method for solving nonlinear fractional differential equations. Comput Methods Appl Mech Eng 197(49–50):4173–4179
50. He JH (1998) Approximate analytical solution for seepage flow with fractional derivatives in porous media. Comput Methods Appl Mech Eng 167:57–68
51. He JH (1999) Variational iteration method - a kind of non-linear analytical technique: some examples. Int J Non Linear Mech 34:699–708

52. He JH (2006) Some asymptotic methods for strongly nonlinear equations. Int J Mod Phys B 20:1141–1199
53. He JH (2012) Asymptotic methods for solitary solutions and compactons. Abstr Appl Anal 2012, Article ID 916793, 130 pp
54. He JH, Wu XH (2007) Variational iteration method: new development and applications. Comput Math Appl 54:881–894
55. Inc M (2008) The approximate and exact solutions of the space- and time-fractional Burgers equations with initial conditions by variational iteration method. J Math Anal Appl 345: 476–484
56. Jafari H, Daftardar-Gejji V (2006) Solving a system of nonlinear fractional differential equations using Adomian decomposition. J Comput Appl Math 196:644–651
57. Jafari H, Daftardar-Gejji V (2006) Positive solutions of nonlinear fractional boundary value problems using Adomian decomposition method. Appl Math Comput 180:700–706
58. Jafari H, Kadem A, Baleanu D, Yilmaz T (2012) Solutions of the fractional davey-stewartson equations with variational iteration method. Rom Rep Phys 64:337–346
59. Khuri SA, Sayfy A (2012) A Laplace variational iteration strategy for the solution of differential equations. Appl Math Lett 25:2298–2305
60. Kilbas AA, Srivastava HM, Trujillo JJ (2006) Theory and applications of fractional differential equations. Elsevier, Amsterdam
61. Klafter J, Lim SC, Metzler R (2011) Fractional dynamics: recent advances. World Scientific, Singapore
62. Lesnic D (2008) The decomposition method for nonlinear, second-order parabolic partial differential equations. Int J Comput Math Numer Simul 1:207–233
63. Li CP, Wang YH (2009) Numerical algorithm based on Adomian decomposition for fractional differential equations. Comput Math Appl 57:1672–1681
64. Mainardi F (1996) Fractional relaxation oscillation and fractional diffusion-wave phenomena. Chaos Solitons Fractals 7:1461–1477
65. Mainardi F (2010) Fractional calculus and waves in linear viscoelasticity. Imperial College, London
66. Momani S, Odibat Z (2007) Numerical comparison of methods for solving linear differential equations of fractional order. Chaos Solitons Fractals 31:1248–1255
67. Momani S, Shawagfeh N (2006) Decomposition method for solving fractional Riccati differential equations. Appl Math Comput 182:1083–1092
68. Nawaz Y (2011) Variational iteration method and homotopy perturbation method for fourth-order fractional integro-differential equations. Comput Math Appl 61:2330–2341
69. Odibat ZM, Momani S (2006) Application of variational iteration method to Nonlinear differential equations of fractional order. Int J Non Linear Sci Numer Simul 7:27–34
70. Podlubny I (1999) Fractional differential equations. Academic Press, San Diego
71. Rach R (1984) A convenient computational form for the Adomian polynomials. J Math Anal Appl 102:415–419
72. Rach R (1987) On the Adomian (decomposition) method and comparisons with Picard's method. J Math Anal Appl 128:480–483
73. Rach R (2008) A new definition of the Adomian polynomials, Kybernetes 37:910–955
74. Rach R (2012) A bibliography of the theory and applications of the Adomian decomposition method, 1961–2011. Kybernetes 41:1087–1148
75. Rach R, Duan JS (2011) Near-field and far-field approximations by the Adomian and asymptotic decomposition methods. Appl Math Comput 217:5910–5922
76. Rach R, Adomian G, Meyers RE (1992) A modified decomposition. Comput Math Appl 23: 17–23
77. Ray SS, Bera RK (2005) An approximate solution of a nonlinear fractional differential equation by Adomian decomposition method. Appl Math Comput 167:561–571
78. Shawagfeh NT (2002) Analytical approximate solutions for nonlinear fractional differential equations. Appl Math Comput 131:517–529

79. Wang YH, Wu GC, Baleanu D (2013) Variational iteration method-a promising technique for constructing equivalent integral equations of fractional order. Cent Eur J Phys. doi:10.2478/s11534-013-0207-3
80. Wazwaz AM (1999) A reliable modification of Adomian decomposition method. Appl Math Comput 102:77–86
81. Wazwaz AM (2000) A new algorithm for calculating Adomian polynomials for nonlinear operators. Appl Math Comput 111:53–69
82. Wazwaz AM (2000) The modified Adomian decomposition method for solving linear and nonlinear boundary value problems of tenth-order and twelfth-order. Int J Non Linear Sci Numer Simul 1:17–24
83. Wazwaz AM (2009) Partial differential equations and solitary waves theory. Higher Education, Beijing/Springer, Berlin
84. Wazwaz AM (2011) Linear and nonlinear integral equations: methods and applications. Higher Education, Beijing/Springer, Berlin
85. Wazwaz AM, El-Sayed SM (2001) A new modification of the Adomian decomposition method for linear and nonlinear operators. Appl Math Comput 122:393–405
86. Wazwaz AM, Rach R (2011) Comparison of the Adomian decomposition method and the variational iteration method for solving the Lane-Emden equations of the first and second kinds. Kybernetes 40:1305–1318
87. Wu GC (2012) Variational iteration method for solving the time-fractional diffusion equations in porous medium. Chin Phys B 21:120504
88. Wu GC (2012) Laplace transform overcoming principle drawbacks in application of the variational iteration method to fractional heat equations. Therm Sci 16:1357–1361
89. Wu GC, Băleanu D (2013) Variational iteration method for fractional calculus - a universal approach by Laplace transform. Adv Diff Equ 2013:18
90. Wu GC, Băleanu D (2013) New applications of the variational iteration method-from differential equations to q-fractional difference equations. Adv Diff Equ 2013:21
91. Wu GC, Băleanu D (2013) Variational iteration method for the Burgers' flow with fractional derivatives-New Lagrange mutipliers. Appl Math Model 37:6183–6190
92. Wu GC, Shi YG, Wu KT (2011) Adomian decomposition method and non-analytical solutions of fractional differential equations. Rom J Phys 56:873–880
93. Yang SP, Xiao AG, Su H (2010) Convergence of the variational iteration method for solving multi-order fractional differential equations. Comput Math Appl 60:2871–2879
94. Yaslan HC (2012) Variational iteration method for the time-fractional elastodynamics of 3D Quasicrystals. Comput Model Eng Sci 86:29–38

Chapter 4
Application of the Local Fractional Fourier Series to Fractal Signals

Xiao-Jun Yang, Dumitru Baleanu, and J. A. Tenreiro Machado

Abstract Local fractional Fourier series is a generalized Fourier series in fractal space. The local fractional calculus is one of useful tools to process the local fractional continuously non-differentiable functions (fractal functions). Based on the local fractional derivative and integration, the present chapter is devoted to the theory and applications of local fractional Fourier analysis in generalized Hilbert space. We recall the local fractional Fourier series, the Fourier transform, the generalized Fourier transform, the discrete Fourier transform and fast Fourier transform in fractal space.

X.-J. Yang (✉)
Department of Mathematics and Mechanics, China University of Mining and Technology, Xuzhou Campus, Xuzhou, Jiangsu 221008, China

Institute of Software Science, Zhengzhou Normal University, Zhengzhou 450044, China

Institute of Applied mathematics, Qujing Normal University, Qujing 655011, China
e-mail: dyangxiaojun@163.com

D. Baleanu
Faculty of Engineering, Department of Chemical and Materials Engineering, King Abdulaziz University, PO Box 80204, Jeddah 21589, Saudi Arabia

Faculty of Arts and Sciences, Department of Mathematics and Computer Sciences, Cankaya University, 06530 Ankara, Turkey

Institute of Space Sciences, Magurele-Bucharest, Romania
e-mail: dumitru@cankaya.edu.tr

J.A.T. Machado
Department of Electrical Engineering, Institute of Engineering, Polytechnic of Porto, Rua Dr. Antonio Bernardino de Almeida, 431, 4200-072 Porto, Portugal
e-mail: jtm@isep.ipp.pt

J.A.T. Machado et al. (eds.), *Discontinuity and Complexity in Nonlinear Physical Systems*, Nonlinear Systems and Complexity 6, DOI 10.1007/978-3-319-01411-1_4,

Keywords Local fractional Fourier series • Local fractional calculus • Local fractional Fourier transform • The generalized local fractional Fourier transform • Discrete local fractional Fourier transform • Fast local fractional Fourier transform • Fractal space

4.1 Introduction

Fourier analysis [1–6] is a mathematical method applied to transform a periodic function with many applications in physics and engineering. It had been used to a wider variety of field in the sciences and in engineering, image and signal processing, containing electrical engineering, quantum mechanics, neurology, optics, acoustics, oceanography, and so on, and after improved and expanded upon it, its general field was come to be known as the field of harmonic analysis [7, 8].

In mathematics, in the area of harmonic analysis, the fractional Fourier transform (FRFT) [9] is a linear transformation generalizing the Fourier transform. The FRFT [10–18] can be used to define fractional convolution, correlation, and other operations, and can also be further generalized into the linear canonical transformation (LCT).

However, the above referred results can't process the non-differentiable time-frequency functions on a fractal set (also local fractional continuous functions). The theory of local fractional calculus (also called fractal calculus [19–51]) is one of the useful tools to handle the fractal and continuously non-differentiable functions, and was successfully applied in describing physical phenomena [22, 23, 26–28, 31–33, 35, 38–40, 42, 44, 45, 47]. Local fractional Fourier analysis [36, 37] that is derived from the local fractional calculus is a generalization of the Fourier analysis in fractal space. Local fractional calculus has played an important role in handling non-differentiable functions.

The aim of this chapter is investigated the theory and applications of the local fractional Fourier series. The organization of this work is as follows. In Sect. 4.2, the preliminary results for the local fractional calculus are investigated. The theory of local fractional Fourier series is presented in Sect. 4.3. Section 4.4 is devoted to theory of the local FRFT in fractal space. Theory of the generalized local FRFT in fractal space is considered in Sect. 4.5. The discrete local FRFT in fractal space is studied in Sect. 4.6. The fast local FRFT in fractal space is considered in Sect. 4.7. The conclusion is in Sect. 4.8.

4.2 Preliminary Results

4.2.1 *Local Fractional Continuity of Functions*

Definition 1. If there is [36–41]

$$|f(x) - f(x_0)| < \epsilon^{\alpha} \tag{4.1}$$

with $|x - x_0| < \delta$, for $\varepsilon, \delta > 0$ and $\varepsilon, \delta \in R$. Now $f(x)$ is called local fractional continuous at $x = x_0$, denote by $\lim_{x \to x_0} f(x) = f(x_0)$. Then $f(x)$ is called local fractional continuous on the interval (a,b), denoted by [36–41, 43]

$$f(x) \in C_\alpha(a,b). \tag{4.2}$$

Lemma 1. Let F be a subset of the real line and be a fractal. If $f : (F,d) \to (\Omega,d)$ is a bi-Lipschitz mapping, then there is for constants $\rho, \tau > 0$ and $F \subset R$,

$$\rho^s H^s(F) \leq H^s(f(F)) \leq \tau^s H^s(F) \tag{4.3}$$

such that for all $x_1, x_2 \in F$,

$$\rho^\alpha |x_1 - x_2|^\alpha \leq |f(x_1) - f(x_2)| \leq \tau^\alpha |x_1 - x_2|^\alpha. \tag{4.4}$$

Hence, we have

$$|f(x_1) - f(x_2)| \leq \tau^\alpha |x_1 - x_2|^\alpha \tag{4.5}$$

such that

$$|f(x_1) - f(x_2)| < \epsilon^\alpha. \tag{4.6}$$

Notice that α is fractal dimension. This result is directly deduced from fractal geometry [38, 44].

4.2.2 Local Fractional Derivative and Integration

Definition 2. Setting $f(x) \in C_\alpha(a,b)$, local fractional derivative of $f(x)$ of order α at $x = x_0$ is defined by [35–44, 46–49]

$$f^{(\alpha)}(x_0) = \frac{d^\alpha f(x)}{dx^\alpha}\Big|_{x=x_0} = \lim_{x \to x_0} \frac{\Delta^\alpha(f(x) - f(x_0))}{(x - x_0)^\alpha}, \tag{4.7}$$

where $\Delta^\alpha(f(x) - f(x_0)) \cong \Gamma(1+\alpha)\Delta(f(x) - f(x_0))$. For any $x \in (a, b)$, there exists

$$f^{(\alpha)}(x) = D_x^{(\alpha)} f(x), \tag{4.8}$$

denoted by

$$f(x) \in D_x^{(\alpha)}(a,b). \tag{4.9}$$

Definition 3. Setting $f(x) \in C_\alpha(a,b)$, local fractional integral of $f(x)$ of order α in the interval $[a,b]$ is defined through [35–48]

$$
\begin{aligned}
&{}_aI_b^{(\alpha)} f(x) \\
&= \frac{1}{\Gamma(1+\alpha)} \int_a^b f(t)(dt)^\alpha \\
&= \frac{1}{\Gamma(1+\alpha)} \lim_{\Delta t \to 0} \sum_{j=0}^{j=N-1} f(t_j)(\Delta t_j)^\alpha,
\end{aligned}
\tag{4.10}
$$

where $\Delta t_j = t_{j+1} - t_j$, $\Delta t = \max\{\Delta t_1, \Delta t_2, \Delta t_j, \ldots\}$ and $[t_j, t_{j+1}]$, $j = 0, \ldots, N-1$, $t_0 = a$, $t_N = b$, is a partition of the interval $[a,b]$. For any $x \in (a,\ b)$, there exists [35–38]

$$
{}_aI_x^{(\alpha)} f(x), \tag{4.11}
$$

denoted by

$$
f(x) \in I_x^{(\alpha)}(a,b). \tag{4.12}
$$

Here, it follows that [35–38]

$$
{}_aI_a^{(\alpha)} f(x) = 0, a = b, \tag{4.13}
$$

$$
{}_aI_b^{(\alpha)} f(x) = -{}_bI_a^{(\alpha)} f(x), a < b, \tag{4.14}
$$

and

$$
{}_aI_a^{(0)} f(x) = f(x). \tag{4.15}
$$

We notice that we have [35–38]

$$
f(x) \in C_\alpha(a,b), \tag{4.16}
$$

if $f(x) \in D_x^{(\alpha)}(a,b)$, or $I_x^{(\alpha)}(a,b)$.

4.2.3 Complex Number of Fractional Order

Definition 4. Fractional-order complex number is defined by [36, 37, 43, 44, 46]

$$
I^\alpha = x^\alpha + i^\alpha y^\alpha, x, y \in R, 0 < \alpha \le 1, \tag{4.17}
$$

where its conjugate of complex number shows that

$$\overline{I^\alpha} = x^\alpha - i^\alpha y^\alpha, \tag{4.18}$$

and where the fractional modulus is derived as

$$|I^\alpha| = I^\alpha \overline{I^\alpha} = \overline{I^\alpha} I^\alpha = \sqrt{x^{2\alpha} + y^{2\alpha}}. \tag{4.19}$$

Definition 5. Complex Mittag–Leffler function in fractal space is defined by [36, 37, 43, 44, 46]

$$E_\alpha\left(z^\alpha\right) := \sum_{k=0}^{\infty} \frac{z^{\alpha k}}{\Gamma\left(1 + k\alpha\right)}, \tag{4.20}$$

for $z \in C$ (complex number set) and $0 < \alpha \leq 1$.

The following rules hold [36, 37, 43, 44]:

$$E_\alpha\left({z_1}^\alpha\right) E_\alpha\left({z_2}^\alpha\right) = E_\alpha\left(\left(z_1 + z_2\right)^\alpha\right); \tag{4.21}$$

$$E_\alpha\left({z_1}^\alpha\right) E_\alpha\left(-{z_2}^\alpha\right) = E_\alpha\left(\left(z_1 - z_2\right)^\alpha\right); \tag{4.22}$$

$$E_\alpha\left(i^\alpha {z_1}^\alpha\right) E_\alpha\left(i^\alpha {z_2}^\alpha\right) = E_\alpha\left(i^\alpha \left({z_1}^\alpha + {z_2}^\alpha\right)^\alpha\right). \tag{4.23}$$

When $z^\alpha = i^\alpha x^\alpha$, the complex Mittag–Leffler function is [36, 37, 43, 44, 46]

$$E_\alpha\left(i^\alpha x^\alpha\right) = \cos_\alpha x^\alpha + i^\alpha \sin_\alpha x^\alpha \tag{4.24}$$

with

$$\cos_\alpha x^\alpha = \sum_{k=0}^{\infty} (-1)^k \frac{x^{2\alpha k}}{\Gamma\left(1 + 2\alpha k\right)}$$

and

$$\sin_\alpha x^\alpha = \sum_{k=0}^{\infty} (-1)^k \frac{x^{\alpha(2k+1)}}{\Gamma\left[1 + \alpha\left(2k + 1\right)\right]},$$

for $x \in R$ and $0 < \alpha \leq 1$, we have that [36, 37]

$$E_\alpha\left(i^\alpha x^\alpha\right) E_\alpha\left(i^\alpha y^\alpha\right) = E_\alpha\left(i^\alpha (x + y)^\alpha\right) \tag{4.25}$$

and

$$E_\alpha\left(i^\alpha x^\alpha\right) E_\alpha\left(-i^\alpha y^\alpha\right) = E_\alpha\left(i^\alpha (x - y)^\alpha\right). \tag{4.26}$$

4.2.4 Generalized Hilbert Space

Definition 6. A generalized Hilbert space is a complete generalized inner-product space [36, 37].

Definition 7. A scalar (or dot) product of two T-periodic functions $f(t)$ and $g(t)$ is defined by [36, 37, 43]

$$\langle f, g\rangle_\alpha = \int_0^T f(t)\overline{g(t)}(dt)^\alpha. \tag{4.27}$$

Suppose $\{e^\alpha{}_n\}$ is an orthonormal system in an inner-product space X. The following results are equivalent [36, 37, 43]:

1. $span\{e_1{}^\alpha, \ldots, e_n{}^\alpha\} = X$, i.e., $\{e_n{}^\alpha\}$ is a basis;
2. (**Pythagorean theorem in fractal space**)
   The equation

$$\sum_{k=1}^{\infty} |a_k{}^\alpha|^2 = \| f \|_\alpha^2 \tag{4.28}$$

   for all $f \in X$, where $a_k{}^\alpha = \langle f, e_k{}^\alpha\rangle_\alpha$;
3. (**Generalized Pythagorean theorem in fractal space**)
   Generalized equation

$$\langle f, g\rangle = \sum_{k=1}^{n} a_k{}^\alpha \overline{b_k{}^\alpha} \tag{4.29}$$

   for all $f, g \in X$, where
   $a_k{}^\alpha = \langle f, e_n{}^\alpha\rangle_\alpha$ and $b_k{}^\alpha = \langle g, e_k{}^\alpha\rangle_\alpha$;
4. $f = \sum_{k=1}^{n} a_k{}^\alpha e_k{}^\alpha$ with sum convergent in X for all $f \in X$.

For more details, we see [4–44].

Here we can take any sequence of T-periodic local fractional continuous functions ϕ_k, $k = 0, 1, \ldots$ that are [36, 37, 43, 44]

Orthogonal:

$$\langle \phi_k, \phi_j\rangle_\alpha = \int_0^T \phi_k(t)\overline{\phi_j(t)}(dt)^\alpha = 0, k \neq j; \tag{4.30}$$

Normalized:

$$\langle \phi_k, \phi_k\rangle_\alpha = \int_0^T \phi_k{}^2(t)(dt)^\alpha = 1; \tag{4.31}$$

Complete: If a function $x(t)$ is such that

$$\langle x, \phi_k \rangle_\alpha = \int_0^T x(t)\phi_k(t)(dt)^\alpha = 0 \tag{4.32}$$

for all i, then $x(t) \equiv 0$.

4.2.5 *Local Fractional Fourier Series in Generalized Hilbert Space*

4.2.5.1 Local Fractional Fourier Series in Generalized Hilbert Space

Definition 8. Let $\{\phi_k(t)\}_{k=1}{}^{\infty}$ be a complete, orthonormal set of functions. Then any T-periodic fractal signal $f(t)$ can be uniquely represented as an infinite series [36, 37, 43]

$$f(t) = \sum_{k=0}^{\infty} \varphi_k \phi_k(t). \tag{4.33}$$

This is called the local fractional Fourier series representation of $f(t)$ in the generalized Hilbert space. The scalars φ_i are called the local fractional Fourier coefficients of $f(t)$.

4.2.5.2 Local Fractional Fourier Coefficients

To derive the formula for φ_k, we write [36, 37, 43]

$$f(t)\phi_k(t) = \sum_{i=0}^{\infty} \varphi_j \phi_j(t)\phi_k(t), \tag{4.34}$$

and integrate over one period by using the generalized Pythagorean theorem in fractal space [36, 37, 43]

$$\begin{aligned}
\langle f, \phi_k \rangle_\alpha &= \int_0^T f(t)\phi_k(t)(dt)^\alpha \\
&= \int_0^T \sum_{j=0}^{\infty} \varphi_j \phi_j(t)\phi_k(t)(dt)^\alpha \\
&= \sum_{j=0}^{\infty} \left(\varphi_j \left(\int_0^T \phi_j(t)\phi_k(t)(dt)^\alpha \right) \right) \\
&= \sum_{j=0}^{\infty} \varphi_j \langle \phi_j, \phi_k \rangle_\alpha = \varphi_k
\end{aligned} \tag{4.35}$$

Because the functions $\phi_k(t)$ form a complete orthonormal system, the partial sums of the local fractional Fourier series

$$f(t)=\sum_{k=0}^{\infty}\varphi_k\phi_k(t) \tag{4.36}$$

converge to $f(t)$ in the following sense:

$$\lim_{N\to\infty}\left(\frac{1}{\Gamma(1+\alpha)}\int_0^T\left(f(t)-\sum_{k=1}^{\infty}\varphi_k\phi_k(t)\right)\overline{\left(f(t)-\sum_{k=1}^{\infty}\varphi_k\phi_k(t)\right)}(dt)^{\alpha}\right)=0. \tag{4.37}$$

Therefore, we can use the partial sums

$$f_N(t)=\sum_{k=1}^{N}\varphi_k\phi_k(t) \tag{4.38}$$

to approximate $f(t)$.

Hence, we have that

$$\int_0^T f^2(t)(dt)^{\alpha}=\sum_{k=1}^{\infty}{\varphi_k}^2. \tag{4.39}$$

The sequence of T-periodic functions in fractal space $\{\phi_k(t)\}_{k=0}{}^{\infty}$ defined by

$$\phi_0(t)=\left(\frac{1}{T}\right)^{\frac{\alpha}{2}}$$

and

$$\phi_k(t)=\begin{cases}\left(\frac{2}{T}\right)^{\frac{\alpha}{2}}\sin_{\alpha}\left(k^{\alpha}{\omega_0}^{\alpha}t^{\alpha}\right), if\, k\geq 1 is\, odd\\ \left(\frac{2}{T}\right)^{\frac{\alpha}{2}}\cos_{\alpha}\left(k^{\alpha}{\omega_0}^{\alpha}t^{\alpha}\right), if\, k>1 is\, even\end{cases} \tag{4.40}$$

are complete and orthonormal, where $\omega_0=\frac{2\pi}{T}$.

Another useful complete orthonormal set is furnished by the Mittag–Leffler functions [36, 37]:

$$\phi_k(t)=\left(\frac{1}{T}\right)^{\frac{\alpha}{2}}E_{\alpha}\left(i^{\alpha}k^{\alpha}{\omega_0}^{\alpha}t^{\alpha}\right), k=0,\pm1,\pm2,\ldots \tag{4.41}$$

where $\omega_0=\frac{2\pi}{T}$.

4.3 Local Fractional Fourier Series

4.3.1 Notations

Definition 9. Local fractional trigonometric Fourier series of $f(t)$ is given by [36, 37, 44]

$$f(t) = a_0 + \sum_{i=1}^{\infty} a_k \sin_\alpha \left(k^\alpha {\omega_0}^\alpha t^\alpha\right) + \sum_{i=1}^{\infty} b_k \cos_\alpha \left(k^\alpha {\omega_0}^\alpha t^\alpha\right). \tag{4.42}$$

Then the local fractional Fourier coefficients can be computed by

$$\begin{cases} a_0 = \dfrac{1}{T^\alpha}\displaystyle\int_0^T f(t)(dt)^\alpha, \\ a_k = \left(\dfrac{2}{T}\right)^\alpha \displaystyle\int_0^T f(t)\sin_\alpha \left(k^\alpha {\omega_0}^\alpha t^\alpha\right)(dt)^\alpha, \\ b_k = \left(\dfrac{2}{T}\right)^\alpha \displaystyle\int_0^T f(t)\cos_\alpha \left(k^\alpha {\omega_0}^\alpha t^\alpha\right)(dt)^\alpha. \end{cases} \tag{4.43}$$

When $\omega_0 = 1$, we get the short form

$$f(t) = a_0 + \sum_{i=1}^{\infty} a_k \sin_\alpha \left(k^\alpha t^\alpha\right) + \sum_{i=1}^{\infty} b_k \cos_\alpha \left(k^\alpha t^\alpha\right).$$

Then the local fractional Fourier coefficients can be computed by

$$\begin{cases} a_0 = \dfrac{1}{T^\alpha}\displaystyle\int_0^T f(t)(dt)^\alpha, \\ a_k = \left(\dfrac{2}{T}\right)^\alpha \displaystyle\int_0^T f(t)\sin_\alpha \left(k^\alpha t^\alpha\right)(dt)^\alpha, \\ b_k = \left(\dfrac{2}{T}\right)^\alpha \displaystyle\int_0^T f(t)\cos_\alpha \left(k^\alpha t^\alpha\right)(dt)^\alpha. \end{cases}$$

The Mittag–Leffler functions expression of local fractional Fourier series is given by [36, 37, 44]

$$f(x) = \sum_{k=-\infty}^{\infty} C_k E_\alpha \left(\frac{\pi^\alpha i^\alpha (kx)^\alpha}{l^\alpha}\right), \tag{4.44}$$

where the local fractional Fourier coefficients is

$$C_k = \frac{1}{(2l)^\alpha} \int_{-l}^{l} f(x) E_\alpha \left(\frac{-\pi^\alpha i^\alpha (kx)^\alpha}{l^\alpha} \right) (dx)^\alpha, k \in Z. \tag{4.45}$$

For local fractional Fourier series (4.45), the weights of the Mittag–Leffler functions are written in the form [44]

$$C_k = \frac{\frac{1}{(2l)^\alpha} \int_{-l+t_0}^{l+t_0} f(x) E_\alpha \left(\frac{-\pi^\alpha i^\alpha (kx)^\alpha}{l^\alpha} \right) (dx)^\alpha}{\frac{1}{(2l)^\alpha} \int_{-l+t_0}^{l+t_0} E_\alpha \left(\frac{-\pi^\alpha i^\alpha (kx)^\alpha}{l^\alpha} \right) \overline{E_\alpha \left(\frac{-\pi^\alpha i^\alpha (kx)^\alpha}{l^\alpha} \right)} (dx)^\alpha}. \tag{4.46}$$

Above is generalized to calculate local fractional Fourier series.

4.3.2 Properties of Local Fractional Fourier Series

The following results are valid [36, 37].

Property 2 (Linearity). Suppose that local fractional Fourier coefficients of $f(x)$ and $g(x)$ are f_n and g_n, respectively, then we has for two constants a and b

$$af(x) + bg(x) \leftrightarrow af_n + bgf_n. \tag{4.47}$$

Property 3 (Conjugation). Suppose that C_n is Fourier coefficients of $f(x)$. Then we have

$$\overline{f(x)} \leftrightarrow \overline{C_{-n}}. \tag{4.48}$$

Property 4 (Shift in time). Suppose that C_n is Fourier coefficients of $f(x)$. Then we have

$$f(x - x_0) \leftrightarrow E_\alpha (-i^\alpha (nx)^\alpha) C_n. \tag{4.49}$$

Property 5 (Time reversal). Suppose that C_n is Fourier coefficients of $f(x)$. Then we have

$$f(-x) \leftrightarrow C_{-n}. \tag{4.50}$$

For proofs of the above, we see [36, 37].

4.3.3 The Basic Theorems of Local Fractional Fourier Series

The following results are valid [36, 37].

Theorem 6 (Local fractional Bessel inequality). Suppose that $f(t)$ is 2π-periodic, bounded and local fractional integral on $[-\pi, \pi]$. If both a_n and b_n are Fourier coefficients of $f(t)$, then there exists the inequality

$$\frac{a_0{}^2}{2} + \sum_{k=1}^{n}\left(a_k{}^2 + b_k{}^2\right) \leq \frac{1}{\pi^\alpha}\int_{-\pi}^{\pi} f^2(t)(dt)^\alpha. \tag{4.51}$$

Theorem 7 (Local fractional Riemann–Lebesgue theorem). Suppose that $f(x)$ is 2π-periodic, bounded and local fractional integral on $[-\pi, \pi]$. Then we have

$$\lim_{n\to+\infty}\frac{1}{(2\pi)^\alpha}\int_{-\pi}^{\pi} f(t)\sin_\alpha(nt)^\alpha(dt)^\alpha = 0 \tag{4.52}$$

and

$$\lim_{n\to+\infty}\frac{1}{(2\pi)^\alpha}\int_{-\pi}^{\pi} f(t)\cos_\alpha(nt)^\alpha(dt)^\alpha = 0. \tag{4.53}$$

Theorem 8. Suppose that $T_{n,\alpha}(x) \sim \frac{a_0}{2^\alpha} + \sum_{n=1}^{n}\left(a_n\cos_\alpha(nx)^\alpha + b_n\sin_\alpha(nx)^\alpha\right)$, then we have that

$$T_{n,\alpha}(x) = \frac{1}{\pi^\alpha}\int_{-\pi}^{\pi} T_{n,\alpha}\,(x+t)\, D_{n,\alpha}(t)(dt)^\alpha, \tag{4.54}$$

where

$$D_{n,\alpha}(t) = \frac{1}{2} + \sum_{k=1}^{n}\cos_\alpha(nx)^\alpha = \frac{\sin_\alpha\left(\frac{(2n+1)x}{2}\right)^\alpha}{\sin_\alpha\left(\frac{x}{2}\right)^\alpha}. \tag{4.55}$$

Theorem 9. Suppose that $f(t)$ is 2π-periodic, bounded and local fractional integral on $[-\pi, \pi]$.

If $f(t) \sim \frac{a_0}{2} + \sum_{k=1}^{\infty}\left(a_k\cos_\alpha(kt)^\alpha + b_k\sin_\alpha(kt)^\alpha\right)$, then we have

$$\frac{1}{\pi^\alpha}\int_{-\pi}^{\pi} f^2(t)(dt)^\alpha = \frac{a_0{}^2}{2} + \sum_{k=0}^{\infty}\left(a_k{}^2 + b_k{}^2\right). \tag{4.56}$$

Theorem 10 (Convergence theorem for local fractional Fourier series). Suppose that $f(t)$ is 2π-periodic, bounded and local fractional integral on $[-\pi, \pi]$. The local fractional series of $f(t)$ converges to $f(t)$ at $t \in [-\pi, \pi]$, and

$$\frac{f(t+0)+f(t-0)}{2} = \frac{a_0}{2} + \sum_{k=1}^{\infty} \left(a_k \cos_\alpha (kt)^\alpha + b_k \sin_\alpha (kt)^\alpha\right), \tag{4.57}$$

where

$$a_0 = \frac{1}{\pi^\alpha} \int_{-\pi}^{\pi} f(x)(dt)^\alpha, \tag{4.58}$$

$$a_n = \frac{1}{\pi^\alpha} \int_{-\pi}^{\pi} f(x) \cos_\alpha (nx)^\alpha (dt)^\alpha \tag{4.59}$$

and

$$b_n = \frac{1}{\pi^\alpha} \int_{-\pi}^{\pi} f(x) \sin_\alpha (nx)^\alpha (dt)^\alpha. \tag{4.60}$$

For proofs of the above, we see [36, 37].

4.3.4 Applications of Local Fractional Fourier Series

4.3.4.1 Applications of Local Fractional Fourier Series to Fractal Signals

Now we consider the applications of local fractional Fourier series to fractal signals.

Expand fractal signal $X(t) = t^\alpha + 1 (-\pi < t \le \pi)$ in local fractional Fourier series.

Now we find the local fractional Fourier coefficients

$$\begin{aligned} a_0 &= \frac{1}{\pi^\alpha} \int_{-\pi}^{\pi} X(t)(dt)^\alpha \\ &= \frac{1}{\pi^\alpha} \int_{-\pi}^{\pi} (t^\alpha + 1)(dt)^\alpha \\ &= \left.\frac{\Gamma^2(1+\alpha)\, t^{2\alpha}}{\pi^\alpha \Gamma(1+2\alpha)}\right|_{-\pi}^{\pi} + \left.\frac{\Gamma(1+\alpha)\, t^\alpha}{\pi^\alpha \Gamma(1+\alpha)}\right|_{-\pi}^{\pi} \\ &= 2, \end{aligned} \tag{4.61}$$

$$\begin{aligned} a_n &= \frac{1}{\pi^\alpha}\int_{-\pi}^{\pi} X(t)\cos_\alpha(nt)^\alpha(dt)^\alpha = \frac{1}{\pi^\alpha}\int_{-\pi}^{\pi}(t^\alpha+1)\cos_\alpha(nt)^\alpha(dt)^\alpha \\ &= \frac{\Gamma(1+\alpha)\,t^\alpha \sin_\alpha(nt)^\alpha}{n^\alpha\pi^\alpha}\Bigg|_{-\pi}^{\pi} - \frac{1}{\pi^\alpha n^\alpha}\int_{-\pi}^{\pi}\sin_\alpha(nt)^\alpha(dt)^\alpha \\ &= \frac{\Gamma(1+\alpha)\,t^\alpha \sin_\alpha(nt)^\alpha}{n^\alpha\pi^\alpha}\Bigg|_{-\pi}^{\pi} + \frac{\Gamma(1+\alpha)\cos_\alpha(nt)^\alpha}{n^{2\alpha}\pi^\alpha}\Bigg|_{-\pi}^{\pi} = 0 \end{aligned} \tag{4.62}$$

and

$$\begin{aligned} b_n &= \frac{1}{\pi^\alpha}\int_{-\pi}^{\pi} X(t)\sin_\alpha(nt)^\alpha(dt)^\alpha = \frac{1}{\pi^\alpha}\int_{-\pi}^{\pi}(t^\alpha+1)\sin_\alpha(nt)^\alpha(dt)^\alpha \\ &= \left[-\frac{\Gamma(1+\alpha)\,t^\alpha \cos_\alpha(nt)^\alpha}{n^\alpha\pi^\alpha}\right]\Bigg|_{-\pi}^{\pi} - \frac{1}{\pi^\alpha n^\alpha}\int_{-\pi}^{\pi}\cos_\alpha(nt)^\alpha(dt)^\alpha \\ &= \left[-\frac{\Gamma(1+\alpha)\,t^\alpha \cos_\alpha(nt)^\alpha}{n^\alpha\pi^\alpha}\right]\Bigg|_{-\pi}^{\pi} + \frac{\Gamma(1+\alpha)\sin_\alpha(nt)^\alpha}{n^{2\alpha}\pi^\alpha}\Bigg|_{-\pi}^{\pi} \\ &= \frac{2\Gamma(1+\alpha)(-1)^{n+1}}{n^\alpha\pi^\alpha}. \end{aligned} \tag{4.63}$$

Therefore, for $-\pi < t \le \pi$ we have local fractional Fourier series representation of fractal signal

$$X(t) = 1 + \sum_{n=1}^{\infty}\left(\frac{2\Gamma(1+\alpha)(-1)^{n+1}}{n^\alpha\pi^\alpha}\sin_\alpha(nt)^\alpha\right). \tag{4.64}$$

4.4 The Local FRFT in Fractal Space

4.4.1 Notations

Definition 10 (The local FRFT in fractal space). Suppose that $f(x) \in C_\alpha(-\infty,\infty)$, the local FRFT, denoted by $F_\alpha\{f(x)\} \equiv f_\omega^{F,\alpha}(\omega)$, is written in the form [37, 38, 45–48]

$$\begin{aligned} F_\alpha\{f(x)\} &= f_\omega^{F,\alpha}(\omega) \\ &= \frac{1}{\Gamma(1+\alpha)}\int_{-\infty}^{\infty} E_\alpha(-i^\alpha\omega^\alpha x^\alpha)\, f(x)(dx)^\alpha, \end{aligned} \tag{4.65}$$

where the latter converges.

Definition 11. If $F_\alpha\{f(x)\} \equiv f_\omega^{F,\alpha}(\omega)$, its inversion formula is written in the form [37, 38, 45–48]

$$\begin{aligned} f(x) &= F_\alpha^{-1}\left(f_\omega^{F,\alpha}(\omega)\right): \\ &= \frac{1}{(2\pi)^\alpha}\int_{-\infty}^{\infty} E_\alpha\left(i^\alpha \omega^\alpha x^\alpha\right) f_\omega^{F,\alpha}(\omega)\,(d\omega)^\alpha, x > 0. \end{aligned} \tag{4.66}$$

For the proofs of the above, we see [37, 38].

4.4.2 The Basic Theorems of Local FRFT

The following results are valid [37, 38].

Theorem 11. Let $F_\alpha\{f(x)\} = f_\omega^{F,\alpha}(\omega)$ and $F_\alpha\{g(x)\} = g_\omega^{F,\alpha}(\omega)$, and let a, b be two constants. Then we have

$$F_\alpha\{af(x) + bg(x)\} = aF_\alpha\{f(x)\} + bF_\alpha\{g(x)\}. \tag{4.67}$$

Theorem 12. Let $F_\alpha\{f(x)\} = f_\omega^{F,\alpha}(\omega)$. If $\lim_{|x|\to\infty} f(x) = 0$, then we have

$$F_\alpha\left\{f^{(\alpha)}(x)\right\} = i^\alpha \omega^\alpha F_\alpha\{f(x)\}. \tag{4.68}$$

As a direct result, repeating this process, when

$$f(0) = f^{(\alpha)}(0) = \cdots = f^{((k-1)\alpha)}(0) = 0$$

we have

$$F_\alpha\left\{f^{(k\alpha)}(x)\right\} = i^{k\alpha}\omega^{k\alpha} F_\alpha\{f(x)\}. \tag{4.69}$$

Theorem 13. Let $F_\alpha\{f(x)\} = f_\omega^{F,\alpha}(\omega)$ and $\lim_{x\to\infty} {}_{-\infty}I_x^{(\alpha)} f(x) \to 0$, then we have

$$F_\alpha\left\{{}_{-\infty}I_x^{(\alpha)} f(x)\right\} = \frac{1}{i^\alpha \omega^\alpha} F_\alpha\{f(x)\}. \tag{4.70}$$

Theorem 14. If $F_\alpha\{f(x)\} = f_\omega^{F,\alpha}(\omega)$ and $a > 0$, then we have

$$F_\alpha\{f(ax)\} = \frac{1}{a^\alpha} f_\omega^{F,\alpha}\left(\frac{\omega}{a}\right). \tag{4.71}$$

Theorem 15. If $F_\alpha\{f(x)\} = f_\omega^{F,\alpha}(\omega)$ and c is a constant, then we have

$$F_\alpha\{f(x-c)\} = E_\alpha\left(-i^\alpha c^\alpha \omega^\alpha\right) F_\alpha\{f(x)\}. \tag{4.72}$$

Theorem 16. If $F_\alpha\{f(x)\} = f_\omega{}^{F,\alpha}(\omega)$ and c is a constant, then we have

$$F_\alpha\{f(x)E_\alpha(-i^\alpha x^\alpha {\omega_0}^\alpha)\} = f_\omega^{F,\alpha}(\omega - \omega_0). \tag{4.73}$$

Theorem 17. Let $F_\alpha\{f_1(x)\} = f_{\omega,1}{}^{F,\alpha}(\omega)$ and $F_\alpha\{f_2(x)\} = f_{\omega,2}{}^{F,\alpha}(\omega)$, then we have

$$F_\alpha\{f_1(x) * f_2(x)\} = f_{\omega,1}^{F,\alpha}(\omega)\, f_{\omega,2}^{F,\alpha}(\omega). \tag{4.74}$$

Theorem 18. If $F_\alpha\{f(x)\} = f_\omega{}^{F,\alpha}(\omega)$, then

$$\frac{1}{\Gamma(1+\alpha)}\int_{-\infty}^{\infty} |f(x)|^2 (dx)^\alpha = \frac{1}{(2\pi)^\alpha}\int_{-\infty}^{\infty} \left|f_\omega^{F,\alpha}(\omega)\right|^2 (d\omega)^\alpha. \tag{4.75}$$

Theorem 19. If $F_\alpha\{f(x)\} = f_\omega{}^{F,\alpha}(\omega)$ and $F_\alpha\{g(x)\} = g_\omega{}^{F,\alpha}(\omega)$, then

$$\frac{1}{\Gamma(1+\alpha)}\int_{-\infty}^{\infty} f(x)\overline{g(x)}(dx)^\alpha = \frac{1}{(2\pi)^\alpha}\int_{-\infty}^{\infty} f_\omega^{F,\alpha}(\omega)\,\overline{g_\omega^{F,\alpha}(\omega)}(d\omega)^\alpha. \tag{4.76}$$

For the proofs of the above, we see [37, 38].

4.4.3 Applications of Local FRFT to Fractal Signals

We now consider applications of local FRFT to fractal signals.

Let a nonperiodic signal $X(t)$ be defined by the relation

$$X(t) = \begin{cases} 1, -t_0 \le t < t_0, \\ 0, else. \end{cases} \tag{4.77}$$

Taking the local FRFTs, we have

$$\begin{aligned} X_\omega^{F,\alpha}(\omega) &= \frac{1}{\Gamma(1+\alpha)}\int_{-\infty}^{\infty} X(t)E_\alpha(-i^\alpha \omega^\alpha x^\alpha)(dx)^\alpha \\ &= \frac{1}{\Gamma(1+\alpha)}\int_{-t}^{t} E_\alpha(-i^\alpha \omega^\alpha x^\alpha)(dx)^\alpha \\ &= \frac{E_\alpha(-i^\alpha \omega^\alpha x^\alpha)}{-i^\alpha \omega^\alpha}\Bigg|_{-t_0}^{t_0}. \end{aligned} \tag{4.78}$$

Taking into account

$$E_\alpha(-i^\alpha x^\alpha) = \cos_\alpha x^\alpha - i^\alpha \sin_\alpha x^\alpha,$$

we get

$$X_{\omega}^{F,\alpha}(\omega) = \frac{2\sin_{\alpha}\omega^{\alpha}t_0{}^{\alpha}}{\omega^{\alpha}} = 2t^{\alpha}\sin_{\alpha C}\omega^{\alpha}t_0{}^{\alpha}, \tag{4.79}$$

where $\sin_{\alpha C}\omega^{\alpha}t_0{}^{\alpha} = \frac{\sin_{\alpha}\omega^{\alpha}t_0{}^{\alpha}}{\omega^{\alpha}t_0{}^{\alpha}}$.

4.5 The Generalized Local FRFT in Fractal Space

4.5.1 Definitions and Notations

Definition 12 (Generalized local FRFT). The generalized local FRFT is written in the form [37, 38, 49]

$$\begin{aligned} F_{\alpha}\{f(x)\} &= f_{\omega}^{F,\alpha}(\omega) \\ &= \frac{1}{\Gamma(1+\alpha)}\int_{-\infty}^{\infty} f(x)E_{\alpha}\left(-i^{\alpha}h_0x^{\alpha}\omega^{\alpha}\right)(dx)^{\alpha}, \end{aligned} \tag{4.80}$$

where $h_0 = \frac{(2\pi)^{\alpha}}{\Gamma(1+\alpha)}$ with $0 < \alpha \leq 1$.

Definition 13. The inverse formula of the generalized local FRFT is written in the form [37, 38, 49]

$$\begin{aligned} F_{\alpha}^{-1}\left(f_{\omega}^{F,\alpha}(\omega)\right) &= f(x) \\ &= \frac{1}{\Gamma(1+\alpha)}\int_{-\infty}^{\infty} f_{\omega}^{F,\alpha}(\omega)\,E_{\alpha}\left(i^{\alpha}h_0x^{\alpha}\omega^{\alpha}\right)(d\omega)^{\alpha}, \end{aligned} \tag{4.81}$$

where $h_0 = \frac{(2\pi)^{\alpha}}{\Gamma(1+\alpha)}$ with $0 < \alpha \leq 1$.

4.5.2 Some Basic Theorems of Local FRFT in Fractal Space

The following result is valid [37, 38, 49].

Theorem 20. Let $F_{\alpha}\{f(x)\} = f_{\omega}{}^{F,\alpha}(\omega)$, then we have

$$f(x) = F_{\alpha}^{-1}\left(f_{\omega}^{F,\alpha}(\omega)\right). \tag{4.82}$$

Theorem 21. Let $F_{\alpha}\{f(x)\} = f_{\omega}{}^{F,\alpha}(\omega)$ and $F_{\alpha}\{g(x)\} = g_{\omega}{}^{F,\alpha}(\omega)$, and let a, b be two constants. Then we have

$$F_{\alpha}\{af(x) + bg(x)\} = aF_{\alpha}\{f(x)\} + bF_{\alpha}\{g(x)\}. \tag{4.83}$$

Theorem 22. Let $F_\alpha\{f(x)\} = f_\omega{}^{F,\alpha}(\omega)$. If $\lim_{|x|\to\infty} f(x) = 0$, then we have

$$F_\alpha\left\{f^{(\alpha)}(x)\right\} = i^\alpha h_0 \omega^\alpha F_\alpha\{f(x)\}. \tag{4.84}$$

As a direct result, repeating this process, when

$$f(0) = f^{(\alpha)}(0) = \cdots = f^{((k-1)\alpha)}(0) = 0$$

we have

$$F_\alpha\left\{f^{(k\alpha)}(x)\right\} = i^{k\alpha} h_0{}^k \omega^{k\alpha} F_\alpha\{f(x)\}. \tag{4.85}$$

Theorem 23. Let $F_\alpha\{f(x)\} = f_\omega{}^{F,\alpha}(\omega)$ and $\lim_{x\to\infty} {}_{-\infty}I_x{}^{(\alpha)} f(x) \to 0$, then we have

$$F_\alpha\left\{{}_{-\infty}I_x{}^{(\alpha)} f(x)\right\} = \frac{1}{i^\alpha h_0 \omega^\alpha} F_\alpha\{f(x)\}. \tag{4.86}$$

Theorem 24. If $F_\alpha\{f(x)\} = f_\omega{}^{F,\alpha}(\omega)$, and $a > 0$, then we have

$$F_\alpha\{f(ax)\} = \frac{1}{a^\alpha} f_\omega^{F,\alpha}\left(\frac{\omega}{a}\right). \tag{4.87}$$

Theorem 25. If $F_\alpha\{f(x)\} = f_\omega{}^{F,\alpha}(\omega)$ and c is a constant, then we have

$$F_\alpha\{f(x-c)\} = E_\alpha\left(-i^\alpha h_0 c^\alpha \omega^\alpha\right) F_\alpha\{f(x)\}. \tag{4.88}$$

Theorem 26. If $F_\alpha\{f(x)\} = f_\omega{}^{F,\alpha}(\omega)$ and c is a constant, then we have

$$F_\alpha\{f(x) E_\alpha\left(-i^\alpha h_0 x^\alpha \omega_0{}^\alpha\right)\} = f_\omega^{F,\alpha}(\omega - \omega_0). \tag{4.89}$$

Theorem 27. Let $F_\alpha\{f_1(x)\} = f_{\omega,1}{}^{F,\alpha}(\omega)$ and $F_\alpha\{f_2(x)\} = f_{\omega,2}{}^{F,\alpha}(\omega)$, then we have

$$F_\alpha\{f_1(x) * f_2(x)\} = f_{\omega,1}^{F,\alpha}(\omega)\, f_{\omega,2}^{F,\alpha}(\omega). \tag{4.90}$$

Theorem 28. If $F_\alpha\{f(x)\} = f_\omega{}^{F,\alpha}(\omega)$, then

$$\frac{1}{\Gamma(1+\alpha)} \int_{-\infty}^{\infty} |f(x)|^2 (dx)^\alpha = \frac{1}{\Gamma(1+\alpha)} \int_{-\infty}^{\infty} \left|f_\omega^{F,\alpha}(\omega)\right|^2 (d\omega)^\alpha. \tag{4.91}$$

Theorem 29. If $F_\alpha\{f(x)\} = f_\omega{}^{F,\alpha}(\omega)$ and $F_\alpha\{g(x)\} = g_\omega{}^{F,\alpha}(\omega)$, then

$$\frac{1}{\Gamma(1+\alpha)}\int_{-\infty}^{\infty} f(x)\overline{g(x)}(dx)^{\alpha} = \frac{1}{\Gamma(1+\alpha)}\int_{-\infty}^{\infty} f_{\omega}^{F,\alpha}(\omega)\,\overline{g_{\omega}^{F,\alpha}(\omega)}(d\omega)^{\alpha}. \tag{4.92}$$

For the proofs of the above, we see [37, 38, 49].

4.6 Discrete Local FRFT in Fractal Space

4.6.1 Definitions and Notations

Definition 14 (Discrete local FRFT). Suppose that $f(n)$ be a periodic discrete-time fractal signal with period N. The N-point discrete local FRFT (DYFT) of $F(n)$ is written in the form [50]

$$\begin{aligned} F(k) &= \sum_{n=0}^{N-1} f(n)E_{\alpha}\left(-i^{\alpha}(2\pi)^{\alpha}n^{\alpha}k^{\alpha}/N^{\alpha}\right) \\ &= \sum_{n=0}^{N-1} f(n)W_{N,\alpha}{}^{-nk}, \end{aligned} \tag{4.93}$$

with $W_{N,\alpha}{}^{-nk} = E_{\alpha}\left(-\frac{i^{\alpha}n^{\alpha}k^{\alpha}(2\pi)^{\alpha}}{N^{\alpha}}\right)$. This is called N-point discrete local FRFT of $F(n)$, denoted by

$$f(n) \leftrightarrow F(k). \tag{4.94}$$

Definition 15 (Inverse discrete local FRFT). The inverse discrete local FRFT (IDYFT) is given by is rewritten as [50]

$$\begin{aligned} f(n) &= \frac{1}{\Gamma(1+\alpha)}\frac{1}{N^{\alpha}}\sum_{k=0}^{N-1} F(k)E_{\alpha}\left(i^{\alpha}n^{\alpha}k^{\alpha}(2\pi)^{\alpha}/N^{\alpha}\right) \\ &= \frac{1}{\Gamma(1+\alpha)}\frac{1}{N^{\alpha}}\sum_{k=0}^{N-1} F(k)W_{N,\alpha}{}^{kn}. \end{aligned} \tag{4.95}$$

with $W_{N,\alpha}{}^{kn} = E_{\alpha}\left(\frac{i^{\alpha}n^{\alpha}k^{\alpha}(2\pi)^{\alpha}}{N^{\alpha}}\right)$.

Taking into account the relation [50]

$$E_{\alpha}\left(i^{\alpha}(2\pi)^{\alpha}(n+1)^{\alpha}\right) = E_{\alpha}\left(i^{\alpha}(2\pi)^{\alpha}n^{\alpha}\right),$$

we deduce that

$$E_\alpha\left(i^\alpha(2\pi)^\alpha n^\alpha\left(\frac{k+N}{N}\right)^\alpha\right)=E_\alpha\left(i^\alpha(2\pi)^\alpha\frac{n^\alpha k^\alpha}{N^\alpha}\right) \tag{4.96}$$

for all $n\in Z$. That is to say,

$$W_{1,\alpha}{}^{(n+N)}=W_{1,\alpha}{}^{n}$$

and

$$W_{N,\alpha}{}^{(k+N)n}=W_{N,\alpha}{}^{kn}.$$

4.6.2 Some Basic Theorems of Discrete Local FRFT in Fractal Space

The following results are valid [50]:

Theorem 30. Suppose that $F(k)=\sum_{n=0}^{N-1}f(n)W_{N,\alpha}{}^{-nk}$, then we have

$$f(n)=\frac{1}{\Gamma(1+\alpha)}\frac{1}{N^\alpha}\sum_{k=0}^{N-1}F(k)W_{N,\alpha}{}^{nk}. \tag{4.97}$$

Theorem 31. Suppose that $f(n)$ be periodic discrete-time signals with period N, then we have

$$\sum_{n=0}^{N-1}f(n)=\sum_{n=j}^{j+N-1}f(n). \tag{4.98}$$

Theorem 32. Suppose that $f_1(n)\leftrightarrow F_1(k)$ and $f_2(n)\leftrightarrow F_2(k)$, then we have

$$af_1(n)+bf_2(n)\leftrightarrow aF_1(k)+bF_2(k). \tag{4.99}$$

Corollary 33.

$$F(n)\leftrightarrow N^\alpha\Gamma(1+\alpha)\,f(-k)\,. \tag{4.100}$$

Corollary 34 (Time reversal rule for DLFFT).

$$f(-n)\leftrightarrow F(-k)\,. \tag{4.101}$$

Definition 16 (Cyclical convolution). The cyclical convolution product of two periodic discrete-time signals $f(n)$ and $g(n)$ with periodic N is the fractal discrete-time signal $(f * g)(n)$ defined by

$$(f * g)(n) = \sum_{l=0}^{N-1} f(l) g(n - l). \tag{4.102}$$

Theorem 35 (Convolution in the n-domain rule for DLFFT). Let $f(n)$ and $g(n)$ be periodic discrete-time signals with period N. Suppose that $f(n) \leftrightarrow F(k)$ and $g(n) \leftrightarrow G(k)$, then

$$(f * g)(n) \leftrightarrow F(k)G(k). \tag{4.103}$$

Theorem 36 (Convolution in the k-domain rule for DLFFT). Let $f(n)$ and $g(n)$ be periodic discrete-time signals with period N. Suppose that $f(n) \leftrightarrow F(k)$ and $g(n) \leftrightarrow G(k)$, then

$$\frac{1}{\Gamma(1+\alpha)} \frac{1}{N^{\alpha}} f(n)g(n) \leftrightarrow (F * G)(k). \tag{4.104}$$

Theorem 37 (Paserval theorem for DLFFT). Let $f(n)$ and $g(n)$ be periodic discrete-time signals with period N. Suppose that $f(n) \leftrightarrow F(k)$ and $g(n) \leftrightarrow G(k)$, then

$$\sum_{n=0}^{N-1} f(n)\overline{g(n)} = \frac{1}{\Gamma(1+\alpha)} \frac{1}{N^{\alpha}} \sum_{k=0}^{N-1} F(k)\overline{G(k)}. \tag{4.105}$$

Corollary 38. Let $f(n)$ and $g(n)$ be periodic discrete-time signals with period N. Suppose that $f(n) \leftrightarrow F(k)$, then

$$\sum_{n=0}^{N-1} |g(n)|^2 = \frac{1}{\Gamma(1+\alpha)} \frac{1}{N^{\alpha}} \sum_{k=0}^{N-1} |G(k)|^2. \tag{4.106}$$

Theorem 39. Let $f(n)$ and $g(n)$ be periodic discrete-time signals with period N. Suppose that $f(n) \leftrightarrow F(k)$ and $g(n) \leftrightarrow G(k)$, then

$$\sum_{n=0}^{N-1} f(n)G(n) = \sum_{k=0}^{N-1} F(k)g(k). \tag{4.107}$$

For the proofs of the above, we see [50].

4.6.3 Application of Discrete Local FRFT to Fractal Signals

Let us consider the discrete local FRFT of the fractal signal

$$f(n) = E_\alpha \left(-i^\alpha (2\pi)^\alpha n^\alpha F_0\right). \tag{4.108}$$

Taking the discrete local FRFT of (4.108) we have

$$\begin{aligned} F(k) &= \sum_{n=0}^{N-1} f(n) E_\alpha \left(-i^\alpha (2\pi)^\alpha n^\alpha k^\alpha / N^\alpha\right) \\ &= \sum_{n=0}^{N-1} E_\alpha \left(-i^\alpha (2\pi)^\alpha n^\alpha F_0\right) E_\alpha \left(-i^\alpha (2\pi)^\alpha n^\alpha k^\alpha / N^\alpha\right) \\ &= \sum_{n=0}^{N-1} E_\alpha \left[-i^\alpha \frac{(2\pi)^\alpha n^\alpha}{N^\alpha} \left(N^\alpha F_0 - k^\alpha\right)\right] \\ &= E_\alpha \left[-i^\alpha \frac{\pi^\alpha (N-1)^\alpha n^\alpha}{N^\alpha} \left(N^\alpha F_0 - k^\alpha\right)\right] \frac{\sin_\alpha \left[\pi^\alpha \left(N^\alpha F_0 - k^\alpha\right)\right]}{\sin_\alpha \left[\pi^\alpha \left(N^\alpha F_0 - k^\alpha\right) / N^\alpha\right]}. \end{aligned} \tag{4.109}$$

4.7 Fast Local FRFT in Fractal Space

4.7.1 Fast Local FRFT of DLFFT

The relations [51]

$$[F_N]^\alpha_{-n,k+1} = \frac{1}{N^\alpha} {W_{N,\alpha}}^{-(k+1)n} = \frac{1}{N^\alpha} {W_{N,\alpha}}^{-kn} {W_{N,\alpha}}^{-n} = [F_N]^\alpha_{-n,k} {W_{N,\alpha}}^{-n} \tag{4.110}$$

and

$$[F_N]^\alpha_{n,k+1} = \frac{1}{N^\alpha} {W_{N,\alpha}}^{(k+1)n} = \frac{1}{N^\alpha} {W_{N,\alpha}}^{kn} {W_{N,\alpha}}^{n} = [F_N]^\alpha_{n,k} {W_{N,\alpha}}^{n} \tag{4.111}$$

are the component formulas for the local FRFT.

Suppose that $\{V_0, V_1, V_2, \ldots, V_{N-1}\}$ is the N_{th} order discrete local FRFTs of $\{v_0, v_1, v_2, \ldots, v_{N-1}\}$. Starting with the component formulas for the discrete local FRFT, we obtain that, for $n = 0, 1, 2, \ldots, N-1$,

$$
\begin{aligned}
V_n &= \sum_{k=0}^{N-1} {W_{N,\alpha}}^{-(k+1)n} v_k \\
&= \sum_{\substack{k=0\\ k-even}}^{N-1} {W_{N,\alpha}}^{-(k+1)n} v_k + \sum_{\substack{k=0\\ k-odd}}^{N-1} {W_{N,\alpha}}^{-(k+1)n} v_k \\
&= \frac{1}{2^\alpha}\left(\sum_{j=0}^{M-1} {W_{2M,\alpha}}^{-n(2j)} v_{2j} + \sum_{j=0}^{M-1} {W_{2M,\alpha}}^{-n(2j+1)} v_{2j+1}\right) \\
&= \frac{1}{2^\alpha}\left(\sum_{j=0}^{M-1} {W_{2M,\alpha}}^{-n(2j)} v_{2j} + {W_{M,\alpha}}^{\frac{n}{2}} \sum_{j=0}^{M-1} {W_{2M,\alpha}}^{-n(2j)} v_{2j+1}\right).
\end{aligned}
\tag{4.112}
$$

and we have the following relation

$$
[F_N V]_n^\alpha = \frac{1}{2^\alpha}\left([F_N V_E]_n^\alpha + {W_{M,\alpha}}^{-\frac{n}{2}} [F_N V_O]_n^\alpha\right), \tag{4.113}
$$

where V is the sequence vector corresponding to $\{V_0, V_1, V_2, \ldots, V_{N-1}\}$, V_E is the M - th order sequence of even-index v_k ' s $\{V_0, V_2, \ldots, V_{N-2}\}$ and V_O is the M - th order sequence of odd-index v_k ' s $\{V_1, V_3, \ldots, V_{N-1}\}$.

Here we can deduce that

$$
\begin{aligned}
{W_{M,\alpha}}^{-(M+l)} &= E_\alpha\left(-i^\alpha\left(\frac{2\pi}{M}\right)^\alpha (M+l)^\alpha\right) \\
&= E_\alpha\left(-i^\alpha\left(\frac{2\pi}{M}\right)^\alpha l^\alpha\right) \\
&= {W_{M,\alpha}}^{-l}
\end{aligned}
\tag{4.114}
$$

and

$$
\begin{aligned}
{W_{M,\alpha}}^{-\frac{M+l}{2}} &= E_\alpha\left(-i^\alpha\left(\frac{\pi}{M}\right)^\alpha (M+l)^\alpha\right) \\
&= -E_\alpha\left(-i^\alpha\left(\frac{\pi}{M}\right)^\alpha l^\alpha\right) \\
&= {W_{M,\alpha}}^{-\frac{l}{2}}.
\end{aligned}
\tag{4.115}
$$

Hence for $l = 0, 1, 2, \ldots, m-1$, we have [51]

$$\begin{aligned} V_l &= \frac{1}{2^\alpha}\left(\sum_{j=0}^{M-1} W_{M,\alpha}{}^{-lj} v_{2j} + W_{M,\alpha}{}^{-\left(\frac{l}{2}\right)j} \sum_{j=0}^{M-1} W_{M,\alpha}{}^{-lj} v_{2j+1}\right) \\ &= \frac{1}{2^\alpha}\left(\left[F_{M V_E{}^{-1}}\right]_l^\alpha + W_{M,\alpha}{}^{-\left(\frac{l}{2}\right)j}\left[F_{M V_0{}^{-1}}\right]_l^\alpha\right) \end{aligned} \tag{4.116}$$

and

$$\begin{aligned} V_{M+l} &= \frac{1}{2^\alpha}\left(\sum_{j=0}^{M-1} W_{M,\alpha}{}^{-lj} v_{2j} - W_{M,\alpha}{}^{\left(\frac{l}{2}\right)j} \sum_{j=0}^{M-1} W_{M,\alpha}{}^{-lj} v_{2j+1}\right) \\ &= \frac{1}{2^\alpha}\left(\left[F_{M V_E{}^{-1}}\right]_l^\alpha - W_{M,\alpha}{}^{-\left(\frac{l}{2}\right)j}\left[F_{M V_0{}^{-1}}\right]_l^\alpha\right). \end{aligned} \tag{4.117}$$

Here, formulas (4.116) and (4.117) contain common elements that can be computed once for each l and then used to compute both V_l and V_{M+l}. Hence we can obtain the total number of computations to find all the V_n' s. That is to say, this process of increasing levels to our algorithm can be continued to the Kth level provided to $N = 2^K N_0$ for some integer N_0. Moreover, that integer, $N_0 = 2^{-K} N$ will also be the order of the discrete Yang–Fourier transforms and inverse discrete local FRFTs. If $N = 2^K$, it is this final K^{th} level algorithm, fully implemented and refined, that is called a fast local FRFT of the discrete local FRFTs.

4.7.2 *Fast Local FRFT of Inverse DLFFT*

Suppose that $\{V_0{}^{-1}, V_1{}^{-1}, \ldots, V_{N-1}{}^{-1}\}$ is the N_{th} order discrete local FRFTs of $\{v_0{}^{-1}, v_1{}^{-1}, \ldots, v_{N-1}{}^{-1}\}$, starting with the component formulas for the inverse discrete local FRFT, we obtain that, for $n = 0, 1, 2, \ldots, N-1$,

$$
\begin{aligned}
V_n{}^{-1} &= \frac{1}{\Gamma(1+\alpha)}\frac{1}{N^\alpha}\sum_{k=0}^{N-1} W_{N,\alpha}{}^{(k+1)n} v_k{}^{-1} \\
&= \frac{1}{\Gamma(1+\alpha)}\frac{1}{N^\alpha}\left(\sum_{\substack{k=0\\ k-even}}^{N-1} W_{N,\alpha}{}^{(k+1)n} v_k{}^{-1} + \sum_{\substack{k=0\\ k-odd}}^{N-1} W_{N,\alpha}{}^{(k+1)n} v_k{}^{-1}\right) \\
&= \frac{1}{\Gamma(1+\alpha)}\frac{1}{(2M)^\alpha}\left(\sum_{j=0}^{M-1} W_{2M,\alpha}{}^{n(2j)} v_{2j}{}^{-1} + \sum_{j=0}^{M-1} W_{2M,\alpha}{}^{n(2j+1)} v_{2j+1}{}^{-1}\right) \\
&= \frac{1}{\Gamma(1+\alpha)}\frac{1}{(2M)^\alpha}\left(\sum_{j=0}^{M-1} W_{2M,\alpha}{}^{n(2j)} v_{2j}{}^{-1} + W_{M,\alpha}{}^{\frac{n}{2}}\sum_{j=0}^{M-1} W_{2M,\alpha}{}^{n(2j)} v_{2j+1}{}^{-1}\right).
\end{aligned}
\tag{4.118}
$$

and we have the following relation

$$
[F_N V]_n^\alpha = \frac{1}{\Gamma(1+\alpha)}\frac{1}{(2M)^\alpha}\left(\left[F_N V_E{}^{-1}\right]_n^\alpha + W_{M,\alpha}{}^{\frac{n}{2}}\left[F_N V_O{}^{-1}\right]_n^\alpha\right), \tag{4.119}
$$

where V^{-1} is the sequence vector corresponding to $\{V_0{}^{-1}, V_1{}^{-1}, V_2{}^{-1}, \ldots, V_{N-1}{}^{-1}\}$, $V_E{}^{-1}$ is the M - th order sequence of even-index $v_k{}^{-1}$'s $\{V_0{}^{-1}, V_2{}^{-1}, \ldots, V_{N-2}{}^{-1}\}$ and $V_O{}^{-1}$ is the M - th order sequence of odd-index $v_k{}^{-1}$'s $\{V_1{}^{-1}, V_3{}^{-1}, \ldots, V_{N-1}{}^{-1}\}$.

Here we can deduce that

$$
\begin{aligned}
W_{M,\alpha}{}^{M+l} &= E_\alpha\left(i^\alpha\left(\frac{2\pi}{M}\right)^\alpha (M+l)^\alpha\right) \\
&= E_\alpha\left(i^\alpha\left(\frac{2\pi}{M}\right)^\alpha l^\alpha\right) \\
&= W_{M,\alpha}{}^{l}
\end{aligned}
\tag{4.120}
$$

and

$$
\begin{aligned}
W_{M,\alpha}{}^{\frac{M+l}{2}} &= E_\alpha\left(i^\alpha\left(\frac{\pi}{M}\right)^\alpha (M+l)^\alpha\right) \\
&= -E_\alpha\left(i^\alpha\left(\frac{\pi}{M}\right)^\alpha l^\alpha\right) \\
&= W_{M,\alpha}{}^{\frac{l}{2}}.
\end{aligned}
\tag{4.121}
$$

Hence for $l = 0, 1, 2, \ldots, m-1$, we have [51]

$$V_l^{-1} = \frac{1}{\Gamma(1+\alpha)}\frac{1}{(2M)^\alpha}\left(\sum_{j=0}^{M-1} W_{M,\alpha}{}^{lj} v_{2j} + W_{M,\alpha}{}^{\left(\frac{l}{2}\right)j}\sum_{j=0}^{M-1} W_{M,\alpha}{}^{lj} v_{2j+1}\right)$$
$$= \frac{1}{\Gamma(1+\alpha)}\frac{1}{2^\alpha}\left([F_M v_E]_l^\alpha + W_{M,\alpha}{}^{\left(\frac{l}{2}\right)j}[F_M v_0]_l^\alpha\right) \tag{4.122}$$

and

$$V_{M+l}{}^{-1} = \frac{1}{\Gamma(1+\alpha)}\frac{1}{(2M)^\alpha}\left(\sum_{j=0}^{M-1} W_{M,\alpha}{}^{lj} v_{2j} - W_{M,\alpha}{}^{\left(\frac{l}{2}\right)j}\sum_{j=0}^{M-1} W_{M,\alpha}{}^{lj} v_{2j+1}\right)$$
$$= \frac{1}{\Gamma(1+\alpha)}\frac{1}{2^\alpha}\left([F_M v_E]_l^\alpha - W_{M,\alpha}{}^{\left(\frac{l}{2}\right)j}[F_M v_0]_l^\alpha\right). \tag{4.123}$$

It is shown that, formulas (4.122) and (4.123) contain common elements that can also be computed once for each l and then used to compute both $V_l{}^{-1}$ and $V_{M+l}{}^{-1}$. These can also yield the total number of computations to find all the $V_n{}^{-1}$'s. That is to say, this process of increasing levels to our algorithm of inverse discrete Yang–Fourier transforms is similar to that of the discrete local FRFTs. Taking into account the relation $N = 2^K$, it is also this final K^{th} level algorithm, fully implemented and refined, that is called a fast local FRFT of the inverse discrete local FRFTs.

4.8 Conclusions

In this chapter we recall some of the theory of local fractional Fourier analysis containing the local fractional Fourier series, the local fractional Fourier transform, the generalized local fractional Fourier transform, the discrete local fractional Fourier transform and fast local fractional Fourier transform. We briefly presented some of their applications. Our attention is devoted to the analytical technique of the local fractional Fourier series for treating with some real world fractal problems in a way accessible to applied scientists and engineers.

References

1. Wiener N (1933) The Fourier integral and certain of its applications. Cambridge University Press, Cambridge

2. KÖrener TW (1988) Fourier analysis. Cambridge University Press, Cambridge
3. Folland B (1992) Fourier analysis and its application. Wadsworth, California
4. Howell KB (2001) Principles of Fourier analysis. Chapman & Hall/CRC, New York
5. Stein M, Weiss G (1971) Introduction to Fourier analysis on Euclidean spaces. Princeton University Press, Princeton
6. Rudin W (1962) Fourier analysis on groups. Wiley, New York
7. Loomis L (1953) Abstract harmonic analysis. Van Nostrand, New York
8. Katznelson Y (1968) An introduction to harmonic analysis. Wiley, New York
9. Namias V (1980) The fractional order Fourier transform and its application to quantum mechanics. IMA J Appl Math 25:241–265
10. Mcbride AC, Kerr FH (1987) On Namias's fractional Fourier transforms. IMA J Appl Math 39(2):159–175
11. Bailey DH, Swarztrauber PN (1991) The fractional Fourier transform and applications. SIAM Rev 33:389–404
12. Lohmann AW (1993) Image rotation, Wigner rotation and the fractional Fourier transform. J Opt Soc Am A 10:2181–2186
13. Almeida LB (1994) The fractional Fourier transform and time-frequency representations. IEEE Trans Signal Process 42(11):3084–3091
14. Candan C, Kutay MA, Ozaktas HM (2000) The discrete fractional Fourier transform. IEEE Trans Signal Process 48(5):1329–1337
15. Pei S-C, Ding J-J (2001) Relations between fractional operations and time-frequency distributions, and their applications. IEEE Trans Signal Process 49(8):1638–1655
16. Saxena R, Singh K (2005) Fractional Fourier transform: a novel tool for signal processing. J Indian Inst Sci 85:11–26
17. Tao R, Deng B, Zhang W-Q, Wang Y (2008) Sampling and sampling rate conversion of band limited signals in the fractional Fourier transform domain. IEEE Trans Signal Process 56(1):158–171
18. Bhandari A, Marziliano P (2010) Sampling and reconstruction of sparse signals in fractional Fourier domain. IEEE Signal Process Lett 17(3):221–224
19. Kolwankar KM, Gangal AD (1996) Fractional differentiability of nowhere differentiable functions and dimensions. Chaos 6(4):505–513
20. Parvate A, Gangal AD (2009) Calculus on fractal subsets of real line - I: formulation. Fractals 17(1):53–81
21. Adda FB, Cresson J (2001) About non-differentiable functions. J Math Anal Appl 263: 721–737
22. Carpinteri A, Chiaia B, Cornetti P (2001) Static-kinematic duality and the principle of virtual work in the mechanics of fractal media. Comput Methods Appl Mech Eng 191:3–19
23. Carpinteri A, Cornetti P (2002) A fractional calculus approach to the description of stress and strain localization in fractal media. Chaos Solitons Fractals 13:85–94
24. Babakhani A, Daftardar-Gejji V (2002) On calculus of local fractional derivatives. J Math Anal Appl 270(1):66–79
25. Chen Y, Yan Y, Zhang K (2010) On the local fractional derivative. J Math Anal Appl 362:17–33
26. Chen W (2006) Time-space fabric underlying anomalous diffusion. Chaos Solitons Fractals 28:923–929
27. Chen W, Sun HG, Zhang XD, Koro D (2010) Anomalous diffusion modeling by fractal and fractional derivatives. Comput Math Appl 59:1754–1758
28. Chen W, Zhang XD, Korosak D (2010) Investigation on fractional and fractal derivative relaxation-oscillation models. Int J Nonlinear Sci Numer Simulat 11:3–9
29. Jumarie G (2005) On the representation of fractional Brownian motion as an integral with respect to $(dt)^a$. Appl Math Lett 18:739–748
30. Jumarie G (2006) Lagrange characteristic method for solving a class of nonlinear partial differential equations of fractional order. Appl Math Lett 19:873–880
31. He JH (2011) A new fractal derivation. Therm Sci 15:S145–S147

32. Fan J, He JH (2012) Fractal derivative model for air permeability in hierarchic porous media. Abstr Appl Anal 2012:354701
33. Li ZB, Zhu WH, He JH (2012) Exact solutions of time-fractional heat conduction equation by the fractional complex transform. Therm Sci 16(2):335–338
34. He JH, Elagan SK, Li ZB (2012) Geometrical explanation of the fractional complex transform and derivative chain rule for fractional calculus. Phys Lett A 376:257–259
35. Yang XJ (2009) Research on fractal mathematics and some applications in mechanics. M.S. thesis, China University of Mining and Technology
36. Yang XJ (2011) Local fractional integral transforms. Prog Nonlinear Sci 4:1–225
37. Yang XJ (2011) Local fractional functional analysis and its applications. Asian Academic publisher, Hong Kong
38. Yang XJ (2012) Advanced local fractional calculus and its applications. World Science Publisher, New York
39. Hu MS, Baleanu D, Yang XJ (2013) One-phase problems for discontinuous heat transfer in fractal media. Math Probl Eng 2013:358473
40. Yang XJ, Baleanu D, Zhong WP (2013) Approximate solutions for diffusion equations on Cantor time-space. Proc Rom Acad A 14(2):127–133
41. Zhong WP, Yang XJ, Gao F (2013) A Cauchy problem for some local fractional abstract differential equation with fractal conditions. J Appl Funct Anal 8(1):92–99
42. Yang XJ, Baleanu D (2013) Fractal heat conduction problem solved by local fractional variation iteration method. Therm Sci 17(2):625–628
43. Liao MK, Yang XJ, Yan Q (2013) A new viewpoint to Fourier analysis in fractal space. In: Anastassiou GA, Duman O (eds) Advances in applied mathematics and approximation theory, chapter 26. Springer, New York, pp 399–411
44. Hu MS, Agarwal RP, Yang XJ (2012) Local fractional Fourier series with application to wave equation in fractal vibrating string. Abstr Appl Anal 2012:567401
45. He JH (2012) Asymptotic methods for solitary solutions and compactons. Abstr Appl Anal 2012:916793
46. Guo Y (2012) Local fractional Z transform in fractal space. Adv Digit Multimedia 1(2):96–102
47. Yang XJ, Liao MK, Chen JW (2012) A novel approach to processing fractal signals using the Yang–Fourier transforms. Procedia Eng 29:2950–2954
48. Zhong WP, Gao F, Shen XM (2012) Applications of Yang–Fourier transform to local fractional equations with local fractional derivative and local fractional integral. Adv Mater Res 461: 306–310
49. Yang XJ (2012) A generalized model for Yang–Fourier transforms in fractal space. Adv Intell Transport Syst 1(4):80–85
50. Yang XJ (2012) The discrete Yang–Fourier transforms in fractal space. Adv Electr Eng Syst 1(2):78–81
51. Yang XJ (2012) Fast Yang–Fourier transforms in fractal space. Adv Intell Transport Syst 1(1):25–28

Chapter 5
Parameter Optimization of Fractional Order $PI^{\lambda}D^{\mu}$ Controller Using Response Surface Methodology

Beyza Billur İskender, Necati Özdemir, and Aslan Deniz Karaoglan

Abstract This chapter presents optimization of fractional order $PI^{\lambda}D^{\mu}$ control parameters by using response surface methodology. The optimization process is observed on a fractional order diffusion system subject to input hysteresis which is defined with Riemann–Liouville fractional derivative. The system is transferred to a fractional order state space model by using eigenfunction expansion method and then Grünwald–Letnikov approximation is applied to solve the system numerically. The necessary data for response surface analysis are read from the obtained numerical solution. Finally, second-order polynomial response surface mathematical model for the experimental design is presented and the optimum control parameters are predicted from this response surface model. The proposed optimization method is compared with the technique of minimization of integral square error by means of settling time and the results are discussed.

Keywords Fractional order controller • Response surface methodology • Integral square error • Hysteresis • Riemann-Liouville fractional derivative • Grünwald-Letnikov approximation

5.1 Introduction

Fractional order system has been drawn great interest recently because of their advantages to model systems more accurately than integer order models. Improvement of fractional order systems in different areas of science and technology brought

B.B. İskender (✉) • N. Özdemir
Department of Mathematics, Faculty of Science and Arts, Balikesir University, Balikesir, Turkey
e-mail: biskender@balikesir.edu.tr; nozdemir@balikesir.edu.tr

A.D. Karaoglan
Department of Industrial Engineering, Faculty of Engineering and Architecture, Balikesir University, Balikesir, Turkey
e-mail: deniz@balikesir.edu.tr

J.A.T. Machado et al. (eds.), *Discontinuity and Complexity in Nonlinear Physical Systems*, Nonlinear Systems and Complexity 6, DOI 10.1007/978-3-319-01411-1_5,

about a new control tool which is called fractional order controllers. CRONE is one of the prior fractional order controller designs which was presented by Oustaloup [20]. Then, Podlubny [24] generalized the classical *PID* controller to the fractional calculus by replacing order of the integral and the derivative controllers with fractional orders λ and μ, respectively. It is called as fractional order $PI^{\lambda}D^{\mu}$ controller. This controller has five parameters to tune while classical *PID* has only three parameters. Therefore, it is more flexible and advantageous. The researches on fractional order controllers were extended to other classical control types, for example fractional order optimal control [1–3, 22] or fractional order sliding mode control [10].

PID controller is the frequently preferred type of controllers due from its ease implementation to industrial systems. Thus, fractional order $PI^{\lambda}D^{\mu}$ controller is also preferable for both integer and fractional order control systems. Many methods have been presented for tuning problem of $PI^{\lambda}D^{\mu}$ controller which is more complex according to tuning problem of classical *PID* [5, 7, 14, 26, 27]. Recently, in [13] response surface methodology is used to tune such type of controllers. This method is a collection of mathematical and statistical techniques where a response of interest is influenced by several variables and the objective is to optimize this response [17]. The optimization of the controller parameters using response surface method is achieved by simultaneous testing of limited number of experiments read from the system under control. In this chapter, it is aimed to improve the response surface design in [13] in terms of settling time where the controlled system is fractional order and is subjected to input hysteresis. Because the considered fractional order system is mathematical, the necessary data are obtained by solving the partial fractional differential equation. The solution is acquired by using eigenfunction expansion and Grünwald–Letnikov numerical methods [4]. Finally, this design is compared with the previous design given in [13] and the method of minimizing integral square error presented in [21].

5.2 Preliminaries

According to the system under consideration it would be possible to correspond different types of fractional derivative, for example Riemann–Liouville, Caputo, Grünwald–Letnikov, Weyl, Marchaud, and Riesz fractional derivatives [16, 19, 23]. In this chapter Riemann–Liouville fractional derivative is used to formulate the system which is defined for a time-dependent function $x(\cdot)$ as

$$ {}_0D_t^{\alpha}x(t) = \frac{1}{\Gamma(n-\alpha)}\left(\frac{d}{dt}\right)^n \int_0^t (t-\tau)^{n-\alpha-1}\, x(\tau)\, d\tau, \tag{5.1}$$

where α is order of derivative such that $n-1 \leq \alpha < n$, n is a nonnegative natural number, and $\Gamma(.)$ is Euler's gamma function. The Riemann–Liouville fractional integral is also defined as

$$ {}_0D_t^{-\alpha}x(t) = \frac{1}{\Gamma(\alpha-1)}\int_0^t (t-\tau)^{\alpha}x(\tau)\,d\tau, \tag{5.2}$$

There is a link between the Riemann–Liouville and Grünwald–Letnikov fractional derivatives and utilizing this link the Riemann–Liouville fractional derivative can be approximated numerically by Grünwald–Letnikov definition which is

$$ {}_0^{GL}D_t^{\alpha}x(t) = \lim_{h\to 0}\frac{1}{h^{\alpha}}\sum_{k=0}^{\left[\frac{t}{h}\right]}(-1)^k\binom{\alpha}{k}x(t-kh), \tag{5.3}$$

where h represents the time increment, $[\frac{t}{h}]$ means the integer parts of $\frac{t}{h}$ and

$$\binom{\alpha}{k} = \frac{\Gamma(\alpha+1)}{\Gamma(k+1)\,\Gamma(\alpha-k+1)}. \tag{5.4}$$

5.3 Fractional Order $PI^{\lambda}D^{\mu}$ Controller

PID controller which is the combination of proportional, integral, and derivative actions represents a basic control structure. It is defined by the following equation

$$u(t) = k_p e(t) + k_i\int e(t)\,dt + k_d\frac{d}{dt}e(t), \tag{5.5}$$

where t is time variable, $u(t)$ is control and $e(t)$ is error functions, k_p, k_i, and k_d are gains of the proportional, integral, and derivative controllers, respectively. The error function is defined as the difference between a desired reference value $r(t)$ and the system output $y(t)$. The response of a system can be optimized by tuning of the coefficients of k_p, k_i and k_d. Since this rule can be easily applied to most of system the controller is still preferable. Therefore, it is generalized for fractional order systems which is known as fractional order $PI^{\lambda}D^{\mu}$ controller which involves an integrator of order λ and a differentiator of order μ. This controller can also be applied to integer order systems. The $PI^{\lambda}D^{\mu}$ is defined by

$$u(t) = k_p e(t) + k_i I^{\lambda}e(t) + k_d D^{\mu}e(t). \tag{5.6}$$

It can be clearly seen that selection of $\lambda = 1$ and $\mu = 1$ gives the classical *PID* controller. As it is seen between the above equations the $PI^{\lambda}D^{\mu}$ controller has five parameters which are the coefficients of k_p, k_i, and k_d, and the orders of λ and μ while the integer order *PID* has only three parameters. Thus, it is deduced that $PI^{\lambda}D^{\mu}$ controller is more flexible than the integer *PID* controller. Moreover, it is less sensitive to change of parameters of controlled system, see [23].

Several tuning strategies have been introduced for $PI^{\lambda}D^{\mu}$ in the literature. In this paper we present response surface method and compare this method with minimization of the integral square error.

5.4 Fractional Order Diffusion Systems Subject to Input Hysteresis

A fractional diffusion process on the one-dimensional spatial domain [0, 1], with diffusion coefficient ν and nonlinear control action applied at point $x_b \in (0, 1)$ via the SSSL (Su, Stepanenko, Svoboda, Leung) hysteresis operator Φ which is a special type of Duhem hysteresis, is given by the following partial fractional differential equation:

$$\frac{\partial^{\alpha} z(t,x)}{\partial t^{\alpha}} = \nu \frac{\partial^2 z(t,x)}{\partial x^2} + \delta(x - x_b)\,\Phi(u(t)) \tag{5.7}$$

with the Dirichlet boundary conditions

$$z(t,0) = z(t,1) = 0, \tag{5.8}$$

and zero initial condition

$$z(0,x) = 0. \tag{5.9}$$

The system is observed at a point $x_c \in (x_b, 1)$ such that

$$y(t) = z(t, x_c). \tag{5.10}$$

The SSSL operator $\omega = \Phi(u)$ is defined by the following differential equation:

$$\frac{d\omega}{dt} = \rho[\zeta u - \omega]\left|\frac{du}{dt}\right| + \eta\frac{du}{dt}. \tag{5.11}$$

In (5.11), the input u and the output ω are real valued functions of time t with piecewise continuous derivatives u and ω, $\left|\frac{du}{dt}\right|$ is the absolute value of $\frac{du}{dt}$. ρ, ζ, and η are some constants satisfying the condition $\zeta > \eta$, see [25].

Solution of the system is obtained by using separation of variables. For this purpose, let (5.7) have a solution of the form:

$$z(t,x) = \overline{T}(t)\,\overline{X}(x). \tag{5.12}$$

Firstly, homogeneous part of (5.7) is considered. Substituting (5.12) in (5.7) gives

$$\frac{1}{\overline{T}}\frac{d^{\alpha}\overline{T}}{dt^{\alpha}} = \nu\frac{1}{\overline{X}}\frac{d^{2}\overline{X}}{dx^{2}}. \tag{5.13}$$

The right-hand side of (5.13) holds if it equals a separation constant shown by χ as in the following equation:

$$\frac{1}{\overline{T}}\frac{d^{\alpha}\overline{T}}{dt^{\alpha}} = \nu\frac{1}{\overline{X}}\frac{d^{2}\overline{X}}{dx^{2}} = -\nu\chi^{2}. \tag{5.14}$$

Using (5.9), the solution of the second part of (5.14) is obtained as

$$\overline{X}(x) = \sin(k\pi x)\,,\ \ k = 1,2,\ldots \tag{5.15}$$

which is called eigenfunctions. The general solution of (5.7) is

$$z(t,x) = \sum_{k=1}^{\infty} q_k(t)\sin(k\pi x)\,. \tag{5.16}$$

Since the higher order terms do not contribute much, it could be of interest to keep only a finite number of terms denoted by m. Substituting (5.16) into (5.7) gives

$$\sum_{k=1}^{m}\frac{d^{\alpha}q_k(t)}{dt^{\alpha}}\sin(k\pi x) = -\nu\sum_{k=1}^{m}q_k(t)\sin(k\pi x) + \delta(x - x_b)\,\Phi(u)\,. \tag{5.17}$$

Multiplying both sides of (5.17) by $\sin(j\pi x)$, $1 \le j \le m$, then integrating from 0 to 1 via variable x, and using the orthogonality property gives

$$\frac{d^{\alpha}q_k(t)}{dt^{\alpha}} = -\nu k^{2}\pi^{2}q_k(t) + 2\sin(k\pi x_b)\,\Phi(u)\,,\ \ k = 1,2,\ldots,m, \tag{5.18}$$

subject to initial conditions $q_k(0) = 0$. Note that the initial conditions are calculated from (5.29). Equation (5.18) can be presented by the state space form

$$\begin{aligned} {}_0D_t^{\alpha}q(t) &= Aq(t) + B\Phi(u(t)) \\ y(t) &= Cq(t) \end{aligned} \tag{5.19}$$

where $q(t) = \left[q_1(t)\, q_2(t) \ldots q_m(t) \right]^T$ is the state variable, $A \in R^{m\times m}$, $B \in R^m$ and $C \in R^{1\times m}$ are the matrices given by

$$\begin{aligned} A &= diagonal\left[-\nu k^2\pi^2\right], \\ B &= \left[b_1\, b_2 \ldots b_m \right]^T, \\ C &= \left[c_1\, c_2 \ldots c_m \right], \end{aligned}$$

in which $k = 1, 2, \ldots, m$, $b_k = 2\sin(k\pi x_b)$, and $c_k = \sin(k\pi x_c)$. The solution of System (5.19) is obtained numerically by Grünwald–Letnikov approximation. For this purpose, the time interval $[0, T]$ is divided N equal parts with size of $h = \frac{1}{N}$ and the nodes are labeled as $0, 1, 2, \ldots, N$. The Grünwald–Letnikov approximation of the Riemann–Liouville fractional derivative at node M is

$$ {}_0D_t^{\alpha} q(hM) = \frac{1}{h^{\alpha}} \sum_{j=0}^{M} w_j^{(\alpha)} q(hM - jh), \tag{5.20}$$

where the coefficients $w_j^{(\alpha)}$ are computed by the following recurrence relationships

$$\begin{aligned} w_0^{(\alpha)} &= 1; \\ w_j^{(\alpha)} &= \left(1 - \frac{\alpha+1}{j}\right) w_{j-1}^{(\alpha)} \end{aligned}$$

for $j = 1, 2, \ldots, N$. Using (5.20), numerical solution of System (5.19) is obtained as

$$ q(hM) = \left(\frac{1}{h^{\alpha}} w_0^{(\alpha)} I - A\right)^{-1} \left(B\Phi(u(hM)) - \frac{1}{h^{\alpha}} \sum_{j=1}^{M} w_j^{(\alpha)} q(hM - jh) \right). \tag{5.21}$$

Similarly, the $PI^{\lambda}D^{\mu}$ controller can be computed by the Grünwald–Letnikov approximation. Note that the integrator of order λ is also approximated with (5.20) by replacing α with $-\lambda$. Therefore, the $PI^{\lambda}D^{\mu}$ controller at node M can be numerically calculated as

$$\begin{aligned} u(Mh) = k_p e(Mh) &+ k_i \frac{1}{h^{-\lambda}} \sum_{j=0}^{M} w_j^{(-\lambda)} e(Mh - jh) \\ &+ k_d \frac{1}{h^{\mu}} \sum_{j=0}^{M} w_j^{(\mu)} e(Mh - jh). \end{aligned} \tag{5.22}$$

Control objective of the system is to get the desired output $y(t) = 1$ with a minimum settling time and no overshoot. This purpose has been achieved by the method of minimizing integral square error in [21] and by response surface methodology in [13]. The details of these methods are given in the following sections and also the previous results obtained in [13] by response surface methodology are improved with additional experimental designs. These methods are applied to the system by using the numerical solutions (5.21) and (5.22) whose parameters are chosen as $\alpha = 0.8$, $\nu = 1$, $x_b = 0.25$, $x_c = 0.375$.. The hysteresis parameters $\rho = 1$, $\zeta = 3.1635$, and $\eta = 0.345$ and the parameters of numerical calculations $m = 15$ and $h = 0.05$ are taken.

5.5 Integral Square Error Method

To adjust the parameters of the $PI^{\lambda}D^{\mu}$ controller with integral square error method the objective function is chosen as

$$J(p) = \int_{0}^{\infty} [e(t, p)]^2 dt, \tag{5.23}$$

where p is the vector of control parameters:

$$p = \left[k_p \; k_i \; k_d \; \lambda \; \mu \right], \tag{5.24}$$

and $e(t, p)$ is the error function between reference input function $r(t)$ and the system response $y(t)$. Then the following algorithm is used to minimize the performance index (5.23) which has been presented by Moradi and Johnson [18].

Step 1. Initialization

- *Choose time interval,*
- *Choose convergent tolerance* $\overline{\epsilon}$,
- *Set loop counter* $k = 0$,
- *Choose the initial controller parameter vector* $p(k)$.

Step 2. Gradient Calculation

- *Calculate gradient of* J. *If the gradient satisfies the following condition*

$$\left| \frac{\partial J}{\partial p}(k) \right| < \overline{\epsilon},$$

then stop.

Step 3. Update calculation

- *Compute the update parameters γ_k and R_k, and compute*

$$p(k+1) = p(k) - \gamma_k R_k^{-1} \frac{\partial J}{\partial p}(k), \tag{5.25}$$

- *Update $k = k + 1$ and go to Step 2.*

Here, R_k^{-1} is chosen as Hessian of J and γ_k is a positive real scalar that determines the step size. Using this algorithm the optimum control parameters have been obtained in [21] as $k_p = 0.2022, k_i = 0.1915, k_d = 0.1958, \lambda = 0.1921$, and $\mu = 0.1904$ via convergence tolerance $\overline{\epsilon} = 0.1$ and initial control parameter vector $p = 1.195\left[0.1\ 0.1\ 0.1\ 0.1\ 0.1\right]$. According to these controller parameters the output of the system reaches $y(t) = 0.997$ with the settling time $t = 13.8$.

5.6 Response Surface Methodology

At the optimization stage of a process, it is important to know the mathematical model that represents the relation between the factors (controllable input variables) and the responses (measured output). By using these mathematical models researchers may perform prediction for not experimented combinations of the factors or may perform optimization by determining the input factor levels that provides the desired results. Design of experiment techniques, which are the combination of statistical and mathematical methods, provide researchers to model the mathematical relations between the factors and the responses by using the results of experimental runs of different combination of factor levels, with minimum number of trials which is important for time saving. Response surface methodology, Taguchi method, and factorial design are the widely used and well-known design of experiment techniques. Response surface is used for modeling systems especially including nonlinear relations. In a response surface model there are quadratic, linear, and interaction terms while factorial design only includes linear and interaction terms. So if there are quadratic relations between the factors and the responses, and also it is important to get the mathematical model with quadratic, linear, and interaction terms (full quadratic model), it is appropriate to use response surface methodology but this information is not a generalized rule. Response surface methodology and factorial design use matrix multiplications and least square estimators while Taguchi uses logarithmic calculations and signal-to-noise ratios which is basically different from response surface methodology and factorial design. By using Taguchi method it is possible to obtain only the optimal parameter combination of determined factor levels at the design of experiment stage while other methods give the optimal solution with decimals. Although this can be seen as a disadvantage of Taguchi method, Taguchi Method requires less experimental runs if the number of factors are quite much [8, 12, 15].

Table 5.1 Initial factor levels

Factors	Minimum	High
k_p	0.15	0.30
k_i	0.15	0.45
k_d	0.25	0.30
λ	0.15	0.45
μ	0.05	0.25

To determine the desired values of the output y and the settling time t for the system given by (5.7) the optimum controller parameters of k_p, k_i, k_d, λ, and μ are calculated by using response surface methodology. First of all the mathematical relationships between the responses (y and t) and the tuning parameters (k_p, k_i, k_d, λ, and μ) are established. By using this mathematical equation optimum parameters are determined for $y(t) = 1$ and minimum t. The general second-order polynomial response surface mathematical model (full quadratic model) for the experimental design presented in the present study ([6, 9, 11, 17]) is

$$Y = \beta_0 + \sum_{i=1}^{n} \beta_i X_i + \sum_{i=1}^{n} \beta_{ii} X_i{}^2 + \sum_{j=1}^{n} \sum_{j<i} \beta_{ij} X_i X_j + \epsilon, \tag{5.26}$$

in which Y is the response (y, t) and the β's are parameters whose values are to be determined. X_i and X_j are the factors and the ϵ is the random error term (residuals). The model in terms of the observations may be written in matrix notation as

$$Y = \beta X + \epsilon, \tag{5.27}$$

where Y is the output matrix and X is the input matrix. The least square estimator of β matrix that composes of coefficients of the regression equation calculated by the given formula:

$$\beta = (X^T X)^{-1} X^T Y. \tag{5.28}$$

To reduce the number of tests, an L_{32} orthogonal array that only needs 32 experimental runs was adopted. Because of using nonrandom system one center point is used in the design of experiment and by this way the number of experiments is reduced to 27 runs. MINITAB 16 statistical package is used to establish mathematical models for achieving the target value of 1 for y, while minimizing t at a desired confidence interval (95%). The experimental design is realized to get the optimum factor levels. The initial factor levels and the experimental design are given in Tables 5.1 and 5.2, respectively.

According to the results of the experiments given in Table 5.2, mathematical models based on response surface method for correlating responses such as the y and t have been established which are represented by (5.29) and (5.30).

Table 5.2 Experimental design

Ex.no	k_p	k_i	k_d	λ	μ	y	t
1	0.150	0.150	0.250	0.150	0.250	1.130	17.75
2	0.300	0.150	0.250	0.150	0.050	1.028	15.95
3	0.150	0.450	0.250	0.150	0.050	1.052	10.20
4	0.300	0.450	0.250	0.150	0.250	1.224	28.50
5	0.150	0.150	0.300	0.150	0.050	0.930	24.40
6	0.300	0.150	0.300	0.150	0.250	1.290	16.35
7	0.150	0.450	0.300	0.150	0.250	1.320	23.20
8	0.300	0.450	0.300	0.150	0.050	1.147	3.90
9	0.150	0.150	0.250	0.450	0.050	0.805	29.00
10	0.300	0.150	0.250	0.450	0.250	1.280	30.00
11	0.150	0.450	0.250	0.450	0.250	1.048	26.65
12	0.300	0.450	0.250	0.450	0.050	1.023	24.00
13	0.150	0.150	0.300	0.450	0.250	1.117	29.15
14	0.300	0.150	0.300	0.450	0.050	1.024	6.50
15	0.150	0.450	0.300	0.450	0.050	0.988	10.30
16	0.300	0.450	0.300	0.450	0.250	1.089	29.15
17	0.150	0.300	0.275	0.300	0.150	1.036	6.35
18	0.300	0.300	0.275	0.300	0.150	1.108	30.00
19	0.225	0.150	0.275	0.300	0.150	1.056	12.20
20	0.225	0.450	0.275	0.300	0.150	1.085	27.60
21	0.225	0.300	0.250	0.300	0.150	1.067	5.45
22	0.225	0.300	0.300	0.300	0.150	1.097	27.60
23	0.225	0.300	0.275	0.150	0.150	1.142	29.80
24	0.225	0.300	0.275	0.450	0.150	1.048	17.30
25	0.225	0.300	0.275	0.300	0.050	1.007	7.20
26	0.225	0.300	0.275	0.300	0.250	1.148	30.00
27	0.225	0.300	0.275	0.300	0.150	1.085	4.55

$$
\begin{aligned}
y = {} & 0.0876 + 4.7234k_p + 1.3262k_i - 1.2102k_d + 0.2525\lambda \\
& +2.3687\mu - 0.7383{k_p}^2 - 0.2512{k_i}^2 + 9.3552{k_d}^2 \\
& +0.8376\lambda^2 + .1347\mu^2 - 3.1383k_pk_i - 10.8300k_pk_d \\
& +1.1117k_p\lambda - 1.4882k_p\mu + 1.3317k_ik_d - 1.2308k_i\lambda \\
& -2.3288k_i\mu - 3.1850k_d\lambda\mu - 1.1725k_d\mu - 0.4746\lambda
\end{aligned} \tag{5.29}
$$

$$
\begin{aligned}
t = {} & -300.83 + 341.23k_p - 90.10k_i + 2267.08k_d + 0.38\lambda \\
& -26.29\mu - 87.68{k_p}^2 + 54.75{k_i}^2 - 3429.09{k_d}^2 \\
& +216.97\lambda^2 - 6.82\mu^2 + 259.44k_pk_i - 1533.33k_pk_d \\
& +256.67k_p\mu - 108.33k_ik_d + 11.39k_i\lambda + 173.75k_i \\
& \lambda + 30.00k_p\lambda - 500k_d\lambda + 725k_d\mu + 57.50\lambda\mu
\end{aligned} \tag{5.30}
$$

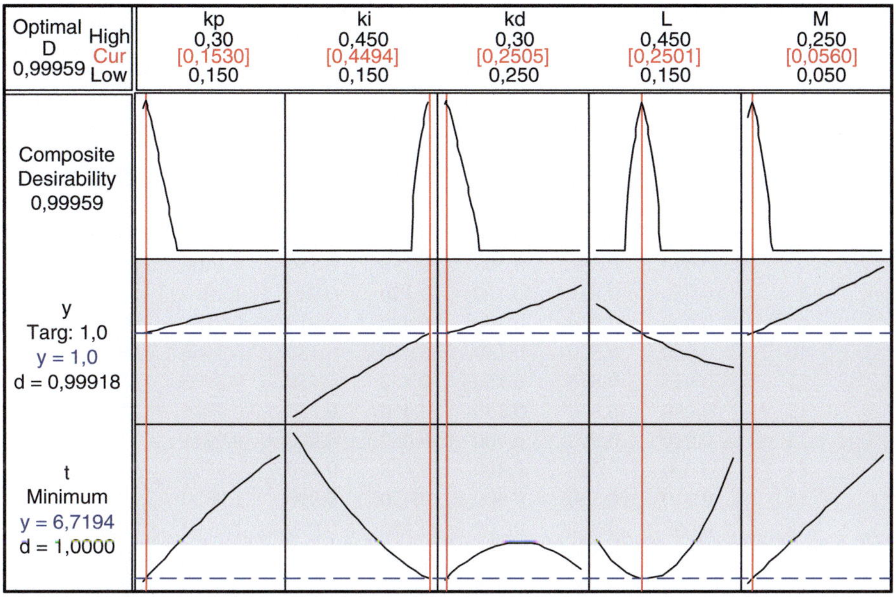

Fig. 5.1 Optimum levels of parameters for initial design obtained from response optimizer module of MINITAB Package

Table 5.3 Factor levels for the new design

Factors	Minimum	High
k_p	0.01	0.05
k_i	0.55	0.85
k_d	0.15	0.30
λ	0.15	0.45
μ	0.01	0.15

By using the response optimizer module of MINITAB the optimum parameter levels are determined as $k_p = 0.1530$, $k_i = 0.4494$, $k_d = 0.2505$, $\lambda = 0.2501$, and $\mu = 0.0560$. By using the given parameters combination y is predicted as 1.00 while t is predicted as 6.7194 which can be found in Fig. 5.1. After the confirmation tests for the given optimum parameter levels by using MATLAB 7.1, $y = 1.00$ and $t = 6.15$ are obtained. Therefore it can be concluded that the settling time is decreased by the response surface method via integral square error method given in [21].

Because the levels of k_p, k_i, and k_d are obtained at the boundary levels which can be seen in Fig. 5.1, current k_p, k_i, and k_d factor levels are rearranged. New factor levels and experimental design are given in Tables 5.3 and 5.4, respectively.

From the experiments given in Table 5.4 the mathematical models based on response surface method for correlating responses have been established by (5.31) and (5.32).

Table 5.4 The new experimental design

Ex.no	k_p	k_i	k_d	λ	μ	y	t
1	0.010	0.550	0.150	0.150	0.150	0.9582	29.95
2	0.050	0.550	0.150	0.150	0.010	0.9246	30.00
3	0.010	0.850	0.150	0.150	0.010	1.0360	5.20
4	0.050	0.850	0.150	0.150	0.150	1.0850	29.40
5	0.010	0.550	0.300	0.150	0.010	1.0110	13.20
6	0.050	0.550	0.300	0.150	0.150	1.1300	28.65
7	0.010	0.850	0.300	0.150	0.150	1.1460	26.15
8	0.050	0.850	0.300	0.150	0.010	1.1010	28.85
9	0.010	0.550	0.150	0.450	0.010	0.9161	99.40
10	0.050	0.550	0.150	0.450	0.150	0.9156	399.4
11	0.010	0.850	0.150	0.450	0.150	0.5900	15.90
12	0.050	0.850	0.150	0.450	0.150	0.9946	21.15
13	0.010	0.550	0.300	0.450	0.010	0.9946	4.95
14	0.050	0.550	0.300	0.450	0.150	0.9704	99.65
15	0.010	0.850	0.300	0.450	0.010	1.0030	18.40
16	0.050	0.850	0.300	0.450	0.010	1.0190	28.05
17	0.010	0.700	0.225	0.300	0.150	0.9796	54.05
18	0.050	0.700	0.225	0.300	0.080	0.9919	5.40
19	0.030	0.550	0.225	0.300	0.080	0.9421	98.30
20	0.030	0.850	0.225	0.300	0.080	1.0150	21.15
21	0.030	0.700	0.150	0.300	0.080	0.9592	95.80
22	0.030	0.700	0.300	0.300	0.080	1.0230	18.65
23	0.03	0.7	0.225	0.150	0.080	1.0640	28.05
24	0.03	0.7	0.225	0.450	0.080	0.9797	75.90
25	0.03	0.7	0.225	0.300	0.010	0.9825	81.45
26	0.03	0.7	0.225	0.300	0.150	1.0150	5.40
27	0.03	0.7	0.225	0.300	0.080	0.9883	78.64

$$\begin{aligned} y = {} & 0.3912 - 2.1518k_p + 1.4834k_i - 0.2448k_d - 0.1235\lambda \\ & +0.8493\mu - 28.2576{k_p}^2 - 0.8224{k_i}^2 - 1.0583{k_d}^2 \\ & +1.1021\lambda^2 + 0.3463\mu^2 + 7.4146k_pk_i - 15.0708k_pk_d \\ & +6.2187k_p\lambda - 19.8348k_p\mu + 0.9106k_ik_d - 1.5064k_i\lambda \\ & -2.7554k_i\mu + 0.9917k_d\lambda + 6.3655k_d\mu - 3.5923\lambda\mu \end{aligned} \tag{5.31}$$

$$\begin{aligned} t = {} & 207.3 + 7539k_p - 776.6k_i + -1540.4k_d + 1248.1\lambda \\ & 1338.9\mu - 45088.3{k_p}^2 + 531.8{k_i}^2 + 1682.6{k_d}^2 \\ & +187.3\lambda^2 - 884.8\mu^2 - 7675.0k_pk_i - 8625.0k_pk_d \\ & +7650.0k_p\mu - 16299.1k_ik_d + 2456.1k_i\lambda - 1410.3k_i \\ & \lambda - 1159.5k_p\lambda - 2150k_d\lambda - 4657.1k_d\mu + 1028.6\lambda\mu \end{aligned} \tag{5.32}$$

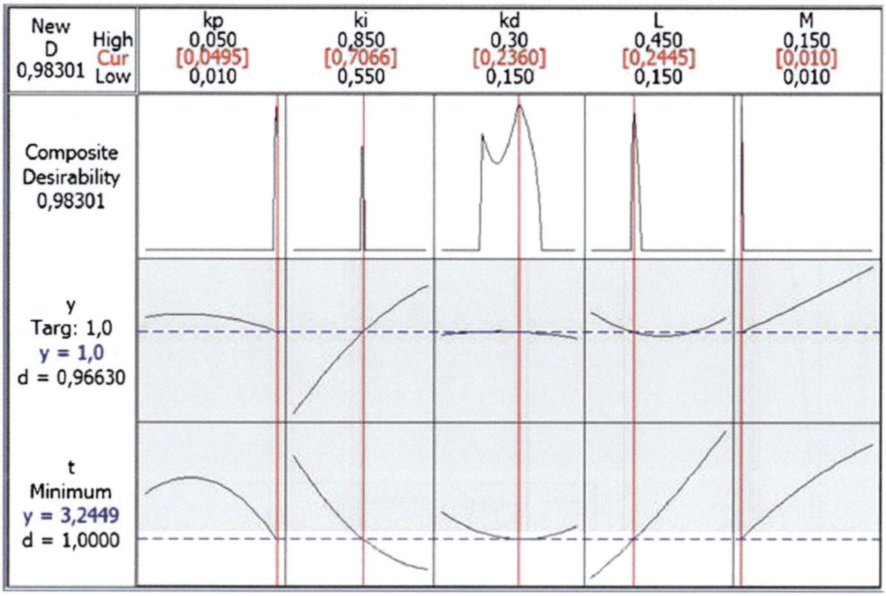

Fig. 5.2 Optimum levels of parameters for new design obtained from response optimizer module of MINITAB Package

The same process is used and the optimum parameter levels are determined as $k_p = 0.0495$, $k_i = 0.7066$, $k_d = 0.2360$, $\lambda = 0.2445$, and $\mu = 0.0100$. The coefficients of determination (R^2) for y and t are calculated as 0.9787 and 0.9248, respectively. As shown is Fig. 5.2, y is predicted as 1.00 while t is predicted as 3.2449 via the parameters combination. After the confirmation tests for the given optimum parameter levels by using MATLAB 7.1, $y = 1.00$ and $t = 3.35$ are obtained.

All of the results that are obtained from minimizing integral square error, first and second designs with response surface methodology are plotted in Fig. 5.3. Therefore, it can be concluded that the settling time is decreased more and more. If the both of response surface designs are compared, it can be easily seen that the optimum values of k_p, k_i, k_d, and λ control parameters are found inside in the levels while the optimum value of μ is found at the boundary of its level. However, it does not need to seek another factor level for μ parameter since this situation leads to further reduce the effect of derivative control. Finally, it can be pointed out the plotted outputs obtained from fitted model and the real results of experiments calculated by using MATLAB are close to each other. Real system results are better for the response t when it is compared with its expected value from MINITAB.

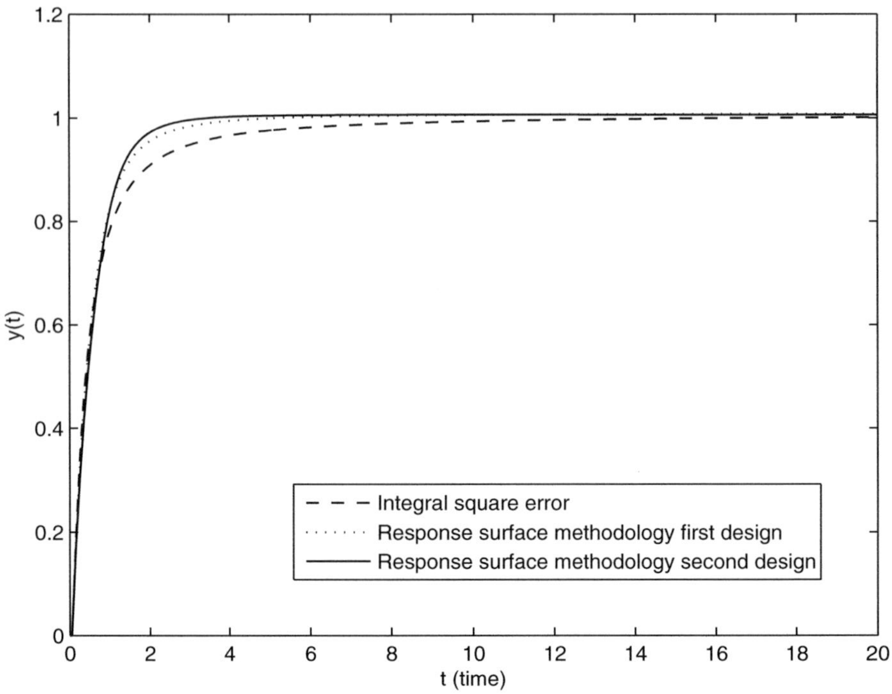

Fig. 5.3 Response of the system with tuning parameters via response surface method

5.7 Conclusions

The tuning strategies of fractional order $PI^{\lambda}D^{\mu}$ controller for a fractional order diffusion process subject to input hysteresis is developed and improved by response surface methodology. To reach the fixed desired output with minimum time 27 experiment data are read on numerical solution of the system and so the orthogonal design of experiment matrix is constructed. The mathematical relation between the response values y and t and the fractional order controller parameters k_p, k_i, k_d, λ, and μ are obtained by a full quadratic model. When comparing output of the system according to response surface method and minimizing integral square error strategy, it can be concluded that the settling time is decreased.

References

1. Agrawal OP (2004) A general formulation and solution scheme for fractional optimal control problems. Nonlinear Dyn 38:323–337
2. Agrawal OP, Defterli Ö, Baleanu D (2010) Fractional optimal control problems with several state and control variables. J Vib Control 16:1967–1976

3. Baleanu D, Defterli Ö, Agrawal OP (2009) A central difference numerical scheme for fractional optimal control problems. J Vib Control 15:583–597
4. Baleanu D, Diethelm K, Scalas E, Trujillo JJ (2012) Fractional calculus models and numerical methods. Series on complexity, nonlinearity and chaos. World Scientific, Singapore
5. Barbosa RS, Silva MF, Machado JAT (2009) Tuning and application of integer and fractional order PID controllers. Intell Eng Syst Comput Cybern 5:245–255
6. Box GEP, Wilson KB (1951) On the experimental attainment of optimum conditions. J R Stat Soc Series B 13:1–38
7. Cao JY, Liang J, Cao BG (2005) Optimization of fractional order PID controllers based on genetic algorithms. Proc Int Conf Mach Learn Cybern 9:5686–5689
8. Castillo ED (2007) Process optimization - a statistical approach. Springer, New York
9. Demirtaş M, Karaoglan AD (2012) Optimization of PI parameters for DSP-based permanent magnet brushless motor drive using response surface methodology. Energy Convers Manag 56:104–111
10. Efe MÖ (2010) Fractional order sliding mode controller design for fractional order dynamic systems. New Trends Nanotechnol Fract Calc Appl 5:463–470
11. Ekren O, Ekren BY (2008) Size optimization of a PV/wind hybrid energy conversion system with battery storage using response surface methodology. Appl Energy 85:1086–1101
12. Giesbrecht FG, Gumpertz ML (2004) Planning, construction, and statistical analysis of comparative experiments. Wiley, New Jersey
13. İskender BB, Özdemir N, Karaoglan AD (2012) Tuning of fractional order $PI^{\lambda}D^{\mu}$ controller with response surface methodology. In: Proceedings of IEEE 4th international conference on nonlinear science and complexity, pp 145–149, 6–11 August 2012
14. Lino P, Maione G (2007) New tuning rules for fractional PI^{α} controllers. Nonlinear Dyn 49:251–257
15. Mason RL, Gunst RF, Hess JL (2003) Statistical design and analysis of experiments, 2nd edn. Wiley, New Jersey
16. Miller KS, Ross B (1993) An introduction to the fractional calculus and fractional differential equations. Wiley, New York
17. Montgomery DC (2005) Design and analysis of experiments: response surface method and designs. Wiley, New Jersey
18. Moradi MH, Johnson MA (2005) PID control. Springer, London
19. Oldham KB, Spanier J (1974) The fractional calculus. Academic, New York
20. Oustaloup A (1995) La Derivation Non Enteire. HERMES, Paris
21. Özdemir N, İskender BB (2010) Fractional order control of fractional diffusion systems subject to input hysteresis. J Comput Nonlinear Dyn 5:021002(1–5)
22. Özdemir N, Karadeniz D İskender BB (2009) Fractional optimal control problem of a distributed system in cylindrical coordinates. Phys Lett A 373:221–226
23. Podlubny I (1999) Fractional differential equations. Academic, San Diego
24. Podlubny I (1999) Fractional-order systems and $PI^{\lambda}D^{\mu}$ -controllers. IEEE Trans Automat Contr 44:208–214
25. Su CY, Stepanenko Y, Svoboda J, Leung TP (2000) Robust and adaptive control of a class of nonlinear systems with unknown backlash-like hysteresis. IEEE Trans Automat Contr 45:2427–2432
26. Valerio D, Costa JS (2006) Tuning of fractional PID controllers with Ziegler-Nichols-type rules. J Signal Process 86:2771–2784
27. Zhao C, Xue D, Chen YQ (2005) A fractional order PID tuning algorithm for a class of fractional order plants. In: Proceedings of the IEEE international conference on mechatronics & automation, Niagara Falls, vol 1, pp 216–221

Chapter 6
Dynamical Response of a Van der Pol System with an External Harmonic Excitation and Fractional Derivative

Arkadiusz Syta and Grzegorz Litak

Abstract We examined the Van der Pol system with external forcing and a memory possessing fractional damping term. Calculating the basins of attraction we showed broad spectrum of nonlinear behaviour connected with sensitivity to the initial conditions. To quantify dynamical response of the system we propose the statistical 0–1 test. The results have been confirmed by bifurcation diagrams, phase portraits and Poincare sections.

Keywords Van der Pol system • Fractional derivative • 0–1 test • Chaos detection

6.1 Introduction

The system with fractional damping dependent on the velocity history has focused a lot of interest and was extensively studied in the last decade [1–6]. To model complex energy dissipation with minimum number of parameters in presence of hysteresis and memory effect, the fractional order derivative in the damping term is proposed. In such systems the damping force is proportional to a fractional derivative of the displacement instead of the classical case (first order derivative of the displacement). The memory of the system was noted to be important factor in different areas [5, 6]. Van der Pol systems, describing relaxation-oscillations are characterized by a non-viscous composite damping term [7, 8] which is small value, negative for small amplitude oscillations and changes the sign to positive

A. Syta (✉) • G. Litak
Faculty of Mechanical Engineering, Lublin University of Technology, Nadbystrzycka 36, PL-20-618 Lublin, Poland
e-mail: a.syta@pollub.pl; g.litak@pollub.pl

J.A.T. Machado et al. (eds.), *Discontinuity and Complexity in Nonlinear Physical Systems*, Nonlinear Systems and Complexity 6, DOI 10.1007/978-3-319-01411-1_6,

for increasing amplitude. This system property is reflected by dynamical response of limit cycle [9]. Comparing to viscous nonlinear systems this implies type of bifurcations and transition to chaos including hopf bifurcations [10, 11].

Recently, Van der Pol systems have been studied in a series of papers [12–15]. Pinto and Machado proposed the complex order van der Pol oscillator [12] reporting the changes in the system response spectrum with varying the fractional order of derivative in the damping term. Attari et al. [13] focused on periodic solutions and studied system parameters for their stability. Suchorsky and Rand [14] investigated the synchronization by a fractional coupling of two Van der Pol systems. Finally, Chen and Chen [15] studied a fractionally damped van der Pol equation with harmonic external forcing. They focus on the effect of fractional damping influence on the dynamic quasi-periodic and chaotic responses. In particular, the transition from quasi-periodic to chaotic motion was demonstrated.

In the present paper we continue the analysis of chaotic motion proposing an efficient method for chaotic solution identification by means of the 0–1 test [16, 17]. The main idea of this method is to use the statistical asymptotics which can distinguish the periodic and non-periodic response by studying a single coordinate of system response.

6.2 Van Der Pol System with a Fractional Damping

The van Der Pol system with external excitation is described by equation:

$$\frac{d^2x}{dt^2} + \epsilon(x^2 - 1)\frac{d^qx}{dt^q} + x = f\cos(\omega t), \tag{6.1}$$

where the fractional order derivative can be described using the Grünwald–Letnikov definition [18, 19]:

$$\frac{d^qx}{dt^q} \equiv_a D_t^q x(t) = \lim_{h\to\infty} \frac{1}{h^q} \sum_{j=0}^{\left[\frac{t-a}{h}\right]} (-1)^j \binom{q}{j} x(t - jh), \tag{6.2}$$

where binomial coefficients can be extended to complex numbers by Euler Gamma function

$$\binom{q}{j} = \frac{q!}{j!(q-j)!} = \frac{\Gamma(q+1)}{\Gamma(j+1)\Gamma(q-j+1)}, \tag{6.3}$$

here a pair of square brackets [.] appearing in the upper limit of the sum denotes the integer part, while a the length of the memory, respectively.

Note that (6.1) can be decomposed into set of equations of lower degree:

$$
\begin{aligned}
{}_LD_t^1 x(t) &= y(t) \\
{}_LD_t^q x(t) &= w(t) \\
{}_LD_t^1 y(t) &= -x(t) - \epsilon(x^2(t) - 1)w(t) + f\cos(\omega t), \qquad (6.4)
\end{aligned}
$$

where w is defined as a fractional time derivative of displacement, while y coincides with velocity ($y = \dot{x}$).

6.3 Test 0–1

To quantify obtained results which can be expressed in the time series of each coordinate we use the 0–1 test for chaos detection ([16, 17, 20–22]). This test combines both spectral and statistical properties of the system and can distinguish different types of dynamic of the system by value $K \in \{0, 1\}$. Below, one can find description of the method.

First of all, we change the coordinates from $(x, \dot{x})$ to the new set (p, q) defined as follows

$$
p(n) = \sum_{j=1}^{n} \tilde{x}_j \cos(jc), \quad q(n) = \sum_{j=1}^{n} \tilde{x}_j \sin(jc), \qquad (6.5)
$$

where $\tilde{x} = [\tilde{x}_1, \tilde{x}_2, \tilde{x}_3, \ldots]$ is a time series sampled from the original simulated series x using and one forth of excitation period [23]. The time interval $T/4$ ($T = 2\pi/\omega$) corresponds to the nodal autocorrelation function of excitation harmonic term $\delta cos(\omega t)$. Note that relevant sampling can make shorter the length of time series used in calculations leading consequently to reduction of computation time. Finally, c is a constant, $c \in (0, \pi)$. One can see that (6.5) resembles the Fourier transform for chosen frequency (in the limit of larger n).

In the next step, one computes the mean square displacement (MSD) of p and q:

$$
\begin{aligned}
MSD(c, j) = \frac{1}{n-j} \sum_{i=1}^{n-j} &\{[p(i+j) - p(i)]^2 \\
&+ [q(i+j) - q(i)]^2\}, \qquad (6.6)
\end{aligned}
$$

where $0 \ll j \ll n$ (in practice $n/100 \leq j \leq n/10$). The main criterion which is based on the trends of $MSD(c, j)$ in higher j limit. It is bounded for regular dynamics or unbounded for chaotic dynamics[16, 17, 20, 22, 24, 25]

The final quantity K is calculated as a asymptotic growth rate of MSD (here given by the correlation method):

$$
K(c) = \frac{\mathrm{Cov}[j, MSD(c, j)]}{\sqrt{\mathrm{Cov}[j, j] \cdot \mathrm{Cov}[MSD(c, j), MSD(c, j)]}}, \qquad (6.7)
$$

where j is based on series of natural numbers: $j = n/100, n/100 + 1 \ldots, n/10$, and Cov[$x_1$,$x_2$] denotes corresponding covariance of two series which for the same arguments $x_1 = x_2$ means variance while for chosen pair of two different series: $x_1 = j$ and $x_2 = MSD(c, j)$, it can be expressed in terms of the expectation value E[.]:

$$\text{Cov}[j, MSD(c, j)] = \text{E}[[j - \text{E}[j]] \cdot [MSD(c, j) - \text{E}[MSD(c, j)]]]. \quad (6.8)$$

6.4 Simulation Results

In our investigations we set $\epsilon = 8.0$, $f = 1.0$, $\omega = 3/10$, and $(x, \dot{x}) = (0.5, 0.0)$ for various q values ($q \in [0.8, 1.2]$).

Figure 6.1 shows the results of the bifurcation diagram of the x coordinate s (red points) versus order of the derivative q. The characteristic broad distributions of points imply the chaotic behaviour while the countable few points (1 to 3 points per q value noticeable in Fig. 6.1) correspond to a periodic solution.

On the other hand the full black line corresponds to parameter K defined for the 0–1 test versus q. Note, different q-parameter regions. $K \approx 0$ correspond to regular (periodic motion) while $K \approx 1$ to chaotic solutions. Note that the $K \approx 0$ regions ideally match the broad distributions in bifurcation diagram. One can also notice some intermediate value of K (for $q = 1.05$) which could tell that reaching the asymptotic limit of K needs longer time series of $\tilde{x}$.

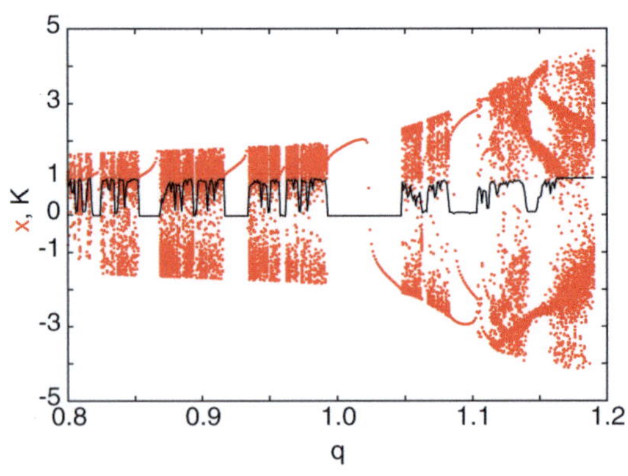

Fig. 6.1 The *red points* indicate the bifurcation (stroboscopic) diagram of the x coordinate versus order of the derivative $q \in [0.8, 1.2]$, initial conditions for each q were $(x, \dot{x}) = (0.5, 0.0)$. Other system parameters: $\epsilon = 8.0$, $f = 1.0$, $\omega = 3/10$. The *full black line* corresponds to parameter K defined for the 0–1 test versus q. Note, different q-parameter regions. $K \approx 0$ correspond to regular (periodic motion) while $K \approx 1$ to chaotic solution. The parameters used for K estimation were as follows: $n = 400$, $j = 4, .., 40$

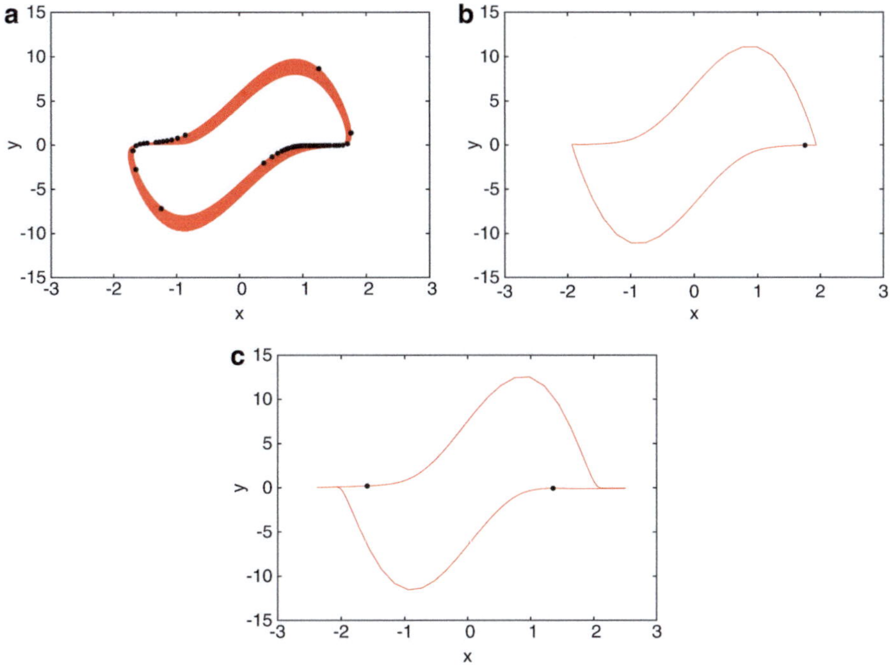

Fig. 6.2 Phase portraits and Poincare points for $q = 0.9$ (**a**), $q = 1.0$ (**b**), and $q = 1.063$ (**c**), respectively. All other system parameters as in Fig. 6.1. The corresponding results for K: 0.91, −0.02, 0.06

For better clarity we show the phase portraits with corresponding Poincare sections in Fig. 6.2a–c. The results also confirm the 0–1 test analysis (see Fig. 6.1).

6.5 Conclusions

We have examined dynamics of the Duffing model with fractional damping term. Using nonlinear methods (phase diagrams, Poincare sections and bifurcation diagrams) we have showed significant different system response while varying the order of the derivative (from non-integer to integer). We also quantified the type of motion by 0–1 test which is based on statistical properties of phase coordinate. Note that the Lyapunov exponent could be difficult to estimate as the phase space dimension is undetermined due to the memory effect. In such a situation the embedding dimension should be estimated for each q value [26].

Acknowledgements The authors gratefully acknowledge the support of the 7th Framework Programme FP7- REGPOT-2009-1, under Grant Agreement No. 245479. The authors are grateful prof. Stefano Lenci for discussions.

References

1. Padovan J, Sawicki JT (1998) Nonlinear vibration of fractionally damped systems. Nonlinear Dyn 16:321–336
2. Seredynska M, Hanyga A (2005) Nonlinear differential equations with fractional damping with application to the 1dof and 2dof pendulum. Acta Mech 176:169–183
3. Gao X, Yu J (2005) Chaos in the fractional order periodically forced complex Duffing's systems. Chaos Solitons Fractals 24:1097–1104
4. Sheu LJ, Chen HK, Tam LM (2007) Chaotic dynamics of the fractionally damped Duffing equation. Chaos Solitons Fractals 32:1459–1468
5. Rossikhin YA, Shitikova MV (2010) Application of fractional calculus for dynamic problems of solid mechanics: novel trends and recent results. Appl Mech Rev 63:010801
6. Machado JAT, Silva MF, Barbosa RS, Jesus IS, Reis CM, Marcos MG, Galhano AF (2010) Some applications of fractional calculus in engineering. Math Probl Eng 2010, 639801
7. Van der Pol B (1926) On relaxation-oscillations. Philos Mag 2:978–992
8. Van der Pol B, Van der Mark J (1928) The heartbeat considered as a relaxation oscillation and an electrical model of the heart. Philos Mag Suppl 6:763–775
9. Steeb W-H, Kunick A (1987) Chaos in system with limit cycle. Int J Nonlinear Mech 22: 349–361
10. Kapitaniak T, Steeb W-H (1990) Transition to chaos in a generalized van der Pol's equation. J Sound Vib 143:167–170
11. Litak G, Spuz-Szpos G, Szabelski K, Warminski J (1999) Vibration analysis of a self-excited system with parametric forcing and nonlinear stiffness. Int J Bifurcat Chaos 9:493–504
12. Pinto CMA, Machado JAT (2011) Complex order van der Pol oscillator. Nonlinear Dyn 65:247–254
13. Attari, M., Haeri, M., Tavazoei MS (2010) Analysis of a fractional order Van der Pol-like oscillator via describing function method. Nonlinear Dyn 61:265–274
14. Suchorsky MK, Rand RH (2012) A pair of van der Pol oscillators coupled by fractional derivatives. Nonlinear Dyn 69:313–324
15. Chen J-H Chen W-C (2008) Chaotic dynamics of the fractionally damped van der Pol equation. Chaos Solitons Fractals 35:188–198
16. Gottwald GA, Melbourne I (2004) A new test for chaos in deterministic systems. Proc R Soc A 460:603–611
17. Gottwald GA, Melbourne I (2005) Testing for chaos in deterministic systems with noise. Physica D 212:100–110
18. Podlubny I (1999) Fractional differential equations. Academic, San Diego
19. Petras I (2010) Fractional-order nonlinear systems: modeling, analysis and simulation. Springer, New York
20. Falconer I, Gottwald GA, Melbourne I, Wormnes K (2007) Application of the 0–1 test for chaos to experimental data. SIAM J Appl Dyn Syst 6:95–402
21. Litak G, Syta A, Wiercigroch M (2009) Identification of chaos in a cutting process by the 0–1 test. Chaos Solitons Fractals 40:2095–2101
22. Litak G, Syta A, Budhraja M, Saha LM (2009) Detection of the chaotic behaviour of a bouncing ball by the 0–1 test. Chaos Solitons Fractals 42:1511–1517
23. Bernardini D, Rega G, Litak G, Syta A (2013) Identification of regular and chaotic isothermal trajectories of a shape memory oscillator using the 0–1 test. Proc IMechE Part K J Multi-body Dyn 227:17–22
24. Krese B, Govekar E (2012) Nonlinear analysis of laser droplet generation by means of 0–1 test for chaos. Nonlinear Dyn 67:2101–2109
25. Litak G, Schubert S, Radons G (2012) Nonlinear dynamics of a regenerative cutting process. Nonlinear Dyn 69:1255–1262
26. Kantz H (1994) A robust method to estimate the maximal Lyapunov exponent of a time series. Phys Lett A 185:77–87

Chapter 7
Fractional Calculus: From Simple Control Solutions to Complex Implementation Issues

Cristina I. Muresan

Abstract Fractional calculus is currently gaining more and more popularity in the control engineering world. Several tuning algorithms for fractional order controllers have been proposed so far. This chapter describes a simple tuning rule for fractional order PI controllers for single-input–single-output processes and an extension of this method to the multivariable case. The implementation of a fractional order PI on an FPGA target for controlling the DC motor speed, as well as the implementation of a multivariable fractional order PI controller for a time delay system is presented. Experimental results are given to show the efficiency and robustness of the tuning algorithm.

Keywords Fractional calculus • Control algorithm • Multivariable processes DC motor speed control • Multivariable fractional order controller • Decoupling FPGA implementation • Micro-controller implementation • Time delays Experimental results • Robustness

7.1 Introduction

Fractional calculus represents the generalization of the integration and differentiation to an arbitrary order. The beginning of fractional calculus dates back to the early days of classical differential calculus, although its inherent complexity postponed its use and application to the engineering world [1]. Nowadays, its use in control engineering has been gaining more and more popularity in both modeling and identification, as well as in the controller tuning. The approach of fractional calculus to modeling is based on the concepts of viscoelasticity, diffusion, and

C.I. Muresan (✉)
Department of Automatic Control, Technical University of Cluj-Napoca, Baritiu str. 26-28, Cluj-Napoca, Romania
e-mail: Cristina.Pop@aut.utcluj.ro

J.A.T. Machado et al. (eds.), *Discontinuity and Complexity in Nonlinear Physical Systems*, Nonlinear Systems and Complexity 6, DOI 10.1007/978-3-319-01411-1_7,

fractal structures that several processes may exhibit, which are more easily and accurately described using fractional order models [2–6].

In terms of controller tuning, the fractional order $PI^{\mu}D^{\lambda}$ controller is in fact a generalization of the classical integer order PID controller. It is generally accepted that the fractional order $PI^{\mu}D^{\lambda}$ controller, due to the two supplementary tuning variables, μ and λ, is able to meet more performance criteria and behave more robustly than the traditional PID controller [7–11]. Several approaches to tuning fractional order $PI^{\mu}D^{\lambda}$ exist, with some notable works that use the theory of fractional calculus in controlling both integer order and fractional order dynamical systems [12–15].

Usually, the design of the fractional order controllers is done by imposing various performance criteria that restrict the open loop system to a certain gain crossover frequency, a given phase margin, a boundary on open loop amplification at certain frequencies, or a robustness to open loop gain variations. Several techniques to find a suitable solution for the fractional order controller parameters that meet all pre-specified closed loop conditions have been developed, ranging from simple optimization routines to more complex genetic algorithms or graphical methods [16, 17].

The complexity of the tuning procedure even for simple, single-input–single-output processes has restricted the application of such fractional order controllers to these types of systems. Very few results are given for the multivariable case [18, 19]. The research of Chenikher et al. [18] proposes in fact a rather complex solution based on an H_{∞} problem with a controller structure constraint, while the controller parameters are optimized to achieve both user-specified robust stability and performance, the controller obtained being tested for controlling systems with multiple delays. The method described in this chapter proposes instead a very simple method for tuning multivariable fractional order controllers, by extending the single-input–single-output version. The method is also used for the general case scenario in which the multivariable system may be non-square and with multiple time delays.

The tuning method is based on a steady state decoupling, followed by several individual designs of fractional order controllers for the decoupled process and a final computation of the multivariable controller. The tuning of the individual fractional order controllers is similar to the single-input–single-output approach and consists in imposing a given gain crossover frequency, a given phase margin, and a gain robustness condition to the open loop system.

The chapter also presents some of the implementation steps and problems to be solved prior to the actual implementation of fractional order controllers on dedicated devices. Two case studies are presented. The first one consists in an FPGA (Field Programmable Gate Array) implementation of a fractional order PI controller for a DC motor, while the second case study presents the microcontroller implementation of the multivariable fractional order PI controller. The experimental results are also given to show the accuracy, efficiency, and robustness of the tuning method.

The paper is structured into five parts. Immediately after the Introduction section, the next subchapter details one of the simplest tuning procedures for computing a fractional order PI controller. The method can easily be applied to both integer order and fractional order single-input–single-output processes. Next, the approach presented for the case of single-input–single-output processes is extended to the multivariable case. The proposed control method for multivariable systems is presented, including the decoupling procedure and the method to compute the final multivariable fractional order PI controller. The next subchapter presents the implementation issues associated with these types of controllers. A single-input–single-output, as well as a multivariable case study is presented. Two different devices are used for implementation purposes, an FPGA and a microcontroller. The final section of this chapter contains the concluding remarks.

7.2 The simplest PI Tuning Algorithm

The most generally used transfer function for a fractional order PI controller (FO-PI) is:

$$H_{FO\text{-}PI}(s) = k_p\left(1 + \frac{k_i}{s^\mu}\right) \tag{7.1}$$

where the fractional order is denoted by μ and is an arbitrary real number. Several tuning algorithms for such FO-PI controllers exist, employing optimization mechanisms for computing the final values of the three controller parameters—k_p, k_i, and μ—starting with some prescribed performance criteria. These performance specifications most often refer to (a) an imposed gain crossover frequency of the open loop system—ω_{gc}, (b) an imposed phase margin of the open loop system—ϕ_m, and (c) a robustness condition.

In frequency domain, the transfer function of the FO-PI controller may be written as:

$$H_{FO\text{-}PI}(j\omega) = k_p\left[1 + k_i\omega^{-\mu}\left(\cos\frac{\pi\mu}{2} - j\sin\frac{\pi\mu}{2}\right)\right] \tag{7.2}$$

where $j^{-\mu} = \cos\frac{\pi\mu}{2} - j\sin\frac{\pi\mu}{2}$.

Considering the process transfer function $H_p(s)$, the open loop system may be written as:

$$H_{open\text{-}loop}(s) = H_{FO\text{-}PI}(s)\, H_P(s) \tag{7.3}$$

The first condition, imposing a gain crossover frequency, leads to the modulus equation:

$$\left|H_{open\text{-}loop}\left(j\omega_{gc}\right)\right| = 1 \tag{7.4}$$

If the process transfer function is written in a complex form as:

$$H_P(j\omega) = \frac{1}{\Re(H_p) + j\Im(H_p)} \tag{7.5}$$

where $\Re(H_p)$ stands for the real part and $\Im(H_p)$ denotes the imaginary part of $H_P(j\omega_{gc})$, then (7.4) results in the following:

$$\left|\frac{1}{\Re(H_p) + j\Im(H_p)}\right|_{\omega_{gc}} \left|k_p\left[1 + k_i\omega_{gc}^{-\mu}\left(\cos\frac{\pi\mu}{2} - j\sin\frac{\pi\mu}{2}\right)\right]\right| = 1 \tag{7.6}$$

Imposing a phase margin for the open loop system translates to:

$$\angle H_{\text{open-loop}}(j\omega_{gc}) = -\pi + \varphi_m \tag{7.7}$$

which may further be written as:

$$a\tan\left(-\frac{k_i\omega_{gc}^{-\mu}\sin\frac{\pi\mu}{2}}{1 + k_i\omega_{gc}^{-\mu}\cos\frac{\pi\mu}{2}}\right) - a\tan\left(\frac{\Im(H_p)}{\Re(H_p)}\right) = -\pi + \varphi_m \tag{7.8}$$

resulting in:

$$\frac{k_i\sin\left(\frac{\pi\cdot\mu}{2}\right)}{\omega_{gc}^{\mu} + k_i\cos\left(\frac{\pi\cdot\mu}{2}\right)} = tg\left(\pi - \varphi_m - a\tan\left(\frac{\Im(H_p)}{\Re(H_p)}\right)\right) \tag{7.9}$$

In (7.9), the only unknown parameters are now k_i and μ, while in (7.6) there is also k_p. A third equation is then used to yield a system of three equations with three unknown parameters. In this chapter, the third equation refers to robustness to gain changes in the open loop system. To ensure such a performance specification, the phase of the open loop system around the gain crossover frequency should be flat, which implies that the derivative of the open loop system would be zero at the gain crossover frequency:

$$\frac{d\left(\angle H_{\text{open-loop}}(j\omega_{gc})\right)}{d\omega_{gc}} = 0 \tag{7.10}$$

Using the phase of the open loop system, (7.10) leads to:

$$\frac{\mu k_i\omega_{gc}^{-\mu-1}\sin\frac{\pi\mu}{2}}{1 + 2k_i\omega_{gc}^{-\mu}\cos\frac{\pi\mu}{2} + k_i^2\omega_{gc}^{-2\mu}} - \frac{\dot{\Im}(H_p)\Re(H_p) - \Im(H_p)\dot{\Re}(H_p)}{\Im(H_p)^2 + \Re(H_p)^2} = 0 \tag{7.11}$$

which is again an equation with only two unknown parameters: k_i and μ, as the case of (7.9). Thus, using (7.9) and (7.11) and applying optimization routines k_i and μ can be uniquely determined given any phase margin, φ_m. Then, using (7.6) and the previously determined values for k_i and μ, the last parameter of the FO-PI controller, k_p, can also be uniquely determined.

7.3 Multivariable Approach to Fractional Order Control

The multivariable approach to fractional order control is based on a steady state decoupling of the process, followed by the tuning of the fractional order controllers for the decoupled process and lastly, the final computation of the multivariable fractional order controller. The approach presented is suitable for square and non-square systems and also for multivariable systems that exhibit multiple time delays.

For a general process transfer function matrix:

$$G_p(s) = \begin{bmatrix} g_{11}e^{-\tau_{11}s} & \dots & g_{1m}e^{-\tau_{1m}s} \\ \vdots & \vdots & \vdots \\ g_{n1}e^{-\tau_{n1}s} & \dots & g_{nm}e^{-\tau_{nm}s} \end{bmatrix} \tag{7.12}$$

having m inputs and n outputs, the steady state decoupling is achieved using the Moore–Penrose pseudo-inverse of the steady state gain matrix [20–22]:

$$G_m^{\#} = G_m(0)^H \cdot \left(G_m(0) \cdot G_m(0)^H\right)^{-1} = \begin{pmatrix} g_{11}^{\#} & \cdots & g_{1n}^{\#} \\ \vdots & \vdots & \vdots \\ g_{m1}^{\#} & \cdots & g_{mn}^{\#} \end{pmatrix} \tag{7.13}$$

where $G_m{}^{\#}$ is the pseudo-inverse, and $(\dots)^H$ is the Hermitian matrix of the steady state matrix $G_m(0)$ computed as:

$$G_m(0) = \begin{bmatrix} g_{110} & \dots & g_{1m0} \\ \vdots & \vdots & \vdots \\ g_{n10} & \dots & g_{nm0} \end{bmatrix} \tag{7.14}$$

The decoupled process transfer function matrix is given by:

$$G_D(s) = G_m(s) \cdot G_m^{\#} = \begin{bmatrix} g_{d11} & \dots & g_{d1n} \\ \vdots & \vdots & \dots \\ g_{dn1} & \dots & g_{dnn} \end{bmatrix} \tag{7.15}$$

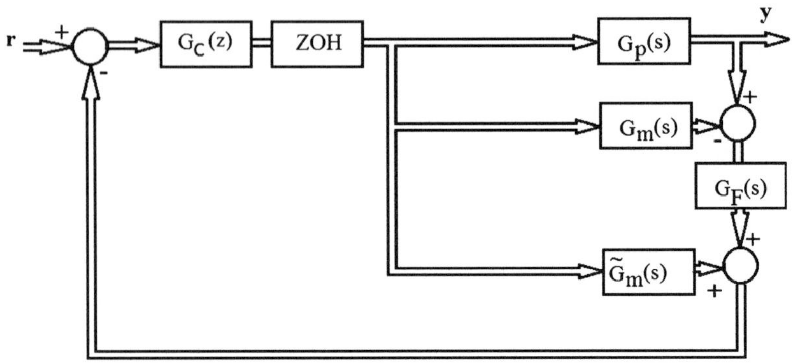

Fig. 7.1 Closed loop control structure in the case of multivariable time delay processes

where

$$G_m(s) = \begin{bmatrix} g_{11}e^{-\tau_{11}s} & \dots & g_{1m}e^{-\tau_{1m}s} \\ : & : & : \\ g_{n1}e^{-\tau_{n1}s} & \dots & g_{nm}e^{-\tau_{nm}s} \end{bmatrix} \tag{7.16}$$

is the model of the multivariable process.

In steady state, applying a step input to the decoupled process in (7.15) leads to output responses that have zero values for all the non-diagonal terms and unitary values for all terms on the main diagonal. Thus, to compute the fractional order controller, all non-diagonal elements in $G_D(s)$ may be discarded. The diagonal elements in $G_D(s)$ would be further used to design the fractional order controller. Each diagonal element has, however, a complex form resulting from the multiplication of the original model $G_m(s)$ with the pseudo-inverse $G_m{}^{\#}$:

$$g_{dii}(s) = \sum_{i=1,j=1}^{i=n,j=m} g_{ij}e^{-\tau_{ij}s}g_{ji}^{\#} \tag{7.17}$$

The diagonal forms, as given in (7.17), are difficult to be used in the design of the controllers. Thus, a more simplified form of (7.17) is required. Denoting the approximations obtained [20, 21] in (7.17) with $H_{pi}(s)$ which can be easily written as in (7.5), the procedure described in the previous section may be used to tune several FO-PI controllers for each $H_{pi}(s)$, by imposing n different gain crossover frequencies and n phase margins φ_m for each of the process outputs. Such a design approach is facilitated since the control structure used would be a Smith predictor, as shown in Fig. 7.1, where $G_p(s)$ is the process transfer function matrix from (7.12), $G_m(s)$ is the model of the process from (7.16), $\tilde{G}_m(s)$ is the model of the process without the corresponding time delays, $G_F(s)$ are some feedback filters [20, 21] and finally $G_c(z)$ is the multivariable fractional order controller, in its discrete form.

For processes that do not exhibit time delays, the same tuning procedure can be applied and the final control structure would be a simple, classical negative feedback closed loop scheme.

The multivariable fractional order controller $G_c(s)$ may be easily computed as soon as the individual FO-PI controllers are determined for each $H_{pi}(s)$:

$$G_C(s) = G_m^{\#} \begin{pmatrix} H_{FO\text{-}PI_1}(s) & 0 & \dots & 0 \\ 0 & H_{FO\text{-}PI_2}(s) & \dots & 0 \\ \dots & \dots & \dots & 0 \\ 0 & 0 & \dots & H_{FO\text{-}PI_n}(s) \end{pmatrix} \tag{7.18}$$

7.4 Practical Implementation Issues

To implement the fractional order controllers, both for the single-input-single-output and for the multivariable processes, an analog or discrete form of these controllers is required. First, to ensure the effect on an integer order integrator, both at high and low frequencies [17], the general form of the FO-PI controller in (7.1) is written as:

$$H_{FO\text{-}PI}(s) = k_p \left(1 + \frac{k_i s^{1-\mu}}{s}\right) \tag{7.19}$$

Compared to traditional integer order systems that have limited memory and are finite dimensional, fractional order systems imply the use of unlimited memory (infinite dimensional), which represents one of the main problems in implementing controllers based on fractional orders. For this reason, the approximation of fractional order systems with finite difference equations becomes even more important. Several analog approximations have been proposed [23], nevertheless in practice it is more convenient to use direct discretization methods.

To implement the fractional order controllers, either for the single-input–single-output processes or for multivariable ones, on digital devices, the discrete form of (7.19) is necessary to be obtained. The discretization method used is based on the recursive Tustin method [24, 25]:

$$s^{\mu} = \left(\frac{2}{T}\right)^{\mu} \frac{A\left(z^{-1}, \mu\right)}{A\left(z^{-1}, -\mu\right)} \tag{7.20}$$

where T is the sampling period. The polynomials A have different forms depending on the order of the discretization:

$$A\left(z^{-1}, \mu\right) = -\frac{\mu}{5} z^{-5} + \frac{\mu^2}{5} z^{-4} - \left(\frac{\mu}{3} + \frac{\mu^3}{15}\right) z^{-3} + \frac{2}{5}\mu^2 z^{-2} - \mu z^{-1} + 1 \tag{7.21}$$

for a 5th order approximation, while

$$
\begin{aligned}
A\left(z^{-1},\mu\right) = & -\frac{\mu}{9}z^{-9} + \frac{\mu^2}{9}z^{-8} - \left(\frac{\mu}{7} + \frac{\mu^3}{21}\right)z^{-7} + \left(\frac{34\mu^2}{189} + \frac{2\mu^4}{189}\right)z^{-6} \\
& - \left(\frac{\mu}{5} + \frac{16\mu^3}{189} + \frac{\mu^5}{945}\right)z^{-5} + \left(\frac{17\mu^2}{63} + \frac{\mu^4}{63}\right)z^{-4} \\
& - \left(\frac{\mu}{3} + \frac{1\mu^3}{9}\right)z^{-3} + \frac{4\mu^2}{9}z^{-2} - \mu z^{-1} + 1
\end{aligned}
\tag{7.22}
$$

is the 9th order approximation.

When using the discrete form of the FO-PI controllers, hardware considerations must be also accounted for. Limitations in digital implementation refer to the memory size available, the necessary computational load, the bounds on the execution time, the number of resources available for compiling, and running the fractional order control algorithm.

The first case study considered in the paper consists in an FPGA implementation of a fractional order PI for controlling the DC motor speed. FPGAs are currently used in a wide area of applications, ranging from digital signal processing systems, space and defense, prototyping, medical systems, intelligent traffic systems (ITS) to language recognition, bioinformatics, cryptography, etc. [26]. The choice for the FPGA implementation is based on a series of advantages: low power consumption, increased flexibility that allows the addition of new features to the controller, its update, the implementation of further data post-processing algorithms, dynamic reconfigurability, and in-system programmability capabilities that allow the implementation of mechanisms that increase the overall performance of the system and its reliability [27]. A couple of notable papers describing the implementation of fractional order control algorithms on PLCs exist [28–30], but research covering the problem of FPGA implementation is rather scarce.

The transfer function of the DC motor to be controlled is given by [31]:

$$
H(s) = \frac{27.5}{0.26s + 1} \tag{7.23}
$$

having the DC motor duty cycle as the control input. To design the fractional order PI controller as described in the previous section, for the single-input–single-output case, the following design specifications are imposed: $\omega_{gc} = 15$ rad/s, $\varphi_m = 70^o$ and robustness to open loop gain variations. Using (7.9) and (7.11), the k_i parameter of the FO-PI controller is computed for different values of the fractional order μ. The results are plotted in a graph given in Fig. 7.2.

Using Fig. 7.2, the intersection point of the two plots yields a fractional order $\mu = 0.7371$ and a corresponding $k_i = 7.85$. With these values, the last parameter of the FO-PI controller in (7.1) can be uniquely determined using (7.6): $k_p = 0.09$. The

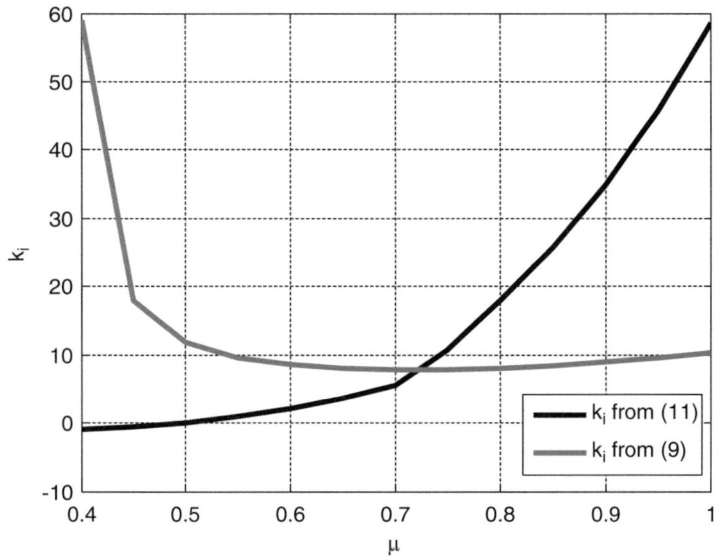

Fig. 7.2 Plots of k_i as a function of the fractional order μ for the DC motor speed control

FO-PI controller is further put into its discrete form using 9th order recursive Tustin method, with a sampling time of 0.015 s.

The main problem with the implementation consists in the data representation. Matlab and PC based programs use floating-point double data representation, whereas FPGAs use fixed point. The implementation of the FO-PI controller in its discrete form implies a recursive equation for the control input, based on its previous values, as well as previous values of the control error. The floating point and the fixed point implementation of the FO-PI controller, on a real time target and an FPGA device, respectively, are given in Fig. 7.3a, b.

For the implementation of the FO-PI controller designed for the DC motor speed control, the optimum data representation was chosen to be 14 bits for the integer word length and 32 bits for the entire word length.

The experimental results are given in Fig. 7.4 and are compared to the Matlab simulation. The results show that the designed FO-PI controller maintains an overshoot of less than 10 % and a settling time of 0.1 s.

The second case study consists in a design of a multivariable FO-PI controller for a process with multiple time delays [21]:

$$G_p(s) = \begin{pmatrix} g_{11}(s)\,e^{-\tau_{11}s} & g_{12}(s)\,e^{-\tau_{12}s} & 0 \\ g_{21}(s)\,e^{-\tau_{21}s} & g_{22}(s)\,e^{-\tau_{22}s} & g_{23}(s) \\ g_{31}(s)\,e^{-\tau_{31}s} & g_{32}(s)\,e^{-\tau_{32}s} & g_{33}(s) \end{pmatrix} \tag{7.24}$$

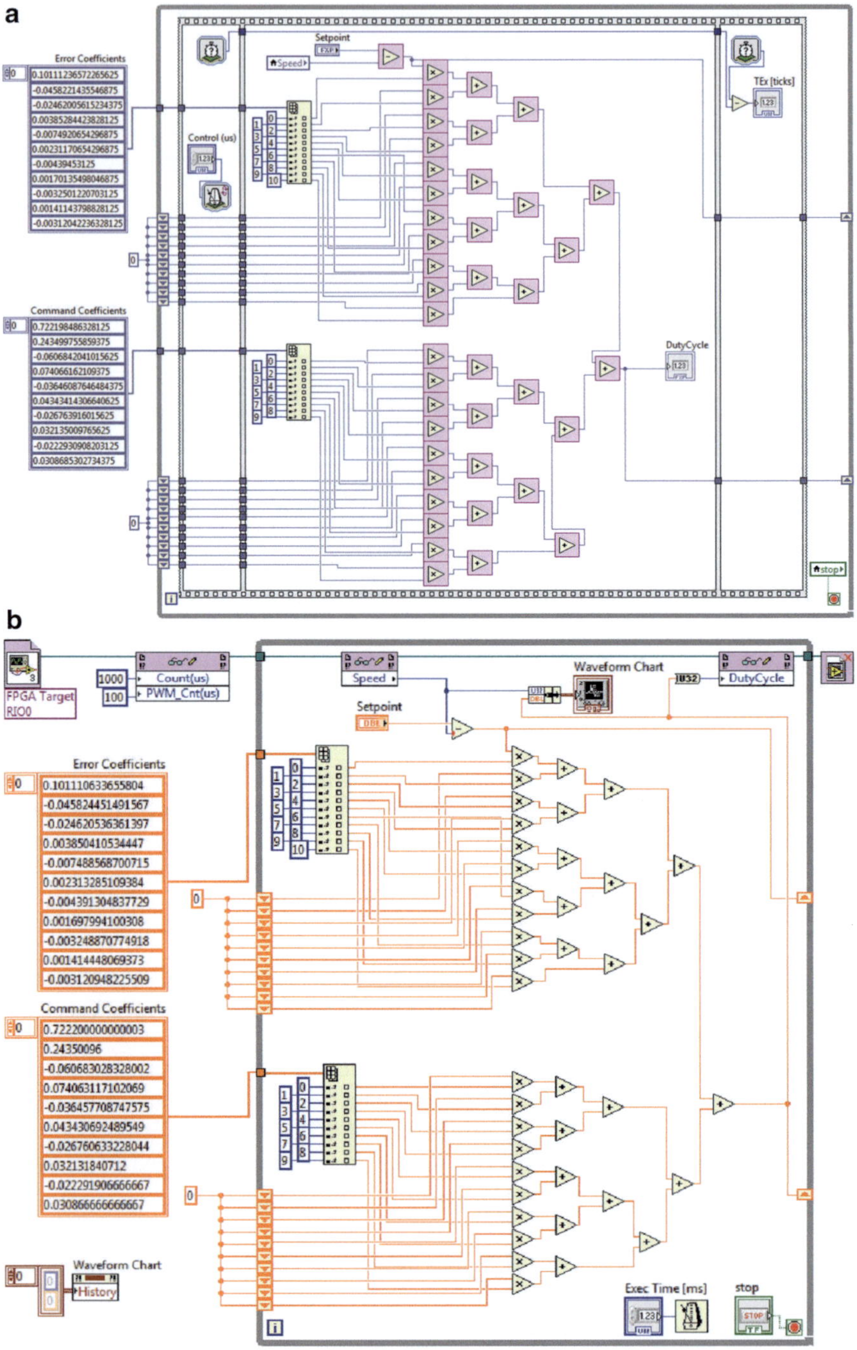

Fig. 7.3 Fractional order control algorithm using (a) fixed-point data running on FPGA and (b) floating-point data running on RT target (© Elsevier, 2013), reprinted with permission

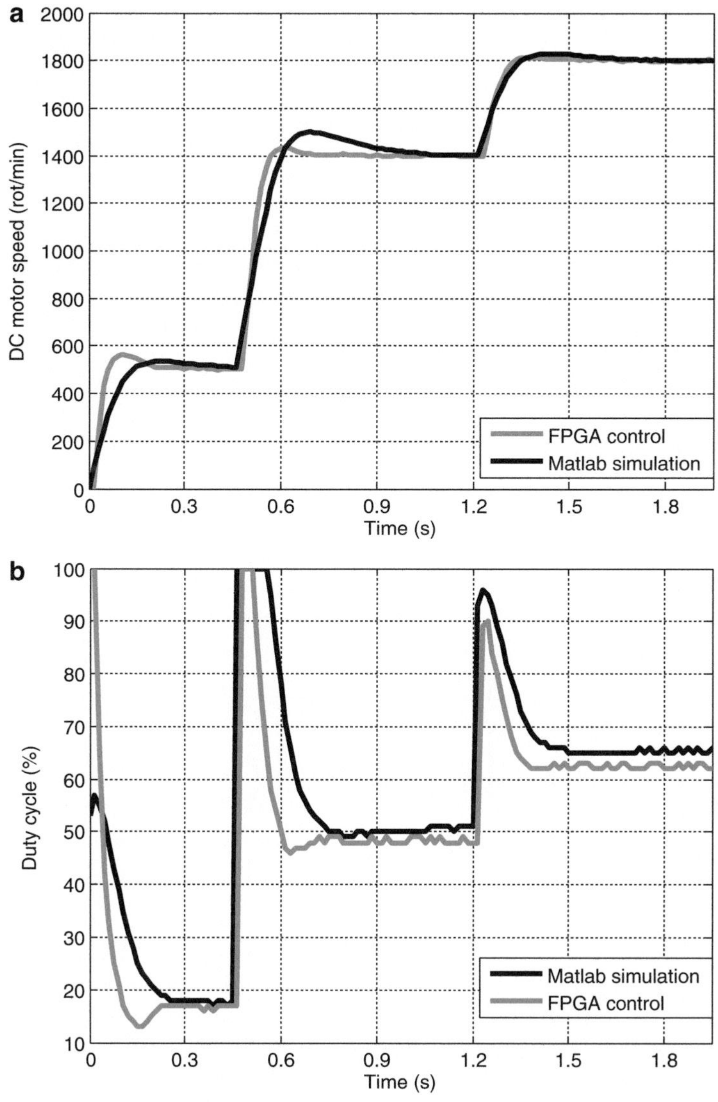

Fig. 7.4 DC motor speed control using FPGA (© Elsevier, 2013), reprinted with permission

with the transfer functions determined using experimental identification methods [21, 32]. The gain matrix of (7.24) is then computed as [21, 32]:

$$G_m(s=0) = \begin{bmatrix} -1.318 & 0.569 & 0 \\ -0.882 & 0.882 & -9.386 \\ -0.140 & 0.063 & 8.585 \end{bmatrix} \qquad (7.25)$$

Since the multivariable process is a square one, the Moore–Penrose pseudo-inverse of the steady state gain matrix would be equal to the actual inverse of (7.25):

$$G_m^{\#} = \begin{bmatrix} -1.432 & 0.856 & 0.936 \\ -1.558 & 1.982 & 2.167 \\ -0.011 & -6.7 \cdot 10^{-4} & 0.115 \end{bmatrix} \quad (7.26)$$

The decoupled process transfer function matrix is computed using (7.26) and (7.24), replaced in (7.15). The diagonal terms of the resulting decoupled process transfer function matrix are given by:

$$g_{d11} = -1.432 \cdot g_{11}(s)\, e^{-\tau_{11}s} - 1.558 \cdot g_{12}(s)\, e^{-\tau_{12}s} \quad (7.27)$$

$$g_{d22} = 0.856 \cdot g_{21}(s)\, e^{-\tau_{21}s} + 1.982 \cdot g_{22}(s)\, e^{-\tau_{22}s} + 6.7 \cdot 10^{-4} \cdot g_{23}(s) \quad (7.28)$$

$$g_{d33} = 0.936 \cdot g_{31}(s)\, e^{-\tau_{31}s} + 2.167 \cdot g_{32}(s)\, e^{-\tau_{32}s} + 0.115 \cdot g_{33}(s) \quad (7.29)$$

To facilitate the design of the controllers, the diagonal terms (7.27), (7.28), and (7.29) are approximated with the following transfer functions [21]:

$$g_{d11}^{*}(s) = \frac{0.022274 \cdot e^{-6s}}{(s^2 + 0.1044s + 0.02223)} \quad (7.30)$$

$$g_{d22}^{*}(s) = \frac{0.0045918 \cdot e^{-8s}}{(s^2 + 0.12s + 0.004592)} \quad (7.31)$$

$$g_{d33}^{*}(s) = \frac{1.0565}{(s + 1.057)} \quad (7.32)$$

For the design of the fractional order controller, due to the Smith predictor structure used, only the delay free parts in (7.30)–(7.32) are considered. The tuning is performed by setting the gain robustness conditions, as well as the additional performance specifications regarding the phase margin and gain crossover frequency for the open loop system: $\omega_{gc1} = 0.01$, $\omega_{gc2} = 0.01$, and $\omega_{gc3} = 0.05$; $\varphi_{m1} = 60^o$, $\varphi_{m2} = 70^o$, and $\varphi_{m3} = 50^o$. Plotting again for each transfer function the k_i parameters computed using (7.9) and (7.11) as a function of the fractional order, as given in Figs. 7.7, 7.5, and 7.6, and using (7.6) to determine the final value for the k_p parameter, yield the following final results: $\mu_1 = 1.327$, $k_{p1} = 0.0424$, and $k_{i1} = 0.0562$, $\mu_2 = 1.19$, $k_{p2} = 0.2213$, and $k_{i1} = 0.02$, and $\mu_3 = 1.436$, $k_{p3} = 0.044$, and $k_{i3} = 0.32$.

To implement the multivariable FO-PI controller on a microcontroller target, the discrete form is obtained using 5th order recursive Tustin method, with a sampling time of 0.3 min. The experimental platform consists in the multivariable process running in a Simulink, Matlab environment, while the multivariable fractional order

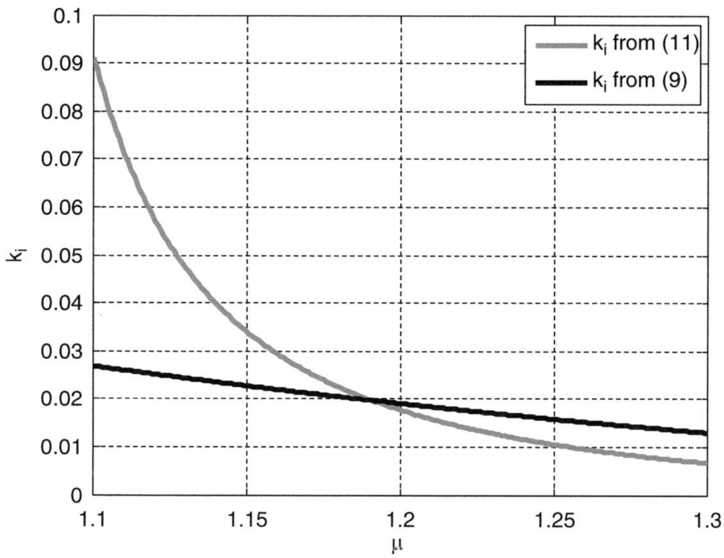

Fig. 7.5 Plots of k_i as a function of the fractional order μ for the second diagonal element of the multivariable decoupled process

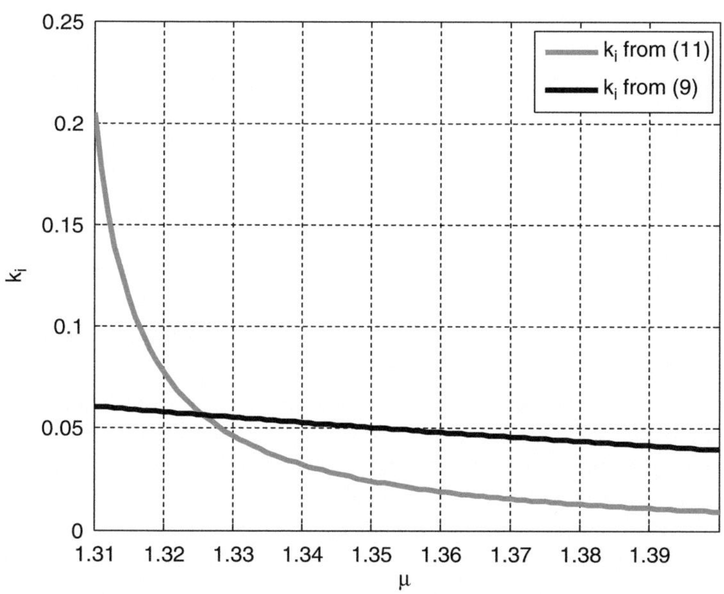

Fig. 7.6 Plots of k_i as a function of the fractional order μ for the third diagonal element of the multivariable decoupled process

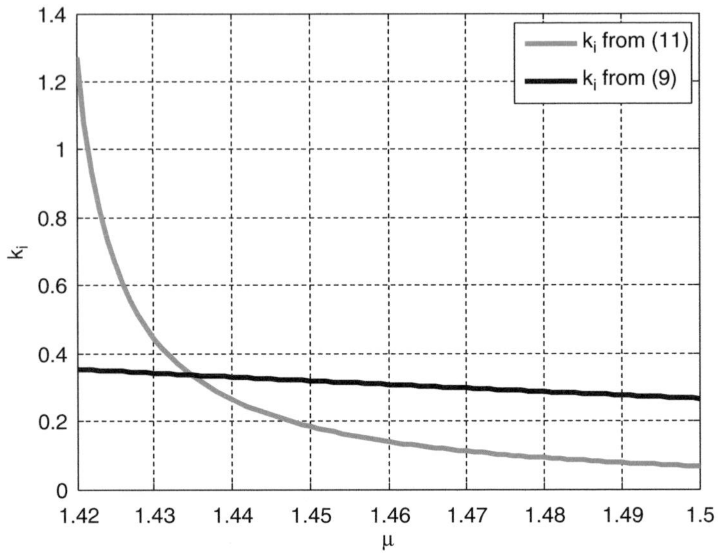

Fig. 7.7 Plots of k_i as a function of the fractional order μ for the first diagonal element of the multivariable decoupled process

control algorithm is developed and runs on a microcontroller. At every sampling time, the microcontroller reads from Simulink the process outputs and generates the control inputs, based on the computed error signals. The communication between the Simulink environment and the microcontroller is a serial one. The multivariable fractional order control algorithm implemented on the microcontroller consists in three recurrent relations that generate the next three control signals based on their previous values, as well as on previous values for the computed tracking errors. The measured process output values are represented on 32 bits, corresponding to single precision, which are then split into 4 groups of 8 bits each, and sent one group at a time to the microcontroller. After the microcontroller computes all of the three control signal values, they would be sent back in the same fashion, in a vector form to the process running in Simulink, Matlab.

Figures 7.8 and 7.9 present the necessary subsystems from Simulink required to connect to the microcontroller. The sending subsystem has three input signals, corresponding to the three output signals of the multivariable process. Since the process is running in a Simulink environment, these outputs are represented in double precision and have to be converted to single precision prior to be sent to the microcontroller. Also, prior to sending the measured output values to the microcontroller, these outputs need to be sampled. This operation is performed in the Single blocks of Simulink, as shown in Fig. 7.8, by providing the sample time.

The multiplexer is used to combine all three measured output values into one vector. This is necessary since the serial communication used is a one-dimensional stream of bits. The vector is then sent by the Serial Send block, maintaining the

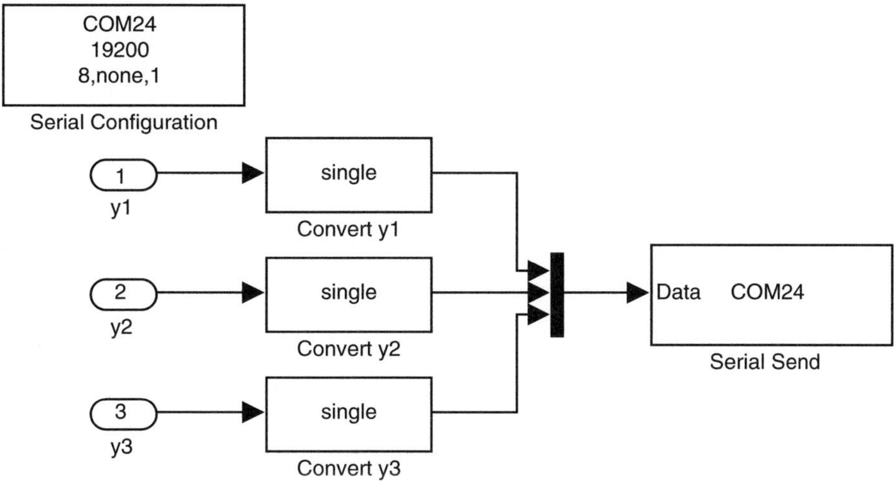

Fig. 7.8 The sending part of the communication subsystem

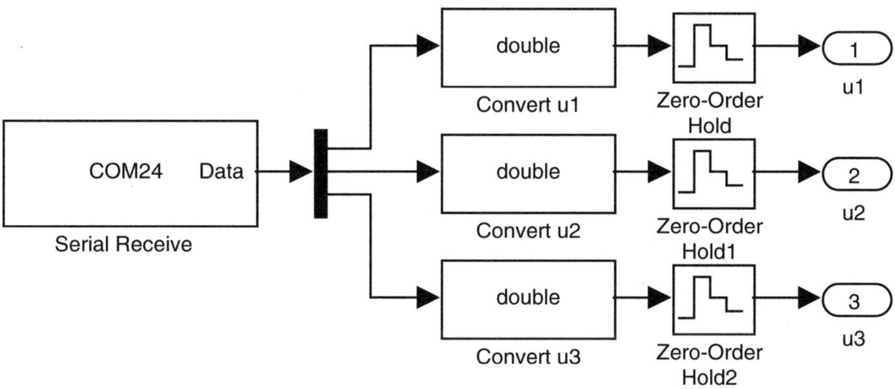

Fig. 7.9 The receiving part of the communication subsystem

order of the three outputs. Each value of the output signals is represented on 32 bits, split into 4 groups of 8 bits each, and sent one group at a time.

The microcontroller computes the three control signals and places them in a vector form which is sent to the Simulink program. The Serial Receive block in Simulink is configured to form a vector dimensional structure, with elements in single precision representation. The demultiplexer in the receiving subsystem performs the inverse operation of the multiplexer and separates the three control signals and converts them in double precision using the Double block of Simulink. Finally, each signal is passed to a zero-order-hold and then to the multivariable process.

```
float in(){
  union u_tag {
    byte b[4]; float fvalue;
  } in;
  in.fvalue=0;

  for (int i=0;i<4;i++)  {
    while (!Serial.available());
    in.b[i]=Serial.read();
  }

  return in.fvalue;
}
```

Fig. 7.10 The microcontroller input function

```
void out(float c){
  union u_tag {
    byte b[4]; float fvalue;
  } out;

  out.fvalue=c;

  Serial.write(out.b[0]);
  Serial.write(out.b[1]);
  Serial.write(out.b[2]);
  Serial.write(out.b[3]);

}
```

Fig. 7.11 The microcontroller output function

In programming the microcontroller, the input function is designed to read from the serial port 8 bits at a time and to save them in a union structure. Once all 32 bits are in place, they will be interpreted as a single precision value, as shown in Fig. 7.10.

The program reads the three measured output values, computes the control signals using the recurrent relations as resulting from the control algorithm, and sends them back, in the same order, corresponding to the order of the output signals.

The output function in the microcontroller receives a single precision floating-point value and sends all its 32 bits, 8 of them at a time, as shown in Fig. 7.11.

The experimental results are given in Figs. 7.12, 7.13, and 7.14 that present the three outputs evolution, considering a step change in the second output reference signal, while Figs. 7.15, 7.16, and 7.17 show the evolution of the three inputs.

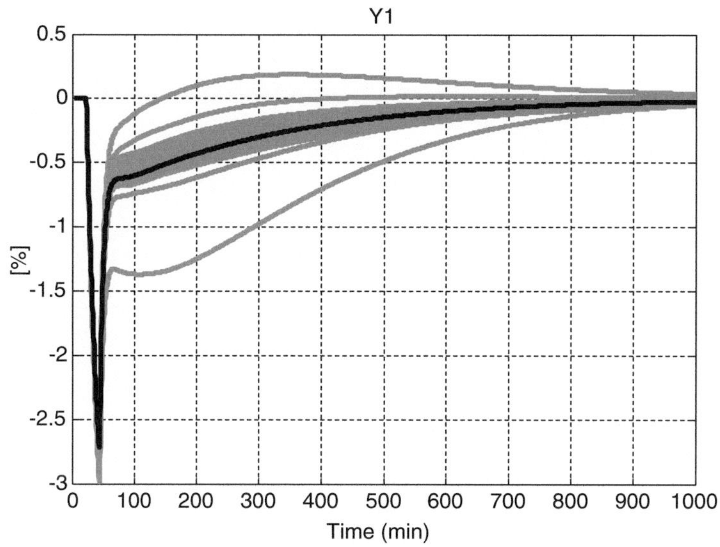

Fig. 7.12 Output y_1 evolution considering a step change in the y_2 reference and a microcontroller implementation of the multivariable FO-PI controller

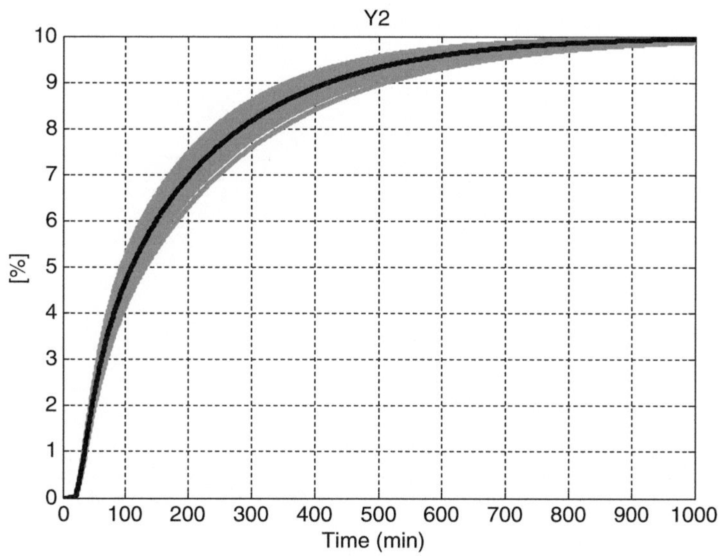

Fig. 7.13 Output y_2 evolution considering a step change in its reference and a microcontroller implementation of the multivariable FO-PI controller

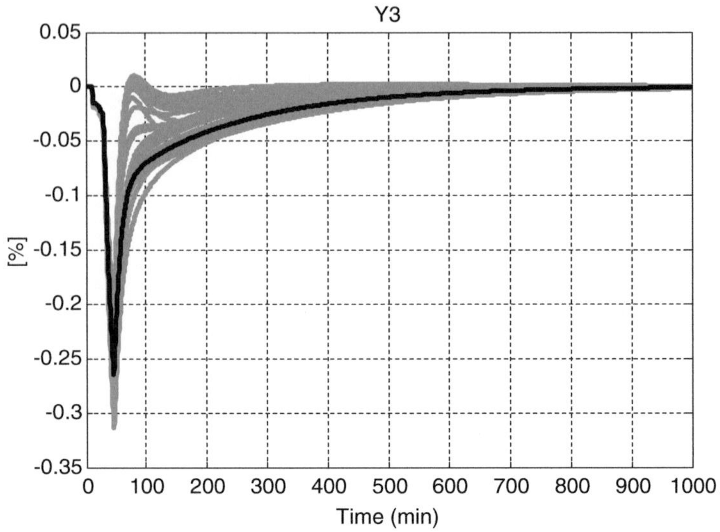

Fig. 7.14 Output y_3 evolution considering a step change in the y_2 reference and a microcontroller implementation of the multivariable FO-PI controller

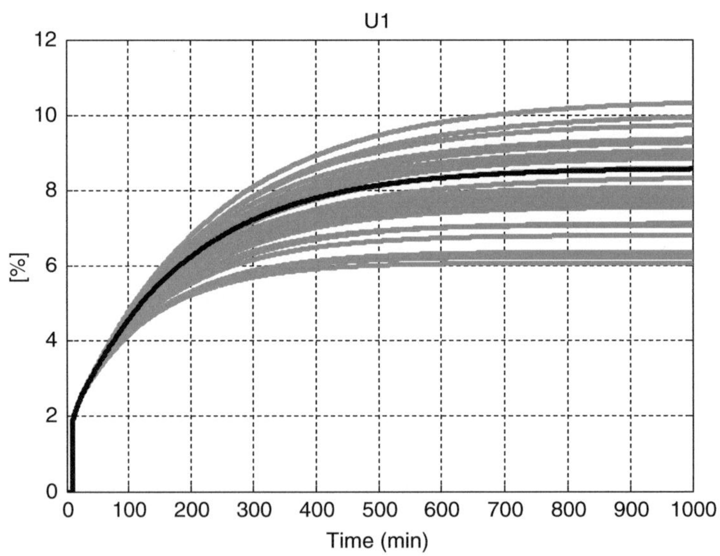

Fig. 7.15 Input u_1 evolution considering a step change in the y_2 reference and a microcontroller implementation of the multivariable FO-PI controller

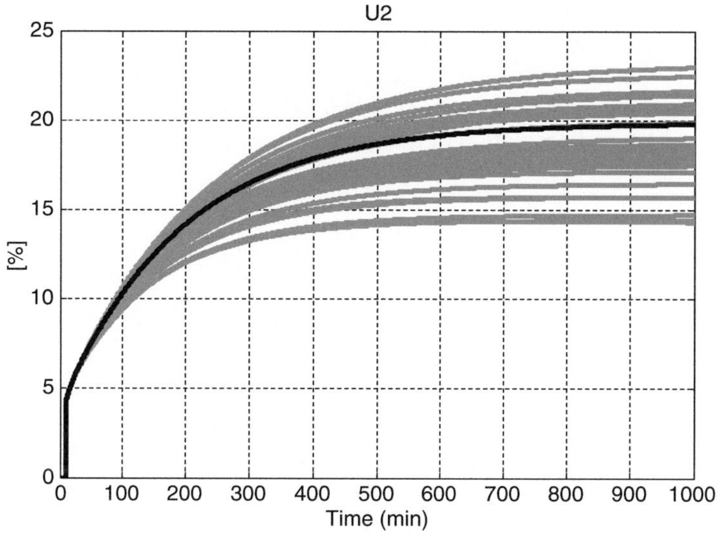

Fig. 7.16 Input u_2 evolution considering a step change in the y_2 reference and a microcontroller implementation of the multivariable FO-PI controller

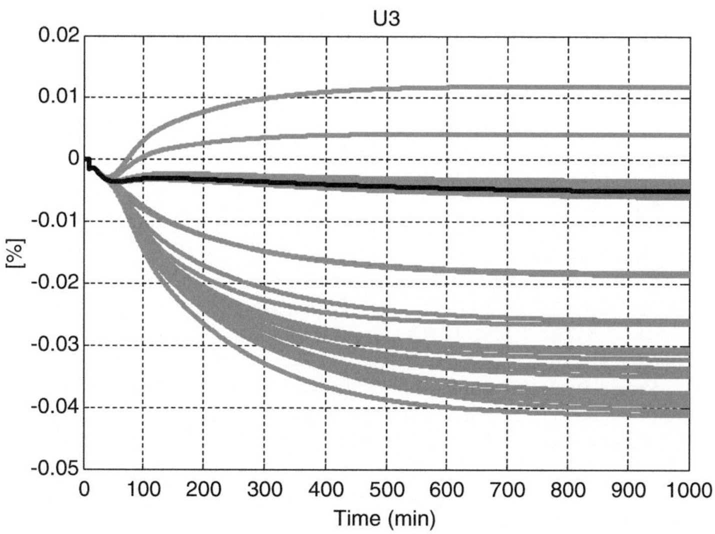

Fig. 7.17 Input u_3 evolution considering a step change in the y_2 reference and a microcontroller implementation of the multivariable FO-PI controller

The nominal case scenarios (black lines), considering $G_p(s) = G_m(s)$, as well as situations considering ±30% gain variations (grey lines) are presented in the figures. The results obtained clearly indicate that the robustness specification regarding gain uncertainties is attained, with zero overshoot as expected from the increased phase margin condition and a variation of the ±15 % in the settling time.

7.5 Conclusions

The main purpose of the chapter is to present a simple tuning rule for FO-PI controllers, both for the single-input–single-output and for the multivariable systems. The tuning algorithm is based on two performance specifications for the open loop system, combined with a robustness condition for gain variations. The case studies presented include a single-input–single-output process, as well as a multivariable one. The implementation steps for both controllers are also presented using two different equipments, an FPGA target and a microcontroller. The experimental results show that the fractional order control algorithm can be easily implemented and represents an adequate control solution for different types of processes.

Acknowledgment This work was supported by a grant of the Romanian National Authority for Scientific Research, CNDI-UEFISCDI, project number 155/2012 PN-II-PT-PCCA-2011-3.2-0591.

References

1. Barbosa RS, Machado JT (2006) Implementation of discrete-time fractional-order controllers based on LS approximations. Acta Polytechn Hung 3(4):5–22
2. Dulf EH, Both R, Dumitrache DC (2010) Fractional order models for a cryogenic separation column. In: International IEEE-TTTC international conference on automation, quality and testing, robotics AQTR 2010 (THETA 17), Cluj-Napoca, 28–30 May 2010, pp 163–169. doi:10.1109/AQTR.2010.5520895
3. Ionescu C, Segers P, De Keyser R (2009) Mechanical properties of the respiratory system derived from morphologic insight. IEEE Trans Biomed Eng 56:949–959
4. Ionescu C, De Keyser R (2009) Relations between fractional order model parameters and lung pathology in chronic obstructive pulmonary disease. IEEE Trans Biomed Eng 56:978–987
5. Mainardi F (2010) Fractional calculus and waves in linear viscoelasticity: an introduction to mathematical models. Imperial College Press, London
6. Zhu L, Knospe CR (2010) Modeling of nonlaminated electromagnetic suspension systems. In: Proceedings of the IEEE-ASME transactions on mechatronics, vol 15, pp 59–69
7. Monje CA, Vinagre BM, Chen YQ, Feliu V (2004) Proposals for fractional PID-tuning. In: Proceedings of the first IFAC symposium on fractional differentiation and its application (FDA04), Bordeaux
8. Oustaloup A, Sabatier J, Lanusse P (1999) From fractional robustness to CRONE control. Fract Calc Appl Anal 2:1–30

9. Oustaloup A (1991) La Commande CRONE: Commande Robuste d'Ordre Non Entiere. Hermes, Paris
10. Pop CI, Ionescu C, De Keyser R, Dulf EH (2012) Robustness evaluation of fractional order control for varying time delay processes. Signal Image Video Process 6(3):453–461. doi:10.1007/s11760-012-0322-4
11. Xue D, Zhao C, Chen YQ (2006) Fractional order PID control a DC-motor with elatic shaft: a case study. In: Proceedings of the 2006 American control conference, Minnesota, pp 3182–3187
12. Luo Y, Chen YQ, Wang CY, Pi YG (2010) Tuning fractional order proportional integral controllers for fractional order systems. J Process Control 20:823–831
13. Oustaloup A, Mathieu B, Lanusse P (1995) The CRONE control of resonant plants: application to a flexible transmission. Eur J Control 1:113–121
14. Podlubny I (1999) Fractional-order systems and $PI^{\lambda}D^{\mu}$-controllers. IEEE Trans Autom Contr 44:208–214
15. Xue D, Chen Y (2002) A comparative introduction of four fractional order controllers. In: Proceedings of the 4th IEEE world congress on intelligent control and automation, Shanghai, pp 3228–3235
16. Cao J-Y, Cao B-G (2006) Design of fractional order controller based on particle swarm optimization. Int J Control Autom Syst 4(6):775–781
17. Monje CA, Chen YQ, Vinagre BM, Xue D, Feliu V (2010) Fractional order systems and controls: fundamentals and applications. Springer, London
18. Chenikher S, Abdelmalek S, Sedraoui M (2012) Control of uncertainly multi-variable system with fractional PID. In: Proceedings of the 2012 16th IEEE Mediterranean electrotechnical conference (MELECON), Yasmine Hammamet, Tunisia, pp 1079–1082
19. Song X, Chen YQ, Tejado I, Vinagre BM (2011) Multivariable fractional order PID controller design via LMI approach. In: Proceedings of the 18th IFAC world congress, Milano
20. Chen J, He ZF, Qi X (2011) A new control method for MIMO first order time delay non-square systems. J Process Control 21:538–546
21. Pop CI, Ionescu CM, De Keyser R (2012) Time delay compensation for the secondary processes in a multivariable carbon isotope separation unit. Chem Eng Sci 80:205–218
22. Pop CI, De Keyser R, Ionescu CM (2011a) A simplified control method for multivariable stable nonsquare systems with multiple time delays. In: Proceedings of the 19th Mediterranean conference on control and automation, Corfu, 20–23 June 2011, pp 382–387. doi:10.1109/MED.2011.5983051
23. Vinagre B, Podlubny I, Hernandez A, Feliu V (2000) Some approximations of fractional order operators used in control theory and applications. Fract Calc Appl Anal 3(3):231–248
24. Chen YQ, Moore KL (2002) Discretization schemes for fractional-order differentiators and integrators. IEEE Trans Circuits Syst I Fundam Theory Appl 49:363–367
25. Vinagre BM, Chen YQ, Petras I (2003) Two direct Tustin discretization methods for fractional-order differentiator/integrator. J Franklin Inst 340:349–362
26. Hulea M, Mois GD, Folea S (2011) Dynamic Wi-Fi reconfigurable FPGA based platform for intelligent traffic systems. In: Folea S (ed) LabVIEW – practical applications and solutions. InTech, pp 377–396
27. Mois GD, Hulea M, Folea S, Miclea L (2011) Self-healing capabilities through wireless reconfiguration of FPGAs. In: Proceedings of the 9th east–west design & test symposium (EWDTS), pp 22–27
28. Lanusse P, Sabatier J (2011) PLC implementation of a CRONE controller. Fract Calc Appl Anal 14(4):505–522
29. Monje CA, Vinagre BM, Santamara GE, Tejado I (2009) Auto-tuning of fractional order PIλDμ controllers using a PLC. In: Proceedings of the 14th IEEE ETFA conference, Palma de Mallorca
30. Petras I, Dorcak L, Podlubny I, Terpak J, O'Leary P (2005) Implementation of fractional-order controllers on PLC B&R 2005. In: Proceedings of the ICCC2005, Miskolc, 24–27 May 2005, pp 141–144

31. Muresan CI, Folea S, Mois G, Dulf EH (2013) Development and implementation of an FPGA based fractional order controller for a dc motor. Elsevier J Mechatron. http://dx.doi.org/10.1016/j.mechatronics.2013.04.001
32. Pop CI, Dulf EH, De Keyser R, Ionescu CM, Muresan B, Festila Cl (2011) Predictive control of the multivariable time delay processes in an isotope separation column. In: Proceedings of the 6th IEEE international symposium on applied computational intelligence and informatics (SACI 2011), Timisoara, 19–21 May 2011, pp 29–34. doi:10.1109/SACI.2011.5872968

Chapter 8
Emerging Tools for Quantifying Unconscious Analgesia: Fractional-Order Impedance Models

Amélie Chevalier, Dana Copot, Clara M. Ionescu, J. A. Tenreiro Machado, and Robin De Keyser

Abstract This paper presents the application of model-based predictive control (MPC) in combination with a sensor for the measurement of analgesia (pain relief) in an unconscious patient in order to control the level of anesthesia. The MPC strategy uses fractional-order impedance models (FOIMs) to model the diffusion process that occurs in the human body when an analgesic drug is taken up. Based on this control strategy an early dawn concept of the pain sensor is developed. The grand challenges that coincide with this development include identification of the patient model, validation of the pain sensor, and validation of the effect of the analgesic drug.

Keywords Analgesia • Pain relief level • Non-invasive pain sensor • Model-based predictive control • Fractional-order impedance model

8.1 Introduction

The last few decades, modern medicine has successfully been influenced by advanced control technologies resulting in applications such as robotic surgery, electro-physiological system life support and image-guided therapy and surgery [1]. An interesting application of control in medicine is clinical pharmacology and in

A. Chevalier (✉) • D. Copot • C.M. Ionescu • R.D. Keyser
Faculty of Engineering and Architecture, Ghent University, Technologiepark 913, 9052 Gent-Zwijnaarde, Belgium
e-mail: Amelie.Chevalier@UGent.be; Dana.Copot@UGent.be; ClaraMihaela.Ionescu@UGent.be; Robain.Dekeyser@UGent.be

J.A.T. Machado
Department of Electrical Engineering, ISEP-Institute of Engineering, Polytechnic of Porto, Rua Dr. Antonio Bernardino de Almeida, 431, 4200-072 Porto, Portugal
e-mail: jtm@isep.ipp.pt

J.A.T. Machado et al. (eds.), *Discontinuity and Complexity in Nonlinear Physical Systems*, Nonlinear Systems and Complexity 6, DOI 10.1007/978-3-319-01411-1_8,

particular the control of general anesthesia during surgery and in the intensive care unit (ICU). Monitoring and controlling the depth of anesthesia for surgical patients poses interesting challenges to the control engineer [2] as it is a multi-variable interaction process that has captured the attention of engineers and clinicians already decades ago [3]. The first designs were expert systems that advised the anesthesiologist upon optimal drug infusion rates during clinical trials [4]. Control of anesthesia has a manifold of challenges, with multi-variable characteristics [5], different dynamics depending on anesthetic substances [6, 7], and stability problems [8].

General anesthesia, where the patient is completely unconscious, has the aim of ensuring sleep, amnesia, loss of pain, relaxation of skeletal muscles, and loss of control of reflexes of the autonomic nervous system. It consists of three components acting simultaneously on the patient's vital signs: hypnosis (ensuring sleep and amnesia), analgesia (ensures loss of pain), and neuromuscular blockade (relaxes the skeletal muscles and the motor reflexes). *Hypnosis* is relatively well characterized and is in standard clinical practice monitored by sensors based on electroencephalogram (EEG) data. *Neuromuscular blockade* immobilizes the patient during surgical procedures or intensive care and is also a relatively well-characterized process with standard sensors, such as motion sensors, available. By contrast, *analgesia* is far from being well characterized and no sensor is available for measuring the pain relief levels that a patient experiences during general anesthesia.

The advantage of automated closed loop control of anesthesia is that it gives a continuous drug delivery, contrary to intermittent control which is nowadays standard practice. A continuous drug delivery ensures that there is no under- or overdose of hypnotic or analgesic drugs that could result in patients that feel pain during surgery but are unable to move. Erroneous feedback information, biased either by the presence of artifacts (e.g., eye movement, leg movement, coughing, sneezing, choking, shivering) or by patient model mismatch, is one of the major problems for the control algorithms [9]. As a result the quality of the measured signals decreases, leading to the need of complex numerical filtering techniques. The latter require longer computation times, hence introducing artificial time delays which vary from one time instant to another, dependent on the signal quality [7]. If not dealt-with appropriately, such varying time-delays are a source of poor feedback control. Advanced control techniques such as model-based predictive control (MPC) can deal successfully with these variable time delays, nonlinearities, input and output constraints [10].

The research presented in this paper merges classical control theory with the young promising field of fractional-order modeling to measure pain relief levels in an unconscious patient and initiate the development of a biosensor for analgesia levels. Few pioneering attempts to measure the analgesic component of general anesthesia have shown that current state of the art is unable to deliver suitable signals and models for optimal regulation. The result is then a high risk of drug over- or under-dosing and unwanted postoperative effects, leading to increased hospitalization and health-care costs for both society and patient [11]. We propose to employ a mathematical tool called fractional-order impedance model (FOIM) to model the pharmacological diffusion process that takes place when the human body takes up an analgesic drug such as remifentanil. These models can be used in an MPC context to control the depth of analgesia in the unconscious patient.

The paper is structured as follows: in Sect. 8.2, we describe analgesia and the coinciding diffusion process. Section 8.3 discusses the control method that will be used in combination with the proposed analgesia sensor and the possible models used in this control. The grand challenges in the development of the sensor are discussed in Sect. 8.4 and the conclusions are summarized in a final section.

8.2 Analgesia as Integrated Part of General Anesthesia

General or complete anesthesia refers to inhibition of sensory, motor and sympathetic nerve transmission at the level of the brain, resulting in unconsciousness and lack of sensation. It consists of three components: hypnosis, analgesia, and neuromuscular blockade. Hypnosis is a general term indicating unconsciousness and absence of postoperative recall of events occurred during surgery. The level of hypnosis is related to the infusion of hypnotic drugs such as propofol and can be monitored by BIS monitoring. Analgesia is defined as an insensibility to pain without loss of consciousness. It is a state in which painful stimuli are not perceived or interpreted as pain and is usually induced by a drug, although trauma or disease may also result in a general or regional analgesia. Neuromuscular blockade is induced to prevent unwanted movement or muscle tone and causes paralysis during surgical procedures. The muscle relaxants are given intravenously (through the bloodstream) and act directly on the muscles.

Hence, analgesia is the amount of pain relief achieved during general anesthesia. The pain relief is obtained by administrating an analgesic drug such as remifentanil to the patient. The effectiveness of the analgesic drugs relies on how they are able to block the neural messages to the brain that are sent by the pain receptors. In the next sections we discuss this process of pain perception and how the analgesic drug is absorbed by the human body, i.e. the drug diffusion process.

8.2.1 Pain Perception

Pain perception is a complicated process where the pain signal is sent from the pain receptors found in the skin to the brain where the signal is interpreted as pain. The signal is transmitted via neurons and synapses, through the spinal cord and then to the brain.

Neurons are the basic cells in the central nervous system (CNS). Classical neurons consist of a cell body, dendrites, and axons (see Fig. 8.1). After the dendrites receive the information from the previous cell, the axon generates an action potential

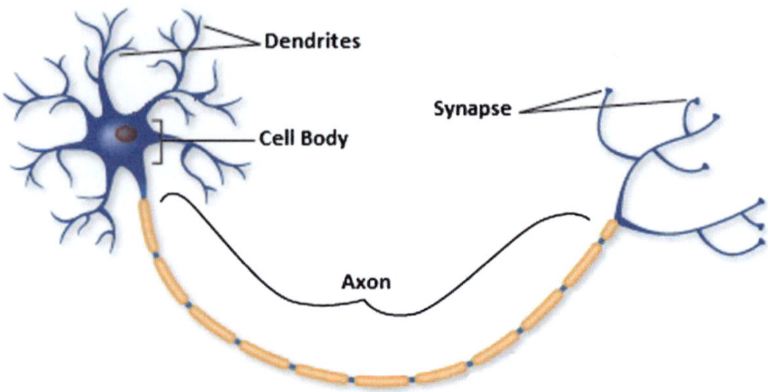

Fig. 8.1 Main parts of a classical neuron

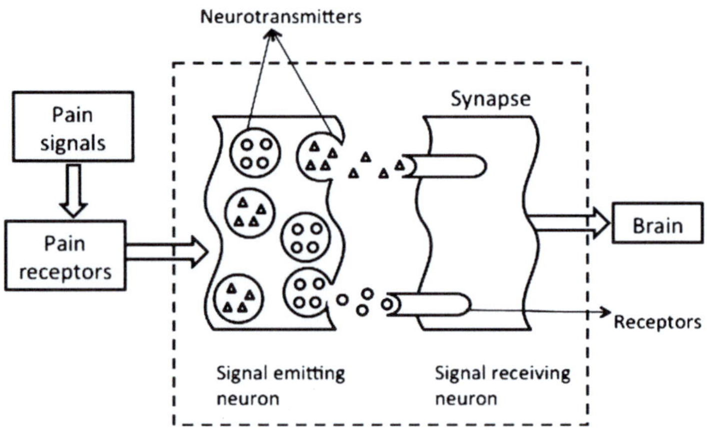

Fig. 8.2 Inner mechanisms of a synapse

(AP), which is an electrical signal, and sends this signal to the next cell via the presynaptic terminals. The presynaptic terminal forms the synapse together with the dendrites of the following cell that receive the information sent through the neuron.

A synapse is the place where two neurons communicate with each other. At this point in the communication we do not have an electrical signal anymore but a chemical signal. The chemicals used in this communication are called neurotransmitters. In Fig. 8.2 we can see a schematic of a synapse between two neurons. When the electric signal reaches the synapse at the side of the signal emitting neuron, it causes the release of chemical messengers, i.e. neurotransmitters from storage vesicles. The neurotransmitters travel across a minute gap between the cells and then interact with protein molecules, i.e. receptors located in the membrane surrounding

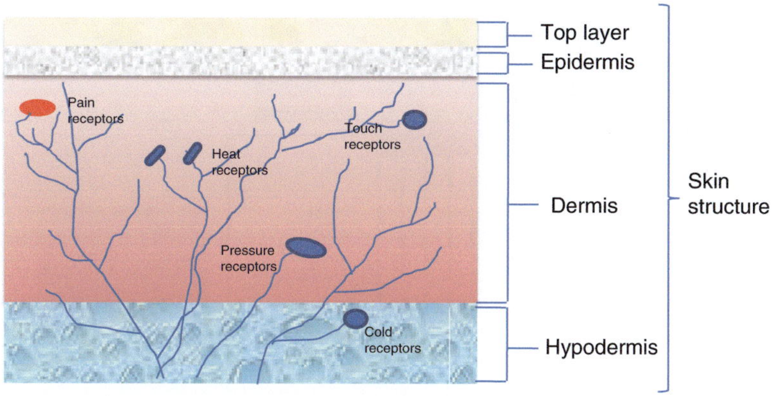

Fig. 8.3 Schematic overview of pain receptors in the skin

the signal receiving neuron. This interaction causes biochemical reactions that result in the generation of a new electrical signal, depending on the type of neuron, neurotransmitter, and receptor involved. Each receptor has a corresponding neurotransmitter. Receptors function much like gates that enable pain signals to pass through and onto the neighboring cells.

Pain receptors, also called nociceptors, are sensory neurons that are found in any area of the body that can sense pain either externally or internally. An external example are the nociceptors in the top layer of the skin (see Fig. 8.3). Internal nociceptors are present in a variety of organs, such as the muscle, joint, bladder, gut and continuing along the digestive tract. Nociceptors can be triggered by exceeding a high threshold that has been reached by either chemical, thermal, or mechanical environments.

Afferent neurons (such as nociceptors), which send information to the CNS, travel back to the spinal cord where they form synapses in its dorsal horn. From the dorsal horn the information is then sent to the thalamus which is located near the brain. The information is then processed in the ventral posterior nucleus and sent to the cerebral cortex, the headquarters for complex thoughts. This is where the signals are interpreted as pain. This entire process of pain transmission is called nociception.

There are many different neurotransmitters in the human body acting in various combinations to produce painful sensations in the body. Some chemicals govern mild pain sensations while others control intense or severe pain. When tissues become injured or inflamed, chemicals are released making nociceptors much more sensitive causing them to transmit pain signals in response to even gentle stimuli such as breeze or a caress. This condition is called allodynia; a state in which pain is produced by harmless stimuli. This can be a major cause for over-dosing of analgesics and should be well understood in order to be avoided.

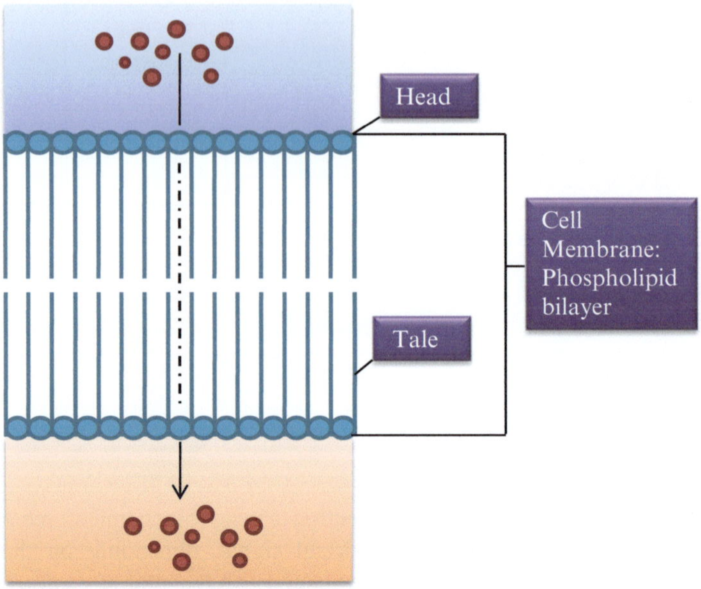

Fig. 8.4 Diffusion of analgesic drug through the phospholipid bilayer of the cell membrane

8.2.2 *Diffusion Process*

An analgesic drug interacts with the CNS to stop the communication between the nociceptors and the brain so that pain cannot be perceived anymore. However, this drug needs first to be taken up by the human body. This is achieved by a complex diffusion process across various cell membranes.

An important function of a biological cell membrane is to serve as a barrier to the outside world. However, membranes are not impenetrable walls. Nutrients must be able to enter the cell and waste products have to leave in order for the cell to survive. For this and many other reasons, it is crucial that membranes be selectively permeable. For example, the movement of ions across membranes is important in regulating vital cell characteristics such as cellular pH and osmotic pressure [12]. Membrane permeability is also a key determinant in the effectiveness of drug absorption, distribution, and elimination. For example, a drug taken orally that targets cells in the CNS must cross several membranes: first the barrier presented by the intestinal epithelium, then the walls of the capillaries that perfuse the gut, then the blood–brain barrier. Some endogenous substances and many drugs easily diffuse across the lipid bilayer. However, the lipid bilayer presents a formidable barrier to larger and more hydrophilic molecules (such as ions). These substances must be transported across the membrane by special protein channels.

Many drugs need to pass through one or more cell membranes (see Fig. 8.4) to reach their site of action. A common feature of all cell membranes is a phospholipid bilayer, about 10 nm thick. Spanning this bilayer or attached to the

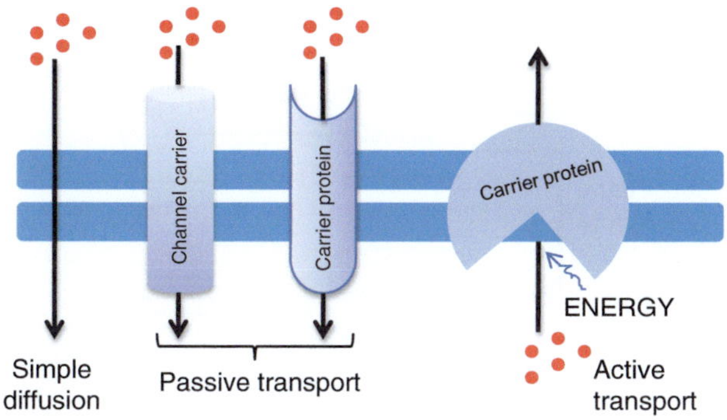

Fig. 8.5 Various mechanisms of diffusion of analgesic drugs across the cell membrane

outer or inner leaflets are glycoproteins, which may act as ion channels, receptors, intermediate messengers (G-proteins), or enzymes. Cells absorb molecules and ions from the extracellular fluid, creating a constant in- and outflow. The interesting thing about cell membranes is that relative concentrations and phospholipid bilayers prevent essential ions from entering the cell. Therefore, in order for drugs to move across the membrane these problems must be addressed. In general, this is completed by facilitated diffusion or active transport. In facilitated diffusion, relative concentrations are used to transport in and out. Active transports use energy, such as ATP (AdenosineTriPhosphate), to transfer molecules and ions in and out of the cell. Cellular signals cross the membrane through a process called signal transduction. This three-step process proceeds when a specific message encounters the outside surface of the cell and makes direct contact with a receptor. A receptor is a specialized molecule that takes information from the environment and passes it throughout various parts of the cell. Next, a connecting switch molecule, called a transducer, passes the message inwards, closer to the cell. Finally, the signal gets amplified, therefore causing the cell to perform a specific function. These functions can include moving, producing more proteins, or even sending out more signals [13].

Diffusion across the lipid bilayer is the spontaneous process where certain molecules can slip between the lipids in the bilayer and cross from one side to the other since the membranes are held together by weak forces (see Fig. 8.5). This process allows molecules that are small and lipophilic (lipid-soluble), including most drugs, to easily enter and exit cells. In order to be able to develop a sensor to measure the analgesic component of general anesthesia, we need to find a way to model this diffusion process.

8.3 Automated Regulation of Depth of Anesthesia

Nowadays, to optimally control the depth of the anesthesia, there is a need for a sensor that can measure the level of analgesia. The nonlinear response profile and inter- and intra-patient variation of the patient's analgesic state to infusion

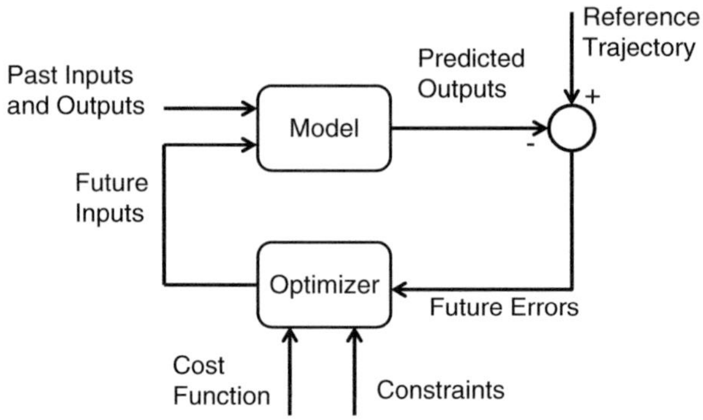

Fig. 8.6 An overview of the MPC strategy

of remifentanil should be handled by a robust controller. From a clinical point of view, an ideal controller would guide the induction of anesthesia in order to reach the target as fast as possible without initial overshoot. Afterwards, the controller would simply maintain the desired target as well as possible. Therefore, from control engineering viewpoint, MPC plays a crucial role in solving such complex problems.

The term model-based predictive control refers to a class of computer-based control strategies. They utilize in real-time an explicit process model to predict at each control interval the future response of the controlled system. The type of models which are currently used in real-life applications are either linear dynamical system models (step response models, transfer function models, linear state space models) or nonlinear dynamical system models. The roots for MPC are dating back to around 1980, when some pioneering institutions started to develop the main ideas and computer algorithms [14]. The MPC strategy can be visualized by the block-scheme in Fig. 8.6.

8.3.1 Non-invasive Sensor

General anesthesia (sedation) is difficult to assess in terms of adequacy because of its subjective nature. Several objective sedation scales such as the Ramsay Sedation Scale and the Sedation-Agitation Scale have been developed [15]. The Ramsay scoring system is one of the most commonly used scales. Even though it is simple, it cannot effectively measure the quality or amount of sedation and has never been objectively validated. Newer sedation scales are reported to show improvements in validity and reliability [16, 17]. Unfortunately, clinical sedation scores do not prevent under- or over-sedation and demand continuous bedside clinical scoring, a task performed by an alert clinical nurse sitting next to the patient.

In order to obtain a correct level of sedation in the patient, continuous monitoring of analgesia (pain relief) is of paramount importance. This demands stand-alone integrated monitoring tools for analgesia. While instrumental tools for the hypnotic component of anesthesia are standardly available and reliable (e.g., the BIS Monitor

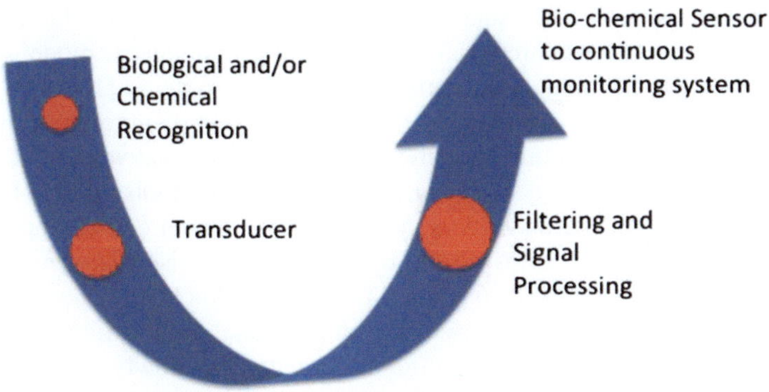

Fig. 8.7 Three main parts of the proposed biosensor

from Aspect Medical [18]), there is a serious lack of available tools to measure the analgesic component in an objective, reliable manner. Hitherto, there exists no sensor that evaluates objectively and continuously the pain relief levels during general anesthesia. The challenge originates from the fact that pain perception in the neural dynamics, and hence in the subsequent biological feedback, is not understood properly since models to characterize this complex biological process are not available.

To date there exists no integrated pain sensor and no information is available on how exactly these nociceptors can be detected. An early dawn conceptual picture of the sensor setup by means of several detection and processing steps is given in Fig. 8.7. The three main parts of a biosensor are presented: the role of biological recognition elements (receptors, enzymes, antibodies, etc.) is to differentiate the target molecules in the presence of various chemicals, the transducer (electrochemical, optical, magnetic, etc.) converts the bio-recognition event into a measurable signal, and the signal processing part converts the signal into a readable form.

8.3.2 A Prediction Model for MPC

MPC is a model-based control strategy. Standard models include step response models, transfer function models, and linear state space models; however, these models do not suffice in modeling the dynamics of the diffusion processes that occur in the human body.

In the past many attempts to model the diffusion process have been made. FOIMs have been shown to well characterize these diffusion processes [19–21], which in essence take place ubiquitously in our body. It is therefore natural to choose these tools in detecting, understanding, and characterizing the process of pain reception at the level of nociceptors.

In medicine, the field of fractional-order calculus has barely been explored. However, this research field promises to serve a whole range of applications with a large impact on the progress of science and welfare. The last decades have shown an increased interest in the research community to employ parametric model structures of fractional-order for analyzing nonlinear biological systems [19]. The concept of fractional-order (FO)—or non-integer order—systems refers to those dynamical systems whose model structure contains arbitrary order derivatives and/or integrals [22]. The fractional-order derivatives and integrals are tools of the Fractional Calculus theory [23]. The dynamical systems whose model can be approximated in a natural way using FO terms exhibit specific features: viscoelasticity, diffusion, and fractal structure. From previous work [24–26], we know now that the multiple scale adaptation of neurons is consistent with fractional order differentiation, such that the neuron's firing rate is a fractional derivative of slowly varying stimulus parameters [27]. The findings of scale-free fluctuations in the activity of neurons and synapses have been used to illustrate the existence of multiple time-scale dynamics in neurons and synapses [28]. Additionally, it has been shown that phase-locking phenomena can be explained by the presence of fractal electrical neuronal networks, which lead to a FOIM of the neural network [29]. However, the theoretical concepts of fractals, chaos, and multi-scale analysis have not yet been employed in the field of anesthesia, where the electrical activity of the brain is altered by the effects of hypnotic (propofol) and analgesic (remifentanil) drugs.

Another option to model the diffusion processes in the human body is to use compartmental models in combination with fractional-order derivatives [30]. Three compartments are used in this diffusion model: blood, muscle, and fat.

Principles of Fractional Calculus

The fractional calculus is a generalization of integration and derivation to non-integer (fractional) order operators. At first, we generalize the differential and integral operators into one fundamental operator D_t^n (n the order of the operation) which is known as *fractional calculus*.

Several definitions of this operator have been proposed. All of them generalize the standard differential–integral operator in two main groups: (a) they become the standard differential–integral operator of any order when n is an integer; (b) the Laplace transform of the operator D_t^n is s^n (provided zero initial conditions), and hence the frequency characteristic of this operator is $(j\omega)^n$. The latter is very appealing for the design of parametric modeling and control algorithms by using specifications in the frequency domain.

A fundamental D_t^n operator, a generalization of integral and differential operators (*differintegration* operator), is introduced as follows:

$$D_t^n = \left\{ \begin{array}{cc} \frac{d^n}{dt^n}, & n > 0 \\ 1, & n = 0 \\ \int_0^t (d\alpha)^{-n}, & n < 0 \end{array} \right\} \tag{8.1}$$

where n is the fractional order and $d\alpha$ is a derivative function. Since this paper will focus on the frequency-domain approach for FO derivatives and integrals, we shall not introduce the complex mathematics for time domain analysis. The Laplace transform for integral and derivative order n are, respectively:

$$L\{D_t^{-n} f(t)\} = s^{-n} F(s) \tag{8.2}$$

$$L\{D_t^{n} f(t)\} = s^{n} F(s) \tag{8.3}$$

where $F(s) = L\{f(t)\}$ and s is the Laplace complex variable. The Fourier transform can be obtained by replacing s by $j\omega$ in the Laplace transform and the equivalent frequency-domain expressions are:

$$\frac{1}{(j\omega)^n} = \frac{1}{\omega^n}\left(\cos\frac{n\pi}{2} - j\sin\frac{n\pi}{2}\right) \tag{8.4}$$

$$(j\omega)^n = \omega^n\left(\cos\frac{n\pi}{2} + j\sin\frac{n\pi}{2}\right) \tag{8.5}$$

Thus, the modulus and the argument of the FO terms are given by:

$$Modulus(\text{dB}) = 20\log\left|(j\omega)^{\mp n}\right| = \mp 20n\log|\omega| \tag{8.6}$$

$$Phase(\text{rad}) = \arg\left((j\omega)^{\mp n}\right) = \mp n\frac{\pi}{2} \tag{8.7}$$

resulting in a straight line with a slope of $\mp 20n$ passing through 0 dB for $\omega = 1$ for the magnitude (dB *vs.* log-frequency), respectively, a horizontal line, thus independent with frequency, with value $\mp n\frac{\pi}{2}$ for the phase (rad *vs.* log-frequency). The respective sketches are given in Fig. 8.8.

Principles of Compartmental Fractional Derivative Models

In this section a two-compartmental fractional derivative model is discussed. The basic idea behind this model can be used to model diffusion processes in the human body by a multi-compartmental model.

The model is formulated so that the mass balance is preserved. In Fig. 8.9, we see a conceptual schematic of a model with two compartments. Assume that $q_i(t) = v_i c_i$, for $i = 1, 2$ denote the amount of a drug in a specific compartment. Here c_i is the concentration of a drug and v_i is the volume of the i-th compartment and K_{ij} is the fractional rate of transfer to compartment i from compartment j.

The first compartment represents the place where the drug is applied, i.e. muscle, subcutaneous tissue, or digestive tract. The second compartment represents the

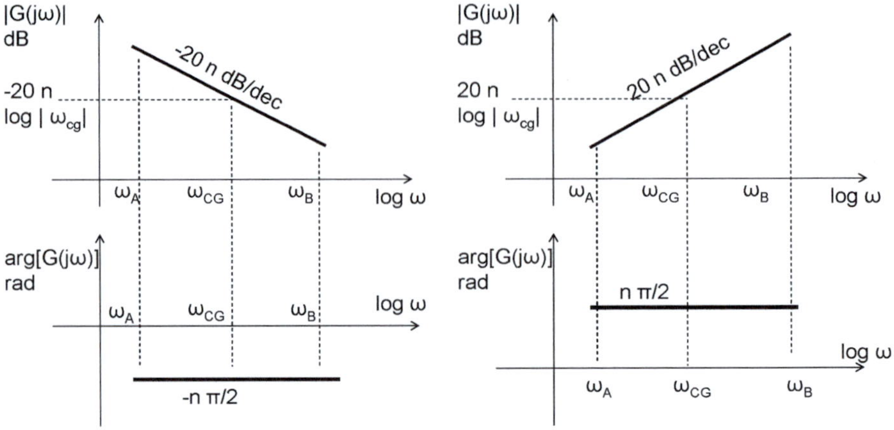

Fig. 8.8 Sketch representation of the FO integral and derivator operators in frequency domain, by means of the Bode plots (magnitude above and phase below)

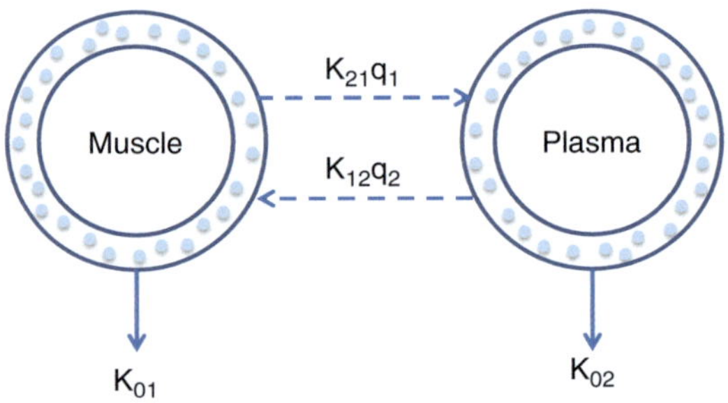

Fig. 8.9 Conceptual schematic of a two-compartment model

plasma or any other region in the body where the kinetics of the drug are uniform. Traditionally, the two compartments are described by a system of differential equations of integer order.

$$\frac{dq_1(t)}{dt} = K_{12}q_2(t) - K_{21}q_1(t) - K_{01}q_1(t) \tag{8.8}$$

$$\frac{dq_2(t)}{dt} = K_{21}q_1(t) - K_{12}q_2(t) - K_{02}q_2(t) \tag{8.9}$$

Recently, the fractional-order models seem to better suit the dynamics of biological systems than other integer models. A simple model of a two-compartmental system is then given by the following equations:

$$\tau_1^{n_1-1}\,{}_0D_t^{n_1}q_1(t) = -K_{21}q_1(t), \tag{8.10}$$

$$\tau_2^{n_2-1}\,{}_0D_t^{n_2}q_2(t) = K_{21}q_1(t) - K_{02}q_2(t), \tag{8.11}$$

where we assumed $K_{01} = 0$, $K_{12} = 0$ and with the initial conditions $q_1(0) = \text{dose}$, and $q_2(0) = 0$. In these equations τ_1 and τ_2 are time constants which represent the speed of diffusion, while n_1 and n_2 represent a non-integer between 0 and 1 and characterize the type of diffusion (sub-, super-, etc).

8.4 Grand Challenges

Several grand challenges are encountered in the development of the proposed analgesic biosensor. We discuss the major challenges in the remainder of this section.

To be able to control the level of analgesia in the unconscious patient, we apply an MPC strategy. This strategy needs a reliable model of the process that it needs to control. In traditional, non-human, systems, this model identification is done by sending excitation signals into the system and analyzing the corresponding output signals of the open-loop system. However, as we are dealing with patients, it is not possible to apply here the same strategy. Output signals to analyze are available only after the nurse administers a certain amount of analgesic drug. As the nurse only administers this drug after examination of the patient, this is no longer an open-loop system and system identification can be compromised by this.

Another difficulty in the model identification is the fact that every person reacts differently to a certain amount or combination of drugs. Therefore, every model differs for every patient i.e. inter-patient variability.

Moreover, the conditions inside the body of every patient are changing as a result of accumulated drug effect, i.e. intra-patient variability. The parameters of the patient model need to be updated regularly.

The pain sensor is supposed to measure a pain signal in an unconscious patient. However, there is no reliable way to validate the pain sensor once it is developed as the patient is the only one who can feel the pain but he/she is not able to indicate it anymore because he/she is unconscious.

Even if you can objectively prove that the pain sensor picks up a pain signal in one patient, it is not certain that the sensor will have the same result in a different patient as the pain threshold for one person can be completely different for another person. This is the result of the inter- and intra-patient variabilities that pose an extra challenge on the development of the pain sensor.

Another major challenge in this research direction is the fact that a combination of drugs is administered to the patient. Therefore, it is difficult to completely separate and validate the effect of the analgesic drug.

8.5 Conclusions

This paper proposes an early dawn pain sensor to measure analgesia level in unconscious patients. The proposed sensor can be used in combination with a MPC strategy to control the level of anesthesia in an unconscious patient. To model the diffusion process in the human body a FOIM is applied. The coinciding challenges in this research direction include identification of the patient model, validation of the pain sensor, and validation of the effect of the analgesic drug.

Acknowledgements Clara M. Ionescu acknowledges the Flanders Research Center (FWO) for its financial support.

References

1. Bailey JM, Haddad WM (2005) Drug dosing control in clinical pharmacology. IEEE Control Syst Mag N Y 25(2):35–51
2. Haddad WM, Hayakawa T, Bailey JM (2003) Nonlinear adaptive control for intensive care unit sedation and operating room hypnosis. Am Control Conf 2:1808–1813
3. O'Hara D, Bogen D, Noordergraaf A (1992) The use of computers for controlling the delivery of anesthesia. Anesthesiology 77:563–581
4. Greenhow SG, Linkens DA, Asbury AJ (1993) Pilot study of an expert system adviser for controlling general anesthesia. Br J Anaesth 71:359–365
5. Petersen-Felix S, Hacisalihzade S, Zbinden AM, Feigenwinter P (1995) Arterial pressure control with isoflurane using fuzzy logic. Br J Anaesth 74:66–72
6. Curatolo M, Derighetti M, Petersen-Felix S, Feigenwinter P, Fisher M, Zbinden AM (1996) Fuzzy logic control of inspired isoflurane and oxygen concentrations using minimal flow anesthesia. Br J Anaesth 76:245–250
7. Struys M, Vereecke H, Moerman A, Jensen EW, Verhaeghen D, De Neve N, Dumortier F, Mortier E (2003) Ability of the bispectral index, autoregressive modelling with exogenous input-derived auditory evoked potentials, and predicted propofol concentrations to measure patient responsiveness during anesthesia with propofol and remifentanil. Anesthesiology 99:802–814
8. Asbury AJ (1997) Feedback control in anesthesia. Int J Clin Monit Comput 14:1–10
9. Northrop RB (2000) Endogenous and exogenous regulation and control of physiological systems. CRC, Boca Raton
10. De Keyser R (2003) Model based predictive control for linear systems. UNESCO Encyclopaedia of Life Support Systems, Article contribution 6.43.16.1, EOLSS Publishers Co Ltd, Oxford, ISBN 0 9542 989 18-26-34, 30p
11. Kress J, Pohlman A, Hall J (2002) Sedation and analgesia in the intensive care unit. Am J Respir Crit Care Med 166:1024–1028
12. Steen-Knudsen O (2002) Biological membranes. Theory of transport, potentials and electric impulses. Cambridge University Press, Cambridge
13. Berg JM (ed) (2002) Biochemistry, 6th edn. W.H. Freeman and Company, New York
14. De Keyser R, Van Cauwenberghe A (1981) A self-tuning multistep predictor application. Automatica 17:167–174

15. Hemmerling TM, Salhab E, Aoun G, Charabati S, Mathieu P (2007) The Analgoscore: a novel score to monitor intraoperative pain and its use for Remifentanil closed loop application. In: Proceedings of the IEEE international conference on systems, man and cybernetics, pp 1494–1499
16. Riker RR, Picard JT, Fraser GL (1999) Prospective evaluation of the sedation-agitation scale for adult critically ill patients. Crit Care Med 27:1325–1329
17. Sessler CN, Gosnell M, Grap MJ, Brophy GT, O'Neal PV, Tesoro E, Elswick RK (2000) A new Agitation-Sedation Scale for critically ill patients: development and testing of validity and inter-rater reliability. Am J Respir Crit Care Med 161:A506
18. Glass PS, Bloom M, Kearse L, Rosow C, Sebel P, Manberg P (1997) Bispectral analysis measures sedation and memory effects of propofol, midazolam, isoflurane, and alfentanil in healthy volunteers. Anesthesiology 86:836–847
19. Losa GA, Merlini D, Nonnenmacher TF, Weibel ER (2005) Fractals in biology and medicine, vol IV. Birkhauser, Berlin
20. West BJ (1990) Fractal physiology and chaos in medicine, studies of nonlinear phenomena in life sciences, vol 1. World Scientific, Singapore
21. Benchellal A, Poinot T, Trigeassou JC (2005) Approximation and identification of diffusive interfaces by fractional models. Signal Process 86:2712–2727
22. Oustaloup A (1996) La derivation non entiere. Hermes, Paris (in French)
23. Podlubny I (1999) Fractional differential equations. Academic, San Diego
24. Ionescu CM, De Keyser R (2009) Relations between fractional order model parameters and lung pathology in chronic obstructive pulmonary disease. IEEE Trans Biomed Eng 56(4): 978–987
25. Ionescu CM, Machado JT, De Keyser R (2011) Modeling of the lung impedance using a fractional order ladder network with constant phase elements. IEEE Trans Biomed Circuits Syst 5(1):83–89
26. Ionescu CM, Hodrea R, De Keyser R (2011) Variable time-delay estimation for anesthesia control during intensive care. IEEE Trans Biomed Eng 58(2):363–369
27. Lundstrom B, Higgs M, Spain W, Fairhall A (2008) Fractional differentiation by neocortical pyramidal neurons. Nat Neurosci 11:1135–1342
28. Drew P, Abbot L (2003) Scale-invariant synaptic dynamics in a computational model of recognition memory. Soc Neurosci Abstr 28:89–99
29. Ionescu CM (2012) Phase constancy in a ladder model of neural dynamics. IEEE Trans Syst Man Cybern A Syst Hum 42(6):1543–1551
30. Dokoumetzidis A, Magin R, Macheras P (2010) A commentary on fractionalization of multi-compartmental models. J Pharmacokinet Pharmacodyn 37(2):203–207

Part II
Chaos and Complexity

Chapter 9
1D Cahn–Hilliard Dynamics: Coarsening and Interrupted Coarsening

Simon Villain-Guillot

Abstract Many systems exhibit a phase where the order parameter is spatially modulated. These patterns can be the result of a frustration caused by the competition between interaction forces with opposite effects.

In all models with local interactions, these ordered phases disappear in the strong segregation regime (low temperature). It is expected, however, that these phases should persist in the case of long-range interactions, which can't be correctly described by a Ginzburg–Landau type model with only a finite number of spatial derivatives of the order parameter.

An alternative approach is to study the dynamics of the phase transition or pattern formation. While, in the usual process of Ostwald ripening, succession of doubling of the domain size leads to a total segregation, or macro-segregation, Misbah and Politi have shown that long-range interactions could cause an interruption of this coalescence process, stabilizing a pattern which then remains in a micro-structured state or super-crystal. We show that this is the case for a modified Cahn–Hilliard dynamics due to Oono which includes a nonlocal term and which is particularly well suited to describe systems with a modulated phase.

Keywords Dynamics of phase transition • Spinodal decomposition • Cahn Hillard equation • Ostwald ripening • Coarsening (interrupted coarsening) • Copolymers • Instability • Pattern formation • Modulated phase • Soliton lattice • Segregation • Interfacial dynamics • Ginzburg–Landau free energy

---

S. Villain-Guillot (✉)
Laboratoire Onde et Matière d'Aquitaine, Université Bordeaux I351, cours de la Libération 33405 Talence Cedex, France
e-mail: simon.villain-guillot@u-bordeaux1.fr

J.A.T. Machado et al. (eds.), *Discontinuity and Complexity in Nonlinear Physical Systems*, Nonlinear Systems and Complexity 6, DOI 10.1007/978-3-319-01411-1_9,

9.1 Introduction

Many systems exhibit phases where the order parameter is spatially modulated and form a pattern [1]. These phases are the result of a frustration caused by the competition between interaction forces with opposite effects.

For example, in a blend of polymers, the difference of interaction energies between homo and heteropolymers generates locally a repulsion between heteropolymers which leads to a macroscopic segregation. But for diblock co-polymers which are built with two heteropolymers A and B which are attached to each other by a chemical bond, such a macroscopic global phase separation is prohibited. They form a disordered phase at high temperature (when the entropic effects prevail), but below a critical temperature, whereas energetic considerations should lead to segregation, this chemical binding prevents separation between A and B heteropolymers over a long distance: the two components A and B self-organized in patterns or domains of finite size (mainly lamellar or hexagonal) in order to minimize nevertheless contacts between heteropolymers en thus the energy of interaction. The relative density in heteropolymers is thus spatially periodically modulated. This spontaneous microstructuration could be helpful to design a new generation of solar cells based on organic semi-conductors [2].

In all models with local interactions, these ordered phases disappear in the strong segregation regime (low temperature). It is expected, however, that these phases should persist in the case of long-range interactions, which can't be correctly described by a Ginzburg–Landau type model with only a finite number of spatial derivatives of an order parameter (which can be defined in our preceding example from the relative density in the two components A and B).

An alternative approach is to study the dynamics of phase transition. While, in the usual process of Ostwald ripening, succession of coarsening events with doubling of the domain size leads to a total segregation, or macro-segregation, Misbah and Politi [3] have shown that long-range interactions could cause an interruption of this coalescence process, stabilizing a pattern that remains consequently in a microstructured pattern or super-crystal.

We show here that this is the case for the equation of Oono [4, 5], which is particularly well suited to describe the dynamics of systems with a modulated phase.

9.2 Dynamics of Phase Transitions

9.2.1 Time-Dependent Ginzburg Landau Equation

9.2.1.1 Derivation of the Model

Different equations can be used to describe the dynamics of a phase transition depending on, for example, if the order parameter is a scalar or a vector, and whether it is conserved by the dynamics or not (for a review see [6, 7]).

As at equilibrium, this order parameter must minimize a free energy, the dynamics out of equilibrium must then involve deviation from this stable order parameter value or function, just like in a simple mechanical system. The simplest dynamics based on Ginzburg–Landau free energy for a scalar order parameter is the TDGL (Time-Dependent Ginzburg Landau or model A in Hohenberg and Halperin classification [6]), which writes

$$\frac{\partial u}{\partial t}(\mathbf{r},t) = -\frac{\delta F_{GL}}{\delta u} = \nabla^2 u - \frac{\varepsilon}{2}u - 2u^3 \tag{9.1}$$

In this equation, $\mathbf{u}(\mathrm{r}, t)$ is a macroscopic order parameter which is a coarse grained of a microscopic order parameter in a small volume around the position $\mathbf{r}$. And ε is the dimensionless control parameter, usually the reduce temperature $\varepsilon = \frac{T-T_c}{T_c}$ where T_c is the critical temperature of the phase transition. This partial differential equation is invariant by the transformations $u \to -u$ and $x_i \to -x_i + a_i$. F_{GL} is the Ginzburg–Landau free energy local density or Lyapunov functional in the context of dynamical systems:

$$F_{GL} = \frac{1}{2}\left((\nabla u)^2 + \frac{\varepsilon}{2}u^2 + u^4\right)$$

The non-local term $(\nabla u)^2$ prevents discontinuity or roughness of the order parameter and assigns energetic overcost to its variations in proportion to their sharpness. When looking at the temporal evolution of the free energy $\int F_{GL}(r,t)dr$:

$$\frac{d}{dt}\int F_{GL}dr = \int \frac{\delta F_{GL}}{\delta u}.\frac{\partial u}{\partial t}dr = \int \frac{\delta F_{GL}}{\delta u}.(-\frac{\delta F_{GL}}{\delta u})dr = -\int (\frac{\delta F_{GL}}{\delta u})^2 dr < 0$$

One notices from (9.1) that the dynamics will induce a change of $u(\mathbf{r})$ as long as it hasn't reached a minimum of the free energy density F_{GL}. If one looks for homogeneous states (where the order parameter is independent of the spatial coordinates) to be stationary states of this equation, they will be the extrema of the Landau potential $V(u) = \frac{\varepsilon}{2}u^2 + u^4$ which is plotted in Fig. 9.1 for two different signs of the control parameter. For $\varepsilon > 0$, the only extremum is $u = 0$, so there is only one homogenous solution, which is stable, being a minimum of the Landau potential (which is a convex function as long as $\varepsilon > 0$). When $\varepsilon < 0$, this potential is now concave in a neighborhood of $u = 0$, which is now a maximum and thus is now linearly instable. Two other symmetric solutions $u = \pm\frac{\sqrt{-\varepsilon}}{2}$ have now appeared due to this pitchfork bifurcation. They are the new stable homogeneous solutions and correspond to a minimum of the potential $V_{\min} = -\varepsilon^2/32$.

9.2.1.2 Linear Stability Analysis

Linear stability analysis consists in computing the growth rate of small fluctuations of a solution. When linearizing equation (9.1) around $u = 0$ (i.e., when neglecting the nonlinear term u^3) one gets

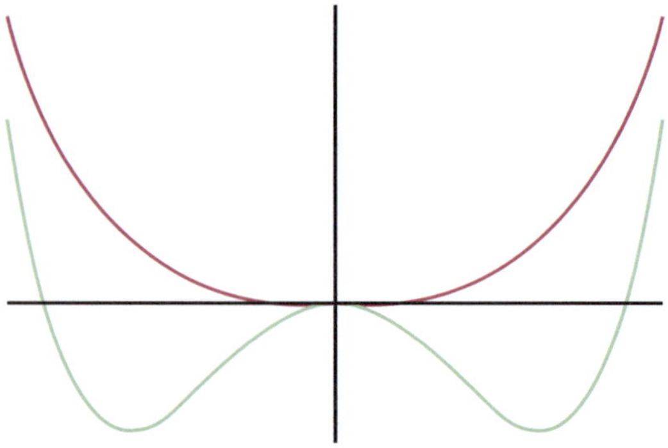

Fig. 9.1 Landau potential as a function of u, the amplitude of the order parameter. We have plotted the profil of this potential above and below the pitchfork bifurcation at $\varepsilon = 0$. For $\varepsilon > 0$, the potential is a convex function and there is only one minimum, $u = 0$. For $\varepsilon < 0$, the Landau potential is a concave function around $u = 0$, which is now a maximum; two other solutions have now appeared as minimum of the potential, symmetric one each other

$$\frac{\partial u}{\partial t}(\mathbf{r}, t) = -\frac{\varepsilon}{2}u + \nabla^2 u$$

Considering this equation in the Fourier space we can decompose u in Fourier series in the case of a finite size problem or Fourier transform in the infinite case:

$$u(\mathbf{r}, t) = \sum_{\mathbf{q}} u_{\mathbf{q}} e^{i\mathbf{q}\cdot\mathbf{r}+\sigma t} \tag{9.2}$$

where u_q is the amplitude of the Fourier mode at $t = 0$. For example, it can be the thermal fluctuations proportional to T. This mode decomposition enables to compute the q-dependence of the amplification factor $\sigma(q)$ (or growth rate or imaginary part of $k = q - i\sigma$):

$$\sigma(\mathbf{q}) = -(q^2 + \frac{\varepsilon}{2}) \tag{9.3}$$

$\sigma(\mathbf{q})$ is negative for $\varepsilon > 0$, and thus the homogeneous solution $u = 0$ is unstable with respect to fluctuations of the order parameter. The whole band $0 < q < \sqrt{(-\varepsilon/2)}$ is linearly unstable as $(\sigma(\mathbf{q}) > 0)$ (see Fig. 9.2).

9.2.1.3 Symmetry Breaking and Conservation Law

The linear stability analysis enables to conclude that the most instable mode is for $q = 0$: it is thus a long wave instability, which will give rise to large homogeneous domains and imply spontaneous symmetry breaking. This is the case, for example, in magnetic systems.

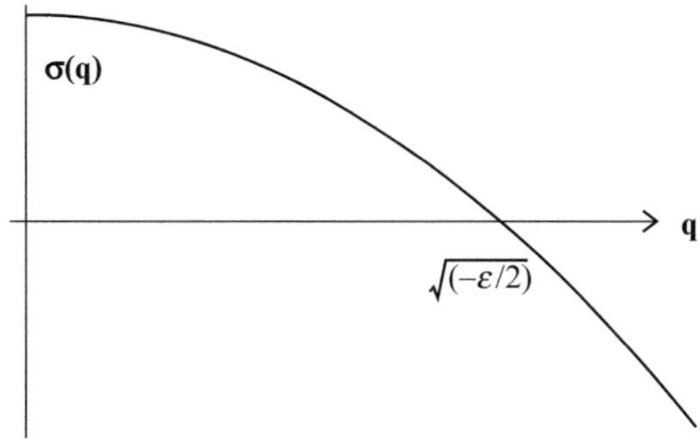

Fig. 9.2 Amplification factor $\sigma(q)$ computed via linear stability analysis of the time-dependent Ginzburg–Landau equation (TDGL). It is positive (growth of the modulations) for all the modes $q < \sqrt{\frac{-\varepsilon}{2}}$

But if there is a conservation law, as for example a conservation of mass, such an instantaneous symmetry breaking is prohibited: the matter, or the different species diffuse with a finite characteristic time. Hillert [8], Cahn and Hilliard [9] have proposed a model to describe segregation in a binary mixture. This equation, later on denoted C–H for Cahn–Hilliard, corresponds to model B in the Hohenberg and Halperin classification [6]. Cahn–Hilliard dynamics is the minimal equation describing phase transition for a conserved scalar order parameter. As this conservation law prevents global symmetry breaking, it will generate numerous domains and interfaces separating them. This dynamic governs a whole class of first order phase transition like the Fréedericksz transition in liquid crystals [11], segregation of granular media in a rotating drum[12, 13], or formation of ripple due to hydrodynamic oscillations [14, 15].

9.2.2 *Model B or Cahn–Hilliard Equation*

9.2.2.1 Derivation of the Model

Cahn–Hilliard dynamics is a modified diffusion equation for a scalar order parameter u, which writes:

$$\frac{\partial u}{\partial t}(\mathbf{r}, t) = \nabla^2(\frac{\varepsilon}{2}u + 2u^3 - \nabla^2 u) = \nabla^2(\frac{\delta F}{\delta u}) \tag{9.4}$$

In the original work of Cahn and Hilliard, $u\,(\mathbf{r}, t)$ represents the concentration of one of the components of a binary alloy. But it can also be the fluctuation of density of a fluid around its mean value, or concentration of one chemical component of a binary mixture, or the height of a copolymer layer [16].

As in model A, this equation is invariant by the transformations $u \to -u$ and $x_i \to -x_i + a_i$ and when looking at the time evolution of the local quantity $F(t)$, we still have:

$$\frac{dF}{dt} = \frac{\delta F}{\delta \Phi u}.\frac{\partial u}{\partial t} = \frac{\delta F}{\delta u}.\nabla^2(\frac{\delta F}{\delta u}) = -(\nabla\frac{\delta F}{\delta u})^2 < 0$$

In order to derive a conservative dynamics, such that $\int \Phi(x,t)dx = cste$, one can start from a detail balance [17], or from a conservation equation for the order parameter Φ.

$$\frac{\partial u}{\partial t} = -\nabla \cdot \mathbf{j}$$

where $\mathbf{j}$ is a matter current associated with u. This current is related to the gradient of the chemical potential μ via the Hartley–Fick law : $\mathbf{j} = -\nabla\mu$). And this chemical potential is itself related to the functional derivative of the free energy $\mu = \frac{\delta F}{\delta \Phi}$. This phenomenological approach enables to recover the C–H equation (9.4).

If one looks globally at the quantity $\int u(x,t)dx =< u >$, the Cahn–Hilliard gives

$$\frac{d < u >}{dt} = \int \frac{\partial u}{\partial t}(x,t)dx = \int \nabla^2(\frac{\delta F}{\delta u}(x,t))dx = \left[-(\nabla \frac{\delta F}{\delta u})\right]$$

So, apart from boundary terms, the order parameter is indeed a conserved quantity.

9.2.2.2 Linear Stability Analysis

Stationary states of the (C–H) are again the extrema of the Landau potential $V(u) = \frac{\varepsilon}{2}u^2 + u^4$. And after a quench, the system undergoes a first order phase transition associated with the pitchfork bifurcation from the $u = 0$ solution to the symmetric solutions $u = \pm\frac{\sqrt{-\varepsilon}}{2}$. But due to the conservation law, the dynamics is different as Cahn and Hilliard have shown via the linear stability analysis of (9.4) around $u = 0$.

$$\frac{\partial u}{\partial t}(\mathbf{r}, t) = \nabla^2 \frac{\varepsilon}{2} u - \nabla^4 u \tag{9.5}$$

one gets for the amplification factor in the Fourier space $\sigma(q)$:

$$\sigma(\mathbf{q}) = -(q^2 + \frac{\varepsilon}{2})q^2 \tag{9.6}$$

So, as $\sigma(\mathbf{q})$ is negative for $\varepsilon > 0$, the $u = 0$ solution is stable with respect to small fluctuations of the order parameter. For negative ε, Fig. 9.3 shows a band of instable Fourier modes, as $\sigma(\mathbf{q}) > 0$ for $0 < q < \sqrt{(-\varepsilon/2)}$. Moreover, linear stability analysis of C–H predicts that the most instable mode is not anymore for $q = 0$ but for $q_{C-H} = \sqrt{-\varepsilon}/2$ (for which $\sigma_{\max} = \frac{\varepsilon^2}{16}$). This wave number of maximum amplification factor will dominate the first stage of the dynamics which is called the spinodal decomposition; this explains in particular why the homogeneous domains appear at length scales close to $L = \lambda_{C-H}/2 = \pi/q_{C-H}$, half the wave length

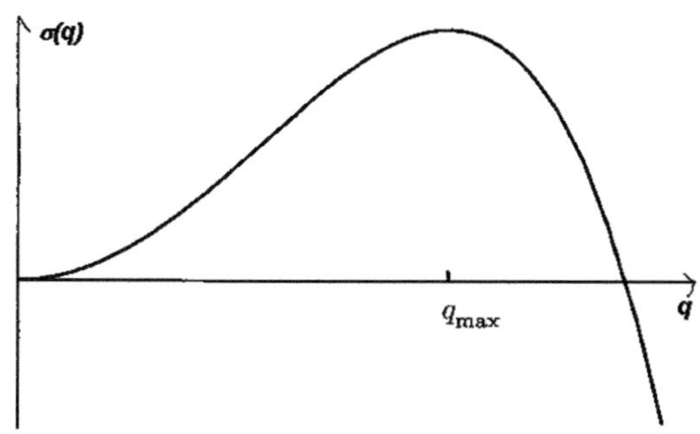

Fig. 9.3 Amplification factor $\sigma(q)$ computed from the linear stability analysis of Cahn and Hilliard equation

associated with the instability. For longer times, interfaces separating each domain interact through Ostwald ripening or coarsening, causing $< L >$ to change slowly toward higher values.

9.3 Cahn–Hilliard Equation

9.3.1 On the Periodic Solutions of Cahn–Hilliard Equation

When the equation is studied for a constant negative ε, via a rescaling of u (as $\sqrt{-\varepsilon u}$), position $\mathbf{r}$ (as $\mathbf{r}/\sqrt{-\varepsilon}$) and time (as $t/|\varepsilon|^2$), we observe that we could restrict the dynamics to the case $\varepsilon = -1$. So later on, we will study the equation

$$\frac{\partial u}{\partial t}(\mathbf{r}, t) = \nabla^2(-\frac{1}{2}u + 2u^3 - \nabla^2 u) \tag{9.7}$$

In 1D, a family of stationary solution of this nonlinear dynamics is the so-called interface-lattice solutions (or soliton-lattice), which writes:

$$U_{k,\varepsilon}(x) = k\Delta \mathrm{Sn}(\frac{x}{\xi}, k)\mathrm{with}\xi = \Delta^{-1} = \sqrt{2\,(k^2+1)} \tag{9.8}$$

where $\mathrm{Sn}(x, k)$ is the Jacobian elliptic function sine-amplitude, or cnoidal mode. This family of solutions is parametrized by the Jacobian modulus $k \in [0, 1]$, or "segregation parameter." These solutions describe periodic patterns of period

$$\lambda = 4K(k)\xi, \mathrm{where} K(k) = \int_0^{\frac{\pi}{2}} \frac{\mathrm{d}t}{\sqrt{1 - k^2 \sin^2 t}} \tag{9.9}$$

is the complete Jacobian elliptic integral of the first kind. $K(k)$ together with k, characterize the segregation, defined as the ratio between the size of the homogeneous domains, $L = \lambda/2$, and the width of the interface separating them,

2ξ. Equation (9.9) and the relation $\xi = \Delta^{-1}$ enable to rewrite this family as:

$$U_{k,\lambda}(x) = \frac{4K(k)\cdot k}{\lambda}\mathrm{Sn}(\frac{4K(k)}{\lambda}x, k). \tag{9.10}$$

and using equations (9.8) and (9.9), we find that for a stationary solution, λ, and k have to be related one another through the following implicit equation (or the state equation):

$$\lambda^2 = 2(1+k^2)\,(4K(k))^2\,. \tag{9.11}$$

Using (9.10) we can compute the free energy per unit length

$$\mathrm{F}_{GL}(k,\lambda) = \left(\frac{4K}{\lambda}\right)^2\left[\frac{-\varepsilon}{4}\left(1-\frac{E}{K}\right)+\left(\frac{1+2k^2}{6}-\frac{E}{6K}(1+k^2)\right)\left(\frac{4K}{\lambda}\right)^2\right]$$

where $E(k)$is the complete Jacobian elliptic integral of the second kind. The absolute minimum for $\mathrm{F}_{GL}(k,\lambda)$ is for $k = 1$ and $\lambda = \infty$, i.e. for complete segregation with a single interface.

9.3.2 Stationary States of the Cahn–Hilliard Dynamics

The dynamics starts initially with $k = 0$, for which $U(x)$ describes a sinusoidal modulation of almost vanishing amplitude around the high temperature homogenous stationary solution $u = 0$

$$\begin{aligned} U_{k\to 0,\varepsilon}(x) &= k\sqrt{\frac{1}{2}}\sin\left(\sqrt{\frac{1}{2}}x\right) \\ &= k\frac{2\pi}{\lambda_{C-H}}\sin\left(\frac{2\pi}{\lambda_{C-H}}x\right) = kq_{C-H}\sin(qx) \end{aligned} \tag{9.12}$$

The spinodal decomposition dynamics will saturate and reach a stationary state which is a periodic pattern with a finite domain length (weak segregation regime) for which $\lambda = \lambda_{C-H}$, and $k = k_0^s$ =0.687 so as to satisfy (9.11), i.e k is solution of the implicit equation :

$$2(1+k_0^{s2})K(k_0^s)^2 = -\frac{\varepsilon_0\lambda_{C-H}^2}{16} = \pi^2. \tag{9.13}$$

The amplitude of the modulation is then $k_0^s\Delta_0^s = 0.400\sqrt{-\varepsilon_0}$, which is different from u_b.

Using linear stability analysis, Langer has shown that the stationary profile thus obtained, $u_0(x) = U_{k_0^s, \lambda_{C-H}}(x)$, is destroyed by stochastic thermal fluctuations [17]. He has identified the most instable mode as an "antiferro" mode, leading to an infinite cascade of period doubling [18]. Disorder of the pattern is also a cause of Ostwald ripening: if the periodicity of the interface-lattice is broken, either when the distance between theses interfaces or when the bulk value in the different domains become non-constant, coarsening is triggered by diffusion of matter between neighboring domains: big domains will then absorb smaller ones [19, 20].

9.3.3 Coarsening

When considering the C–H equation 9.4 as a diffusion equation, Politi and Misbah have shown that there should be coarsening as long as $\mathrm{d}\nu/\mathrm{d}\lambda$ is positive, where ν is the amplitude of the modulation and λ its [3]. As in Cahn–Hilliard dynamics

$$\nu = k\Delta = k\sqrt{\frac{-\varepsilon}{2(k^2+1)}} \quad \text{and} \quad \lambda = 4K(k)\xi = 4K(k)\sqrt{\frac{2(k^2+1)}{-\varepsilon}}$$

are two growing functions of the parameter k, this diffusion coefficient will always remain positive and coarsening will proceed until $\lambda \to \infty$ (as in Fig. 9.4 Left).

When looking at Fig. 9.1, one can see that the bulk energy is decreasing when the amplitude varies from $\nu = 0$ to $\nu = \pm\frac{\sqrt{-\varepsilon}}{2}$, that is, when the segregation increase. As the interfacial energy is proportional, the interfacial energy is proportional to the period, we finally get that the total energy decreases when the period of the stationary solutions gets longer and longer. But for other dynamics (as in Fig. 9.4 Right), $\mathrm{d}\nu/\mathrm{d}\lambda$ can change of sign as we will see in the following: segregation then remains partial. Politi and Misbah speak then of interrupted coarsening.

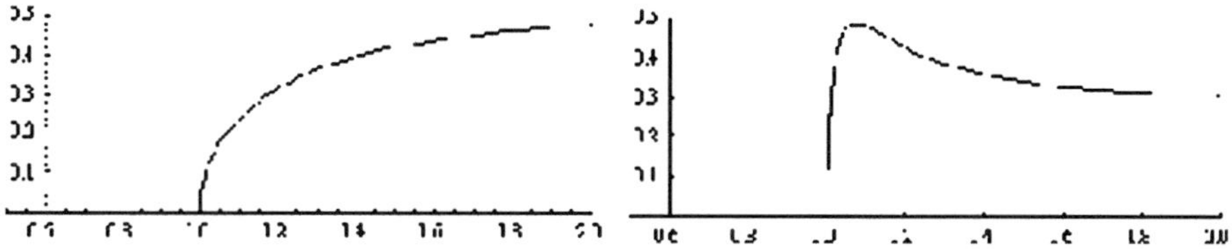

Fig. 9.4 *Left*: evolution of the amplitude of the modulation of the stationary states as a function of the period, in the cases of a Cahn–Hilliard dynamics. As $\mathrm{d}\,\nu/\,\mathrm{d}\,\lambda$ is always positive, the pattern will ripen until all the interfaces disappear but one (note that as $\mathrm{d}\,\nu/\,\mathrm{d}\,\lambda \to 0$, there is a slowing down of the coarsening process). *Right*, a model where $\mathrm{d}\,\nu/\,\mathrm{d}\,\lambda$ changes sign: the coarsening will then be interrupted

9.4 Oono's Model

9.4.1 Derivation of the Model

We would like to work out the period of modulated phase systems for which there is a competition between two types of interactions: a short-range interaction which tends to make the system more homogeneous together with a long-range one, or a non-local one, which prefers proliferation of domain walls. This competition results in a microphase separation with a preferred mesoscopic length scale. These systems forming a super-crystal can be studied using a modified Landau–Ginzburg approach, derived from Cahn–Hilliard equation and of practical use for numerical simulations [4, 5]:

$$\frac{\partial u}{\partial t} = (\nabla^2 \frac{\delta F_{GL}(u)}{\delta u}) - \beta^2 u = \nabla^2 (\frac{-1}{2} u + 2u^3 - \nabla^2 u) - \left(\frac{\beta}{4}\right)^2 u. \qquad (9.14)$$

The $-\beta^2 u$ term models in the Cahn–Hilliard equation the long-range interactions, which prevents the formation of macroscopic domains and favors the modulation. We will see that the inclusion of such a term, following Oono, enables to describe the behavior of modulated systems at T much lower than T_c. If we suppose, for example, that in a 3D problem, the long-range interaction decreases like $\frac{1}{r}$, the full free energy density writes

$$\begin{aligned} F(u) &= F_{GL} + F_{int} \\ &= \frac{1}{2}(\nabla u(r))^2 + \frac{-1}{4} u^2(r) + \frac{1}{2} u^4(r) + \int u(r') g(r', r) u(r) \mathrm{d}r', \qquad (9.15) \end{aligned}$$

where $g(r', r) = 4\pi \frac{\left(\frac{\beta}{4}\right)^2}{|r'-r|}$ in D = 3, or $|x' - x|$ in D = 1. The long-range interaction $g(r', r)$ corresponds to a repulsive interaction when $u(r')$ and $u(r)$ are of the same sign: thus it favors the formation of interphases. If we want to study the dynamic of this phase separation, we use the Cahn–Hilliard equation:

$$\begin{aligned} \frac{\partial u}{\partial t} &= \nabla_r^2 \left(\frac{\delta F(u)}{\delta u} \right) \\ &= \nabla_r^2 \left(\frac{-1}{2} u + 2u^3 - \nabla^2 u + \int u(r') g(r', r) \mathrm{d}r' \right). \qquad (9.16) \end{aligned}$$

If one recalls that $\frac{-1}{|r'-r|}$ is the Green's function associated with the Laplacian operator ∇_r^2 in 3D, the preceding equation then transforms into

$$\begin{aligned} \nabla_r^2 \left(\int u(r') g(r', r) \mathrm{d}r' \right) &= \int u(r') \nabla_r^2 g(r', r) \mathrm{d}r' \\ &= -\left(\frac{\beta}{4}\right)^2 \int u(r') \delta(r', r) \mathrm{d}r' = -\left(\frac{\beta}{4}\right)^2 u(r). \qquad (9.17) \end{aligned}$$

which leads to (9.14). Note that, even with the new term added by Oono to the usual Cahn-Hilliard dynamics, this equation remains in the class of the conservative models, as it derives from a equation of conservation. Note also that the free energy F_{int} is infinite if $u(r)$ is of the same sign in a macroscopic domain.

9.4.2 *Linear Stability Analysis for Oono's Model*

If we look at the linear stability analysis of the homogenous solution $u = 0$, we found almost the same results as in the original work of Cahn and Hilliard, except that the amplification factor $\sigma(\mathbf{q})$ now write:

$$\sigma(\mathbf{q}) = (\frac{1}{2} - \mathbf{q}^2)\mathbf{q}^2 - \left(\frac{\beta}{4}\right)^2$$

This shows immediately that $u = 0$ is linearly instable if $\beta < 1$, with a band of unstable Fourier modes $0.5\sqrt{1-\sqrt{1-\beta^2}} < q < 0.5\sqrt{1+\sqrt{1-\beta^2}}$ (for which $\sigma(\mathbf{q}) > 0$). The most unstable mode is for $q_{C-H} = 0.5$ like in the simplest Cahn–Hilliard model (9.4). Therefore, during the initial stage of the dynamics, the spinodal decomposition the homogeneous domains appear at length scales close to $L = 2\pi$, as in the usual Cahn Hilliard dynamics. But one sees that, contrary to the simple Cahn–Hilliard case, the long wave length modulations are now stable as $\sigma(\mathbf{q}) < 0$ for $q < 0.5\sqrt{1-\sqrt{1-\beta^2}}$. This explains qualitatively why, for any finite value of β, the dynamics will end in a micro segregated regime, as it is observed numerically and as we will discuss quantitatively below.

It has been noticed in different models [21] that if the interaction responsible of the modulation is local, i.e. described in the free energy by local terms only, like $-(\nabla u)^2$ in the Swift Hohenberg model, then for low temperature or small β, the macrosegregated regime (one unique interface) will be energetically favored compared to the microphase separation.

However, in this model by Oono, because the interaction is long range (i.e., non-local), no matter how small is β, there will always be a finite region around $q = 0$ where $\sigma(\mathbf{q}) < 0$. Indeed, $\sigma(\mathbf{0}) = -\left(\frac{\beta}{4}\right)^2$. Consequently, a modulated phase should always end the dynamics[22].

9.4.3 *Direct Minimization of the Free Energy*

For D = 1, the contribution of the long-range interaction to the free energy per unit length is [23]

$$\mathcal{F}_{int} = \frac{1}{\lambda}\int_0^\lambda F_{int}\mathrm{d}r = \frac{-\beta^2}{2\lambda}\int_0^{\frac{\lambda}{2}}\int_0^{\frac{\lambda}{2}} \Psi(r')\left|r'-r\right|\Psi(r)\mathrm{d}r\mathrm{d}r'.$$

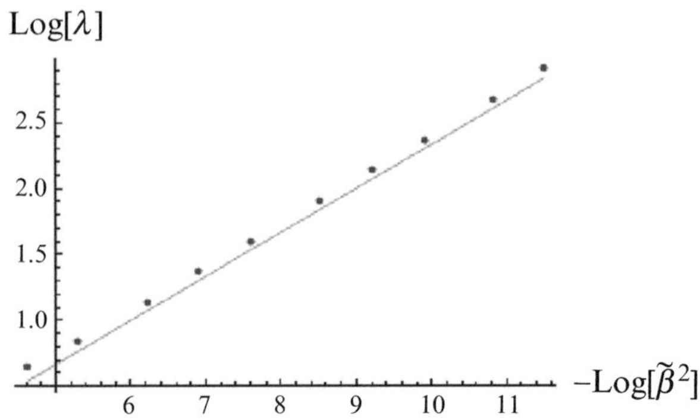

Fig. 9.5 Graph of the stable period $\lambda\left(\beta^2\right)$ computed by minimizing the free energy $F_{GL}(k,\lambda(k)) + F_{int}(k,\beta^2)$ with respect to k. The result scales like $\left(\beta^2\right)^{1/3}$.

When using as ansatz the family of interface-lattice solutions $U_{k,\lambda}(x)$, we then obtain

$$\mathcal{F}_{int} = \frac{-\beta^2}{2\lambda}\int_0^{\frac{\lambda}{2}}\int_0^{\frac{\lambda}{2}} k^2(\frac{4K}{\lambda})^2\left|r'-r\right|\mathrm{Sn}(\frac{4K(k)}{\lambda}r,k)\mathrm{Sn}(\frac{4K(k)}{\lambda}r',k)\mathrm{d}r\mathrm{d}r'$$
$$= \frac{\pi}{K}\frac{-\beta^2}{8}\int_0^{2K}\int_0^{2K} k^2\left|x'-x\right|\mathrm{Sn}(x,k)\mathrm{Sn}(x',k)\mathrm{d}x\mathrm{d}x'.$$

Thus, this contribution is independent of λ and the only minimization is with respect to k. Consequently, the minimization with respect to λ concerns only F_{GL} and enables to find λ as a function of k : $\lambda(k) = 8K\sqrt{\frac{1+k^2}{3} + \frac{k^2}{3(1-\frac{E}{K})}}$. And the minimization of the free energy $F_{GL}(k,\lambda(k)) + F_{int}(k)$ is simply with respect to a single variable k, which can be done numerically for different values of the interaction strength β^2.

Figure 9.5 presents $\lambda\left(\beta^2\right)$ which scales like $\left(\beta^2\right)^{1/3}$.

9.4.4 Stationary Microsegregated Patterns

The family (9.10) is not anymore an exact stationary solution of the dynamics (9.14) because of its last term. Nevertheless, it is a good candidate for an approximate solution (especially in the case of small β) and thus can be used as a tool for calculation using a solvability condition or Fredholm's alternative.

Indeed, we can write deviation from a given periodic stationary profile of period λ as $u(x,t) = u_0(\phi(x,t)) + \varepsilon u_1(\phi(x,t)) + \ldots$ where ε is a small parameter and u_0 is a periodic function of the phase $\phi(x,t)$. For a steady state solution $\phi(x,t) = qx$ with $q = 2\pi/\lambda$. In the general case $\phi(x,t) = q(X,T)x$ where $X = \epsilon x$ and $T = \epsilon^2 t$, i.e. $q = \frac{\partial\phi}{\partial x}$ is now a slowly varying function of x and t.

$$\frac{\partial u}{\partial t} = \frac{\partial u}{\partial \phi}\frac{\partial \phi}{\partial t} = \frac{\partial u}{\partial \phi}\frac{\partial \phi}{\partial T}\frac{dT}{dt} = \epsilon^2 \frac{\partial \phi}{\partial T}\frac{\partial u}{\partial \phi}$$
$$\frac{\partial u}{\partial x} = \frac{\partial u}{\partial \phi}\frac{\partial \phi}{\partial x} = \frac{\partial u}{\partial \phi}(q + \frac{\partial \phi}{\partial X}\frac{dX}{dx}) = q\frac{\partial u}{\partial \phi} + \epsilon\frac{\partial u}{\partial X}$$

If we denote $\Psi(X,T) = \epsilon\phi(x,t)$, then the local wave number is $q(X,T) = \frac{\partial\phi}{\partial x} = \frac{\partial\Psi}{\partial X}$ and

$$\frac{\partial}{\partial t} = \epsilon\partial_T\Psi\partial_\phi$$
$$\frac{\partial}{\partial x} = q\partial_\phi + \epsilon\frac{\partial q}{\partial X}\frac{\partial}{\partial q} = q\partial_\phi + \epsilon\frac{\partial^2\Psi}{\partial X^2}\partial_q$$
$$\frac{\partial^2}{\partial x^2} = q\frac{\partial}{\partial\phi}\left(q\partial_\phi + \epsilon\partial^2_{XX}\Psi\partial_q\right) + \epsilon\partial^2_{XX}\partial_q\left(q\partial_\phi + \epsilon\partial^2_{XX}\Psi\partial_q\right)$$
$$\frac{\partial^2}{\partial x^2} = q^2\partial_{\phi\phi} + \epsilon\partial^2_{XX}\Psi\partial_\phi + 2\epsilon\partial^2_{XX}\Psi q\partial_q\partial_\phi$$
$$\frac{\partial^2}{\partial x^2} = q^2\partial_{\phi\phi} + \partial^2_{XX}\Psi\left(1 + 2q\partial_q\right)\partial_\phi$$

where we have kept only the first order terms in ϵ.

If we consider a stationary profile u_0 which satisfies (zero order equation):

$$q^2\frac{\partial^2}{\partial\phi^2}\left(\frac{-1}{2}u_0 + 2u_0^3 - q^2\frac{\partial^2}{\partial\phi^2}u_0\right) - \left(\frac{\beta}{4}\right)^2 u_0 = 0$$

$$\text{i.e. } \frac{\partial}{\partial\phi}\left(\frac{-1}{2}u_0 + 2u_0^3 - q^2\frac{\partial^2}{\partial\phi^2}u_0\right) = \left(\frac{\beta}{4}\right)^2 w \text{where} \partial_\phi w = q^{-2}u_0 \tag{9.18}$$

Oono's equation (9.14) becomes then at order one in ϵ

$$\epsilon\partial_T\Psi\partial_\phi u_0 = \epsilon\mathcal{N}_0(u_1) + \epsilon\mathcal{N}_1(u_0)\text{where}$$
$$\mathcal{N}_0(u_1) = q^2\frac{\partial^2}{\partial\phi^2}(\frac{-1}{2}u_1 + 6u_0^2u_1 - q^2\frac{\partial^2}{\partial\phi^2}u_1) - \left(\frac{\beta}{4}\right)^2 u_1$$
$$= q^2\frac{\partial^2}{\partial\phi^2}\mathcal{L}(u_1) - \left(\frac{\beta}{4}\right)^2 u_1\text{and}$$
$$\mathcal{N}_1(u_0) = \partial^2_{XX}\Psi\left(1 + 2q\partial_q\right)\partial_\phi\left(\frac{-1}{2}u_0 + 2u_0^3 - q^2\frac{\partial^2}{\partial\phi^2}u_0\right)$$
$$- q^2\frac{\partial^2}{\partial\phi^2}\left(\partial^2_{XX}\Psi\left(1 + 2q\partial_q\right)\partial_\phi u_0\right)$$
$$= \left(\frac{\beta}{4}\right)^2\partial^2_{XX}\Psi\left(1 + 2q\partial_q\right)w - q^2\frac{\partial^2}{\partial\phi^2}\left(\partial^2_{XX}\Psi\left(1 + 2q\partial_q\right)\partial_\phi u_0\right)$$

where we have used $\partial_\phi w = q^{-2}u_0$ and equation (9.18) to simplify $\mathcal{N}_1(u_0)$. So Oono's equation (9.14) writes

$$\epsilon\partial_T\Psi\partial_\phi u_0 - \left(\frac{\beta}{4}\right)^2 \partial^2_{XX}\Psi\left(1+2q\partial_q\right)w$$
$$+q^2\partial^2_{XX}\Psi\frac{\partial^2}{\partial\phi^2}\left(\left(1+2q\partial_q\right)\partial_\phi u_0\right) = q^2\frac{\partial^2}{\partial\phi^2}\mathcal{L}(u_1) - \left(\frac{\beta}{4}\right)^2 u_1 \quad (9.19)$$

9.4.5 *Stability of Stationary Microsegregated Patterns*

A necessary condition for a solution to exist is that the left-hand side of the system is orthogonal to the kernel of the adjoint operator $\mathcal{N}_0^\dagger = \left(q^2\partial_{\phi\phi}\mathcal{L} - \left(\frac{\beta}{4}\right)^2 Id\right)^\dagger$; if $v\ \in \mathrm{Ker}\left(q^2\partial_{\phi\phi}\mathcal{L} - \left(\frac{\beta}{4}\right)^2 Id\right)^\dagger$, then the solvability condition (or Fredholm alternative) writes:

$$< v|\partial_T\Psi\partial_\phi u_0 - \mathcal{N}_1(u_0) >=< v|\mathcal{N}_0(u_1) >= 0$$

As for any v we have

$$< v|q^2\frac{\partial^2}{\partial\phi^2}\left(\frac{-1}{2}u_1 + 6u_0^2u_1 - q^2\frac{\partial^2}{\partial\phi^2}u_1\right) - \left(\frac{\beta}{4}\right)^2 u_1 >$$
$$=< q^2\frac{\partial^2}{\partial\phi^2}v|\frac{-1}{2} + 6u_0^2 - q^2\frac{\partial^2}{\partial\phi^2})u_1 > - \left(\frac{\beta}{4}\right)^2 < v|u_1 >$$
$$=< q^2\left(\frac{-1}{2} + 6u_0^2 - q^2\frac{\partial^2}{\partial\phi^2}\right)\partial_{\phi\phi}v|u_1 > - \left(\frac{\beta}{4}\right)^2 < v|u_1 >$$

this adjoint operator writes:

$$\mathcal{N}_0^\dagger = \left(q^2\partial_{\phi\phi}\mathcal{L} - \left(\frac{\beta}{4}\right)^2 Id\right)^\dagger = q^2\left(\frac{-1}{2} + 6u_0^2 - q^2\frac{\partial^2}{\partial\phi^2}\right)\partial_{\phi\phi} - \left(\frac{\beta}{4}\right)^2$$

If $v\ \in \mathrm{Ker}\mathcal{N}_0^\dagger$, we can define $\tilde{u}$ such that $q^2\partial_{\phi\phi}v = \tilde{u}$ and which satisfies

$$q^2\frac{\partial^2}{\partial\phi^2}\left(\frac{-1}{2}\tilde{u} + 6u_0^2\tilde{u} - q^2\frac{\partial^2}{\partial\phi^2}\tilde{u}\right) = q^2\left(\frac{\beta}{4}\right)^2\partial_{\phi\phi}v = \left(\frac{\beta}{4}\right)^2\tilde{u}. \quad (9.20)$$

So$\tilde{u}$issolutionof$\frac{-1}{2}\tilde{u} + 6u_0^2\tilde{u} - q^2\frac{\partial^2}{\partial\phi^2}\tilde{u} = \left(\frac{\beta}{4}\right)^2 v.$

Using (9.18), we thus find that v defined by $q^2\partial_{\phi\phi}v(=\tilde{u}) = \partial_\phi u_0$ is an element of $\mathrm{Ker}\left(q^2\partial_{\phi\phi}\mathcal{L} - \left(\frac{\beta}{4}\right)^2 Id\right)^\dagger$. As a consequence, the diffusion equation writes

$$\epsilon\partial_T\Psi = \frac{-q^2 < v|\frac{\partial^2}{\partial\phi^2}\left(\left(1+2q\partial_q\right)\partial_\phi u_0\right) > + < v|\left(\frac{\beta}{4}\right)^2\left(1+2q\partial_q\right)w >}{< v|\partial_\phi u_0 >}\partial^2_{XX}\Psi$$

As $q^2\partial_\phi w = u_0$ and $q^2\partial_{\phi\phi}v = \partial_\phi u_0$ we get the equality

$$v = w.$$

$$\text{So} < v|\partial_\phi u_0 >= - < \partial_\phi v|u_0 >= - < \partial_\phi w|u_0 >$$

$$= -q^{-2} < u_0|u_0 >$$

and consequently (9.19) is a diffusion equation

$$\epsilon\partial_T\Psi = D\,\partial^2_{XX}\Psi$$

$$\epsilon\partial_T\Psi = q^2\frac{\partial_q < q\left(\partial_\phi u_0\right)^2 > - \left(\frac{\beta}{4}\right)^2\partial_q < q\,w^2 >}{< u_0^2 >}\partial^2_{XX}\Psi$$

9.5 Conclusion

As long as the diffusion coefficient is negative (due to the $< \partial_k u_0|\left(\left(1+2q\partial_q\right)\partial_\phi u_0\right) >= \partial_q < q\left(\partial_\phi u_0\right)^2 >$ term), the coarsening process goes on, in order to minimize interfacial energy. But, due to its second part in β^2, the diffusion coefficient will vanish and thus the coarsening will be interrupted at a finite length scale.

Acknowledgements The authors would like to thank Dr. Chaouqi Misbah (LIPhy, Grenoble) for fruitful discussions and an invitation in Grenoble where part of this work was done.

References

1. Seul M, Andelman D (1995) Domain shapes and patterns: the phenomenology of modulated phases. Science 267:476
2. Hadziioannou G (2002) Semiconducting block copolymers for self-assembled photovoltaic devices. MRS Bull 27:456
3. Politi P, Misbah C (2004) When does coarsening occur in the dynamics of one-dimensional fronts? Phys Rev Lett 92:090601
4. Oono Y, Puri S (1987) Computationally efficient modeling of ordering of quenched phases. Phys Rev Lett 58:836

5. Oono Y, Shiwa Y (1987) Computationally efficient modeling of block copolymer and Benard pattern formations. Mod Phys Lett B 1:49
6. Hohenberg PC, Halperin BI (1977) Theory of dynamical critical phenomena. Rev Mod Phys 49:435. See also Cross MC, Hohenberg PC (1993) Pattern formation out of equilibrium. Rev Mod Phys 65:851
7. Gunton JD, San Miguel M, Sahni PS (1983) In: Domb C, Lebowitz JL (eds) Phase transition et critical phenomena, vol 8. Academic, London, p 267
8. Hillert M (1961) A solid solution model for inhomogeneous systems. Acta Metall 9:525
9. Cahn JW, Hilliard JE (1958) Free energy of a nonuniform system. I. Interfacial free energy. J Chem Phys 28:258 (1958)
10. Cahn JW (1965) Phase separation by spinodal decomposition in isotropic systems. J Chem Phys 42:93
11. Chevallard C, Clerc M, Coullet P, Gilli JM (2000) Interface dynamics in liquid crystals. Eur Phys J E 1:179
12. Oyama Y (1939) Mixing of solids. Bull Inst Phys Chem Res Rep 5:600. English translation Weidenbaum SS (1958) Adv Chem Eng 2:211
13. Puri S, Hayakawa H (2001) Segregation of granular mixtures in a rotating drum. Adv Complex Syst 4(4)469–479
14. Scherer MA, Melo F, Marder M (1999) Sand ripples in an oscillating annular sand–water cell. Phys Fluids 11:58
15. Stegner A, Wesfreid JE (1999) Dynamical evolution of sand ripples under water. Phys Rev E 60:R3487
16. Joly S, Raquois A, Paris F, Hamdoun B, Auvray L, Ausserre D, Gallot Y (1996) Early stage of spinodal decomposition in 2D. Phys Rev Lett 77:4394
17. Langer JS (1971) Theory of spinodal decomposition in alloys. Ann Phys 65:53
18. Villain-Guillot S (2004) Coalescence in the 1D Cahn–Hilliard model. J Phys A Math Gen 37:6929
19. Calisto H, Clerc MG, Rojas R, Tirapegui E (2000) Bubbles interaction in Canh-Hilliard equation. Phys Rev Lett 85:3805
20. Argentina M, Clerc MG, Rojas R, Tirapegui E (2005) Coarsening dynamics of the one-dimensional Cahn-Hilliard model. Phys Rev E 71:046210
21. Buzdin AI, Kachkachi H (1997) Generalized Ginzburg-Landau theory for nonuniform FFLO superconductors. Phys Lett A 225:341
22. Andelman D, Brochard F, Joanny J-F (1987) Phase transitions in Langmuir monolayers of polar molecules. J Chem Phys 86:3673
23. Liu F, Goldenfeld N (1989) Dynamics of phase separation in block copolymer melts. Phys Rev A 39:4805

Chapter 10
Nonlinear Analysis of Phase-locked Loop-Based Circuits

R.E. Best, N.V. Kuznetsov, G.A. Leonov, M.V. Yuldashev, and R.V. Yuldashev

Abstract Main problems of simulation and mathematical modeling of high-frequency signals for analog Costas loop and for analog phase-locked loop (PLL) are considered. Two approachers which allow to solve these problems are considered. In the first approach, nonlinear models of classical PLL and classical Costas loop are considered. In the second approach, engineering solutions for this problems are described. Nonlinear differential equations are derived for both approaches.

Keywords Phase-locked loop • Nonlinear analysis • Dynamical model • Simulation

The Phase-locked loop (PLL) is a classical circuit widely used in telecommunication and computer architectures. PLL was invented in the 1930s–1940s [5] and then intensive studies of the theory and practice of PLL were carried out [11, 33, 40]. One of the first applications of PLL is related to the problems of wireless data transfer. In radio engineering, PLL-based circuits (e.g., Costas Loop, PLL with squarer) are used for carrier recovery, demodulation, and frequency synthesis (see, e.g., [6, 14, 35]).

R.E. Best
Best Engineering, Oberwil, Switzerland
e-mail: Rolandbest@aol.com

N.V. Kuznetsov (✉) • G.A. Leonov • M.V. Yuldashev • R.V. Yuldashev
Saint Petersburg State University, St Petersburg, Russia

University of Jyväskylä, Jyväskylä, Finland
e-mail: nkuznetsov239@gmail.com; leonov@math.spbu.ru; maratyv@gmail.com; renatyv@gmail.com

J.A.T. Machado et al. (eds.), *Discontinuity and Complexity in Nonlinear Physical Systems*, Nonlinear Systems and Complexity 6, DOI 10.1007/978-3-319-01411-1_10,

Although the PLL is essentially a nonlinear control system, in modern literature, devoted to the analysis of PLL-based circuits, the main direction is the use of simplified linear models, the methods of linear analysis, empirical rules, and numerical simulation (see plenary lecture of Abramovich at American Control Conference 2002 [1]). Rigorous nonlinear analysis of PLL-based circuit models is often a very difficult task [4, 9, 10, 37], so for analysis of nonlinear PLL models, in practice, in numerical simulation is widely used (see, e.g., [6]). However for high-frequency signals, complete numerical simulation of *physical model of PLL-based circuit in signals/time space*, which is described by a nonlinear non-autonomous system of differential equations, is a very challenging task [2, 3] since it is necessary to observe simultaneously *very fast time scale of the input signals* and *slow time scale of signal's phases*. Here the relatively small discretization step in numerical procedure does not allow one to consider phase locking processes for high-frequency signals in reasonable time.

Here two approaches, which allow one to overcome these difficulties, are considered. The first idea is traced back to the works of [40] and consists in construction of mathematical models of PLL-based circuits in phase-frequency space. This approach requires to determine mathematical characteristics of the circuit components and to prove reliability of considered mathematical model. The second idea is traced back to the works of [8] and consists in design of circuit components in such a way that there is no oscillation with a double frequency component in the loop.

10.1 Phase-Frequency Model of Classical PLL

To overcome simulation difficulties for PLL-based circuits it is possible to construct *a mathematical model in phase-frequency/time space* [30], which can be described by a nonlinear dynamical system of differential equations, here only low frequency signals have to be analyzed. That, in turn, requires [1] the computation of phase detector characteristics (nonlinear element used to math reference and controllable signals), which depends on waveforms of the considered signals. Using results of analysis of this mathematical model for conclusions on behavior of the physical model requires rigorous justification.

Consider a classical PLL on the level of electronic realization (Fig. 10.1)

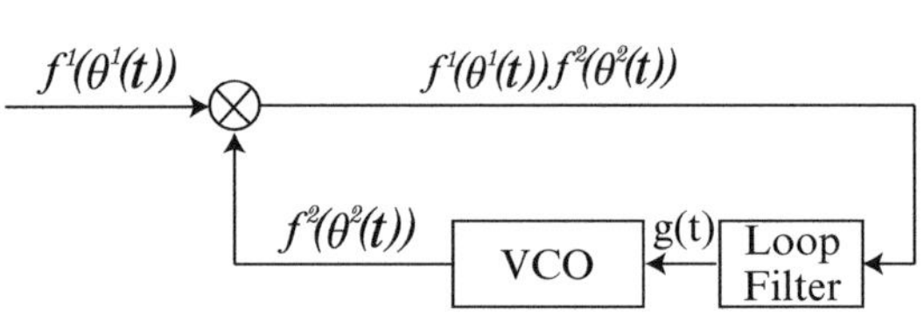

Fig. 10.1 Block diagram of PLL on the level of electronic realization

Here signals $f^p(t) = f^p(\theta^p(t)),\ p = 1, 2$ with $\theta^p(t)$ as phases are oscillations generated by the reference oscillator and the tunable voltage-controlled oscillator (VCO), respectively.

The block $\otimes$ is a multiplier (used as a phase detector) of oscillations $f^1(t)$ and $f^2(t)$, and the signal $f^1(\theta^1(t))f^2(\theta(t))$ is its output. The relation between the input $\xi(t)$ and the output $\sigma(t)$ of linear filter has the form:

$$\sigma(t) = \alpha_0(t) + \int_0^t \gamma(t-\tau)\xi(\tau)\,\mathrm{d}\tau, \tag{10.1}$$

where $\gamma(t)$ is an impulse response function of filter and $\alpha_0(t)$ is an exponentially damped function depending on the initial data of the filter at $t = 0$. By assumption, $\gamma(t)$ is a differentiable function with bounded derivative (this is true for the most considered filters [38]).

10.1.1 High-Frequency Property of Signals

Suppose that the waveforms $f^{1,2}(\theta)$ are bounded 2π-periodic piecewise differentiable functions.[1] Consider Fourier series representation of such functions

$$f^p(\theta) = \sum_{i=1}^{\infty} \big(a_i^p \sin(i\theta) + b_i^p \cos(i\theta)\big),\ p = 1, 2,$$

$$a_i^p = \frac{1}{\pi}\int_{-\pi}^{\pi} f^p(\theta)\sin(i\theta)d\theta, b_i^p = \frac{1}{\pi}\int_{-\pi}^{\pi} f^p(\theta)\cos(i\theta)d\theta.$$

A high-frequency property of signals can be reformulated in the following way. By assumption, the phases $\theta^p(t)$ are smooth functions (this means that frequencies are changing continuously, which corresponds to classical PLL analysis [6, 14]). Suppose also that there exists a sufficiently large number ω_{min} such that the following conditions are satisfied on a fixed time interval $[0, T]$:

$$\dot\theta^p(\tau) \geq \omega_{min} > 0,\ \ p = 1, 2, \tag{10.2}$$

where T is independent of ω_{min} and $\dot\theta^p(\tau) = \frac{d\theta^p(\tau)}{d\tau}$ denotes frequencies of signals. The frequencies difference is assumed to be uniformly bounded

$$\big|\dot\theta^1(\tau) - \dot\theta^2(\tau)\big| \leq \Delta\omega,\ \forall\tau \in [0, T]. \tag{10.3}$$

[1]The functions with a finite number of jump discontinuity points differentiable on their continuity intervals

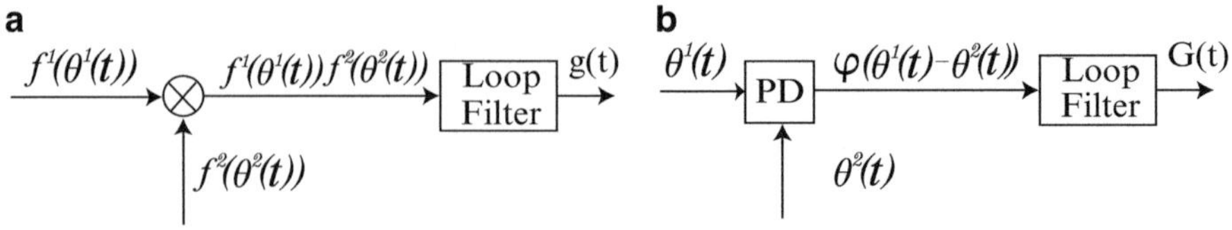

Fig. 10.2 Phase detector models. (**a**) Multiplier and filter, (**b**) Phase detector and filter

Requirements (10.2) and (10.3) are obviously satisfied for the tuning of two high-frequency oscillators with close frequencies. Denote $\delta = \omega_{min}^{-\frac{1}{2}}$. Consider the following relations

$$|\dot{\theta}^p(\tau) - \dot{\theta}^p(t)| \le \Delta\Omega, \ p = 1, 2,$$
$$|t - \tau| \le \delta, \quad \forall \tau, t \in [0, T], \tag{10.4}$$

where $\Delta\Omega$ is independent of δ. Conditions (10.2)–(10.4) mean that the functions $\dot{\theta}^p(\tau)$ are almost constant and the functions $f^p(\theta^p(\tau))$ are rapidly oscillating on small intervals $[t, t + \delta]$.

The boundedness of derivative of $\gamma(t)$ implies

$$|\gamma(\tau) - \gamma(t)| = O(\delta), \quad |t - \tau| \le \delta, \quad \forall \tau, t \in [0, T]. \tag{10.5}$$

10.1.2 Phase Detector Characteristic Computation for Classical PLL

Consider two block diagrams shown in Fig. 10.2a, b. Here, PD is a nonlinear block with characteristic $\varphi(\theta)$. The phases $\theta^p(t)$ are PD block inputs and the output is a function $\varphi(\theta^1(t) - \theta^2(t))$. The PD characteristic $\varphi(\theta)$ depends on waveforms of input signals.

The signal $f^1(\theta^1(t))f^2(\theta^2(t))$ and the function $\varphi(\theta^1(t) - \theta^2(t))$ are the inputs of the same filters with the same impulse response function $\gamma(t)$ and with the same initial state. The outputs of filters are the functions $g(t)$ and $G(t)$, respectively. By (10.1) one can obtain $g(t)$ and $G(t)$:

$$g(t) = \alpha_0(t) + \int_0^t \gamma(t - \tau) f^1\big(\theta^1(\tau)\big) f^2\big(\theta^2(\tau)\big) d\tau,$$
$$G(t) = \alpha_0(t) + \int_0^t \gamma(t - \tau)\varphi\big(\theta^1(\tau) - \theta^2(\tau)\big) d\tau. \tag{10.6}$$

Using the approaches outlined in [20, 22, 24, 26], the following result can be proved.

Theorem 1. [23, 29, 30] *Let conditions* (10.2)–(10.5) *be satisfied and*

$$\varphi(\theta) = \frac{1}{2}\sum_{l=1}^{\infty}\Big((a_l^1 a_l^2 + b_l^1 b_l^2)\cos(l\theta) + (a_l^1 b_l^2 - b_l^1 a_l^2)\sin(l\theta)\Big). \tag{10.7}$$

Then the following relation

$$|G(t) - g(t)| = O(\delta), \quad \forall t \in [0, T]$$

is valid.

See Appendix for a proof of this theorem.

Broadly speaking, this theorem separates the low-frequency error-correcting signal from parasitic high-frequency oscillations. This theorem allows one to compute a phase detector characteristic for various typical waveforms of signals.

10.1.3 Description of classical Costas Loop

Nowadays BPSK and QPSK modulation techniques are used in telecommunication. For these techniques different modifications of the PLL are used: e.g. a circuit with a squaring device, or the Costas Loop [6, 11, 33]. However, the realization of some parts of PLL with squarer, used in analog circuits, can be quite difficult [6]. In the digital circuits, maximum data rate is limited by the speed of analog-to-digital converter (ADC) [12, 13]. Here, we will consider analog Costas Loops, which are easy for implementation and effective for demodulation.

Various methods for analysis of Costas loop are well developed by engineers and considered in many publications (see, e.g., [11, 14, 32]). However, the problems of construction of adequate nonlinear models and nonlinear analysis of such models are still far from being resolved. Further we will consider only classical BPSK Costas loops, but a similar analysis could be done for QPSK Costas Loop.

Consider the physical model of classical Costas Loop (Fig. 10.3).

Here $f^1(t)$ is a carrier and $m(t) = \pm 1$ is data signal. Hilbert transform block shifts phase of input signal by $-\frac{\pi}{2}$.

In the simplest case when

$$\begin{aligned}
&f^1(\theta^1(t)) = \cos\big(\omega^1 t\big), f^2(\theta^2(t)) = \sin\big(\omega^2 t\big) m(t) f^1(\theta^1(t)) f^2(\theta^2(t)) \\
&= \frac{m(t)}{2}\big(\sin(\omega^2 t - \omega^1 t) - \sin(\sin(\omega^2 t + \omega^1 t))\big)\, m(t) f^1(\theta^1(t)) f^2(\theta^2(t) - \frac{\pi}{2}) \\
&= \frac{m(t)}{2}\big(\cos(\omega^2 t - \omega^1 t) + \cos(\sin(\omega^2 t + \omega^1 t))\big)
\end{aligned} \tag{10.8}$$

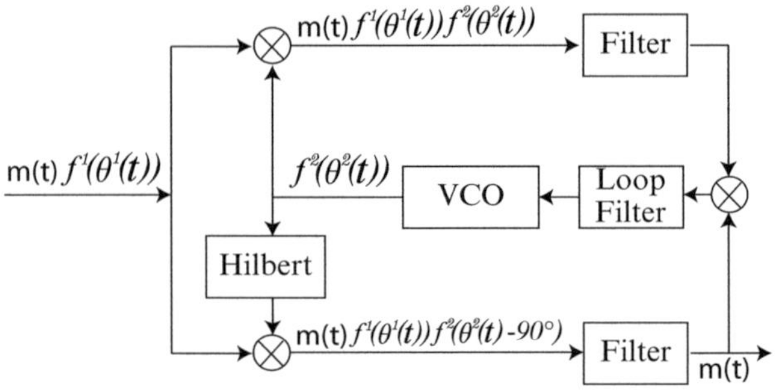

Fig. 10.3 Block diagram of Costas Loop at the level of electronic realization

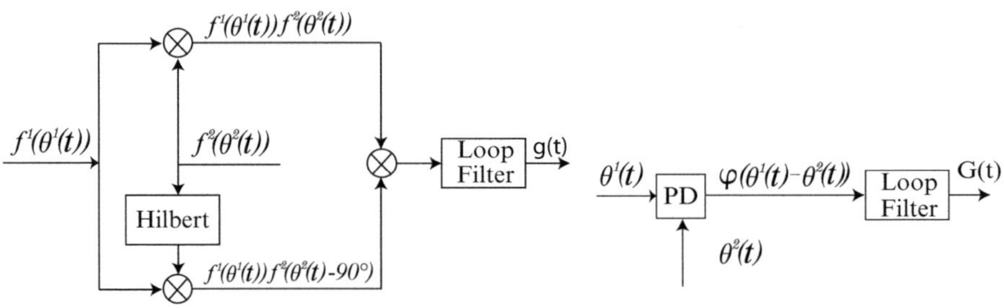

Fig. 10.4 Equivalent block diagrams of Costas loop in signals/time and phase-frequency spaces

standard engineering assumption is that a low pass filter removes the upper sideband having a frequency of about twice the carrier frequency but leaves the lower sideband unchanged. Thus, after synchronization one gets demodulated the data $m(t)\cos((\omega^1 - \omega^2)t) = m(t)\cos(0) = m(t)$ at the output of the lower filter (see Fig. 10.3).

Further, to avoid this assumption, a rigorous mathematical approach for the analysis of Costas loop will be demonstrated.

10.1.4 Computation of Phase Detector Characteristic for Costas Loop

From a theoretical point of view, since two lowpass filters in Fig. 10.3 are used for demodulation, for analysis of synchronization processes one can study the Costas loop with only Loop filter. Also since $m(t)^2 = 1$, the transmitted data $m(t)$ do not affect the operation of VCO. Thus one can consider the following equivalent block diagrams of the Costas loop in signals/time and phase-frequency spaces (Fig. 10.4).

In both diagrams the filters are the same and have the same impulse transient function $\gamma(t)$ and the same initial data. The filter outputs are the functions $g(t)$ and $G(t)$, respectively.

Consider a case of non-sinusoidal piecewise-differentiable carrier oscillation $f^1(\theta^1(t))$ and tunable harmonic oscillation

$$f^1(\theta) = \sum_{i=1}^{\infty} \big(a_i^1 \cos(i\theta) + b_i^1 \sin(i\theta)\big),$$
$$f^2(\theta) = b_1^2 \sin(\theta). \tag{10.9}$$

The following assertion is valid.

Theorem 2. [24, 31] *If conditions* (10.2)–(10.5) *are satisfied and*

$$\varphi(\theta) = \frac{(b_1^2)^2}{8}\Big[(a_1^1)^2 \sin(2\theta) + 2\sum_{q=1}^{\infty} a_q^1 a_{q+2}^1 \sin(2\theta)$$
$$-2a_1^1 b_1^1 \cos(2\theta) + 2\sum_{q=1}^{\infty} a_{q+2}^1 b_q^1 \cos(2\theta) - 2\sum_{q=1}^{\infty} a_q^1 b_{q+2}^1 \cos(2\theta)$$
$$-(b_1^1)^2 \sin(2\theta) + 2\sum_{q=1}^{\infty} b_q^1 b_{q+2}^1 \sin(2\theta)\Big)\Big]. \tag{10.10}$$

then the following relation

$$G(t) - g(t) = O(\delta), \quad \forall t \in [0, T] \tag{10.11}$$

is valid.

In general, the proof of this result repeats the proof of Theorem 1. The details of the proof can be found in [24, 31]. Note that this result could be easily extended to the case of two non-sinusoidal signals.

10.2 Engineering Solutions for Elimination of High-Frequency Oscillations

Consider engineering solution for elimination of high-frequency oscillations on the output of PD for harmonic signals. Further it is considered special analog PLL and analog Costas loop implementations, which allow one to effectively solve this problem.

10.2.1 Two-Phase PLL

Consider a special modification of the PLL (two phase PLL) suggested in [8] (Fig. 10.5).

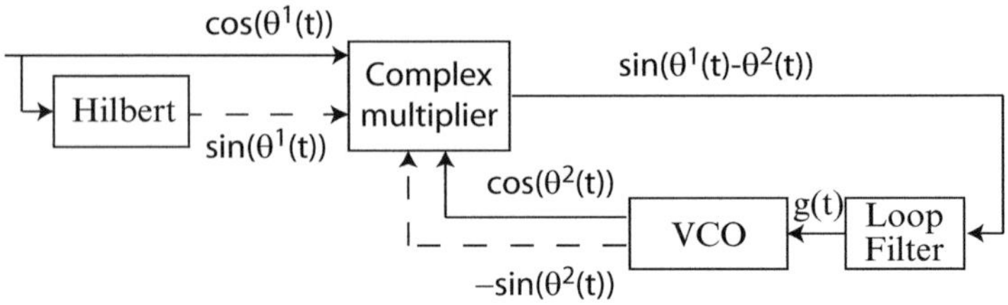

Fig. 10.5 Two-phase PLL

Fig. 10.6 Phase detector in two-phase PLL

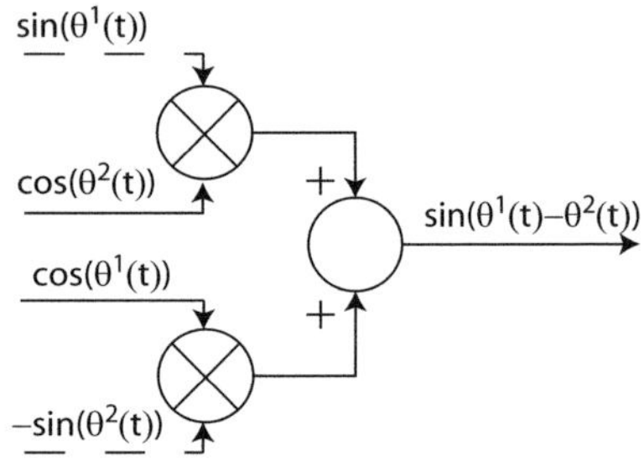

Here, a carrier is $\sin(\theta^1(t))$ with $\theta^1(t)$ as a phase and the output of Hilbert block is $\cos(\theta^1(t))$. The VCO generates oscillations $-\sin(\theta^2(t))$ and $\cos(\theta^2(t))$ with $\theta^2(t)$ as a phase. Figure 10.6 shows the structure of phase detector—complex multiplier. The phase detector consists of two analog multipliers and an analog subtractor. Here

$$\sin(\theta^1(t))\cos(\theta^2(t)) - \cos(\theta^1(t))\sin(\theta^2(t)) = \sin(\theta^1(t) - \theta^2(t))$$

In this case there is no high-frequency component at the output of the phase detector. Thus the block diagram in Fig. 10.6 is equivalent to the block diagram in Fig. 10.2b, where phase detector characteristic is $\varphi(\theta) = \sin(\theta)$.

10.2.2 Two-Phase Costas Loop

Consider now an engineering solution [39] for the problem of elimination of high-frequency oscillations in the Costas Loop (Fig. 10.7).

Here the carrier is $\cos(\theta^1(t))$ with $\theta^1(t)$ as a phase. The VCO generates the oscillations $\cos(\theta^2(t))$ and $-\sin(\theta^2(t))$ with $\theta^2(t)$ as a phase, and $m(t) = \pm 1$ is a relatively slowly varying data signal (carrier period is several orders of magnitude smaller than the symbol duration). In Fig. 10.8 is shown a structure of phase detector. Here the outputs of phase detector are the following

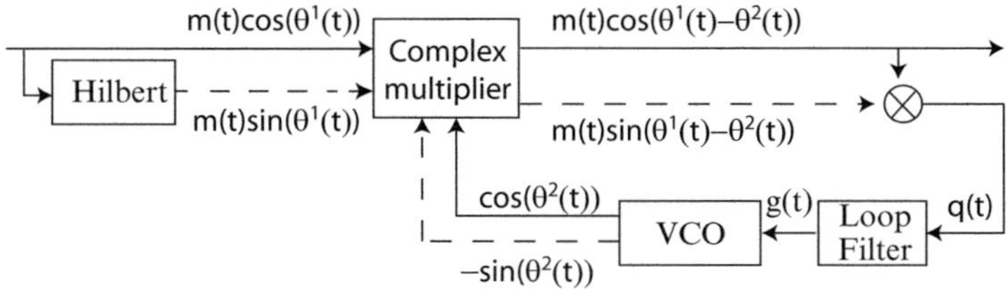

Fig. 10.7 Two-phase Costas loop

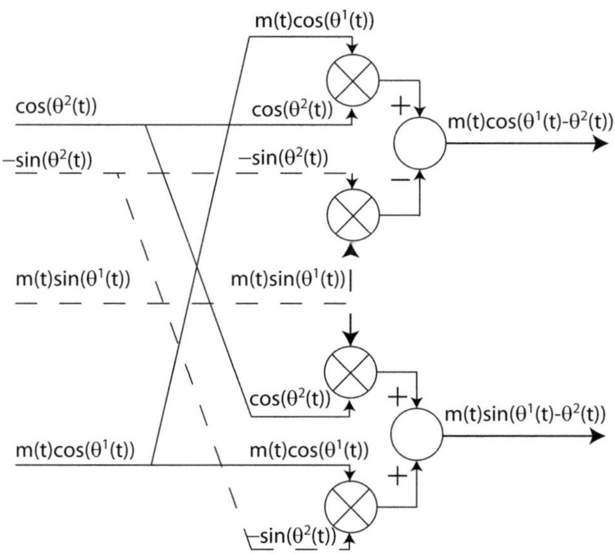

Fig. 10.8 Phase detector in two-phase Costas loop

$$
\begin{aligned}
m(t)\left(\cos(\theta^1(t))\cos(\theta^2(t)) + \sin(\theta^1(t))\sin(\theta^2(t))\right) &= m(t)\cos(\theta^1(t) - \theta^2(t)) \\
m(t)\left(\sin(\theta^1(t))\cos(\theta^2(t)) - \cos(\theta^1(t))\sin(\theta^2(t))\right) &= m(t)\sin(\theta^1(t) - \theta^2(t))
\end{aligned}
\tag{10.12}
$$

If oscillators are synchronized (i.e., $\theta^1(t) = \theta^2(t)$), one of the outputs of phase detector contains only data signal $m(t)$. Therefore, taking into account $m(t) = \pm 1$, the input of the Loop filter takes the form

$$
m(t)\cos(\theta^1(t) - \theta^2(t))m(t)\sin(\theta^1(t) - \theta^2(t)) = \frac{1}{2}\sin(2(\theta^1(t) - \theta^2(t)))
$$

and it depends only on the phase difference of VCO and carrier. Thus the block diagram in Fig. 10.8 is equivalent to block-scheme in Fig. 10.2b, where the phase detector characteristic is $\varphi(\theta) = \frac{1}{2}\sin(2(\theta))$.

10.3 Differential Equation for PLL and Costas Loop

Here differential equations for the considered PLL-based circuits are derived.

From a mathematical point of view, a linear low-pass filter can be described by a system of linear differential equations

$$\dot{x} = Ax + p\xi(t),\ \sigma = c^* x, \tag{10.13}$$

a solution of which takes the form (10.1). Here, A is a constant matrix, $x(t)$ is a state vector of filter, b and c are constant vectors.

The model of the tunable generator is usually assumed to be linear [6, 14]:

$$\dot{\theta}^2(t) = \omega^2_{free} + LG(t),\ t \in [0, T]. \tag{10.14}$$

where ω^2_{free} is a free-running frequency of the tunable generator and L is an oscillator gain. Here it is also possible to use nonlinear models of VCO; see, e.g., [7, 36].

Suppose that the frequency of the master generator is constant $\dot{\theta}^1(t) \equiv \omega^1$. Equation of the tunable generator (10.14) and equation of the filter (10.13) yield

$$\dot{x} = Ax + p\xi(t),\ \dot{\theta}^2 = \omega^2_{free} + Lc^* x. \tag{10.15}$$

For a classical PLL circuit

$$\xi(t) = f^1(\theta^1(t)) f^2(\theta^2(t)), \tag{10.16}$$

for a classical Costas loop

$$\xi(t) = f^1(\theta^1(t)) f^2(\theta^2(t) - \frac{\pi}{2}) f^1(\theta^1(t)) f^2(\theta^2(t)), \tag{10.17}$$

for the two-phase PLL

$$\xi(t) = \sin(\theta^1(t) - \theta^2(t)), \tag{10.18}$$

and for two-phase Costas loop

$$\xi(t) = \frac{1}{2} \sin\left(2(\theta^1(t) - \theta^2(t))\right), \tag{10.19}$$

While for two-phase PLLs and Costas loops, system (10.15) is autonomous, for classical PLL and Costas loop, system (10.15) is nonautonomous and rather difficult for investigation [16, 34]. Here, Theorems 1 and 2 allow one to study more simple autonomous system of differential equations

$$\dot{x} = Ax + p\varphi(\Delta\theta),\ \Delta\dot{\theta} = \omega^2_{free} - \omega^1 + Lc^* x,$$
$$\Delta\theta = \theta^2 - \theta^1, \tag{10.20}$$

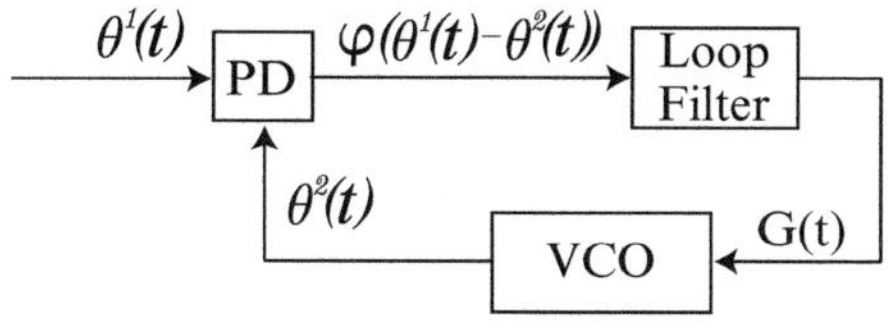

Fig. 10.9 Block scheme of phase-locked loop in phase-frequency space

where $\varphi(\theta)$ is the corresponding characteristic of the phase detector. By well-known averaging methods [15] one can show that solutions of (10.15) and (10.20) are close under some assumptions. Thus, by Theorems 2 and 1, the block diagrams of PLL and Costas Loop in signals/time space (Figs. 10.1 and 10.3) can be asymptotically replaced [for high-frequency generators, see conditions (10.2)–(10.4)] for the block-scheme in phase-frequency space (Fig. 10.9).

The methods of nonlinear analysis for system (10.20) are well developed (see, e.g., [17–19, 25, 27, 28]). The simulation approach for PLL analysis and design, based on the obtained analytical results, is discussed in [21].

It should be noted that instead of conditions (10.3) and (10.5) for simulations of real system, it is necessary to consider the following conditions

$$|\Delta\omega| << \omega_{min}, \ |\lambda_A| << \omega_{min},$$

where λ_A is the largest (in modulus) eigenvalue of matrix A. Also, for correctness of transition from (10.21) to (10.25) it is necessary to consider $T << \omega_{min}$. It is easy to see that for sinusoidal waveforms operations of classical PLL and two-phase PLL are very similar because the phase detector characteristic and corresponding phase-frequency models are the same. Theoretical results are justified by simulation of classical PLL and two-phase PLL (Fig. 10.10).

Unlike the filter output for the phase-frequency model of classical and two-phase PLLs, for signals/time space model of classical PLL the outputs of filter and phase detector contains additional high-frequency oscillations. These high-frequency oscillations interfere with qualitative analysis and efficient simulation of PLL. The filter output of two-phase PLL is delayed compared to the classical one because of the non-ideality of the Hilbert transformer. Similar results can be obtained for the Costas loop (see Fig.10.11).

Appendix

Proof. Suppose that $t \in [0, T]$. Consider a difference

$$g(t) - G(t) = \int_0^t \gamma(t-s)\Big[f^1\big(\theta^1(s)\big)f^2\big(\theta^2(s)\big) - \varphi\big(\theta^1(s) - \theta^2(s)\big)\Big]ds. \tag{10.21}$$

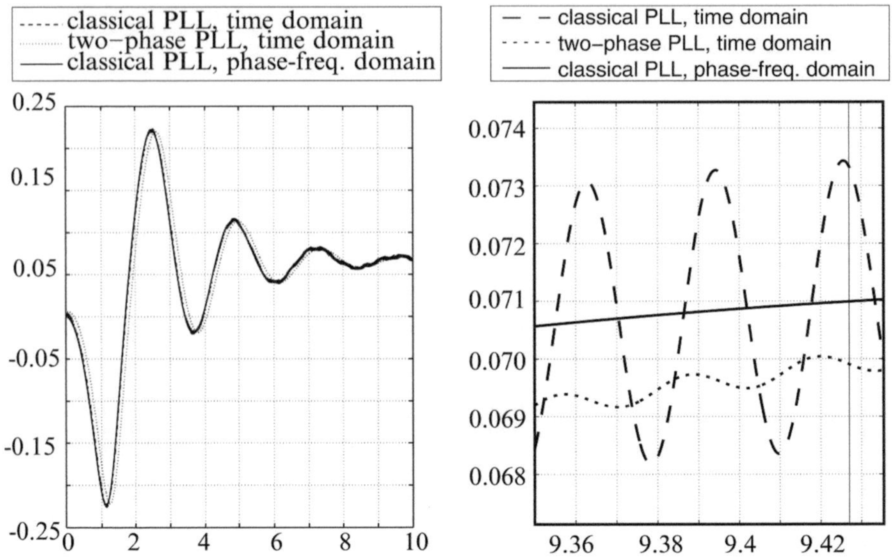

Fig. 10.10 Classical PLL in signals/time space; two-phase and classical PLLs in phase-frequency space, $\omega^2_{free} = 99\,\text{Hz}$, $\omega^1 = 100\,\text{Hz}$, $L = 15$, filter transfer functions $\frac{1}{s+1}$

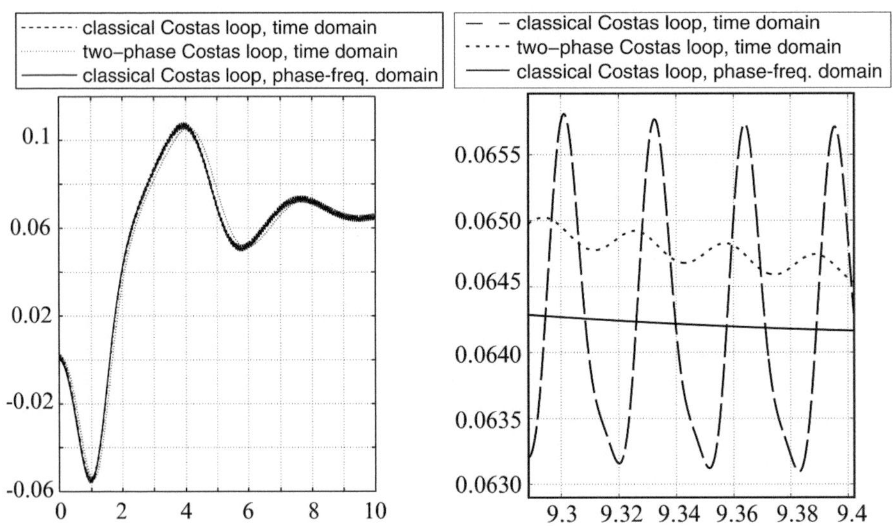

Fig. 10.11 Classical Costas Loop in signals/time space, two-phase and classical Costas Loops in phase-frequency space, $\omega^2_{free} = 99\,\text{Hz}$, $\omega^1 = 100\,\text{Hz}$, $L = 15$, filter transfer functions $\frac{1}{s+1}$

Suppose that there exists $m \in \mathbb{N} \cup \{0\}$ such that $t \in [m\delta, (m+1)\delta]$. By definition of δ, one has $m < \frac{T}{\delta} + 1$. The continuity condition implies that $\gamma(t)$ is bounded on $[0, T]$ and $f^1(\theta)$, $f^2(\theta)$ are bounded on $\mathbb{R}$. Since $f^{1,2}(\theta)$ are piecewise differentiable, one can obtain

$$a_i^p = O(\frac{1}{i}),\ b_i^p = O(\frac{1}{i}). \tag{10.22}$$

Hence $\varphi(\theta)$ converges uniformly and $\varphi(\theta)$ is continuous, piecewise differentiable, and bounded. Then the following estimates

$$\int_t^{(m+1)\delta} \gamma(t-s) f^1\big(\theta^1(s)\big) f^2\big(\theta^2(s)\big)ds = O(\delta),$$

$$\int_t^{(m+1)\delta} \gamma(t-s)\varphi\big(\theta^1(s) - \theta^2(s)\big)ds = O(\delta)$$

are satisfied. It follows that (10.21) can be represented as

$$g(t) - G(t) = \sum_{k=0}^{m} \int_{[k\delta,(k+1)\delta]} \gamma(t-s)$$
$$\Big[f^1\big(\theta^1(s)\big) f^2\big(\theta^2(s)\big) - \varphi\big(\theta^1(s) - \theta^2(s)\big)\Big]ds + O(\delta). \tag{10.23}$$

Prove now that on each interval $[k\delta, (k+1)\delta]$ the corresponding integrals are equal to $O(\delta^2)$.

Condition (10.5) implies that on each intervals $[k\delta, (k+1)\delta]$ the following relation

$$\gamma(t-s) = \gamma(t-k\delta) + O(\delta),\ t > s,\ s,t \in [k\delta, (k+1)\delta] \tag{10.24}$$

is valid. Here $O(\delta)$ is independent of k and the relation is satisfied uniformly with respect to t. By (10.23), (10.24), and the boundedness of $f^1(\theta)$, $f^2(\theta)$, and $\varphi(\theta)$,

$$g(t) - G(t) = \sum_{k=0}^{m} \gamma(t-k\delta) \int\limits_{[k\delta,(k+1)\delta]}$$
$$\Big[f^1\big(\theta^1(s)\big) f^2\big(\theta^2(s)\big) - \varphi\big(\theta^1(s) - \theta^2(s)\big)\Big]ds + O(\delta). \tag{10.25}$$

Denote

$$\theta_k^p(s) = \theta^p(k\delta) + \dot\theta^p(k\delta)(s - k\delta),\ p = 1, 2.$$

Then for $s \in [k\delta, (k+1)\delta]$, condition (10.4) yields

$$\theta^p(s) = \theta_k^p(s) + O(\delta).$$

From (10.3) and the boundedness of the derivative $\varphi(\theta)$ on $\mathbb{R}$ it follows that

$$\int\limits_{[k\delta,(k+1)\delta]} \left|\varphi\big(\theta^1(s)-\theta^2(s)\big)-\varphi\big(\theta_k^1(s)-\theta_k^2(s)\big)\right|ds = O(\delta^2). \tag{10.26}$$

If $f^1(\theta)$ and $f^2(\theta)$ are continuous on $\mathbb{R}$, then for $f^1\big(\theta^1(s)\big)f^2\big(\theta^2(s)\big)$ the following relation holds

$$\int_{[k\delta,(k+1)\delta]} f^1(\theta^1(s))f^2(\theta^2(s))ds$$
$$= \int_{[k\delta,(k+1)\delta]} f^1(\theta_k^1(s))f^2(\theta_k^2(s))ds + O(\delta^2). \tag{10.27}$$

Consider the validity of this estimate for the considered class of piecewise-differentiable waveforms. Since the conditions (10.2) and (10.4) are satisfied and the functions $\theta^{1,2}(s)$ are differentiable and satisfy (10.3), for all $k = 0,\ldots,m$ there exist sets E_k [the union of sufficiently small neighborhoods of discontinuity points of $f^{1,2}(t)$] such that the following relation $\int\limits_{E_k} ds = O(\delta^2)$ is valid, in which case the relation is satisfied uniformly with respect to k. Then from the piecewise differentiability and the boundedness of $f^{1,2}(\theta)$ it is possible to obtain (10.27) (see Corollary 1).

By (10.27) and (10.26), relation (10.25) can be rewritten as

$$\begin{aligned}
g(t)-G(t) &= \sum_{k=0}^{m}\gamma(t-k\delta)\int\limits_{[k\delta,(k+1)\delta]} \\
&\quad [f^1\big(\theta_k^1(s)\big)f^2\big(\theta_k^2(s)\big)-\varphi\big(\theta_k^1(s)-\theta_k^2(s)\big)]ds + O(\delta) \\
&= \sum_{k=0}^{m}\gamma(t-k\delta)\int\limits_{[k\delta,(k+1)\delta]} \\
&\quad \Bigg[\Big(\sum_{i=1}^{\infty} a_i^1\cos\big(i\theta_k^1(s)\big)+b_i^1\sin\big(i\theta_k^1(s)\big)\Big) \\
&\quad \times\Big(\sum_{j=1}^{\infty} a_j^2\cos\big(j\theta_k^2(s)\big)+b_j^2\sin\big(j\theta_k^2(s)\big)\Big) \\
&\quad -\varphi\big(\theta_k^1(s)-\theta_k^2(s)\big)\Bigg]ds + O(\delta).
\end{aligned} \tag{10.28}$$

Since conditions (10.2)–(10.4) are satisfied, it is possible to choose $O(\frac{1}{\delta})$ of sufficiently small time intervals of length $O(\delta^3)$, outside of which the functions

$f^p(\theta^p(t))$ and $f^p(\theta_k^p(t))$ are continuous. It is known that on each interval, which has no discontinuity points, Fourier series of the functions $f^1(\theta)$ and $f^2(\theta)$ converge uniformly. Then there exists a number $M = M(\delta) > 0$ such that outside sufficiently small neighborhoods of discontinuity points of $f^p(\theta^p(t))$ and $f^p(\theta_k^p(t))$, the sum of the first M series terms approximates the original function with accuracy to $O(\delta)$. In this case by relation (10.28) and the boundedness of $f^1(\theta)$ and $f^2(\theta)$ on $\mathbb{R}$, we obtain

$$g(t) - G(t) = \sum_{k=0}^{m} \gamma(t - k\delta) \int\limits_{[k\delta,(k+1)\delta]} \sum_{i=1}^{M}\sum_{j=1}^{M} \left[\mu_{i,j}(s) - \varphi\big(\theta_k^1(s) - \theta_k^2(s)\big)\right] ds + O(\delta), \tag{10.29}$$

where

$$\begin{aligned}\mu_{i,j}(s) = \frac{1}{2}\Big(& (a_i^1 a_j^2 + b_i^1 b_j^2)\cos(i\theta^1 - j\theta^2) \\ & +(-a_i^1 b_j^2 + b_i^1 a_j^2)\sin(i\theta^1 - j\theta^2) \\ & +(-b_i^1 b_j^2 + a_i^1 a_j^2)\cos(i\theta^1 + j\theta^2) \\ & +(a_i^1 b_j^2 + b_i^1 a_j^2)\sin(i\theta^1 + j\theta^2)\Big).\end{aligned}$$

From definition of δ and (10.22) it follows that $\forall i \in \mathbb{N}, j \in \mathbb{N}$ the relation

$$\int\limits_{[k\delta,(k+1)\delta]} \frac{1}{i} \cos\big(j(\omega_{min} s + \theta_0)\big) ds = \frac{O(\delta^2)}{ij} \tag{10.30}$$

is valid. Taking into account (10.30) and (10.2), one obtains the estimate

$$\int\limits_{[k\delta,(k+1)\delta]} b_j^p \cos\big(j\theta_k^p(s)\big) ds = \frac{O(\delta^2)}{j^2}.$$

A similar estimate is also valid for the addends with sin.

Consider the addend involving $\cos(i\theta_k^1(s) + j\theta_k^2(s))$ in $\mu_{i,j}(s)$. By (10.2) one can obtain $i\dot{\theta}^1(k\delta) + j\dot{\theta}^2(k\delta) \geq (i + j)\omega_{min}$. Then (10.30) yields the following relation

$$\int\limits_{[k\delta,(k+1)\delta]} \cos\Big(i\big(\theta^1(k\delta) + \dot{\theta}^1(k\delta)(s - k\delta)\big) + j\big(\theta^2(k\delta) + \dot{\theta}^2(k\delta)(s - k\delta)\big)\Big) ds$$

$$= \int\limits_{[k\delta,(k+1)\delta]} \cos\Big(\big(i\dot\theta^1(k\delta) + j\dot\theta^2(k\delta)\big)s - \big(i\theta^1(k\delta) + j\theta^2(k\delta)\big)k\delta + \big(i\dot\theta^1(k\delta) + j\dot\theta^2(k\delta)\big)k\delta\Big)ds = O\left(\frac{\delta^2}{i+j}\right). \tag{10.31}$$

Then

$$\sum_{i=1}^{M}\sum_{j=1}^{M}\int\limits_{[k\delta,(k+1)\delta]} \frac{-b_i^1 b_j^2 + a_i^1 a_j^2}{2}\cos\Big(i\left(\theta_k^1(s)\right) + j\left(\theta_k^2(s)\right)\Big)ds$$
$$= \sum_{i=1}^{M}\sum_{j=1}^{M}\frac{O(\delta^2)}{ij(i+j)}.$$

The convergence of series $\sum\limits_{i=1}^{\infty}\sum\limits_{j=1}^{\infty}\frac{1}{ij(i+j)}$ implies that the above expression is $O(\delta^2)$. Obviously, a similar relation occurs for the addend $\sin(i\theta_k^1(s) + j\theta_k^2(s))$.

Thus, by (10.29)

$$g(t) - G(t) = \sum_{k=0}^{m}\gamma(t-k\delta)\int\limits_{[k\delta,(k+1)\delta]}\Bigg[\sum_{i=1}^{M}\sum_{j=1}^{M}\Bigg\{\frac{a_i^1 a_j^2 + b_i^1 b_j^2}{2}\cos\Big(i\theta_k^1(s) - j\theta_k^2(s)\Big) + \frac{a_i^1 b_j^2 - b_i^1 a_j^2}{2}\sin\Big(i\theta_k^1(s) - j\theta_k^2(s)\Big)\Bigg\} - \varphi\big(\theta_k^1(s) - \theta_k^2(s)\big)\Bigg]ds + O(\delta).$$

Note that, here, the addends with indices $i = j$ give, in sum, $\varphi\big(\theta_k^1(s) - \theta_k^2(s)\big)$ with accuracy to $O(\delta)$. Consider the addends with indices $i < j$, involving cos (for the addends with indices $i > j$, involving sin, similar relations are satisfied). By (10.3), similar to (10.31), the following relation

$$\sum_{i=2}^{M}\sum_{j=1}^{i-1}\frac{a_i^1 a_j^2 + b_i^1 b_j^2}{2}\int\limits_{[k\delta,(k+1)\delta]}\cos\Big(i\left(\theta_k^1(s)\right) - j\left(\theta_k^2(s)\right)\Big)ds$$
$$= \sum_{i=2}^{M}\sum_{j=1}^{i-1} O(\delta^2)O\left(\frac{1}{ij\,|i-j|}\right) = O(\delta^2)$$

is valid (see Lemma 2). □

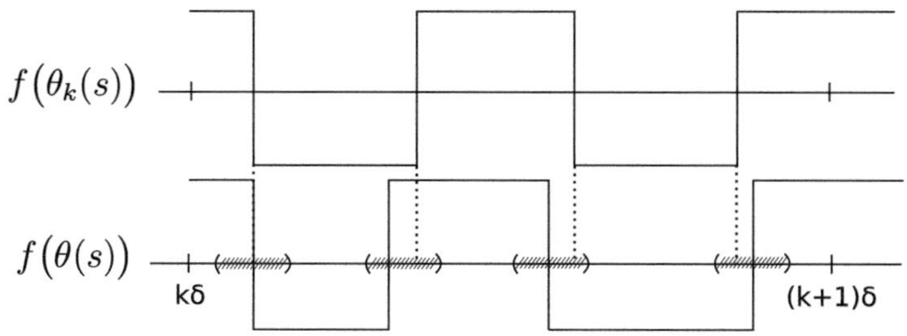

Fig. 10.12 Approximation of function $f(\theta(s))$ with the help of $f(\theta_k(s))$

Let us now proceed to the proof of the used auxiliary lemmas.

One can show that it is possible to choose sufficiently small neighborhoods of points of discontinuity of $f(\theta(t))$, in which there are also points of discontinuity of $f(\theta_k(s))$ (see. Fig. 10.12, indicated neighborhoods are hatched). Then $f(\theta(s))$ can be approximated with the help of $f(\theta_k(s))$ (see. Corollary 1).

Lemma 1. *Suppose, $f(\theta)$ is piecewise-differentiable 2π-periodic bounded function. $\theta(t)$ is a smooth function such that the conditions of high-frequency property* (10.2)–(10.4) *are satisfied. Then there exist sets $E_{\varepsilon,k}$ such that any ε-neighborhood of point of discontinuity of $f(\theta)$ acted by $\theta^{-1}(s)$ and $\theta_k^{-1}(s)$, attains the same interval, completely contained in $E_{\varepsilon,k}$, where*

$$\theta_k(s) = \theta(k\delta) + \dot{\theta}(k\delta)(s - k\delta), \tag{10.32}$$

in which case these sets are small:

$$\int\limits_{E_{\varepsilon,k}} ds = O(\delta^2). \tag{10.33}$$

Proof. By the data, $f(\theta)$ is bounded on $\mathbb{R}$. If $f(\theta)$ is continuous on $\mathbb{R}$, then the assertion of Lemma is obvious. Consider the case when $f(\theta)$ has at least 1 point of discontinuity.

Taking into account (10.4) and a smoothness of θ, it is possible to introduce the following notion

$$\omega_{min} \le m_k = \min_{[k\delta,(k+1)\delta]} \dot{\theta}(s),$$

$$\omega_{min} \le M_k = \max_{[k\delta,(k+1)\delta]} \dot{\theta}(s).$$

Then for $s \in [k\delta, (k+1)\delta]$ one obtains

$$\theta(k\delta) + m_k(s - k\delta) \le \theta(s) \le \theta(k\delta) + M_k(s - k\delta). \tag{10.34}$$

Then (10.32) implies

$$\theta(s) \in [\theta(k\delta), \theta(k\delta) + M_k\delta],$$
$$\theta_k(s) \in [\theta(k\delta), \theta(k\delta) + M_k\delta]. \tag{10.35}$$

Suppose, $a_1, a_2, \ldots a_N$ are discontinuity points of $f(\theta)$ such that

$$a_j \in [\theta(k\delta), \theta(k\delta) + M_k\delta]. \tag{10.36}$$

Here there are altogether $O(\frac{1}{\delta})$ intervals of length δ on $[0, T]$, on each of which the increase of $\dot{\theta}(s)$ is less than $\Delta\Omega$. Thus, $\dot{\theta}(s) \leq \omega_{min} + \Delta\Omega O(\frac{1}{\delta})$, i.e. $M_k = O(\frac{1}{\delta^2})$. If on interval $[0, 2\pi]$ the function $f(\theta)$ has $N_{[0,2\pi]}$ discontinuities, then on interval of the length $M_k\delta$ there are $N = \frac{1}{2\pi}M_k\delta N_{[0,2\pi]}$ discontinuities. However $M_k\delta = O(\frac{1}{\delta^2})\delta = O(\frac{1}{\delta})$. Thus, $N = O(\frac{1}{\delta})$.

Consider ε-neighborhoods

$$V^j_{\varepsilon,k} = (a_j - \varepsilon, a_j + \varepsilon),\ 0 < \varepsilon < \delta.$$

The choice of such neighborhoods becomes clear in proving the Lemma from the latter relations of (10.42).

Introduce the following notion

$$\left(\frac{a_j - \varepsilon - \theta(k\delta) + M_k k\delta}{M_k}, \frac{a_j + \varepsilon - \theta(k\delta) + m_k k\delta}{m_k}\right) = \tilde{E}^j_{\varepsilon,k}, \tag{10.37}$$

$$E^j_{\varepsilon,k} = \tilde{E}^j_{\varepsilon,k} \cap [k\delta, (k+1)\delta]. \tag{10.38}$$

In this case if $s \in [k\delta, (k+1)\delta] \setminus E^j_{\varepsilon,k}$, then $\theta(s), \theta_k(s)$ do not attain ε—neighborhoods of a_j, denoted by $V^j_{\varepsilon,k}$. Denote

$$E_{\varepsilon,k} = \bigcup_{j=1}^{N} E^j_{\varepsilon,k}. \tag{10.39}$$

This implies that condition (10.48) is satisfied.

Further, for $E_{\varepsilon,k}$, it will be proved that property (10.33) is satisfied. The following estimation

$$\int\limits_{E^j_{\varepsilon,k}} ds \leq \left(\frac{a_j + \varepsilon - \theta(k\delta) + m_k k\delta}{m_k} - \frac{a_j - \varepsilon - \theta(k\delta) + M_k k\delta}{M_k}\right) \tag{10.40}$$

is valid. Using (10.34), one obtains $|M_k - m_k| \leq C$. Then

$$\begin{aligned}
&\frac{a_j + \varepsilon - \theta(k\delta) + m_k k\delta}{m_k} - \frac{a_j - \varepsilon - \theta(k\delta) + M_k k\delta}{M_k} \\
&\leq \frac{\big(a_j - \theta(k\delta)\big)\big(M_k - m_k\big)}{M_k m_k} + \varepsilon\Big(\frac{1}{m_k} + \frac{1}{M_k}\Big) + \Big(\frac{m_k k\delta}{m_k} - \frac{M_k k\delta}{M_k}\Big) \\
&\leq \big(a_j - \theta(k\delta)\big)\frac{M_k - m_k}{M_k m_k} + \frac{2\varepsilon}{\omega_{min}} \\
&\leq \big(a_j - \theta(k\delta)\big)\frac{C}{M_k m} + \frac{2\varepsilon}{\omega_{min}}. \qquad (10.41)
\end{aligned}$$

By (10.36)

$$\begin{aligned}
&\big(a_j - \theta(k\delta)\big)\frac{C}{M_k m_k} + \frac{2\varepsilon}{\omega_{min}} \\
&\leq M_k\delta\frac{C}{M_k m_k} + \frac{2\varepsilon}{\omega_{min}} \\
&\leq \frac{C\delta}{m_k} + \frac{2\varepsilon}{\omega_{min}} \\
&\leq \frac{C\delta}{\omega_{min}} + \frac{2\varepsilon}{\omega_{min}} \\
&= O(\delta^3) + \varepsilon O(\delta^2). \qquad (10.42)
\end{aligned}$$

The relations (10.40), (10.41), and (10.42) imply

$$\int\limits_{E^j_{\varepsilon,k}} ds = O(\delta^3). \qquad (10.43)$$

Taking into account that the number of points of discontinuity is equal to $N = O\left(\frac{1}{\delta}\right)$, one proves the assertion of Lemma 1:

$$\int\limits_{E_{\varepsilon,k}} ds = \sum_{j=1}^{N} \int\limits_{E^j_{\varepsilon,k}} ds = O(\delta^2). \qquad (10.44)$$

□

Corollary 1. *Suppose, $f(\theta)$ is a piecewise-differentiable 2π-periodic bounded function. $\theta(t)$ is a smooth function and the conditions of high-frequency property* (10.2)–(10.4) *are satisfied.*

Then

$$\int_{k\delta}^{(k+1)\delta} f(\theta(s))ds = \int_{k\delta}^{(k+1)\delta} f(\theta_k(s)) + O(\delta^2), \tag{10.45}$$

where

$$\theta_k(s) = \theta(k\delta) + \dot{\theta}(k\delta)(s - k\delta). \tag{10.46}$$

Proof. By the data, $f(\theta)$ is bounded on $\mathbb{R}$. If $f(\theta)$ is continuous on $\mathbb{R}$, then the assertion of Lemma is obvious. Consider the case when $f(\theta)$ has at least 1 point of discontinuity. Since the conditions of Lemma 1 are satisfied, there exist sets $E_{\varepsilon,k}$. Then the use of (10.33) and the boundedness of $f(\theta)$ gives

$$\int_{[k\delta,(k+1)\delta]} f(\theta(s))ds = \int_{[k\delta,(k+1)\delta]\setminus E_{\varepsilon,k}} f(\theta(s))ds + O(\delta^2),$$

$$\int_{[k\delta,(k+1)\delta]} f(\theta_k(s))ds = \int_{[k\delta,(k+1)\delta]\setminus E_{\varepsilon,k}} f(\theta_k(s))ds + O(\delta^2). \tag{10.47}$$

In addition, according to assertion of Lemma 1, the functions $f(\theta)$ are differentiable with respect to θ and their derivatives are bounded for

$$\theta \in \{\theta(s)|s \in [k\delta, (k+1)\delta] \setminus E_{\varepsilon,k}\} \cup \{\theta_k(s)|s \in [k\delta, (k+1)\delta] \setminus E_{\varepsilon,k}\}, \tag{10.48}$$

i.e. on this set, $f(\theta)$ is Lipschitzian.

By (10.46) and (10.4)

$$\theta(s) = \theta(k\delta) + \int_{k\delta}^{(k+1)\delta} \dot{\theta}(v)dv = \theta_k(s) + O(\delta). \tag{10.49}$$

Then (10.48) and (10.49) yield

$$\int_{[k\delta,(k+1)\delta]\setminus E_{\varepsilon,k}} \left|f\left(\theta(s)\right) - f\left(\theta_k(s)\right)\right|ds = O(\delta^2),$$

$$\int_{[k\delta,(k+1)\delta]\setminus E_{\varepsilon,k}} f\left(\theta(s)\right)ds = \int_{[k\delta,(k+1)\delta]\setminus E_{\varepsilon,k}} f\left(\theta_k(s)\right)ds + O(\delta^2), \tag{10.50}$$

Then (10.47) implies the assertion of Lemma. □

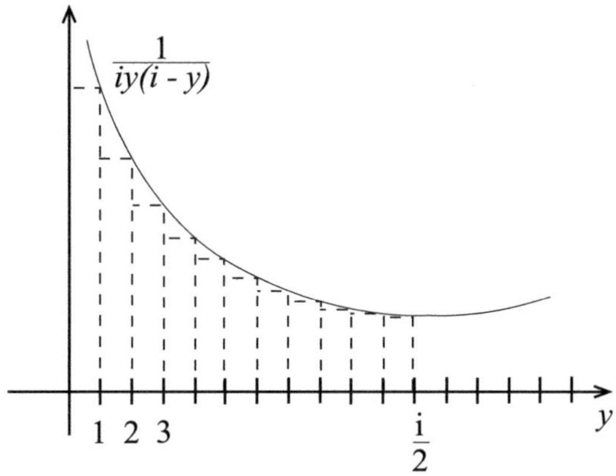

Fig. 10.13 Proof idea for Lemma 2

Lemma 2. *The following series* $\sum_{i=1}^{\infty} \sum_{j=1, j \neq i}^{\infty} \frac{1}{ij|i-j|}$ *converges.*

Proof. For Lemma to be proved, it is sufficient to prove convergence of the following series

$$\sum_{i=2}^{\infty} \sum_{j=1}^{i-1} \frac{1}{ij(i-j)} \tag{10.51}$$

Consider $i = 9$

$$\begin{aligned} &\frac{1}{9}\Big(\frac{1}{1(9-1)} + \frac{1}{2(9-2)} + \frac{1}{3(9-3)} + \frac{1}{4(9-4)} + \frac{1}{5(9-5)} + \frac{1}{6(9-6)} \\ &+ \frac{1}{7(9-7)} + \frac{1}{8(9-8)}\Big) = \frac{1}{9}\Big(\frac{1}{1(9-1)} + \frac{1}{2(9-2)} + \frac{1}{3(9-3)} \\ &+ \frac{1}{4(9-4)} + \frac{1}{(9-4)4} + \frac{1}{(9-3)3} + \frac{1}{(9-2)2} + \frac{1}{(9-1)1}\Big) \end{aligned} \tag{10.52}$$

This implies that it is sufficient to prove convergence of the following series

$$\sum_{i=2}^{\infty} \sum_{j=1}^{\lfloor \frac{i}{2} \rfloor} \frac{1}{ij(i-j)}, \tag{10.53}$$

where $\lfloor x \rfloor = \max\limits_{d \in \mathbb{Z},\ d \leq x} d$. Since

$$\frac{1}{iy(i-y)} \tag{10.54}$$

decreases for $y \in (0, \frac{i}{2}], i \geq 2$, one obtains that (Fig. 10.13)

$$\sum_{j=1}^{\lfloor \frac{i}{2} \rfloor} \frac{1}{ij(i-j)} \leq \frac{1}{i(i-1)} + \int_{1}^{\frac{i}{2}} \frac{1}{iy(i-y)} dy, \ i \geq 2, \tag{10.55}$$

However

$$\int_{1}^{\frac{i}{2}} \frac{1}{iy(i-y)} dy = \frac{1}{i^2} \Big(\log(y) - \log(i-y) \Big) \Big|_{1}^{\frac{i}{2}}$$

$$= \frac{1}{i^2} \Big(\log(\frac{i}{2}) - \log(i - \frac{i}{2}) - \log(1) + \log(i-1) \Big) = \frac{\log(i-1)}{i^2}. \tag{10.56}$$

It follows that a series

$$\sum_{i=2}^{\infty} \frac{\log(i-1)}{i^2} \tag{10.57}$$

converges. □

References

1. Abramovitch D (2002) Phase-locked loops: a control centric tutorial. Proc Am Control Conf 1:1–15
2. Abramovitch D (2008) Efficient and flexible simulation of phase locked loops, part I: simulator design. In: American control conference, Seattle, pp 4672–4677
3. Abramovitch D (2008) Efficient and flexible simulation of phase locked loops, part II: post processing and a design example. In: American control conference, Seattle, pp 4678–4683
4. Banerjee T, Sarkar B (2008) Chaos and bifurcation in a third-order digital phase-locked loop. Int J Electron Commun 62:86–91
5. Bellescize H (1932) La réception synchrone. L'onde Électrique 11:230–340
6. Best RE (2007) Phase-lock loops: design, simulation and application. McGraw-Hill, New York
7. Demir A, Mehrotra A, Roychowdhury J (2000) Phase noise in oscillators: a unifying theory and numerical methods for characterization. IEEE Trans Circuits Syst I 47:655–674
8. Emura T (1982) A study of a servomechanism for nc machines using 90 degrees phase difference method. Progress Report of JSPE, pp 419–421
9. Feely O (2007) Nonlinear dynamics of discrete-time circuits: a survey. Int J Circuit Theory Appl 35:515–531
10. Feely O, Curran PF, Bi C (2012) Dynamics of charge-pump phase-locked loops. Int J Circuit Theory Appl 27. doi:10.1002/cta.1814
11. Gardner F (1966) Phase–lock techniques. Wiley, New York
12. Gardner F (1993) Interpolation in digital modems - part i: fundamentals. IEEE Electron Commun Eng J 41(3):501–507
13. Gardner F, Erup L, Harris R (1993) Interpolation in digital modems - part ii: implementation and performance. IEEE Electron Commun Eng J 41(6):998–1008

14. Kroupa V (2003) Phase lock loops and frequency synthesis. Wiley, New York
15. Krylov N, Bogolyubov N (1947) Introduction to non-linear mechanics. Princeton University Press, Princeton
16. Kudrewicz J, Wasowicz S (2007) Equations of phase-locked loops: dynamics on the circle, torus and cylinder, A, vol 59. World Scientific, Singapore
17. Kuznetsov NV, Leonov GA, Seledzhi SS (2008) Phase locked loops design and analysis. In: Proceedings of ICINCO 2008 - 5th international conference on informatics in control, automation and robotics, vol SPSMC, pp 114–118. doi:10.5220/0001485401140118
18. Kuznetsov NV, Leonov GA, Seledzhi SM (2009) Nonlinear analysis of the Costas loop and phase-locked loop with squarer. In: Proceedings of the IASTED international conference on signal and image processing, SIP 2009, pp 1–7
19. Kuznetsov NV, Leonov GA, Seledzhi SM, Neittaanmäki P (2009) Analysis and design of computer architecture circuits with controllable delay line. In: Proceedings of ICINCO 2009 - 6th international conference on informatics in control, automation and robotics, vol 3 SPSMC, pp 221–224. doi:10.5220/0002205002210224
20. Kuznetsov NV, Leonov GA, Neittaanmäki P, Seledzhi SM, Yuldashev MV, Yuldashev RV (2010) Nonlinear analysis of phase-locked loop. In: IFAC proceedings volumes (IFAC-PapersOnline), vol 4(1), pp 34–38. doi:10.3182/20100826-3-TR-4016.00010
21. Kuznetsov NV, Leonov GA, Seledzhi SM, Yuldashev MV, Yuldashev RV (2011) Method for determining the operating parameters of phase-locked oscillator frequency and device for its implementation. Patent RU2449463 C1
22. Kuznetsov NV, Neittaanmäki P, Leonov GA, Seledzhi SM, Yuldashev MV, Yuldashev RV (2011) High-frequency analysis of phase-locked loop and phase detector characteristic computation. In: ICINCO 2011 - proceedings of the 8th international conference on informatics in control, automation and robotics vol 1, pp 272–278. doi:10.5220/0003522502720278
23. Kuznetsov NV, Leonov GA, Neittaanmäki P, Seledzhi S, Yuldashev MV, Yuldashev RV (2012) Simulation of phase-locked loops in phase-frequency domain. In: International congress on ultra modern telecommunications and control systems, IEEE Press, pp 364–368
24. Kuznetsov NV, Leonov GA, Yuldashev MV, Yuldashev RV (2012) Nonlinear analysis of Costas loop circuit. In: ICINCO 2012 - proceedings of the 9th international conference on informatics in control, automation and robotics 1:557–560. doi10.5220/0003976705570560
25. Leonov GA (2006) Phase-locked loops. Theory and application. Autom Remote Control 10:47–55
26. Leonov GA (2008) Computation of phase detector characteristics in phase-locked loops for clock synchronization. Dokl Math 78(1):643–645
27. Leonov GA, Kuznetsov NV, Seledzhi SM (2006) Analysis of phase-locked systems with discontinuous characteristics. In: IFAC proceedings volumes (IFAC-PapersOnline), vol 1, pp 107–112. doi10.3182/20060628-3-FR-3903.00021
28. Leonov GA, Kuznetsov NV, Seledzhi SM (2009) Nonlinear analysis and design of phase-locked loops. In: Automation control - theory and practice. In-Tech, New York, pp 89–114. doi:10.5772/7900
29. Leonov GA, Kuznetsov NV, Yuldahsev MV, Yuldashev RV (2011) Computation of phase detector characteristics in synchronization systems. Dokl Math 84(1):586–590. doi:10.1134/S1064562411040223
30. Leonov GA, Kuznetsov NV, Yuldahsev MV, Yuldashev RV (2012) Analytical method for computation of phase-detector characteristic. IEEE Trans Circuits Syst II Express Briefs 59(10):633–647. doi:10.1109/TCSII.2012.2213362
31. Leonov GA, Kuznetsov NV, Yuldashev MV, Yuldashev RV (2012) Differential equations of Costas loop. Dokl Math 86(2):723–728. doi:10.1134/S1064562412050080
32. Lindsey W (1972) Synchronization systems in communication and control. Prentice-Hall, New Jersey
33. Lindsey W, Simon M (1973) Telecommunication systems engineering. Prentice Hall, New Jersey

34. Margaris W (2004) Theory of the non-linear analog phase locked loop. Springer, New Jersey
35. Stiffler JP (1964) Bit and subcarrier synchronization in a binary psk communication system Natl Telemetering Conf
36. Suarez A, Quere R (2003) Stability analysis of nonlinear microwave circuits. Artech House, New Jersey
37. Suarez A, Fernandez E, Ramirez F, Sancho S (2012) Stability and bifurcation analysis of self-oscillating quasi-periodic regimes. IEEE Trans Microw Theory Tech 60(3):528–541
38. Thede L (2005) Practical analog and digital filter design. Artech House, New Jersey
39. Tretter SA (2007) Communication system design using DSP algorithms with laboratory experiments for the TMS320C6713TM DSK. Springer, New York
40. Viterbi A (1966) Principles of coherent communications. McGraw-Hill, New York

Chapter 11
Approaches to Defining and Measuring Assembly Supply Chain Complexity

V. Modrak and D. Marton

Abstract The present study examines static complexity of assembly supply chains (ASCs). While static complexity describes the structure of the supply chain, the number and the variety of its components, and interactions between relevant units; the dynamic complexity of supply chains involves the aspects of time and randomness. The aim is to come up with a methodological framework for conceptual modeling of ASC structures. Models of such ASC structures are divided into classes on the basis of the numbers of initial suppliers. Subsequently, we propose to apply different indices for measuring a structural complexity of ASC structures based on specific demand conditions. Special attention is also paid here to so-called Vertex Degree Index. It is a complexity measure originating from information theory and is based on the Shannon entropy. Finally, we outline a reference model for defining levels of parameterized complexity of ASC structures.

Keywords Static complexity • Assembly • Supply chain • Topological classes

11.1 Introduction

The main goal of assembly supply chains (ASCs) is to reduce uncertainty and thus help diminish the volatility of business results. General supply chain frequently involves three segments: upstream, where sourcing or procurement from external suppliers occurs; internal supply chain, where production, assembly, and packaging take place; and downstream, where distribution to customers takes place. Our focus is concentrated on the exploration of convergent ASCs commonly associated with automotive and similar industries. Recently, the studies of ASC systems are

---

V. Modrak (✉) • D. Marton
Faculty of Manufacturing Technologies, Department of Manufacturing Management, Technical University of Kosice, Presov, Slovakia
e-mail: vladimir.modrak@tuke.sk; david.marton@tuke.sk

J.A.T. Machado et al. (eds.), *Discontinuity and Complexity in Nonlinear Physical Systems*, Nonlinear Systems and Complexity 6, DOI 10.1007/978-3-319-01411-1_11,

mainly focused on stochastic models. Instead, our ambition is to examine the static complexity of ASCs. While static complexity describes the structure of the supply chain, the number and the variety of its components, and interactions between relevant units; the dynamic complexity of supply chains involves the aspects of time and randomness. Structural properties of ASCs are assumed to be especially important indicators at the early design stage when making a decision about a suitable networked manufacturing configuration. In this context any reduction of redundant complexity of ASC is considered as a way to increase organizational performance and reduce operational inefficiencies.

In general, high complexity of any nonlinear dynamic system including ASC systems makes it difficult to analyze, because a small change leads to a massive reaction. Nonlinear systems that are unpredictable cannot be solved exactly and need to be approximated. One way of how to approximate a dynamic complexity of such systems is to transform them into simpler ones. Therefore, structural complexity is linked to dynamical complexity. In structural complexity the main focus is on complexity classes, as opposed to the study of systems behavior to be conducted more efficiently. According to [1], "structural complexity investigates both internal structures of complexity classes, and relations that hold between different complexity classes." In this study our intent is to determine topological classes of ASCs and subsequently to determine a parameterized measure of topological complexity of such networks.

The aim is also to come up with a methodological framework for conceptual modeling of ASC structures. Models of such ASC structures are divided into classes on the basis of the numbers of initial suppliers. Subsequently, we propose to apply different indices for measuring a structural complexity of ASC structures based on specific demand conditions. Finally, we outline a reference model for defining levels of parameterized complexity of ASC structures.

11.2 Related Works

Supply chain can be defined in numerous ways. According to [2], supply chain is a network of organizations that are involved, through upstream and downstream linkages, in the different processes and activities that produce value in the form of products and services delivered to the ultimate consumer. The authors [3] add that each functional level of this network is represented by numerous facilities that along with the structure of the material and information flows contribute to the complexity of the chain.

In practical and theoretical approaches to system complexity issues it is useful to remember the formulation by [4] that "every good regulator of a system must be a model of that system." This principle can also be used in the diagnosis of failures of complex systems. Complexity of systems has many facets, some of which are mutually correlated. For example, Kolmogorov complexity [5, 6] is based on algorithmic information theory, which is related to Shannon entropy [7]. Both theories use the same unit the *bit* for measuring information. Shannon's information has been widely used in biological and ecological networks in the form of information indices, characterizing different aspects of chemical structure [8–10].

Another category of intricacy so-called stochastic complexity is defined using the concept of the minimum description length principle [11, 12]. Information theories consider information complexity as the minimum description size of a system [13–15]. Related pertinent findings with regard to the impact of organization size on increasing differentiation have been expressed in literature [16–18]. These authors maintain that increasing differentiation of networks creates a control problem of integrating the differentiated subunits. According to [19], the most basic issues in the study of complex networks are structural properties because structure always affects function. Moreover, he adds that there are missing unifying principles underlying their topology. The lack of such principles makes it difficult to evaluate certain topological aspects of networks including complexity.

There is a rich body of literature studying inventory models with supply uncertainty. Authors [20, 21] assume that the supplier has a random capacity, while authors [22, 23] model supply uncertainty using random yield. There are several papers, e.g., [23, 24], on assembly systems with random yield that determine the optimal assembly target level and the optimal order quantities of components.

Managing an ASC can be very difficult, since various sources of uncertainty are combined in the ASC. Uncertainty may result from customer's demand variability or unreliability in external suppliers [25]. In this context, various deterministic and stochastic models have been developed to study supply chain control and management [26, 27].

11.3 Classification of ASCs

Obviously, supply chains come in all shapes and sizes and can also be very specific. For the purposes of this work, the supply chain structure classification according to Fig. 11.1 has been used. Convergent class of structure that represents assembly-type of supply chains is that one in which each node in the chain has at most one successor, but may have any number of predecessors. This class of SC structures is matter of interest in this study. Convergent supply chains can be divided into two basic groups: Modular SCs and Non-modular SCs [29].

Moreover, it is suggested here to divide the Modular SCs into two specific categories: Modular SCs with minimal number of echelons and Modular SCs with maximal number of echelons. This categorization is conditioned on the requirement that number of initial nodes is the same for these two altered structures. In the modular configuration, the final producer purchases subcomponents from intermediate subassemblers instead of doing all the assembly activities itself. Modular assembly is typical for many industries, such as automotive, agricultural equipment,

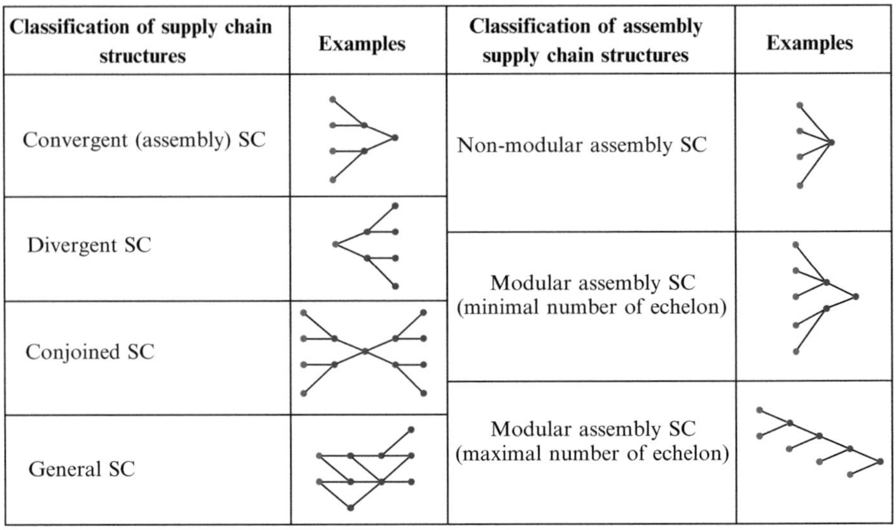

Fig. 11.1 Supply chain structure classification (adapted from [28])

aerospace, and others. In this context, it is proposed here to establish a framework for creating topological classes of ASCs.

11.4 Generating of ASC Classes

Assembly-type of supply chains is that one in which each node in the chain has at most one successor, but may have any number of predecessors. Such supply chain structures are convergent and can be divided into two types, modular and non-modular. In the modular structure, the intermediate sub-assemblers are understood as assembly modules, while the non-modular structure consists only from suppliers (initial nodes) and a final assembler (end node). The framework for creating topological classes of ASC structures follows the work [29] who outlined the way forward to model possible supply chain structures with four original suppliers as shown in Fig. 11.2.

Generating all possible combinations of structures brings enormous combinatorial difficulties. Thus, it is proposed here to establish a framework for creating topological classes of ASCs for non-modular and modular ASC structures based on number of initial nodes respecting the following rules [30]:

1. The initial nodes in topological alternatives are allocated to possible tiers t_l ($l = 1, \ldots, m$), except the tier t_m, in which is situated a final assembler,
2. The minimal number of initial nodes in the first tier t_l equals 2,

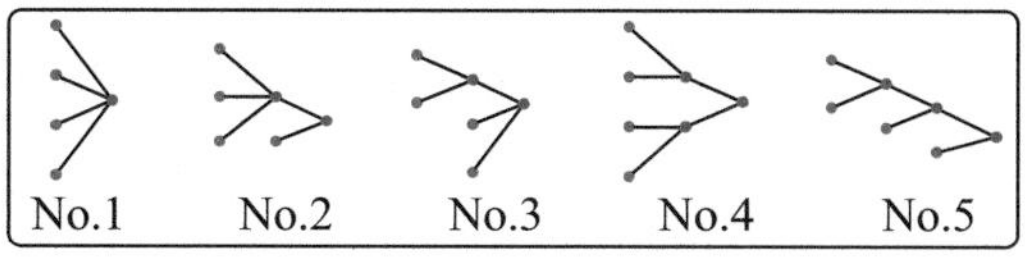

Fig. 11.2 Possible ASC structures with four initial suppliers (adopted from [29])

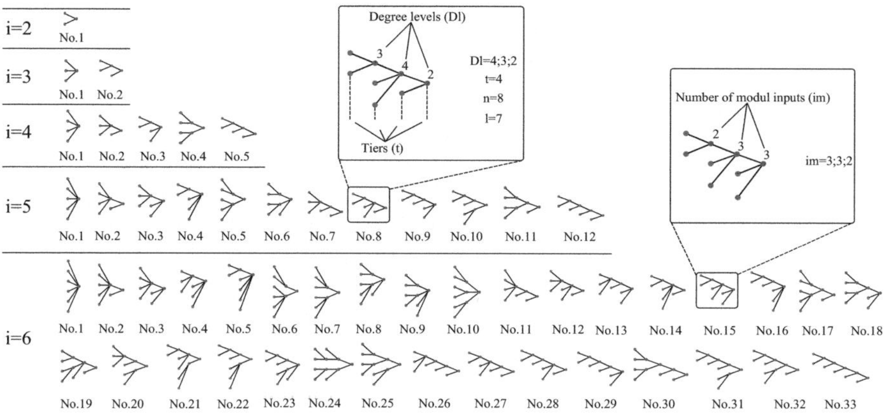

Fig. 11.3 Graphical models of the selected classes of ASC structures

3. In case of non-modular ASC structure, the number of initial nodes in the most upstream echelon is equal to the number of individual assembly parts or inputs (i = 1, . . . ,r).

Then, all possible structures for given number of initial nodes can be created. An example of generating the sets of structures for the classes with numbers of initial nodes from 2 to 6 is shown in Fig. 11.3.

The numbers of all possible ASC structures for arbitrary class of a structure can be determined by the following manner. We first need to calculate the sum of non-repeated combinations for each class of ASC structures through the so-called Cardinal Number [31]. The individual classes are determined by number of initial nodes (inputs) denoted by "i." Then, for any integer $i \geq 2$, we denote by S(i) the finite set consisting of all q-tuples $(i_1, \ldots, i_q)$ of integers $i_1, \ldots, i_q \geq 2$ with $i_1 + \cdots + i_q \leq i$, where q is a nonnegative integer.

The Cardinal Number #S(i) of S(i) is equal to $p(i) - 1$, where p(i) denotes the number of partition of "i," which increases quite rapidly with "i." For instance, for i = 2, 3, 4, 5, 6, 7, 8, 9, 10, the cardinal numbers #S(i) are given, respectively, by 1, 2, 4, 6, 10, 14, 21, 29, 41 [32].

Subsequently, for each non-repeated combination "K," a multiplication coefficient "M" has to be assigned (see Fig. 11.4). Then, $\sum M_i$—the number for all possible combinations of ASC structures for a given class can be obtained. This number is applied in Fig. 11.4.

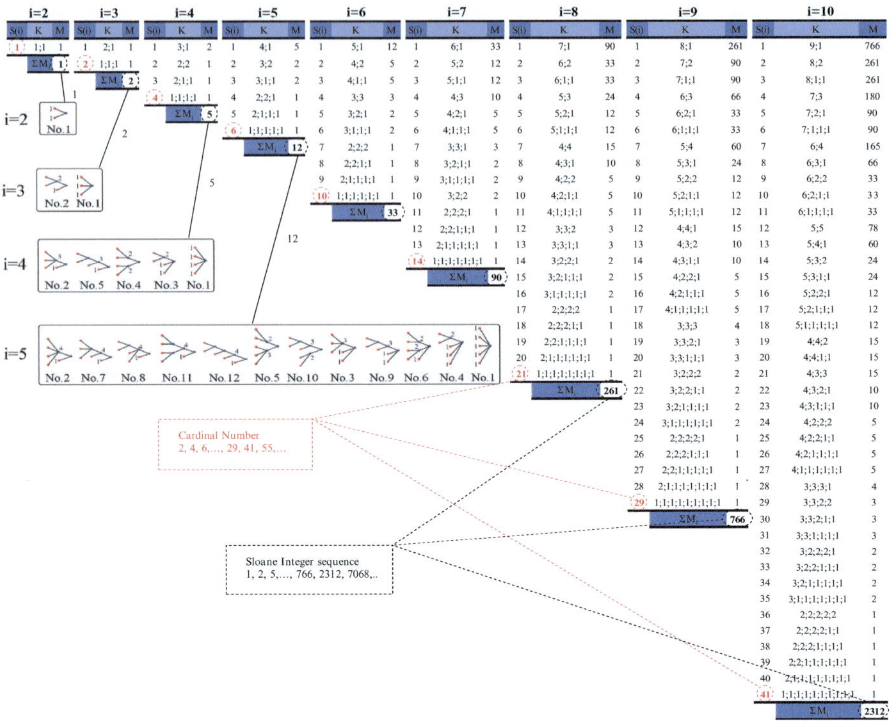

Fig. 11.4 Determination of total combinations of ASC networks related to the given classes

A critical step in determining all possible combinations of ASC structures for a given class (starting with a class for i = 2) is rules by which we can prescribe a multiplication coefficient "M."

In the case when we consider the number of initial nodes equal 2, there is only one numerical combination (1;1) corresponding with appropriate graphical model of ASC structure, and thus M = 1. Similarly, for each numerical combination has to be found exact logic rule. Accordingly we can formulate the following rules:

R1. If the numerical combination "K" consists only of numeric characters (digits), assigned by symbol "n," $n \leq 2$, then $M_{(2)} = 1$.

R2. If the numerical combination "K" consists just of one digit "3" and other digits are < 3 or do not appear respectively, then $M_{(3)} = 2$.

R3. If the numerical combination "K" consists just of one digit "4" and other digits are < 3 or do not appear, respectively, then $M_{(4)} = 5$.

Analogically, we can continue to determine multiplication coefficients "M" for similar cases when numerical combinations "K" consist just of one digit ≥ 5 and other digits are < 3 or do not appear, respectively. Then we would obtain the following multiplication coefficients: $M_{(5)} = 12$; $M_{(6)} = 33$; $M_{(7)} = 90$;

$M_{(8)} = 261$, etc. The multiplication coefficients, in such case, follow the Sloane Integer sequence 1, 2, 5, ..., 261, 766, 2312, 7068, ... [32].

For other cases has to be applied next rules:

R4. If the numerical combination "K" consists just of two digits "3" and other digits are < 3 or do not appear, respectively, then $M_{(3,3)} = 3$. Calculation of this multiplication coefficient can be formally expressed in this manner:

$$M_{(3,3)} = M_{(3)} + \left(M_{(3)} - 1\right) = 2 + 1 \Rightarrow M_{(3,3)} = 3 \tag{11.1}$$

R5. If the numerical combination "K" consists just of two digits "4" and other digits are < 3 or do not appear, respectively, then $M_{(4,4)} = 15$. $M_{(4,4)}$ can be computed similarly as Eq. (11.1):

$$\begin{aligned} M_{(4,4)} &= M_{(4)} + \left(M_{(4)} - 1\right) + \left(M_{(4)} - 2\right) + \left(M_{(4)} - 3\right) + \left(M_{(4)} - 4\right) \\ M_{(4,4)} &= 5 + 4 + 3 + 2 + 1 \Rightarrow M_{(4,4)} = 15 \end{aligned} \tag{11.2}$$

Analogically, we can continue to determine multiplication coefficients "M" for similar cases when numerical combinations "K" consist just of two digits ≥ 5 and other digits are < 3 or do not appear, respectively. For such cases we can calculate the multiplication coefficients by this equation:

$$M_{(n,n)} = M_{(n)} + \left(M_{(n)} - 1\right) + \left(M_{(n)} - 2\right) +, \ldots, + \left[M_n - \left(M_{(n)} - 1\right)\right] \tag{11.3}$$

11.5 Static Structural Complexity Metrics for ASC Structures

11.5.1 Some Terminology and Definitions

The following section consists of theoretical concepts and working definitions for the given research domain. General networks can be properly defined as well as effectively recognized as structural patterns by Graph Theory (GT). GT deals with the mathematical properties of structures as well as with problems of a general nature. In this context, a graph is a network of nodes (vertices) and links (edges) from some nodes to others or to themselves. Graph G consists of a set of *V vertices*, $\{V\} \equiv \{v_1, v_2, \ldots, v_V\}$, and the set of *E edges*, $\{E\} \equiv \{e_1, e_2, \ldots, eE\}$. The edge $\{ij\}$ is the path from Vertex i and ends in vertex j. The number of the nearest-neighbors of a Vertex "i" is termed vertex Degree and denoted $deg(v)$.

The maximum degree of a graph G, denoted by Δ(G), and the minimum degree of a graph, denoted by δ(G), are the maximum and minimum degree of its vertices. For a vertex, the number of head endpoints adjacent to a vertex is called the in-degree of the vertex and the number of tail endpoints is its out-degree. For a directed

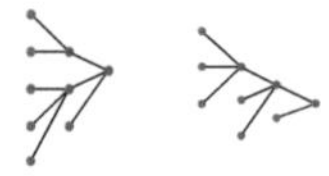

i=6				
No.	t	n	l	m
No.8	3	9	8	3
No.12	4	9	8	3

Fig. 11.5 Example of comparison of ASC structures with same number of links and modules but with different number of tiers (i = 6)

graph, the sum of the vertex in-degree and out-degree is the vertex degree [33].

$$\deg(v) = \deg^+(v)_i + \deg^-(v)_i \tag{11.4}$$

11.5.2 Approaches to Measuring ASC Complexity

Complexity reduction of convergent ASC systems is among topical questions under discussions in a framework of material flow optimization. Our interest in this context will be focused on selected measurable complexity indicators that have a potential to characterize configuration complexity attributes. In the proposed approaches this problem is treated for three different conditions:

- There is only one dominant final product among all the variants determined by a final product portfolio. They also showed that in the first scenario where one variant significantly dominates the demand, the optimal ASC with smallest complexity should be *non-modular* [28].
- Demand shares are equal across all variants determined by a final product portfolio. In the scenario of equal demand shares, the *modular* ASCs are more beneficial than non-modular ones when the product variety is rather large than small [28].
- We do not consider the above-mentioned specifications in such case we study ASC as general networks.

11.5.2.1 The Case When Dominant Demand Exists

Based on the previous premise for these scenarios two propositions can be formulated:

1. For a given class of ASS structures the optimal structure is one with the smallest number of links.
2. When comparing two or more structures with the same number of *links* "l," *nodes* "n," and *modules* "m" but with different number of tiers "t" (see Fig. 11.5), the following argument can be constructed:

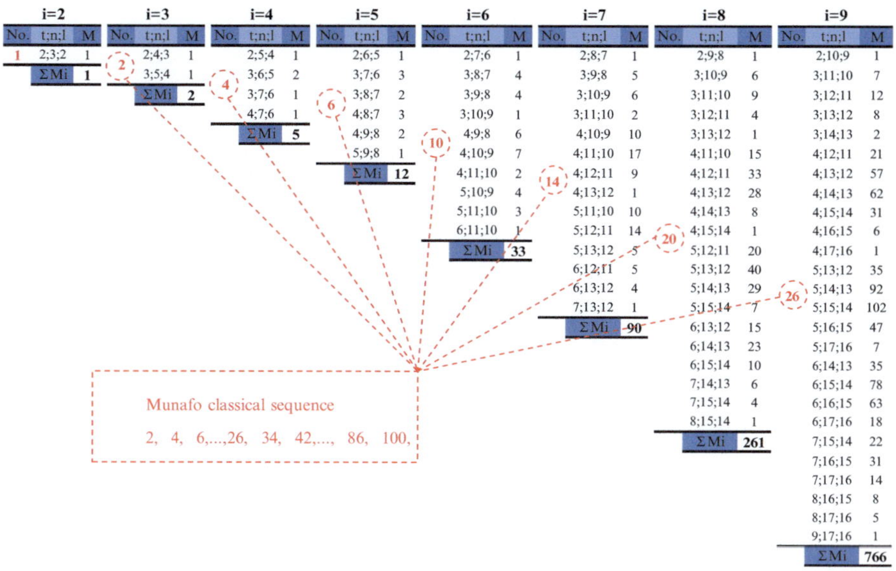

Fig. 11.6 Non-repeated sets of ASC structures based on t,n,l parameters

The structure with the smallest number of tiers is topologically less complex than other one(s). Then, it is proposed to measure structural complexity by formula Links/Tiers Index [34, 35]:

$$LTI = \sum_{j=1}^{p} \sum_{l=1}^{m} l_j . t_l 0, 1 \tag{11.5}$$

In order to have, at our disposal, all non-repeated ASC structures of selective classes we need to clear redundant structures with the identical parameters "t, n, l." To do so, we can determine exact numbers of all non-repeated ASC structures that are shown in Fig. 11.6. These numbers follow a classical sequence MCS6858778 introduced by Munafo and used by, e.g., [36].

When applying Eq. 11.5 to calculate structural complexity measures for all non-repeated ASC structures of the selective classes (for i = 4–10) we obtain values that are depicted in Fig. 11.7. Subsequently, it is possible to use these values for comparison of arbitrary structures from the structural complexity point of view.

Naturally the next question arises about a mechanism of generating all non-repeated ASC structures for the higher relevant classes (from i = 11, 12, . . . ,n). This mechanism is graphically outlined in Fig. 11.8.

The principle of generating sub-sets of non-repeated structures (assigned with blocks in the figure above) from lower classes to higher classes is quite simple, but its formal description would require rather complicated procedures.

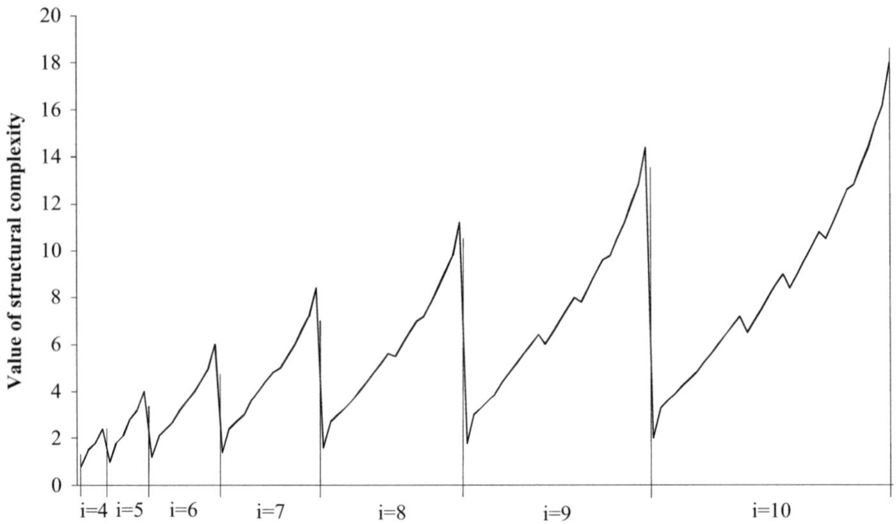

Fig. 11.7 Computational results of the LTI for selected classes of ASC structures

11.5.2.2 The Case When Dominant Demand Does Not Exists

According to the assumption for this scenario, modular ASCs are more beneficial than non-modular ones. Authors [29] of this premise showed that, e.g., for the structures in Fig. 11.9 the following relation can be formulated:

$$\text{Complexity (I)} > \text{Complexity (II)} > \text{Complexity (III)}.$$

Considering this assumption, it is proposed the following parameterization with aim to obtain measures that allow comparing complexity of structures:

1. To split a given structure into substructures which are represented by Non-modular ones, the number of which is just equal to sum of the intermediate subassemblers plus one assembler of final products (see Fig. 11.10),
2. To calculate a structural complexity for each substructure of original structure,
3. To calculate a total structural complexity of an original structure.

For step 2, to measure substructure complexity, the following parameter of *Module Degree* can be formulated:

$$\deg(m)_i = (i_m - 1)^2, \tag{11.6}$$

where i_m presents a number of module inputs ($i_m = 1, \ldots, r$) of given Non-modular structure.

For step 3, to measure the Index of Module Degree, the following formula is used:

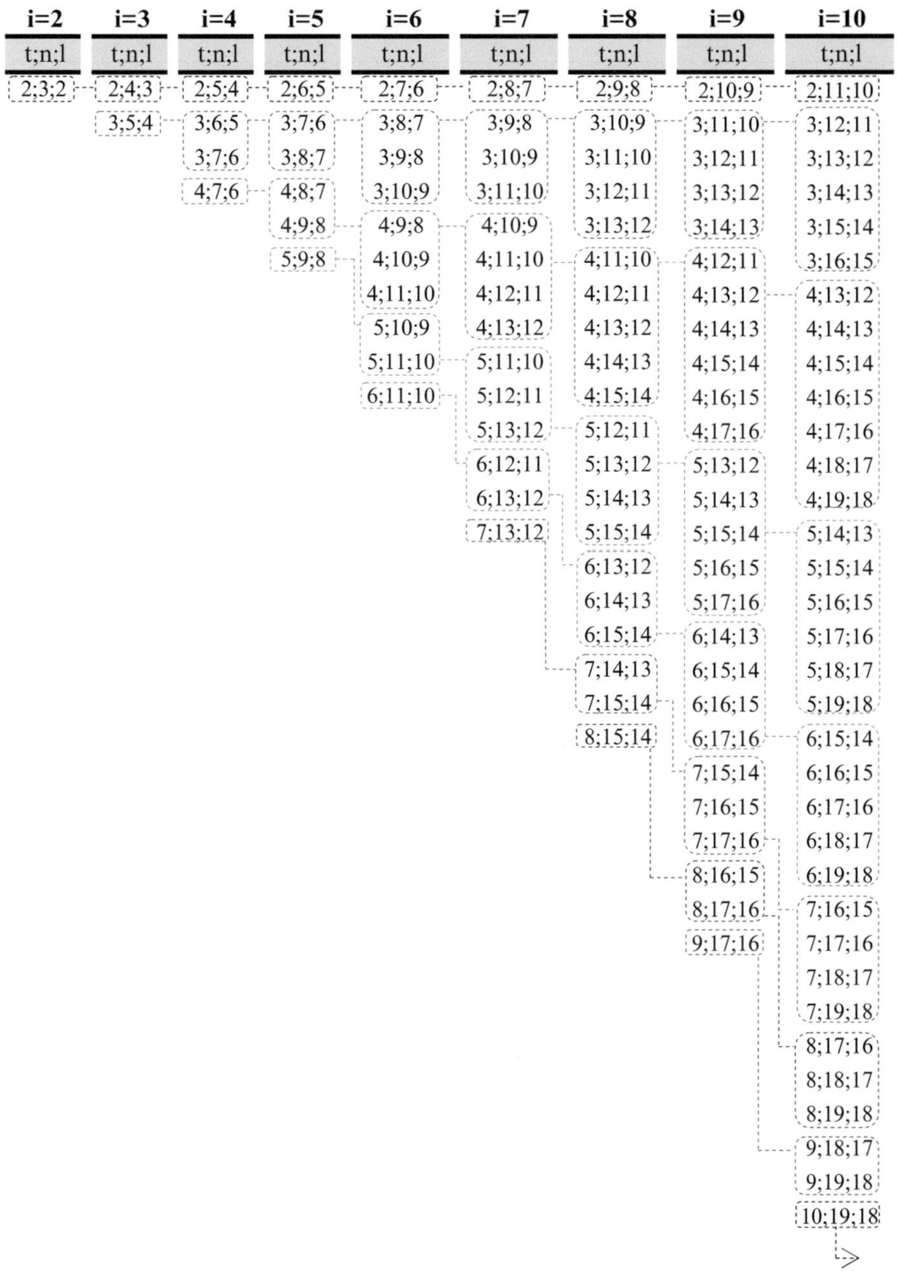

Fig. 11.8 The mechanism of generating subsets of non-repeated structures based on t,n,l parameters

Fig. 11.9 Example of Modular ASC structures (i = 8)

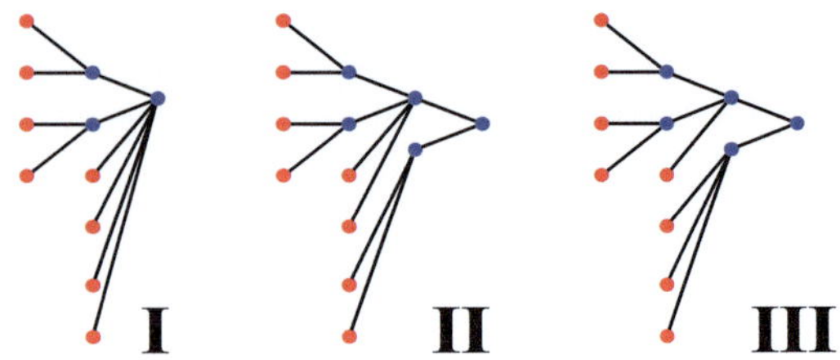

Fig. 11.10 Substructures of the original structure (i = 8)

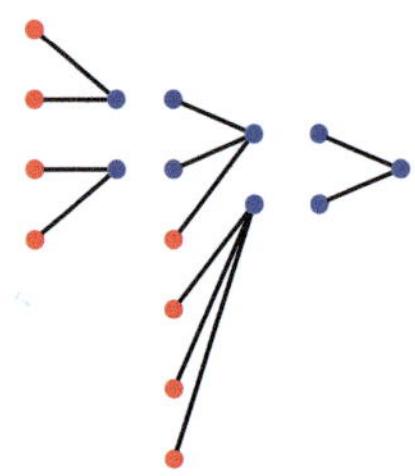

i=2

No.	i_m	M
1	2	1
	ΣM_i	**1**

i=3

No.	i_m	M
2	3	1
	2;2	1
	ΣM_i	**2**

i=4

No.	i_m	M
	4	1
3	3;2	2
	2;2;2	2
	ΣM_i	**5**

i=5

No.	i_m	M
	5	1
	4;2	2
5	3;3	1
	3;2;2	5
	2;2;2;2	3
	ΣM_i	**12**

i=6

No.	i_m	M
	6	1
	5;2	2
	4;3	2
7	4;2;2	5
	3;3;2	5
	3;2;2;2	12
	2;2;2;2;2	6
	ΣM_i	**33**

i=7

No.	i_m	M
	7	1
	6;2	2
	5;3	2
	5;2;2	5
	4;4	1
11	4;3;2	9
	4;2;2;2	12
	3;3;3	2
	3;3;2;2	17
	3;2;2;2;2	28
	2;2;2;2;2;2	11
	ΣM_i	**90**

i=8

No.	i_m	M
	8	1
	7;2	2
	6;3	2
	6;2;2	5
	5;4	2
	5;3;2	9
	5;2;2;2	12
15	4;4;2	5
	4;3;3	5
	4;3;2;2	33
	4;2;2;2;2	29
	3;3;3;2	12
	3;3;2;2;2	55
	3;2;2;2;2;2	66
	2;2;2;2;2;2;2	23
	ΣM_i	**261**

i=9

No.	i_m	M
	9	1
	8;2	2
	7;3	2
	7;2;2	5
	6;4	2
	6;3;2	9
	6;2;2;2	12
	5;5	1
	5;4;2	9
	5;3;3	5
22	5;3;3;2	34
	5;2;2;2;2	28
	4;4;3	5
	4;4;2;2	17
	4;3;3;2	33
	4;3;2;2;2	109
	4;2;2;2;2;2	66
	3;3;3;3	4
	3;3;3;2;2	57
	3;3;2;2;2;2	167
	3;2;2;2;2;2;2	155
	2;2;2;2;2;2;2;2	43
	ΣM_i	**766**

Sloane integer sequence

2, 3, 5,..., 15, 22, 30,..., 101, 135, 176,

Fig. 11.11 Non-repeated sets of ASC structures based on the number of module inputs parameter

$$I_{md} = \sum_{s=1}^{q} \deg{(m)_i} \tag{11.7}$$

In order to have, at our disposal, all non-repeated ASC structures of selective classes we need to clear redundant structures with the identical parameter named as "number of module inputs" (i_m). To do so, we can determine exact numbers of all non-repeated ASC structures that are shown in Fig. 11.11. These numbers follow the integer sequence A000041 introduced by [37].

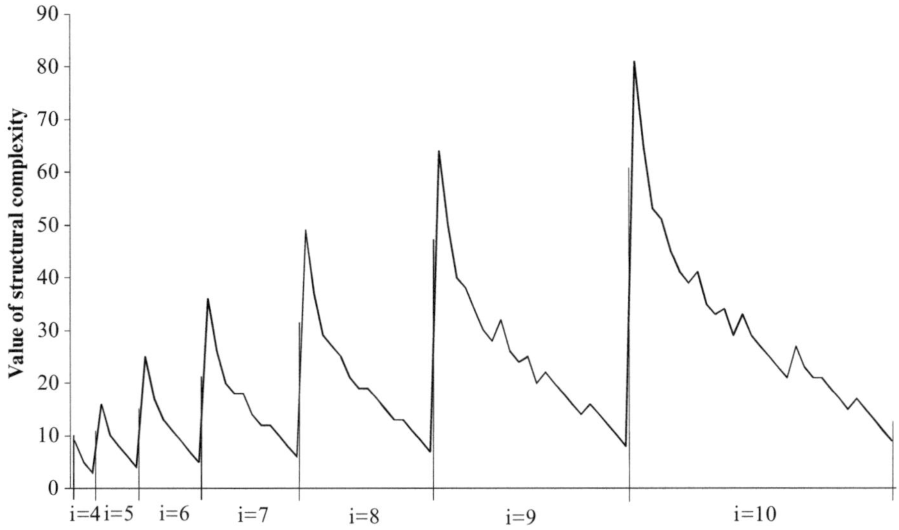

Fig. 11.12 Computational results of the I_{md} for selected classes of ASC structures

When applying Eq. 11.7 to calculate structural complexity measures for all non-repeated ASC structures of the selective classes (for i = 4–10) we obtain values that are depicted in Fig. 11.12. Subsequently, it is possible to use these values for comparison of arbitrary structures from the structural complexity point of view.

Anyway, the concern is obviously about finding a mathematical model (mechanism) of generating all non-repeated ASC structures for the higher relevant classes (from i = 11, 12, . . . ,n). The mechanism is graphically outlined in Fig. 11.13.

11.5.2.3 The Case When We Consider ASC as General Networks

According to Shannon's information theory [7], *the entropy of information* $H(\alpha)$ in describing a message of N system elements (or symbols), distributed according to some equivalence criterion α into k groups of $N_1, N_2, \ldots, N_k$ elements, is calculated by the formula:

$$H\ (\alpha) = -\sum_{i=1}^{k} p_i \log_2 p_i = -\sum_{i=1}^{k} \frac{N_i}{N} \log_2 \frac{N_i}{N} \tag{11.8}$$

where p_i specifies the probability of occurrence of the elements of the ith group.

Since it is of interest to characterize *entropy of information* of a network according to Eq. 11.8, it is possible to substitute symbols or system elements for the vertices.

i=2	i=3	i=4	i=5	i=6	i=7	i=8	i=9	i=10
i_m	i_m	i_m	i_m	i_m	i_m	i_m	i_m	i_m
2	3	4	5	6	7	8	9	10
	2;2	3;2	4;2	5;2	6;2	7;2	8;2	9;2
		2;2;2	3;3	4;3	5;3	6;3	7;3	8;3
			3;2;2	4;2;2	5;2;2	6;2;2	7;2;2	8;2;2
			2;2;2;2	3;3;2	4;4	5;4	6;4	7;4
				3;2;2;2	4;3;2	5;3;2	6;3;2	7;3;2
				2;2;2;2;2	4;2;2;2	5;2;2;2	6;2;2;2	7;2;2;2
					3;3;3	4;4;2	5;5	6;5
					3;3;2;2	4;3;3	5;4;2	6;4;2
					3;2;2;2;2	4;3;2;2	5;3;3	6;3;3
					2;2;2;2;2;2	4;2;2;2;2	5;3;3;2	6;3;3;2
						3;3;3;2	5;2;2;2;2	6;2;2;2;2
						3;3;2;2;2	4;4;3	5;5;2
						3;2;2;2;2;2	4;4;2;2	5;4;3
						2;2;2;2;2;2;2	4;3;3;2	5;4;2;2
							4;3;2;2;2	5;3;3;2
							4;2;2;2;2;2	5;3;2;2;2
							3;3;3;3	5;2;2;2;2;2
							3;3;3;2;2	4;4;4
							3;3;2;2;2;2	4;4;3;2
							3;2;2;2;2;2;2	4;4;2;2;2
							2;2;2;2;2;2;2;2	4;3;3;3
								4;3;3;2;2
								4;3;2;2;2;2
								4;2;2;2;2;2;2
								3;3;3;3;2
								3;3;3;2;2;2
								3;3;2;2;2;2;2
								3;2;2;2;2;2;2;2
								2;2;2;2;2;2;2;2;2

Fig. 11.13 The graphical principle of generating non-repeated structures based on i_m parameter

In order to define the probability for a randomly chosen system element i it is possible to formulate general weight function as $p_i = w_i/\Sigma w_i$, assuming that $\Sigma p_i = 1$.

Author [38] claims that, considering the system elements, the vertices and supposing the weights assigned to each vertex to be the corresponding vertex degrees, one easily distinguishes the null complexity of the totally disconnected graph from the high complexity of the complete graph.

Then, the probability for a randomly chosen vertex i in the complete graph of V vertices to have a certain degree $\deg(v)_i$ can be expressed by the formula:

$$p_i = \frac{\deg(v)_i}{\sum_{i=1}^{V} \deg(v)_i}. \tag{11.9}$$

Based on our previous experiences the most feasible indicator to configuration complexity of general structures seems to be *Vertex degree index* (I_{vd}). The information entropy of a graph with a total weight W and vertex weights w_i can be expressed in the form of the equation:

$$H(W) = W\log_2 W - \sum_{i=1}^{V} w_i \log_2 w_i \tag{11.10}$$

Since the maximum entropy is when all $w_i = 1$, then:

$$H_{\max} = W\log_2 W \tag{11.11}$$

By substituting $W = \sum \deg(v)_i$ and $w_i = \deg(v)_i$, the information content of the vertex degree distribution of a network called as Vertex Degree Index (I_{vd}) is derived by [38] that is expressed as follows:

$$I_{vd} = \sum_{i=1}^{V} \deg(v)_i \log_2 \deg(v)_i \tag{11.12}$$

In order to have, at our disposal, all non-repeated ASC structures of selective classes we need to clear redundant structures with the identical parameter named as “vertex degree” ($\deg(v)_i$). To do so, we can determine exact numbers of all non-repeated ASC structures that are shown in Fig. 11.14. These numbers follow the integer sequence A139582 introduced by [39].

When applying Eq. 11.12 to calculate structural complexity measures for all non-repeated ASC structures of the selective classes (for i = 4–10) we obtain values that are depicted in Fig. 11.15.

Subsequently, it is possible to use these values for comparison of arbitrary structures from the structural complexity point of view.

In order to find a mathematical model of generating all non-repeated ASC structures for the higher relevant classes (from i = 11, 12,…,n) the possible mechanism is graphically outlined in Fig. 11.16.

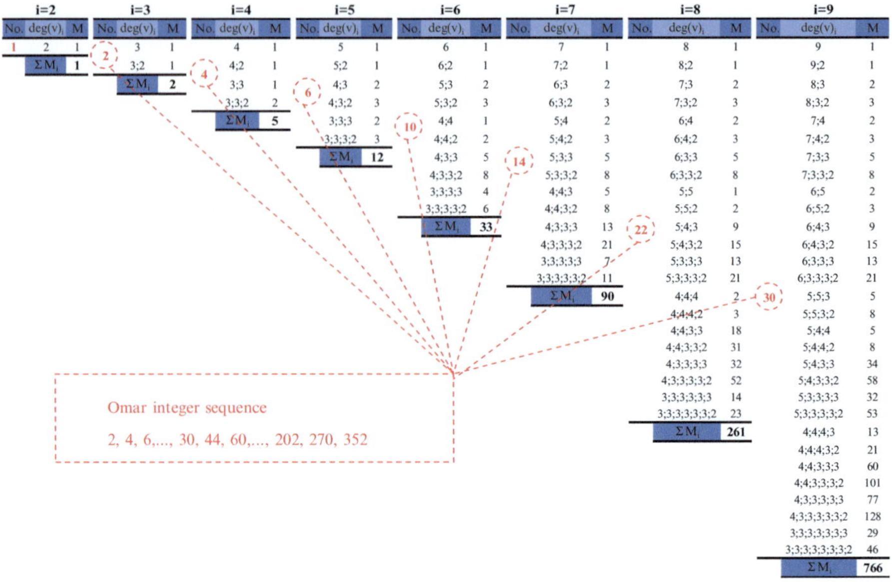

Fig. 11.14 Non-repeated ASC structures based on the "vertex degree" parameter

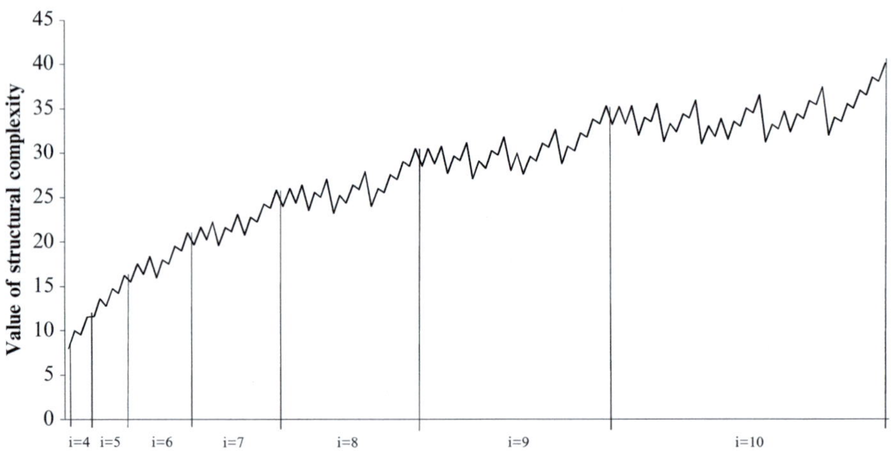

Fig. 11.15 Computational results of the I_{vd} for selected classes of ASC structures

11.6 Reference Model for Defining Complexity Levels of Supply Chain Structures

Basically, the comparison of complexity is of a relative and subjective nature. It is also clear that through a relative complexity metric we can compare the complexity of the existing configuration against the simplest one. Perhaps, the most important

i=2	i=3	i=4	i=5	i=6	i=7	i=8	i=9	i=10
$deg(v)_i$	$deg(v)_i$	$deg(v)_i$	$deg(v)_i$	$deg(v)_i$	$deg(v)_i$	$deg(v)_i$	$deg(v)_i$	$deg(v)_i$
2	3	4	5	6	7	8	9	10
	3;2	4;2	5;2	6;2	7;2	8;2	9;2	10;2
		3;3	4;3	5;3	6;3	7;3	8;3	9;3
		3;3;2	4;3;2	5;3;2	6;3;2	7;3;2	8;3;2	9;3;2
			3;3;3	4;4	5;4	6;4	7;4	8;4
			3;3;3;2	4;4;2	5;4;2	6;4;2	7;4;2	8;4;2
				4;3;3	5;3;3	6;3;3	7;3;3	8;3;3
				4;3;3;2	5;3;3;2	6;3;3;2	7;3;3;2	8;3;3;2
				3;3;3;3	4;4;3	5;5	6;5	7;5
				3;3;3;3;2	4;4;3;2	5;5;2	6;5;2	7;5;2
					4;3;3;3	5;4;3	6;4;3	7;4;3
					4;3;3;3;2	5;4;3;2	6;4;3;2	7;4;3;2
					3;3;3;3;3	5;3;3;3	6;3;3;3	7;3;3;3
					3;3;3;3;3;2	5;3;3;3;2	6;3;3;3;2	7;3;3;3;2
						4;4;4	5;5;3	6;6
						4;4;4;2	5;5;3;2	6;6;2
						4;4;3;3	5;4;4	6;5;3
						4;4;3;3;2	5;4;4;2	6;5;3;2
						4;3;3;3;3	5;4;3;3	6;4;4
						4;3;3;3;3;2	5;4;3;3;2	6;4;4;2
						3;3;3;3;3;3	5;3;3;3;3	6;4;3;3
						3;3;3;3;3;3;2	5;3;3;3;3;2	6;4;3;3;2
							4;4;4;3	6;3;3;3;3
							4;4;4;3;2	6;3;3;3;3;2
							4;4;3;3;3	5;5;4
							4;4;3;3;3;2	5;5;4;2
							4;3;3;3;3;3	5;5;3;3
							4;3;3;3;3;3;2	5;5;3;3;2
							3;3;3;3;3;3;3	5;4;4;3
							3;3;3;3;3;3;3;2	5;4;4;3;2
								5;4;3;3;3
								5;4;3;3;3;2
								5;3;3;3;3;3
								5;3;3;3;3;3;2
								4;4;4;4
								4;4;4;4;2
								4;4;4;3;3
								4;4;4;3;3;2
								4;4;3;3;3;3
								4;4;3;3;3;3;2
								4;3;3;3;3;3;3
								4;3;3;3;3;3;3;2
								3;3;3;3;3;3;3;3
								3;3;3;3;3;3;3;3;2

Fig. 11.16 The graphical principle of generating non-repeated structures based on vertex degree parameter

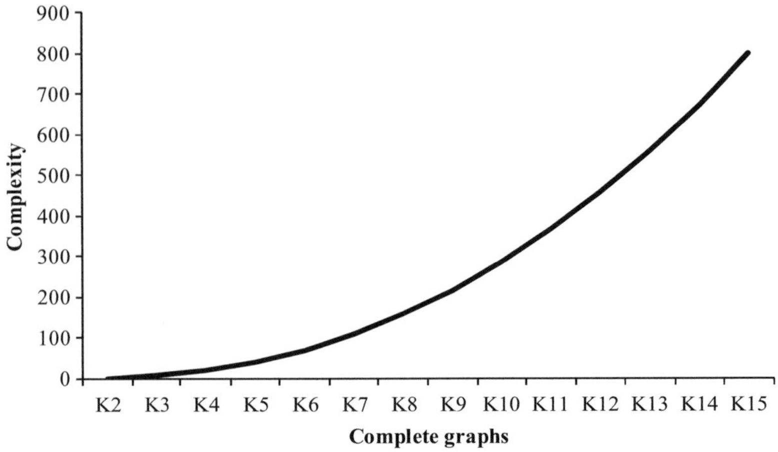

Fig. 11.17 Graph of the complexity measures for the selected complete graphs

feature of the relative complexity metric is that we can generalize it to other areas [40].

Comparing the results of the three structural complexity indicators from graphs (shown in Figs. 11.7, 11.12 and 11.15) the Vertex degree index seems to be suitable as generic approach for use in most structural applications. To test this anticipation we apply this indicator in order to compare complexities of the complete graphs with $v(v-1)/2$ edges (see Fig. 11.17). Subsequently, one can determine upper bounds of configuration complexity for any general supply chain structure with a given number of vertices.

Under an assumption that we will consider upper bound value of a complete graph with V vertices as lower bound for a complete graph with $V+1$ vertices, then we suggest to rate these complexity values of complete graphs complexity as boundaries of levels configuration complexity of general supply chain networks.

Based on these considerations, the reference model for defining levels of parameterized complexity of supply chain networks is outlined in Fig. 11.18.

11.7 Conclusions

The main contributions of this paper we see in the following three aspects:

1. A new framework for creating topological classes of ASC networks under defined specific condition is developed (see Fig. 11.4). This methodological framework is enabling exactly determining all relevant topological graphs for any class of ASC structure. The usefulness of such framework is especially in

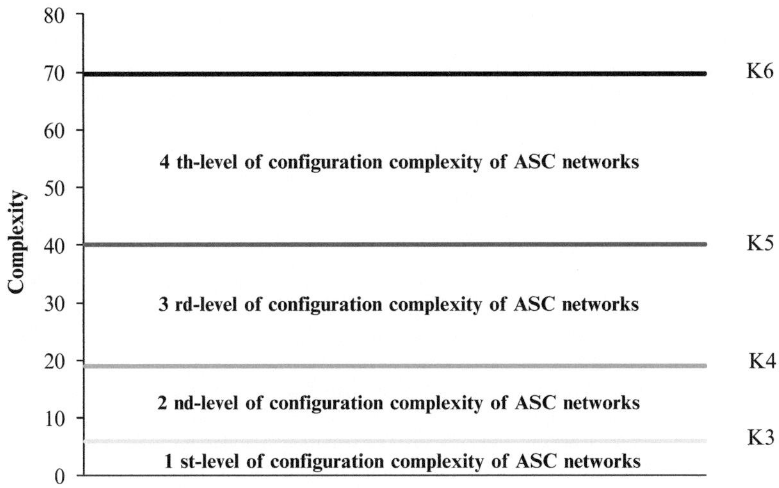

Fig. 11.18 Reference model for defining levels of configuration complexity of SC networks

cases when it is necessary to apply relative complexity metrics to compare the complexity of the existing configuration against the simplest one.

2. As the second contribution of this paper, it is showed that structural complexity of ASC networks can be dependable measured for the three different scenarios through the proposed formulas. Potentially, these structural complexity measures can be used to find or create optimal ASC configurations according to one of the specific criteria.
3. The reference model for defining levels of configuration complexity of general supply chain networks is outlined based on an upper bound concept.

However, this research path requires further independent research to confirm presented results and proposals.

Acknowledgments This paper has been supported by KEGA project "The Development of a Web Learning System to Support an External Form of Education in Study Program Manufacturing Management" (no. 054TUKE-4/2012) and VEGA project "The Development and Application of a Genetic Algorithm for Planning and Scheduling of One piece low Lines" (no. 1/1028/11) granted by the Ministry of Education of the Slovak Republic.

References

1. Hartmanis J (1988) New developments in structural complexity theory. Lect Notes Comput Sci 317:271–286
2. Christopher M (1998) Logistics and supply chain management: strategies for reducing cost and improving services. Financial Times Pitman Publishing, London

3. Beamon BM, Chen VCP (2001) Performance analysis of conjoined supply chains. Int J Prod Res 39:3195–3218
4. Ashby WR (1956) An introduction to cybernetics. Chapman and Hall, London
5. Kolmogorov AN (1965) Three approaches to the quantitative definition of information. Probl Inf Transm 1:1–7
6. Chaitin G (1966) On the length of programs for computing finite binary sequences. J Assoc Comput Mach 13:547–569
7. Shannon CE (1948) A mathematical theory of communication. Bell Syst Tech J 27:379–423
8. Bonchev D, Trinajstic N (1977) Information theory, distance matrix and molecular branching. J Chem Phys 67:4517–4533
9. Bonchev D, Kamenska V (1978) Information theory in describing the electronic structure of atoms. Croat Chem Acta 51:19–27
10. Bonchev D (1979) Information indices for atoms and molecules. MATCH Commum Math Comput Chem 7:65–113
11. Rissanen J (1978) Modeling by shortest data description. Automatica 14:445–471
12. Grünwald P (1986) The minimum description length principle. MIT, Cambridge
13. Grassberger P (1986) Towards a quantitative theory of self-generated complexity. Int J Theor Phys 25:907–938
14. Grassberger P (1991) Information and complexity measures in dynamical systems. In: Atmanspacher H, Scheingraber H (eds) Information dynamics. Plenum, New York, pp 15–33
15. Crutchfield JP, Young K (1989) Inferring statistical complexity. Phys Rev Lett 63:105
16. Ouchi WG (1977) The relationship between organizational structure and organizational control. Adm Sci Q 22:95–113
17. Blau PM, Scott WR (1962) Formal organizations. Chandler, San Francisco
18. Blau PM, Schoenherr RA (1971) The structure of organizations. Basic Books, New York
19. Strogatz SH (2001) Exploring complex networks. Nature 410:268–276
20. Ciarallo FW, Akella R, Morton TE (1994) A periodic reviews, production planning model with uncertain capacity and uncertain demand optimality of extended myopic policies. Manag Sci 40:320–332
21. Wang Y, Gerchak Y (1996) Periodic review production models with variable capacity, random yield, and uncertain demand. Manag Sci 42:130–137
22. Gupta D, Cooper WL (2005) Stochastic comparisons in production yield management. Oper Res 53:377–384
23. Bollapragada R, Rao US, Zhang J (2004) Managing inventory and supply performance in assembly systems with random supply capacity and demand. Manag Sci 50:1729–1743
24. Gurnani H, Akella R, Lehoczky J (2000) Supply management in assembly systems with random yield and random demand. IIE Trans 32:701–714
25. Mahnam M, Yadollahpour MR, Famil-Dardashti V, Hejazi SR (2009) Supply chain modeling in uncertain environment with bi-objective approach. Comput Ind Eng 56:1535–1544
26. Ishii K, Takahashi K, Muramatsu R (1988) Integrated production, inventory and distribution systems. Int J Prod Res 26:473–482
27. Williams JF (1981) Heuristic techniques for simultaneous scheduling of production and distribution in multi-echelon structures: theory and empirical comparisons. Manag Sci 27: 336–352
28. Wang H, Ko J, Zhu X, Hu SJ (2010) A complexity model for assembly supply chain and its application to configuration design. J Manuf Sci Eng 132:1–46
29. Hu SJ, Zhu XW, Wang H, Koren Y (2008) Product variety and manufacturing complexity in assembly systems and supply chains. CIRP Ann Manuf Technol 57:45–48
30. Modrak V, Marton D, Kulpa W, Hricova R (2012) Unraveling complexity in assembly supply chain networks. In: Proceedings of the 4th IEEE international symposium on logistic and industrial informatics, pp 151–155
31. Deiser O (2010) On the development of the notion of a cardinal number. Hist Philos Log 319:123–143
32. Chen BY (2004) What can we do with Nash's embedding theorem? Sooch J Math 30:303–338

33. Barrat A, Barthelemy M, Vespignani A (2008) Dynamical processes on complex networks. Cambridge University Press, Cambridge
34. Modrak V, Marton D (2012) Modelling and complexity assessment of assembly supply chain systems. Proc Eng 48:428–435
35. Modrak V, Marton D (2013) Complexity metrics for assembly supply chains: a comparative study. Adv Mater Res 629:757–762
36. Liu H, Yan J (2010) The direct discontinuous Galerkin (DDG) method for diffusion with interface corrections. Commun Comput Phys 8:541–564
37. Sloane NJA (1973) A handbook of integer sequences. Academic, New York
38. Bonchev D, Buck GA (2005) Quantitative measures of network complexity. In: Bonchev D, Rouvray DH (eds) Complexity in chemistry, biology and ecology. Springer, New York
39. Omar EP (2008) Twice partition numbers. The on-line encyclopedia of integer sequences. http://oeis.org/A139582
40. Munson JC, Khoshgoftaar TM (1990) Applications of a relative complexity metric for software project management. J Syst Softw 12:283–291

Chapter 12
Non-commutative Tomography: Applications to Data Analysis

Françoise Briolle and Xavier Leoncini

Abstract In this chapter, we briefly recall the theory of non-commutative tomography in a pedagogical way. We then consider its applications to signal analysis. The advantages and drawbacks of these techniques to finite samples of data are discussed. Then the method is applied, first to signals originating from reflectometry measurements in magnetized fusion plasmas, and then to data obtained from the advection of tracers in a two-dimensional time-dependent flow generated by three point vortices. In the first case, we show that the tomogram allows to pick a base to represent our signal which has the advantage of isolating the reflection coming from the plasma and then to improve the estimation of the density profile. In the second case, we show how, with a "tricky transformation" the method allows us to detect Lévy flights and extract some of their properties.

Keywords Nuclear fusion • Signal analysis • Non commutative tomogram • Anomalous transport • Chaotic advection • Reflectometry • Lévy flights

F. Briolle (✉)
Centre de Physique Théorique, Campus de Luminy, CReA, BA 701, 13300 Salon de Provence, France

Aix Marseille Université, CNRS, CPT, UMR 7332, 13288 Marseille, France

Université de Toulon, CNRS, CPT, UMR 7332, 83957 La Garde, France
e-mail: Francoise.Briolle@cpt.univ-mrs.fr

X. Leoncini
Aix Marseille Université, CNRS, CPT, UMR 7332, 13288 Marseille, France

Université de Toulon, CNRS, CPT, UMR 7332, 83957 La Garde, France
e-mail: Xavier.Leoncini@cpt.univ-mrs.fr

J.A.T. Machado et al. (eds.), *Discontinuity and Complexity in Nonlinear Physical Systems*, Nonlinear Systems and Complexity 6, DOI 10.1007/978-3-319-01411-1_12,

12.1 Introduction

The notion of time has been throughout the ages of constant debate and reflexion. In physics the emergence of the theory of relativity and its consequences have related in an intertwined manner the notions of space and time, leading, for instance, to the definition of the meter through the speed of light since 1983. From another perspective, the classical mechanical time which is essential to Newton's second law and lead to the rise of the dynamical systems branch of physics and mathematics can be as well challenged, essentially by data analysis. Indeed physics is grounded on experimental relevance of its laws, in order to uncover or verify these, the experimentalist acquires data usually originating from a time-dependent signal. When dealing with time-dependent systems, data acquisition and signal analysis become crucial, not only because, as quantum mechanics taught us, measuring something changes it, but also and quite often on the macroscopic scale because of imperfections, noise, and possible biases. When dealing with almost periodic data, such as the one acquired by looking at our sky and planetary motion, we usually rely on Fourier series, who introduced them in 1822. It took though a long time to develop a full mathematical theory of the basis of this approach, which eventually leads to the notion of functional analysis, with its vector space, basis or generating ensembles, and scalar products useful to define a norm, projections and a distance between functions. In some way, using this approach we try to describe an unknown function (signal) with a set of functions that are well known. Fourier analysis was then able to be deployed using the integral formalism and the full Fourier transform. At the same time, the notion of wave-length and frequency could be seen as dual representation of time and space, and hence the notion of time or its representation could become fuzzier, leading to the notion of time-frequency representations. Following this trend, the switch to numerical treatment, the development of new algorithms such as the Fast Fourier Transform, especially tailored for finite sampled data, lead to the uncovering of some of the shortcomings of Fourier analysis, and most notably for un-stationary signals, for which the time-frequency representations become crucial. This paved the way for the development of new signal processing tools such as for instance wavelet analysis. In this chapter we focus on another approach to signal analysis, it of course comes like most other approaches for the original vein of Fourier analysis, and is as well somewhat inspired from wavelet analysis. It, however, adds a new degree of freedom, in the sense that we use a parametric generating set, that allows us to tune this parameter to "optimize" our signal representation for certain desired tasks, such as isolating some components or signature from an un-stationary signal.

Most of signals are non-stationary with a time-dependent spectral content. Therefore an adequate joint time and frequency representation is desired for a characterization of such signals. Several types of linear transforms, such as Gabor transform or wavelets transforms, are widely used.

The Wigner-Ville quasi distribution is considered to be optimal in the sense that the spread in the time-frequency plane is minimal. But the Wigner-Ville distribution has in general positive and negative values and the interference terms (artifacts) may be nonzero and the interpretation of its representation could be delicate.

Tomograms transforms are recent mathematical techniques, based on group theory. Associated with a linear combination of non-commutative operators, tomograms are quadratic positive signal transforms. Then, in contrast to a time-frequency representation, the tomogram is the exact probability distribution of the signal on the variable X, corresponding to a linear combination of the chosen operators. We may define as a component of the signal any distinct feature (ridge, peak, etc.) of the representation.

In Sect. 12.2 we will give an overview of several transformations as linear transforms, quasi-distributions and tomograms, which can be used to characterize unstationary signals. Non-commutative tomograms, elaborated with the generators of the one-dimensional conformal group will be presented, with a particular emphasis to the time-frequency tomogram. Then, two applications of this transformation will be extensively presented. In Sect. 12.3, tomograms are used for the analysis of measurements of reflectometry on magnetized plasma, allowing to isolate the only reflection on the plasma, and then to estimate with accuracy the density profile. In Sect. 12.4, the anomalous transport of particles in a flow generated by three points vortices will be detected and characterized. After a transformation of the arclength of chaotic trajectories as the instantaneous frequency of a signal, the time-frequency tomogram transformation is used for the detection and characterization of Lévy flights.

12.2 Non-commutative Tomograms

Several types of integral transform [68, 77] are used in signal processing and are applied in different fields such as engineering, acoustic, communications, radar, medicine, we will consider here an analytic signal $f(t) = x(t) + iy(t)$, where $y(t)$ is the Hilbert transform of $x(t)$, where $x(t)$ is the real measured signal.

In addition to the traditional Fourier analysis [36] widely used, other transforms have been developed like the wavelet [20, 30, 31]. Recently the non-commutative tomograms, based on the linear combination of non-commutative operators, were suggested [56, 57]. We will present in this section a unified picture of different methods of signal processing using linear or bilinear transform in the Hilbert space. Mutual relations of the Wavelets, Wigner-Ville and tomographic transformations will be exhibited.

12.2.1 Linear Transforms, Quasi-distributions and Tomograms

A unified framework to characterize linear transforms, quasidistributions and tomograms was developed in [57]. This is briefly summarized here.

Consider:

- a normalized analytic signal $f(t)$ as vectors $| f\rangle$ in a dense nuclear subspace $\mathscr{N}$ of a Hilbert space $\mathscr{H}$ with dual space $\mathscr{N}^*$ (with the canonical identification $\mathscr{N} \subset \mathscr{N}^*$)
- a family of operators $\{U(\alpha) : \alpha \in I, I \subset R^n\}$ defined on $\mathscr{N}^*$, satisfying the completeness conditions (which is the case when $U(\alpha)$ generates a unitary group $U(\alpha) = e^{iB(\alpha)}$).
- a reference vector $\langle h \in \mathscr{N}^*$ be a reference vector chosen in such way that the linear span of $\{U(\alpha)h| \in \mathscr{N}^* : \alpha \in I\}$ is dense in $\mathscr{N}^*$. This means that, out of the set $\{U(\alpha)h\}$, a complete set of vectors can be chosen to serve as a basis.

If $U(\alpha)$ is a unitary operator, there is a self-adjoint operator $B(\alpha)$, such that $U(\alpha) = e^{iB(\alpha)}$.

In this setting three types of integral transforms are constructed.

1. **Linear transform:** $W_f^{(h)}(\alpha) = \langle U(\alpha)h \mid f\rangle$

   - Fourier transform [36] is the representation of the analytic signal as a linear superposition of planes waves which are the eigenvectors $| \omega\rangle$ of the frequency operator $\hat{\omega} = -i\frac{d}{dt}$. The plane wave signals reads

     $$f_\omega(t) = \langle t \mid \omega\rangle = \frac{1}{\sqrt{2\pi}} e^{i\omega t},$$

     and the Fourier transform of the analytic signal is

     $$F_f(\omega) = \langle \omega \mid f\rangle = \frac{1}{\sqrt{2\pi}} \int f(t) e^{-i\omega t} dt.$$

     This transformation is invertible and gives the possibility to reconstruct the signal $f(t)$ by means of the inverse Fourier transform

     $$f(t) = \langle t \mid f\rangle = \frac{1}{\sqrt{2\pi}} \int F(\omega) e^{i\omega t} d\omega.$$

     The main problem with the Fourier transform is that the signal $f(t)$ has a finite duration and the plane waves $f_\omega(t)$ are supposed of infinite duration. And in the case of unstationary signals, this transformation will not give any information of the spectral evolution in time. In fact, it is necessary to use a joint time-frequency description of the signal to get the evolution of the phase derivative (instantaneous frequency) as a function of time.

- Gabor transform [69] or Short-Time Fourier transform [1, 61, 67] gives the possibility to represent the spectral evolution of the signal $f(t)$, using a window function of fixed width. The signal will be projected on "wave packets" of finite duration:

$$h_{\tau,\omega}(t) = h(t-\tau)e^{i\omega t},$$

and the Gabor transform is

$$G_f(\tau,\omega) = \langle h_{\tau,\omega} \mid f \rangle = \frac{1}{\sqrt{2\pi}} \int f(t)h^*(t-\tau)e^{-i\omega t}dt.$$

For each τ, the window $h(t)$ will take only a portion of the signal beforehand the Fourier transform. To get a good resolution in time, the width of the window $h(t)$ should be very small, but then the resolution in frequency is degraded. And to get a good resolution in frequency, the window has to be very large, and then the resolution in time is very bad. However, this transformation, also called the spectrogram, is widely used to represent unstationary signals.

- Wavelet transform [60, 72] is the projection of the signal $f(t)$ on a "basic wavelet" $h(t)$ translated and expanded:

$$h_{s,\tau}(t) = \frac{1}{\sqrt{s}} h(\frac{t-\tau}{s})e^{i\omega t},$$

and the Wavelet transform is

$$W_f(s,\tau) = \langle h_{s,\tau} \mid f \rangle = \int f(t)h^*_{s,\tau}(t)dt.$$

To get a finite integral, the "basic wavelet" should satisfy the eligibility conditions such as $\int h(t)dt = 0$ (zero mean) and $\int |H(\omega)|^2 \frac{d\omega}{\omega} = 1$. A lots of "basic wavelets" can be used as the Mexican hat wavelet

$$h(t) = (1-t^2)e^{-t^2/2},$$

or the Morlet wavelet

$$h(t) = \frac{1}{2\pi} e^{-t^2/2} e^{i\omega_0 t}.$$

Unlike the Short-Time Fourier transform which gives a unique resolution (in time or in frequency) for each point of the time-frequency plane, the wavelet transform will give different resolutions according to the frequency: for low frequency, the resolution will be good in frequency at the cost of a bad localization in time. On the contrary, for high frequency, the compression of the wavelet will allow to a good resolution in time to the detriment of the frequency resolution. This transformation elaborated in the years 1980 by Grossman and Morley [39] is now used in many applications of signal processing.

2. *Quasidistribution transform:* $Q_f(\alpha) = \langle U(\alpha) f | f \rangle$
   - Wigner-Ville transform [75, 76] is a bilinear map of the function $f(t)$

$$W(t,\omega) = \int f(t + \frac{u}{2}) f^*(t - \frac{u}{2}) e^{-i\omega u} du$$

   Wigner-Ville quasidistribution provides information in the joint time-frequency domain with good energy resolution. But the oscillating cross-term makes the interpretation of this transform a difficult matter. Even if the average of the cross-terms is small, their amplitude may be greater than the signal in time-frequency regions that carry no physical information. This is a consequence of the basic fact that the time ($\hat{t}$) and the frequency ($\hat{\omega} = i\frac{d}{dt}$) operators associated with this quasi distribution are a pair of non-commutative operators and then precludes the existence of joint probabilities density in the time-frequency plane. Hence a joint probability density cannot be defined.
   To profit from the time-frequency energy resolution of the bilinear transforms while controlling the cross-terms problem, modifications to the Wigner-Ville transform have been proposed. Transforms in the Cohen class [25, 26] make a two-dimensional filtering of the Wigner-Ville quasidistribution.
   - Ambiguity function: the analytic signal $f(t)$ can also be described by a function called the ambiguity function of two variables

$$AF_f(\tau,\omega) = \int f(t + \frac{\tau}{2}) f^*(t - \frac{\tau}{2}) e^{-i\omega t} dt$$

   This function is the two-dimensional Fourier transform of the Wigner-Ville quasidistribution. Thus, the ambiguity function contains the same information on a signal as the Wigner-Ville transformation $W(t,\omega)$.

3. *Quadratic signal transforms:* $M_f^{(B)}(X) = \langle f \mid \delta(B(\alpha) - X) \mid f \rangle$
   Recently, a new type of strictly positive bilinear transforms has been proposed [56, 57], called *tomograms*, which is a generalization of the Radon transform [32] to non-commutative pairs of operators.
   Let X take values on the spectrum of $B(\alpha)$. Considering a set of generalized eigenstates (in $\mathcal{N}^*$) of $B(\alpha)$, one obtains for the kernel

$$\langle Y \mid \delta(B(\alpha) - X) \mid Y' \rangle = \delta(Y' - X)\,\delta(Y - Y') = \langle Y \mid X \rangle \langle X \mid Y' \rangle$$

   Therefore, we may identify $\delta(B(\alpha) - X)$ with the projector $\mid X \rangle \langle X \mid$

$$\delta(B(\alpha) - X) = \mid X \rangle \langle X \mid = P_X$$

From this, it follows

$$M_f^{(B)} = \langle f \mid \delta\left(B(\alpha) - X\right) \mid f \rangle = \langle f \mid X \rangle \langle X \mid f \rangle = |\langle X \mid f \rangle|^2, \quad (12.1)$$

showing the positivity of the tomogram and its nature as the squared amplitude of the projection on generalized eigenvectors of $B(\alpha)$. For a normalized analytic signal $f(t)$, the tomogram is normalized

$$\int M_f^{(B)}(X)\, dX = 1.$$

Then, the tomogram can be interpreted as the probability distribution of the random variable X corresponding to the observable defined by the operator $B\,(\alpha)$ and provides a full characterization of the signal.

Let us consider the operator $B(\alpha)$ as a linear combination of the operators $\mathcal{O}_1, \mathcal{O}_2$ and its eigeinvectors $\{\Psi_\alpha^X(t)\}$. The B-tomogram, which explores the signal along lines in the plane (O_1, O_2), is the projection of the analytic signal on the eigenvectors:

$$M_f^{(B)}(X) = \langle f, \Psi_\alpha^X \rangle = \int f(t) \Psi_\alpha^X(t) dt$$

Here, we consider one-dimensional conformal group with its generators

$$\hat{t} \qquad \hat{\omega} = -i\frac{d}{dt} \qquad D = \left(\hat{t}\hat{\omega} + \hat{\omega}\hat{t}\right) \qquad K = i\left(\hat{t}^2 \frac{d}{dt} + \hat{t}\right).$$

One may elaborate a linear combination of those non-commutative operators to construct one-dimensional tomograms.

- Time-frequency tomogram

  The operator $B_1(\alpha)$ is a linear combination of the time $\hat{t}$ and frequency $\hat{\omega}$ operators,

  $$B_1(\mu, \nu) = \mu\hat{t} + \nu\hat{\omega}.$$

  The eigenvectors $\Psi_{\mu,\nu}^X(t)$, associated with the eigenvalue X are

  $$\Psi_{\mu,\nu}^X(t) = e^{-i\left(\frac{\mu t^2}{2\nu} - \frac{tX}{\nu}\right)},$$

  and the time-frequency tomogram is the projection of the analytic signal on the eigenvectors

  $$M_1\left(\mu, \nu, X\right) = \frac{1}{2\pi|\nu|} \left| \int e^{i\left(\frac{\mu t^2}{2\nu} - \frac{tX}{\nu}\right)} f(t)\, dt \right|^2. \quad (12.2)$$

This tomogram is studied in more detail in Sect. 12.2.2 and two applications using this transformation are extensively detailed in Sects. 12.3 and 12.4.

- Time-scale tomogram
  For this tomogram, the operator $B_2(\alpha)$ is a linear combination of the time $\hat{t}$ and the dilatation operator $D = (\hat{t}\hat{\omega} + \hat{\omega}\hat{t}) = -i\left(\hat{t}\frac{d}{dt} + \frac{1}{2}\right)$, instead of the operator $\hat{\omega}$ used for the previous operator,

$$B_2(\mu, \nu) = \mu\hat{t} + \nu D.$$

  The time-scale tomogram is defined as the projection of the signal on the eigenvectors of the operator $B_2(\mu, \nu)$ associated with the eigenvalue X,

$$M_2(\mu, \nu, X) = \frac{1}{2\pi|\nu|}\left|\int dt\, \frac{f(t)}{\sqrt{|t|}} e^{\left[i\left(\frac{\mu}{\nu}t - \frac{X}{\nu}\log|t|\right)\right]}\right|^2. \tag{12.3}$$

- Frequency-scale tomogram
  This tomogram is elaborated with the operator $B_3(\alpha)$, a linear combination of the frequency operator $\hat{\omega}$ and the dilatation operator D,

$$B_3(\mu, \nu) = \mu\hat{\omega} + \nu D. \tag{12.4}$$

  and then, the projections of the signal on the eigenvectors will give the frequency-scale tomogram

$$M_3(\mu, \nu, X) = \frac{1}{2\pi|\nu|}\left|\int \frac{F_f(\omega)}{\sqrt{|\omega|}} e^{\left[-i\left(\frac{\mu}{\nu}\omega - \frac{X}{\nu}\log|\omega|\right)\right]} d\omega\right|^2, \tag{12.5}$$

  with $F_f(\omega)$ being the Fourier transform of the analytic signal $f(t)$.

- Time-conformal tomogram
  For this tomogram, the operator $B_4(\mu, \nu)$ is a linear combination of the time $\hat{t}$ and the conformal operator K

$$B_4(\mu, \nu) = \mu\hat{t} + \nu K = \mu\hat{t} + i\nu\left(t^2\frac{d}{dt} + t\right).$$

Then, the tomograms related to this operator is

$$M_4(\mu, \nu, X) = \frac{1}{2\pi|\nu|}\left|\int dt\, \frac{f(t)}{|t|} e^{\left[i\left(\frac{X}{\nu t} + \frac{\mu}{\nu}\log|t|\right)\right]}\right|^2. \tag{12.6}$$

For more details on non-commutative tomograms defined on the one-dimensional conformal group, see [16, 57].

4. *Quantum mechanics formalism*
   The linear and the quasidistribution transforms can be written using group theory formalism.

If $U(\alpha)$ are unitary operators, by Stone's theorem, there are self-adjoint operators $B(\alpha)$ such that $U(\alpha) = e^{iB(\alpha)}$. The linear and quasidistribution transforms can be written as

$$W_f^{(h)}(\alpha) = \langle h|e^{iB(\alpha)}|f\rangle$$

$$Q_f^{(B)}(\alpha) = \langle f|e^{iB(\alpha)}|f\rangle$$

For $B(\boldsymbol{\alpha}) = \alpha_1 \hat{t} + \alpha_2 \hat{\omega}$ and h a generalized eigenvector of the time-translation operator, the linear transform $W_f^{(h)}$ becomes the Fourier transform. For $B(\boldsymbol{\alpha})$ plus the parity operator $\frac{\pi(\hat{t}^2+\hat{\omega}^2-1}{2}$, the $Q_f(\alpha)$ would be the Wigner -Ville transform. Similarly, for $B(\boldsymbol{\alpha}) = \alpha_1 D + \alpha_2 \hat{\omega}$ where D is the dilatation operator $D = \frac{1}{2}(\hat{t}\hat{\omega} + \hat{\omega}\hat{t})$, the linear transform $W_f^{(h)}$ is a wavelet transform and the $Q_f(\alpha)$ the Bertrand transform.
The relations between the transformations are established in [57].

12.2.2 *Time-Frequency Tomogram*

For a signal $f(t)$, the time-frequency tomogram is defined as:

$$M_f(X, \mu, \nu) = \frac{1}{2\pi|\nu|} \left| \int f(t) \exp\left(\frac{i\mu}{2\nu} t^2 - \frac{iX}{\nu} t \right) dt \right|^2, \tag{12.7}$$

For each (μ, ν) pair corresponding to a linear combination of the time and frequency operators the tomogram provides a probability distribution on the variable X [see Eq. (3)]. The tomogram $M_f(X, \mu, \nu)$ is an image in the $(X, (\mu, \nu))$ hyper-plane of the probability flow from the t-description of the signal to the frequency-description, through all the intermediate steps of the linear combination.

For an easy interpretation of the time-frequency tomogram, we consider a particular case $\mu = \cos\theta, \nu = \sin\theta$ with the self-adjoint operator $B(\theta) = \cos\theta\hat{t} + \sin\theta\hat{\omega}$. The tomogram is defined as:

$$M_f(X, \theta) = \frac{1}{2\pi|\sin\theta|} \left| \int f(t) \exp\left(\frac{i\cos\theta}{2\sin\theta} t^2 - \frac{iX}{\sin\theta} t \right) dt \right|^2. \tag{12.8}$$

Then, in the plane (X, θ) the tomogram $M_f(X, \theta)$ can be interpreted as the probability distribution on the variable X. For this particular case, the tomogram $M_f(X, \theta)$ coincides with the Radon transform [37], which has already been used for signal analysis by several authors [7, 78, 79] in a different context.

For $\theta = \frac{\pi}{2}$, the tomogram $M_f(X, \frac{\pi}{2})$ is the frequency-description of the signal,

$$M_f(X, \frac{\pi}{2}) = \frac{1}{2\pi}\left|\int f(t)e^{-iXt}\,dt\right|^2.$$

For $\theta = 0$, the operator $B(\theta) = \hat{t}$ and the tomogram $M_f(X, 0)$ is the time-description of the signal. The limit of the Fresnel tomogram $M_f^F(X, \theta)$ defined for small θ in [33] is

$$\lim_{\theta\to 0} M_f^F(X, \theta) = |s(t)|^2.$$

The variable X is the time for $\theta = 0$, the frequency for $\theta = \pi/2$ and is a generalized variable X, mixture of time and frequency, for other values of θ.

We can make the link between the time-frequency tomogram $M_f(X, \theta)$ and the fractional Fourier transform [66], defined as:

$$\mathscr{F}_s(x, \theta) = C(\theta)e^{\frac{i\pi x^2}{\tan\theta}}\int s(t)exp\left(\frac{i\pi\cos\theta}{\sin\theta}t^2 - \frac{2\pi x}{\sin\theta}t\right)dt. \tag{12.9}$$

Up to a phase factor $\exp(ix^2/2\tan\theta)$ and a normalization constant $C(\theta)$, the fractional Fourier transform is similar to the time-frequency tomogram $M_f(X, \theta)$. They can be both interpreted as the projection of the analytic signal $f(t)$ on a basis of chirp signals [13]

$$\psi_{\theta,x}(t) = e^{i\left[(\pi/2\tan\theta)t^2 - (x/sin\theta)t\right]}.$$

12.2.3 *Time-Frequency Tomogram and Data Analysis*

12.2.3.1 Signal of Finite Duration *T*

For a signal of duration T the time-frequency tomogram can be written as:

$$M_s(x, \theta) = \left|\int s(t)\Psi_x^{\theta,T}(t)\,dt\right|^2 = \left|< s, \Psi_x^{\theta,T} >\right|^2, \tag{12.10}$$

with

$$\Psi_x^{\theta,T}(t) = \frac{1}{\sqrt{T}}\exp\left(\frac{-i\cos\theta}{2\sin\theta}t^2 + \frac{ix}{\sin\theta}t\right). \tag{12.11}$$

The family $\{\Psi_{x_n}^{\theta,T}(t)\}$ is orthogonal and normalized basis: $< \Psi_{x_m}^{\theta,T}, \Psi_{x_n}^{\theta,T} >= \delta_{m,n}$ for a family of values $\{x_n = x_0 + \frac{2n\pi}{T} \sin\theta\}$, where x_0 is freely chosen (in general we take $x_0 = 0$).

The time-frequency tomogram can be written, for each angle $\theta_1, \ldots, \theta_k, \ldots, \theta_P$, as:

$$M(x_n, \theta_k) = \left|c^{\theta_k}(x_n)\right|^2. \tag{12.12}$$

For a digital signal $\{s[n\}_{n=0,\ldots,N-1}$, of length NT, $c^{\theta_k}(x_n)$ is the Fast Fourier Transform of the digital signal:

$$c^{\theta_k}(x_n) = FFT\left(s[n]\exp\left[\frac{i\cos\theta_k}{2\sin\theta_k}n^2\right]\right). \tag{12.13}$$

The fast implementation of the time-frequency tomogram is of complexity $\mathcal{O}(N\log N)$, for each θ_k.

It is then possible, from the projections $c^{\theta_k}(x_n)$ to recover the original signal $s[n]$:

$$s[n] = IFFT\left(c^{\theta_k}(x_n)\right).\exp\left[\frac{-i\cos\theta_k}{2\sin\theta_k}n^2\right] \tag{12.14}$$

12.2.3.2 Density of Magnetized Plasma from Reflectometry Measurements

Reflectometry measurements on magnetized plasma are difficult to analyze. Indeed the signal is a mixture of components such as reflections on the porthole, on the wall of the machine and, that which is of interest, the reflection on the plasma. For this application, we use the time-frequency tomogram as a kind of "chirp filter." For an angle θ_k, the probability distribution of the signal on the variable x allows to separate the three components. Then, from the tomogram projections $c^{\theta_k}(x_n)$, we will "re-synthesize" each component and their phase derivative. We are able to extract the component of interest, the reflection on the plasma, and then to extract information of the plasma density. This application is developed in Sect. 12.3.

12.2.3.3 Detection and Characterization of Lévy Flights

Transport of advected passive particles in two-dimensional flows with coherent structures (vortex) is anomalous when it contains Lévy flights. The arclength of the particle trajectories is characterized by a linear behavior with respect to the time (ballistic motion). The arclength of the trajectory will be transformed as the phase derivative of a new signal to emphasize the linear part of the trajectory. Then, the time-frequency tomogram will be used to detect linear chirps in a two-dimensional time-frequency representation. This application is developed in Sect. 12.4.

12.3 Measurement of the Density Profile of Magnetized Plasma

12.3.1 Context

The energy confinement in ITER is predicted with scaling laws extrapolated from measurements on smaller machines such as Tore Supra and Jet. tokamaks. When rewritten with dimensionless parameters, large uncertainties remain on some parameter dependence such as the ratio of plasma pressure to magnetic pressure. The understanding of the anomalous transport of particles in magnetized plasmas is a key issue for a fusion reactor. The large heat and particle transport is attributed to drift wave turbulence destabilized by temperature and density gradients [40].

Density measurements play an important role in the study of the anomalous transport of magnetically confined plasma for a better understanding of the turbulence. Microwave reflectometry is a radar-like technique, widely used to measure the electronic density profile in tokamak plasmas. Reflectometers have been developed along two main applications: density profile and density fluctuation measurements [43, 62].

In the years 2010, we participated in the analysis of data coming from new reflectometers on Tore Supra [14, 15]. The goal was to extract from a mixture of multi-reflections (reflectometry measurements) the sole reflection on the plasma.

In this part, we will first explain the principle of reflectometry measurements in magnetized plasma and then give some results of tomographic data analysis and its future applications to reflectometers.

12.3.2 Principle of Reflectometry

Derived from radar principles, reflectometry measures the amplitude and the phase variation of a microwave $E_R(t)$ reflected inside the plasma at a cutoff layer where the refractive index n becomes zero, by mixing the reflected wave with the probing wave (reference) $E_0(t)$.

For measuring the density profile, a standard method uses a frequency sweeping of the probing wave.

$$E_0(t) = cos\{\Omega(t).t)\} \qquad with \qquad \frac{\partial \Omega(t)}{\partial t} = a.t + b. \tag{12.15}$$

Then the reflected wave is equal to $E_R(t) = A(t)cos\{\Omega(t).t) + \phi(t)\}$. This signal is multiplied by a pure frequency $cos\{\Omega(t).t)\}$ and low-pass filtered afterwards in order to get, at the output of the mixer, the signal:

$$s(t) = A(t)cos\{\phi(t)\}. \tag{12.16}$$

In the mixer output, the amplitude $A(t)$ of the reflected wave $E_R(t)$ depends on the variation of the reflectivity of the cutoff layer. This is due to geometrical effects like the divergence of the microwave beam or the tilting of the cutoff layer when a large perturbation modifies the flux surfaces. The phase $\phi(t)$, that contains the most reliable information about the plasma density, is the main quantity of interest. In Sect. 12.3.3, the experimental setup, which allows us to get the amplitude and the phase of the reflected wave, is exposed in details.

There are two modes of polarization of the probing wave: the ordinary polarization, where the wave polarization is in the direction of the magnetic field B of the plasma ($E \parallel B$), the so-called the 0–mode, and the extraordinary polarization, X-mode, where the wave polarization is orthogonal to the magnetic field ($E \perp B$). The value of the refractive index is depending on the polarization of the probing wave, as it is shown in Sect. 12.3.2.1.

12.3.2.1 Wave Propagation in a Plasma

With the hypothesis of cold (the particles are static), homogeneous (the characteristics lengths are large in comparison with the wavelength), and stationary plasma (the evolution time is large in comparison with the wave period), it is possible to write the equation of propagation of a plane wave [38, 70, 71]. Then, the dielectric tensor is:

$$\begin{cases} \epsilon_{xx} = \epsilon_{yy} = 1 - \frac{\omega_{pe}^2}{\omega^2 - \omega_{ce}^2} \\ \epsilon_{xy} = \epsilon_{yx} = -\frac{\omega_{ce}}{\omega}\frac{\omega_{pe}^2}{\omega^2 - \omega_{ce}^2} \\ \epsilon_{zz} = 1 - \frac{\omega_{pe}^2}{\omega^2} \\ \epsilon_{xz} = \epsilon_{zx} = \epsilon_{yz} = \epsilon_{zy} = 0 \end{cases} \tag{12.17}$$

where ω is the pulsation of the probing wave, $\omega_{pe} = \sqrt{\frac{e^2 n_e}{\epsilon_0 m_e}}$ the electronic plasma pulsation and $\omega_{ce} = eB/m_e$ the cyclotronic electronic pulsation, n_e is the electron density, e and m_e the electronic charge and mass, ε_0 the permittivity of the vacuum.

The propagation equation of a wave, perpendicular to the direction $0y$, when the magnetic field B is constant and Oz oriented, can be written as:

$$\begin{pmatrix} \epsilon_{xx} & -i\epsilon_{xy} & 0 \\ i\epsilon_{xy} & \epsilon_{xx} - N^2 & 0 \\ 0 & 0 & \epsilon_{zz} - N^2 \end{pmatrix} \begin{pmatrix} E_x \\ E_y \\ E_z \end{pmatrix} = \mathbf{0} \tag{12.18}$$

With N the refractive index, $N = kc/\omega$.

- Ordinary polarization (O-mode)
  In the ordinary polarization (O-mode), the wave polarization is in the direction of the magnetic field of the plasma ($E \parallel B$). In this case $E_x = 0$, and the propagation equation has a unique solution:

$$N_O^2 = 1 - \frac{\omega_{pe}^2}{\omega^2}. \tag{12.19}$$

  Then, the cutoff frequency f_O is equal to the plasma frequency $\omega_{pe}/2\pi$ and depends only on the electron density n_e.

$$f_O = \frac{1}{2\pi}\sqrt{\frac{e^2 n_e}{\epsilon_0 m_e}}. \tag{12.20}$$

  The O-mode is widely used in reflectometry, but the measurements can be done only for density gradient between 0.3 and 0.8. The edge and the center density of the plasma can't be reached with this kind of measurements.
- Extraordinary polarization (X-mode)
  In the extraordinary polarization, the wave polarization is orthogonal to the magnetic field ($E \perp B$). Then, the refractive index is equal to:

$$N_X^2 = 1 - \frac{\omega_{pe}^2 (1 - \frac{\omega_{pe}^2}{\omega^2})}{\omega^2 - \omega_{pe}^2 - \omega_{ce}^2}. \tag{12.21}$$

  If the frequency f of the probing wave is equal to $\frac{1}{2\pi}\sqrt{\omega_{pe}^2 - \omega_{ce}^2}$, then the wave will become evanescent and will be absorbed by the plasma.
  The wave will be reflected when $N_X = 0$. There are two cutoff frequencies, namely the upper f_X^{up} and lower f_X^{low}:

$$f_X^{up} = \frac{1}{2\pi}\frac{\sqrt{\omega_{ce}^2 + 4\omega_{pe}^2} + \omega_{ce}}{2} \quad and \quad f_X^{low} = \frac{1}{2\pi}\frac{\sqrt{\omega_{ce}^2 + 4\omega_{pe}^2} - \omega_{ce}}{2}. \tag{12.22}$$

  The edge density can be probed using the upper cutoff frequency since the frequency is finite. It allows us to measure weak density at the edge of the plasma.

12.3.2.2 Density Profile Reconstruction

Using the WKB approximation along the propagation path (1D approximation), the phase variation between the antenna at $r = 0$ and the reflecting layer at $r = r_{co}$ can be estimated:

$$\phi_p = \frac{4\pi}{c}.f.\int_{r=0}^{r=r_{co}} N(r, f, t)dr - \frac{\pi}{2}, \tag{12.23}$$

where f is the frequency of the probing wave, N(r,f,t) the plasma refractive index at the frequency f. The term $-\frac{\pi}{2}$ indicates that the reflection inside the plasma is nonmetallic.

A variation of the phase ϕ_p can be due either to a variation of the probing frequency or to a variation of the optical path length between the antenna and the cutoff layer along the line of sight. Temporal changes of the phase can thus be written as:

$$\frac{\partial \phi_p}{\partial t} = \frac{4\pi}{c} . \frac{\partial f}{\partial t} . \int_{r=0}^{r=r_{co}} N(r, f, t) dr + \frac{4\pi}{c} . f . \frac{\partial}{\partial t} \left(\int_{r=0}^{r=r_{co}} N(r, f, t) dr \right) . \quad (12.24)$$

The first term is proportional to the optical path length $\int_{r=0}^{r=r_{co}} N(r, f, t) dr$, i.e. the position r_{co} of the reflecting layer, when the frequency f is swept.

The second term describes the phase changes introduced by fluctuations of the optical path length arising from temporal and spatial fluctuations of the electron density.

The beat frequency is defined as:

$$f_b = \frac{1}{2\pi} \frac{\partial \phi_p}{\partial t}, \quad (12.25)$$

and the group delay of the reflected wave, namely the time of flight:

$$\tau_g = \frac{1}{2\pi} \frac{\partial \phi_p}{\partial f} = f_b / \frac{\partial f}{\partial t}. \quad (12.26)$$

- Ordinary polarization (O-mode)
  It is possible to reconstruct a monotonic density profile, with the estimation of the group delay τ_g of the reflected wave. The localization of the reflecting layer $r_c(F_p)$ for the frequency F_p is given by the analytic expression [27]:

$$r_c(F_p) - a = \frac{c}{\pi} \int_0^{F_p} \frac{\tau_g(f)}{\sqrt{F_p^2 - f^2}} df. \quad (12.27)$$

- Extraordinary polarization (X-mode)
  In the extraordinary polarization (X-mode), the density profiles are recovered from the phase using the Bottolier algorithm [12]. Initialization of the profile is the most interesting feature of the X-mode polarization. Contrary to the O-mode polarization, where at zero density the cutoff frequency equals zero, in X mode the edge density profile position can be setup with the rise of the detected amplitude. Assuming that the first cutoff is for a null density, the start of the plasma can be set providing knowledge of the local magnetic field.

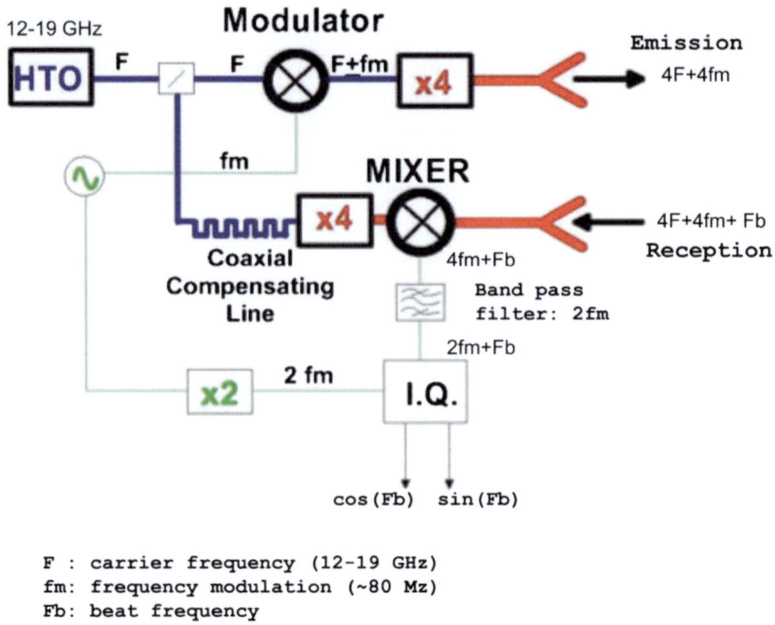

Fig. 12.1 X-mode Reflectometer working in V-Band (50–75 GHz)

12.3.3 Experimental Setup

A broadband reflectometer operating in the frequency range 50–75 GHz (V-Band) in extraordinary mode polarization has been developed on Tore Supra to measure edge density profiles [21–23].

Fast sweeping improves greatly the profile reconstruction. On Tore-Supra, the cutoff layer displacement during the turbulence correlation time (microsecond range) is comparable to the turbulence correlation length (centimeter range). At a sweeping rate of 1 GHz μs^{-1}, the wavelength rate is 30 cm μs^{-1}, which is comparable to the displacement of the cutoff layer. With the experimental setup described in Fig. 12.1, the probing wave operates in the range 50–75 GHz with a sweeping rate of 20 μs.

The output of a Hyperabrupt varactor Tuned Oscillator (HTO) providing fast linear frequency sweeps from 12 to 19 GHz in 20 μs is mixed to a low frequency signal $fm \sim 100$ MHz.

After amplification, the frequency $\omega(t) + fm$ is multiplied by 4 to provide a probing signal with a frequency coverage between 48 and 76 GHz. The probe signal $E_0(t) = cos(4\{\omega(t) + f_m\}.t)$ is then emitted through wave guides.

Emission and reception are done with two separate identical rectangular antennas, one near of the other, outside the vacuum vessel through a porthole, around 120 cm away from the plasma edge, as it is shown in Fig. 12.2.

A sweep is done before every discharge and the reflection on the inner wall of the vessel is used as a reference to correct the dispersion in waveguides and antennas.

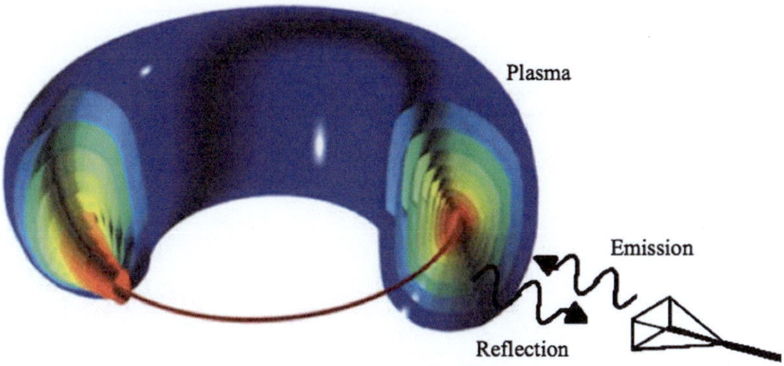

Fig. 12.2 The experimental setup is located outside the vacuum vessel, close to the plasma edge

The output of the HTO, at the frequency $4\omega(t)$, is τ delayed by a delay line, to obtain the signal: $u(t) = cos(4\omega(t).(t+\tau))$. During the time τ, the path of the probing wave is equal to $D = \tau.c$, where D is the distance from the emitting antenna to the inner wall of the vessel. Then, the phase differences between the probing and the reflected waves will be mainly due to the position of the cutoff layer.

The reflected wave, $E_R(t) = A(t)cos\{(4\omega(t) + 4f_m).(t + \tau_R) + \phi(t)\}$, is then mixed to $u(t)$ and band-pass filtered at $4fm \pm 50MHz$ to obtain a low frequency signal $v(t) = A(t)cos\{(4f_m t + \phi(t)\}$.

An heterodyne demodulation at $4f_m$, providing in-phase and 90° phase detection, leaves the reflected wave in the base-band leaving only the cutoff data of the probing frequency $s(t) = A(t)e^{i\phi(t)}$.

The reflectometer can achieve a repetition rate of $5\mu s$ between sweeps, so the dynamic behavior of fast plasma events can be followed.

12.3.4 Data Processing

The goal is to measure the density at the edge of the plasma on the extraordinary mode polarization (X mode) on Tore Supra.

The sweep-frequency reflectometer launches a probing wave in the V band (50–75 GHz). The reflectometry system repeatedly sends sweeps of duration 20 μs. The heterodyne reflectometers, with I/Q detection, provide a good signal-to-noise ratio, up to 40 dB.

As it is described in detail in Sect. 12.3.3, for each sweep, the reflected chirp $E_R(t)$ is mixed with the incident sweep $E_{ref}(t)$ and only the interference term is recorded as an in-phase and a 90° phase shifted sampled signals. Let the reflected signal be:

$$s(t) = x_1(t) + ix_2(t) = A(t)e^{i\varphi(t)}. \tag{12.28}$$

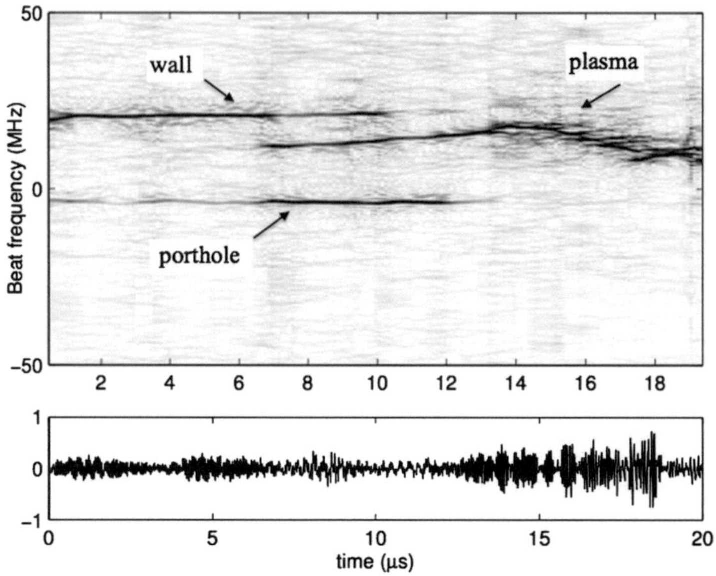

Fig. 12.3 *Up* Time-frequency representation of the base-band downshifted reflected wave $A(t)e^{\phi(t)}$. *Down* Real part of the base-band downshifted reflected wave

For one of the measurements (choc #42824) the Gabor Transform $G_s(t,\omega)$, namely the spectrogram gives a time-frequency representation of the signal $s(t)$ [20].

$$G_s(t,\omega) = \frac{1}{\sqrt{2\pi}} \int_{-\infty}^{+\infty} s(\tau) e^{-\pi(\tau-t)^2} e^{-i\omega\tau} d\tau.$$

The Gabor Transform is obtained with short time Fourier transforms (STFT) of the sampled signal broken up into M windowed chunks, which usually overlap, and Fourier transformed. The spectrums are then "laid side by side" to form the image or a three-dimensional surface. For a better representation, the amplitude of the spectrum is represented by gray scales, to obtain a two-dimensional image where the horizontal axis is the time and the vertical one the beat frequency. Each vertical line represents the spectrum of a trunk (Fig. 12.3).

The signal $s(t)$ is sampled at the frequency of 100 MHz, so we get only 2,000 samples by trial. For a nice time-frequency representation, the length of the chunks is equal to 100 samples with an overlapping equal 90%.

As it can be seen on the time-frequency representation, the base-band reflected wave $s(t)$ is a mixture of different signals: a reflection of the probing wave on the inner wall of the vessel ($0 < t < 10\ \mu s$; beat frequency $\sim$ 20 MHz) a reflection on the porthole, placed in front of the antennas ($0 < t < 12 \mu s$; beat frequency $\sim$-5 MHz) and the reflection on the cutoff layers of the plasma ($7 < t < 20\ \mu s$; beat frequency between 5 to 20 MHz). The reflections on the inner wall and the porthole are represented by straight lines while the plasma reflection is more heckled.

The goal is then to extract only the reflection on the plasma from the base-band reflected wave. The reflections are overlapping in time and in frequency, the reflection on the inner wall is very close to some reflections on the plasma: a band-pass filter will not give good results.

A time-frequency tomographic analysis is therefore used to achieve the separation of different reflections merged in the reflected wave.

12.3.5 Tomographic Analysis

12.3.5.1 Time-Frequency Tomograms (Signal of Finite Duration)

In part I, we described in much detail the time-frequency tomograms. Here, we will describe the method of component separation for the operator:

$$B_\theta^S = cos\theta \; t + sin\theta \; \omega, \tag{12.29}$$

where t and $\omega = i\frac{\partial}{\partial t}$ are, respectively, the time and frequency non-commutative operators.

A probability family of distributions, $M_s(x, \theta)$, is defined from a complex signal $s(t), t \in [0, T]$ by:

$$M_s(x, \theta) = \left| \int s(t) \Psi_x^{\theta,T}(t) \, dt \right|^2 = \left| < s, \Psi_x^{\theta,T} > \right|^2, \tag{12.30}$$

with

$$\Psi_x^{\theta,T}(t) = \frac{1}{\sqrt{T}} e^{i \left(\frac{-\cos\theta}{2\sin\theta} t^2 + \frac{x}{\sin\theta} t \right)}. \tag{12.31}$$

Note that the $\Psi_x^{\theta,T}$ are generalized eigenfunctions for any spectral value x of the operator B_θ^S. Therefore $M_s(x, \theta)$ is a (positive) probability distribution as a function of x for each θ.

A glance at the shape of the functions (12.31) shows that, for fixed θ, the oscillation length at a given t decreases when $|x|$ increase. As a result, the projection of the signal on the $\left\{ \Psi_{x_n}^{\theta,T}(t) \right\}$ basis locally explores different scales. On the other hand the local time scale is larger when θ also becomes larger, in agreement with the uncertainty principle for a non-commuting pair of operators.

Here θ is a parameter that interpolates between the time and the frequency operators, thus running from 0 to $\pi/2$ whereas x is allowed to be any real number. For $\theta = 0$, the tomogram $M_s(x, \theta)$ is the probability distribution of the signal in time $|s(t)|^2$ and for $\theta = \frac{\pi}{2}$, the probability distribution of the signal in frequency $|S(f)|^2$.

Our strategy is to search for intermediate values of θ where a good compromise may be found to separate the components of the signal. For such intermediate values it is possible to pull apart different components of the signal (see Fig. 12.4, a tomographic representation ($0 < \theta < \frac{\pi}{2}$) of the reflected wave).

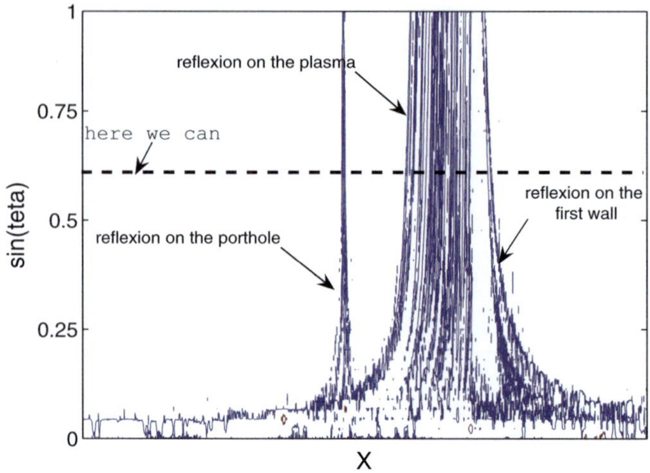

Fig. 12.4 Time-frequency tomographic representation of the base-band reflected wave $A(t)e^{\phi(t)}$. For $\theta = 0$, the tomogram $M_s(x, \theta)$ is probability distribution of $|s(t)|^2$ and for $\theta = \frac{\pi}{2}$, the probability distribution of $|S(f)|^2$

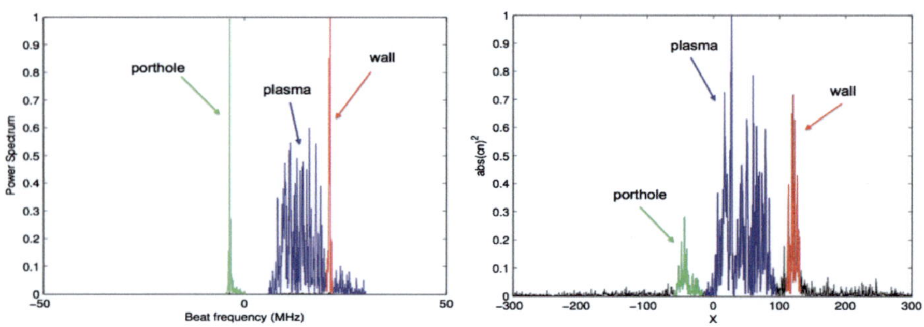

Fig. 12.5 *Left*: Fourier transform of the base-band signal $s(t)$ measured at the output of the reflectometer. *Right*: Tomogram, for $\theta = \frac{\pi}{5}$ of the signal $s(t)$

As it can be seen in Fig. 12.5, an intermediate value of $\sin\theta \sim 0.6$ ($\theta = \frac{\pi}{5}$) allows us to separate the three components, taking into account both time and frequency information.

The Fourier transform of $s(t)$(left part Fig. 12.5) shows that it is impossible to use a band-pass filter to get the only reflection on the plasma. With a tomogram of the signal, for $\theta = \frac{\pi}{5}$, the three components can be distinguished (right part Fig. 12.5).

12.3.5.2 Components Factorization

First we select a subset of numbers $\{x_n\}$ in such a way that the corresponding family $\{\Psi_{x_n}^{\theta,T}(t)\}_n$ is orthogonal and normalized:

$$< \Psi_{x_m}^{\theta,T}, \Psi_{x_n}^{\theta,T} >= \delta_{m,n}. \tag{12.32}$$

This is possible using the sequence

$$x_n = x_0 + \frac{2n\pi}{T} \sin\theta, \tag{12.33}$$

where x_0 is freely chosen (in general we take $x_0 = 0$ but it is possible to make other choices, depending on what is more suitable for the signal under study).

We then consider the projections of the signal $s(t)$ on the orthonormal basis $\{\Psi_{x_n}^{\theta,T}\}$

$$c_{x_n,\theta}^{s} =< s, \Psi_{x_n}^{\theta,T} >, \tag{12.34}$$

and use the coefficients $c_{x_n,\theta}^{s}$ for our signal processing purposes.

As it is shown on the right part of the Fig. 12.4, it is possible, using a threshold, to select three subsets $\mathscr{F}_k$ of the $\{x_n\}$. A multi-component analysis of the signal [15] is done by reconstructing the partial signals:

$$s_k(t) = \sum_{n\in\mathscr{F}_k} c_{x_n,\theta}^{s} \Psi_{x_n}^{\theta,T}(t) \qquad k = 1, 2, 3. \tag{12.35}$$

From the projections of the signal $s(t)$ on the orthonormal basis $\{\Psi_{x_n}^{\theta,T}\}$, for $\theta = \frac{\pi}{5}$, using a threshold ($\epsilon = 0.04$) it is possible to select the spectral projections of three different components (see Fig. 12.6).

First component, the reflection on the porthole

The first component, $\widetilde{s_1}(t)$, corresponds to $-20 \leq x_n \leq 0$ and is therefore defined as:

$$\widetilde{s_1}(t) = \sum_{x_n=-20}^{0} c_{x_n}^{\theta}(y)\Psi_{x_n}^{\theta}(t). \tag{12.36}$$

This component is the reflection of the probing wave on the porthole. The distance from the emitting/reception antenna to the porthole is around 80 cm. It is a constant low frequency signal (see Fig. 12.7): the phase derivative of the reflection is proportional to the distance from the antenna to the reflector. The duration of this signal is around 12 μs.

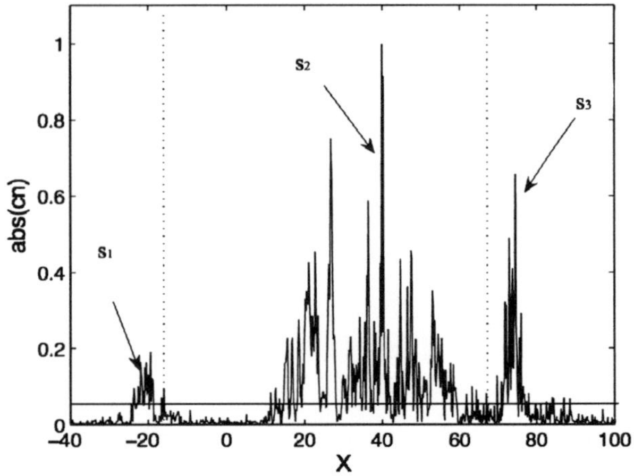

Fig. 12.6 Tomogram of the signal $s(t)$ for $\theta = \frac{\pi}{5}$, $M_s(x, \theta = \frac{\pi}{5})$

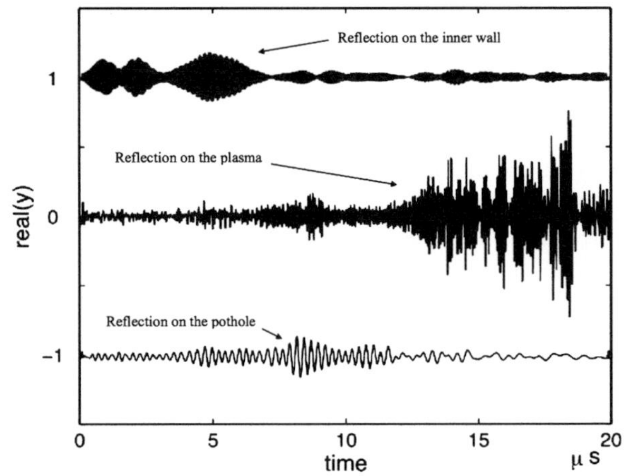

Fig. 12.7 The three components of the reflectometry signal. For visual purposes, the average of $\widetilde{s_1}(t)$ is shifted to 1 and the average of $\widetilde{s_3}(t)$ to -1

Second component, reflection on the plasma

The second component is the reflection on the cutoff frequency of the plasma (see Fig. 12.7). The reflection starts around 10 μs after the reflection on the porthole. The frequency and the amplitude of this reflected wave is quite heckled. This component, $\widetilde{s_2}(t)$, corresponds to $0 \le x_n \le 110$ and is therefore defined as:

$$\widetilde{s_2}(t) = \sum_{x_n=0}^{110} c^{\theta}_{x_n}(y)\Psi^{\theta}_{x_n}(t). \tag{12.37}$$

Third component, reflection on the inner wall of the vessel

The last component corresponds to the reflection of the probing wave on the wall of the vacuum vessel. The frequency is quite constant (Fig. 12.7), and related to the distance antenna-wall. The duration of this signal is around 10 μs. This component, $\widetilde{s_3}(t)$, corresponds to $110 \le x_n \le 140$ and is therefore defined as:

$$\widetilde{s_3}(t) = \sum_{x_n=10}^{140} c^{\theta}_{x_n}(y)\Psi^{\theta}_{x_n}(t). \tag{12.38}$$

12.3.5.3 Estimation of the Phase Derivative

To compute the density profile of the plasma, with reflectometry measurements in the X-mode, it is necessary to estimate the phase derivative of the reflection on the cutoff layer of the plasma. The usual process is to isolate this refection and then to unwrap the phase using a classical gradient procedure. Given a signal $s(t) = A(t)e^{i\phi(t)}$, the time derivative of the phase may be obtained from

$$\frac{\partial}{\partial t}\phi(t) = \mathrm{Im}\left(\frac{\frac{\partial s}{\partial t}}{s(t)}\right). \tag{12.39}$$

Using a tomographic decomposition allows us to get the time derivative of the phase directly. Let us remember that:

$$\widetilde{s_k}(t) = A_k(t)e^{i\phi_k(t)} = \sum_{n\in\mathscr{F}_k} c^{s}_{x_n,\theta}\Psi^{\theta,T}_{x_n}(t) \qquad k = 1,2,3, \tag{12.40}$$

then,

$$\frac{\partial}{\partial t}\widetilde{s_k}(t) = \sum_{x_n} c^{s}_{x_n,\theta}\frac{\partial}{\partial t}\Psi^{\theta,T}_{x_n}(t). \tag{12.41}$$

Notice that an explicit analytic expression for $\frac{\partial}{\partial t}\Psi^{\theta,T}_{x_n}(t)$ is known, namely:

$$\frac{\partial}{\partial t}\Psi^{\theta,T}_{x_n}(t) = i\left(\frac{-\cos\theta}{\sin\theta}t + \frac{x}{\sin\theta}\right)\Psi^{\theta,T}_{x_n}(t). \tag{12.42}$$

Therefore we obtain a direct expression for the phase derivative in terms of the coefficients $c^{s}_{x_n,\theta}$ without having to use the values of s_k for neighboring values of t.

$$\frac{\partial}{\partial t}\phi_k(t) = \mathrm{Im}\left(\frac{\sum_{x_n} c^{s}_{x_n,\theta} i\left(\frac{-\cos\theta}{\sin\theta}t + \frac{x}{\sin\theta}\right)\Psi^{\theta,T}_{x_n}(t)}{\sum_{x_n} c^{s}_{x_n,\theta}\Psi^{\theta,T}_{x_n}(t)}\right) \qquad k = 1,2,3. \tag{12.43}$$

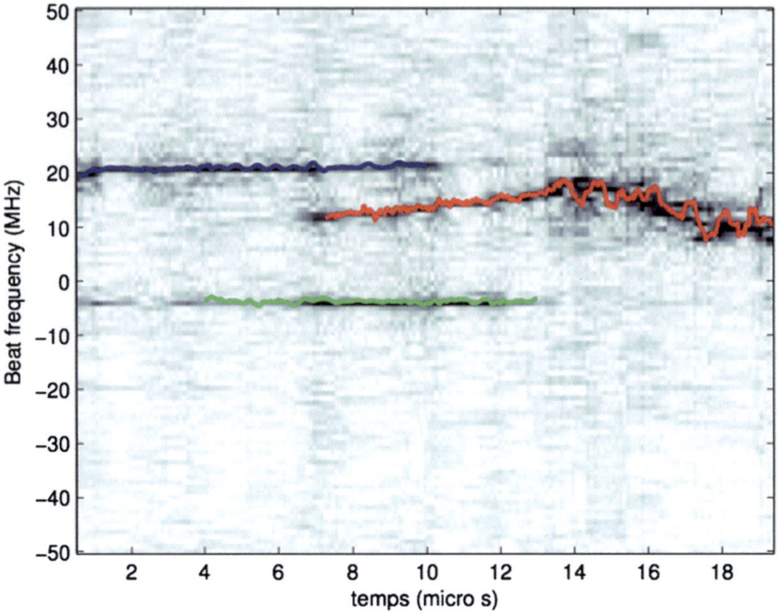

Fig. 12.8 The phase derivative of the three components, estimated by the tomogram method, is plotted on the time-frequency representation of the signal $s(t)$

This provides a more robust method to estimate the derivative. The phase derivative of the three components obtained with this method is plotted on the time-frequency representation of the signal (Fig. 12.8).

The expression of the phase derivative $\frac{\partial}{\partial t}\phi_k(t)$, given by the Eq. (12.43), is true for all $t \in [0, T]$.

As it can be seen in Fig. 12.7, the reflections on the porthole ($\widetilde{s_1}$) and the inner wall ($\widetilde{s_3}$) are very weak $t > 11\mu s$. The reflection on the plasma ($\widetilde{s_2}$) starts only after $7\,\mu s$. The phase derivative will be computed only when the signal exists.

For some values of t, the denominator in Eq. (12.43) could be very small, and then the estimation of the phase derivative is not good. To overcome this problem, we use a low pass filter. More details are given in [14].

The estimation of the phase derivative of the three components, with the method described above, is plotted on the time-frequency representation of the full signal $s(t)$. The method gives good results as it is shown in Fig. 12.8.

The data processing (tomogram) will be used for the new reflectometer on Tore Supra and on Jet [11, 24, 73, 74].

12.4 Detection and Characterization of Lévy Flights

12.4.1 Context: Stickiness and Lévy Flights in Chaotic Advection

In order to detect Lévy flights, we shall consider a specific physical context namely the stickiness phenomenon which leads to the presence of such flights. For this purpose we shall introduce briefly this phenomenon, in the case when it occurs in low-dimensional Hamiltonian systems [45]. To be more explicit stickiness occurs at the border between an island of regular motion and the chaotic sea. This stickiness induces long time correlations and as such memory effects and Lévy flights. We consider a specific physical context for which this phenomenon has been explicitly exhibited. This will allow the reader to get a better intuition on physical mechanisms behind the stickiness and how it affects transport properties. The considered system is the advection of passive tracers by a two-dimensional time dependent flow leading to the phenomenon of chaotic advection. For this purpose and in order to be more explicit we shall consider a specific flow generated by three point vortices (see, for instance, [54]).

12.4.1.1 Chaotic Advection

Let us start by giving some definition and clarifying the background of chaotic advection. Let us consider a flow $\mathbf{v}(\mathbf{r}, t)$ of an incompressible fluid ($\nabla \cdot \mathbf{v} = 0$) and a particle advected by this flow: one can, for instance, picture a small object floating on the surface of a river and transported by the stream. We then need to introduce the notion of passive particle or passive tracer. This notion defines an idealized particle which presence and motion in the fluid imposes no feedback on the flow and thus does not modify it. By definition this would be true for a fluid particle itself, but for other types of particles or tracers this is usually not true. However if the size of particle is small enough with respect to the length scales involved in the system and governing the flow, and other factors such as density and rugosity are more or less those of the considered fluid this ideal hypothesis is a good approximation. We can then derive the equation of motion of a passive particle which transported by the fluid so that its speed equals that of the fluid and hence its motion is governed by:

$$\dot{\mathbf{r}} = \mathbf{v}(\mathbf{r}, t), \tag{12.44}$$

where $\mathbf{r} = (x, y, z)$ refers to the tracer's position, and the $\dot{}$ to the time derivative.

We shall see now how this relates to Hamiltonian chaos. In fact for an incompressible flow, we can define a stream function which resumes to a scalar field for a two-dimensional system, such that the fluid velocity can be written as

$$\mathbf{v} = \nabla \wedge (\Psi\, \mathbf{z}), \tag{12.45}$$

where **z** corresponds to the unit vector perpendicular to the two dimensional space. Using Ψ, we can rewrite the equations governing the motion of a passive tracer Eq. (12.44) projected on each coordinate, as

$$\dot{x} = \frac{\partial \Psi}{\partial y}, \qquad \dot{y} = -\frac{\partial \Psi}{\partial x}. \tag{12.46}$$

And we recognize Hamilton equation of motion, where the space coordinates (x, y) are actually canonically conjugated and the stream function Ψ acts as an Hamiltonian.

When the flow is time independent, then the Hamiltonian Ψ reduces to an autonomous one degree of freedom system and is therefore integrable, which translates into the particular considered case that our passive tracers are following velocity field lines. However, it is possible and likely that the stream function Ψ is actually time dependent. In this case, we have actually a non-autonomous system and we have a time-dependent Hamiltonian system, meaning a system with $1 - \frac{1}{2}$ degrees of freedom. And it is known that generically, such systems generate the so-called Hamiltonian chaos. Note that this chaotic phenomenon can also occur in a stationary incompressible flow, but then the flow has to be three-dimensional, and we talk about chaos of field lines, see, for instance, [55] and references therein.

In the context of the advection of particles in flows, this chaotic nature of trajectories was called as a phenomenon of chaotic advection [3, 4, 64]. One of the major consequences of this phenomenon concerns the mixing of trajectories. Indeed chaotic advection can enhance drastically the mixing properties of the flow, meaning that the mixing process generated by the chaotic motion is much more efficient than the one occurring through molecular diffusion. And this effect is even more patent when the flow is laminar [6, 9, 28, 65, 81]. When dealing with mixing in micro-fluid experiments and devices, chaotic advection becomes crucial. Indeed since the Reynolds number is usually small, chaotic mixing becomes, de facto, an efficient way to mix. There are also numerous domains of physics, displaying chaotic advection-like phenomena, for instance in geophysical flows or magnetized fusion plasmas [2, 10, 17–19, 29, 34, 35, 48].

To detect the Lévy flights we use data coming from the simulation of passive tracers advected by the flow generated by three point vortices. We now shall recall quickly what is a point vortex and how they appear and can be useful in two-dimensional flows.

12.4.1.2 Definition of a Point Vortex

In order to describe the notion of a point vortex it is convenient to start with Euler equation. In fact, when considering a perfect two-dimensional incompressible flow governed by the Euler equation, if we are interested in the dynamics of the vorticity field Ω, we simply take the rotational of the Euler equation. This helps getting rid of the pressure and other potential forces gradients and we end up with the following equation

$$\frac{\partial \Omega}{\partial t} + \{\Omega, \Psi\} = 0, \ \Omega = -\nabla^2 \Psi, \tag{12.47}$$

where $\{\cdot, \cdot\}$ corresponds to the Poisson brackets. In order for the point vortices to "appear," we assume a vorticity field given by a superposition of point concentrated vorticities (Dirac functions) written as

$$\Omega(\mathbf{r}, t) = \sum_{i=1}^{N} k_i \delta\left(\mathbf{r} - \mathbf{r}_i(t)\right), \tag{12.48}$$

where k_i is the vorticity of a point vortex, and the vortex is localized by the point $\mathbf{r}_i(t)$ in the plane. This singular distribution is actually an exact solution of the Euler equation (12.47) when each of the N vortices obeys a specific and prescribed motion [58]. In fact the dynamics of the vortices ends up being equivalent to the one coming from an N-body Hamiltonian dynamics. The form of the Hamiltonian is strongly related to the Green function and therefore depends on the considered boundary conditions. Typically if one considers no specific boundary conditions, meaning that we allow the flow to evolve on the whole plane. In this case, the Hamiltonian is quite simple and writes

$$H = \frac{1}{2\pi} \sum_{i>j} k_i k_j \ln |\mathbf{r}_i - \mathbf{r}_j|, \tag{12.49}$$

where the canonically conjugated variables are $k_i y_i$ and x_i. This is reminiscent of the passive tracer Hamiltonian as the canonical variables are intimely linked to the vortex position $\mathbf{r}_i(t)$ in the plane; however, it is important to recall that the phase space corresponds now to a 2N dimensional space.

The equations of motion derived from Hamiltonian (12.49) just state the fact that each vortex is advected by the velocity field generated by the other vortices. We also can note that since we know in time the positions of the point vortices, we know as well the stream function (the Hamiltonian governing passive tracers) of the flow:

$$\Psi(\mathbf{r}, t) = -\frac{1}{2\pi} \sum_{i=1}^{N} k_i \ln |\mathbf{r} - \mathbf{r}_i(t)|. \tag{12.50}$$

As a last remark and important point concerning point vortex dynamics, it is important to notice that the Hamiltonian (12.49) is invariant by translation and by rotation. There are thus three constant of the motion besides the "energy" associated with these symmetries. However only three integrals are really in involution and Hamiltonian chaos appears in point vortex motion when we have more than $N = 3$ vortices [5, 46, 47, 63]. Note that point vortices can be also useful to model some geophysical flows [49]. And that three vortices can have singular solution, leading to finite time singularities which can lead to interesting properties and considerations [46, 50].

In order to obtain a regular (laminar) and time-dependent flow, the flow generated by three vortices is a good compromise. Indeed the integrable motion of three point vortices shows a large variety of behaviors, quasi-periodic and aperiodic flows are both possible [5, 46, 63], and are more easy to tackle than flows with more vortices see, for instance, [44, 47], as Poincaré maps can be computed [41, 42, 54]. In order to choose among the different possibilities, we would like to point out that usually, to address transport properties, asymptotic (large times) behavior and time translational properties are desired. So in order to achieve a situation where these features exist, we have had to consider the quasi-periodic motion of vortices. Note that these discussions are inspired by the work related to transport of passive tracers in the case of three identical vortices found in [41, 42] and the one reported in [54] corresponding to a situation of vortices with vorticities with different signs.

12.4.1.3 Stickiness and Anomalous Transport

Until now, we have briefly reviewed the notion of chaotic mixing in a flow generated by three point vortices. As a matter of fact transport in these systems is potentially anomalous [54]. In order to emphasize what we mean by anomalous, we would like to remind the reader that the type of transport can be defined by considering the behavior of the second moment of the displacement distribution and, for instance, extracting a value of a characteristic exponent. If we proceed as mentioned, we end up with a rough definition of anomalous transport, meaning that transport is said to be anomalous when it is not Gaussian (diffusive), meaning that

$$\langle X^2 - \langle X \rangle^2 \rangle \sim t^{\mu}, \tag{12.51}$$

with $\mu \neq 1$ and as such:

1. If $\mu < 1$ transport is anomalous and sub-diffusion is present.
2. If $\mu = 1$ transport is Gaussian and we have diffusion.
3. If $\mu > 1$ transport is anomalous and super-diffusion is present.

Going back to our point vortex system, the motion of passive tracers is depicted in the Poincaré section depicted in Fig. 12.9. We can notice that there are islands of regular motion, surrounded by a finite chaotic sea. When measuring transport, we shall consider only initial conditions in the stochastic sea, but since this chaotic region is bounded. Measuring plain dispersion is not convenient, it is, however, possible to circumvent this problem by working instead with length of trajectories and then to measure the dispersion of distance travelled among different trajectories.

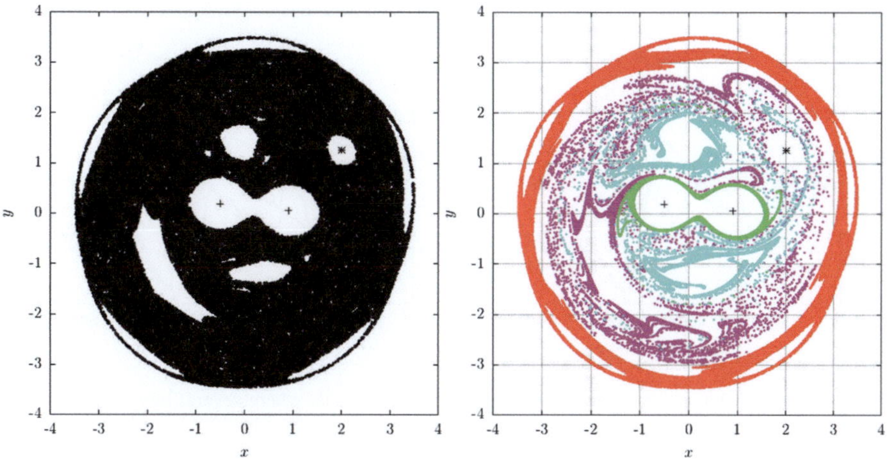

Fig. 12.9 *Left*: Poincaré section of passive particle in a flow generated by three point vortices. *Right*: localization of sticky regions contributing to different types of flights (see [54] for details)

$$s_i(t) = \int_0^t |v_i(\tau)| d\tau, \tag{12.52}$$

where $v_i(\tau)$ denotes the speed of particle i at time τ.

Once we have the length we can compute transport properties by computing the moments of the distribution

$$M_q(t) \equiv \langle |s(t) - \langle s(t) \rangle|^q \rangle, \tag{12.53}$$

where $\langle \ldots \rangle$ corresponds to ensemble averaging (average over different trajectories). Finally once we have the moments, we shall estimate the characteristic exponent of each moment, from its time evolution.

$$M_q(t) \sim t^{\mu(q)}. \tag{12.54}$$

As a result of this analysis the transport properties are found to be super-diffusive and multi-fractal [54], and this is the result of the memory effects engendered by stickiness.

Stickiness is a phenomenon which is found in Hamiltonian systems with mixed phase spaces, meaning phase spaces where regions of regular motion coexists with region of chaotic motion. When this is the case, in the vicinity of an island, trajectories can stay for arbitrary large times, we can think of them mimicking the behavior of the regular trajectories nearby inside the island, these sticky borders act then as pseudo-traps [51, 52, 59, 80].

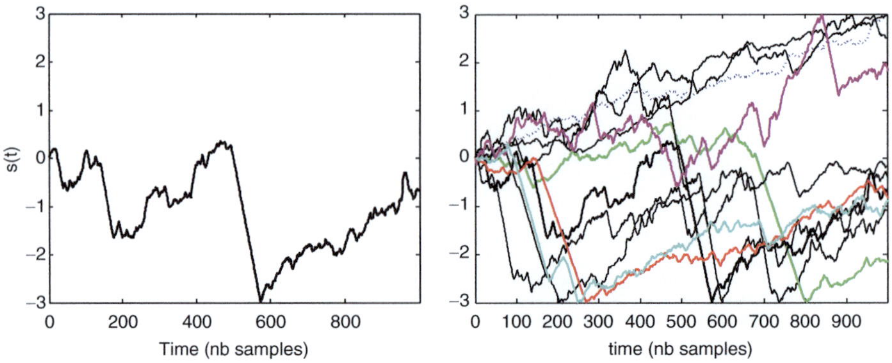

Fig. 12.10 *Left*: Arclength of the trajectory of a single particle, where several Lévy flights can be observed. *Right*: A set of trajectories of advected particles

In the end, stickiness induces memory effects and implies a slow decay of correlations. This affects strongly transport properties and ends up in an anomalous super-diffusive transport.

In order to visualize the effect of stickiness, we have extracted from transport data the points which were corresponding to Lévy flights and then localized them on the Poincaré section. We can see clearly that sticky regions are responsible for these flights and are located near regular islands (note that not all islands are necessary sticky, see, for instance, [53]). The plot is drawn in Fig. 12.9 (see [54,59] for details). To resume, once a trajectory sticks near an island, its length starts to grow almost linearly with time, it does so usually with an average speed generically different from the average peed over the chaotic sea. When looking at the transport data, this statement will imply the presence of Lévy flights in. In Fig. 12.9, we can see that four different sticking regions are present. We can thus expect to have four different types of Lévy flights in our advected data.

12.4.2 *Data Processing*

We shall now introduce the particularities of the data set from a signal processing point of view and describe the first step of the analyzing method.

A typical trajectory s is a one-dimensional signal of $N = 1000$ sampling points $s(t)$, $t \in [1, N]$. An example of such signal is shown in Fig. 12.10 (left) and a set of trajectories in Fig. 12.10 (right). Several parts can be distinguished: a random fluctuation (Brownian motion) and some *almost* linear segments of different length corresponding to Lévy flights.

12.4.2.1 A Time-Frequency Transformation

The robustness of our method relies on an uncertainty principle which is reminiscent of quantum mechanics. It can be shown that one cannot measure exactly both frequency and time of a given signal. We use this latter relation to our advantage. Through an elementary transformation we turn random fluctuations of the signal amplitude into random fluctuations of the frequency of a new signal. When these frequencies are rapidly varying, as it is the case for random behaviors or noise in the signal, the uncertainty principle makes it impossible to have precise information on these variations. In the meantime, coherent behavior is emphasized since it is less fluctuating.

It is then important to notice that thanks to the uncertainty principle:

- random fluctuations in frequency cannot be rendered precisely in the time-frequency plane. It requires to be precise both in time and in frequency, which is forbidden.
- linear parts or more generally slowly varying frequency components are emphasized by the time-frequency representation. Moreover, linear parts, called chirp signals, can be detected efficiently using the fractional Fourier transform.

It is then interesting and natural to take advantage of this fact for the analysis of the data set. To perform our analysis we shall therefore interpret the arclength $s(t)$ as the phase derivative (the fluctuation of the "frequency component") of a new signal $S(t)$. This corresponds to the first step of the process: Let us introduce the phase

$$\varphi(t) = \sum_{\tau=1}^{t} s(\tau), \tag{12.55}$$

and the signal

$$S(t) = e^{i\varphi(t)}. \tag{12.56}$$

The signal $S(t)$ is a non-stationary signal of magnitude one and made of a single frequency component which fluctuations are the one of the initial function $s(t)$.

The time-frequency representation (Gabor transform [20]) of S presented in Fig. 12.11 (right) is the absolute value of the short-time Fourier transform of S. One single frequency component can be seen which mimics the behavior of the signal s plotted on the left. But the important difference is now, because of the uncertainty principle, that brownian fluctuations become diffuse stains in Fig. 12.11 (right).

A consequence of this time-frequency transformation is that the random behavior is blurred even more, spread over a neighborhood zone, whereas the linear parts remain relatively sharp.

Our first objective is attained: the linear behavior has been emphasized over the brownian motion, thanks to the uncertainty principle.

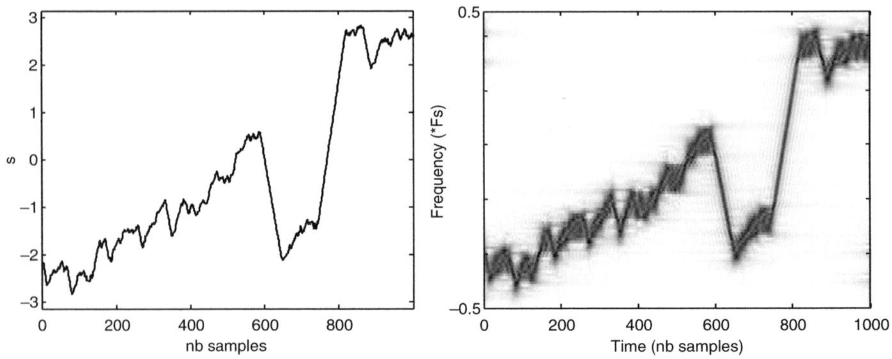

Fig. 12.11 *Left*: tracer trajectory s with fluctuating regions and linear regions (Lévy flights). *Right*: Gabor Transform (spectrogram) of S (absolute value of the short-time Fourier transform of S). *Darker* regions are associated to high values of $|\mathcal{V}_S|$

12.4.3 Tomographic Analysis

For the detection of linear behavior in chaotic signals, we need a method able to detect these straight line patterns. In a 2-dimensional image, one would use techniques such as the Hough transform. In our case, we need a similar tool retrieving straight lines which would appear when a time-frequency decomposition is done (such as the short-time Fourier transform, the Gabor transform, or the Wigner-Ville transform). The appropriate tool for this purpose is based on the time-frequency tomogram.

In Sect. 12.3.5 we describe in detail the time-frequency tomogram applied to a reflectometry signal of finite duration T.

In this application, projections of the reflectometry signal on an orthogonal basis $\left\{\Psi_{x_n}^{\theta,T}(t)\right\}_{x_0,\ldots,x_N}$ are used to extract the different components of the signal (see Sect. 12.3.5.2). Each element of the basis $\Psi_{x_n}^{\theta,T}(t)$ is equal to:

$$\Psi_{x_n}^{\theta,T}(t) = \frac{1}{\sqrt{T}} e^{i\left(\frac{-\cos\theta}{2\sin\theta} t^2 + \frac{x_n}{\sin\theta} t\right)} = \frac{1}{\sqrt{T}} e^{i\alpha(t)}. \tag{12.57}$$

We can notice that the phase derivative $\frac{d\alpha}{dt}$ is linear:

$$\frac{d\alpha}{dt}(t) = -\frac{1}{\tan\theta} t + \frac{x_n}{\sin\theta}. \tag{12.58}$$

That means that the projections of a signal on such basis will be appropriate to detect linear part of its phase derivative. Considering a signal with a linear phase derivative such $s(t) = e^{i\left(\frac{b}{2}t^2 + ct\right)}$, it is easy to demonstrate that the set of projections on an orthogonal basis:

$$c_{x_n,\theta}^{s} = < S, \Psi_{x_n}^{\theta,T} >, \tag{12.59}$$

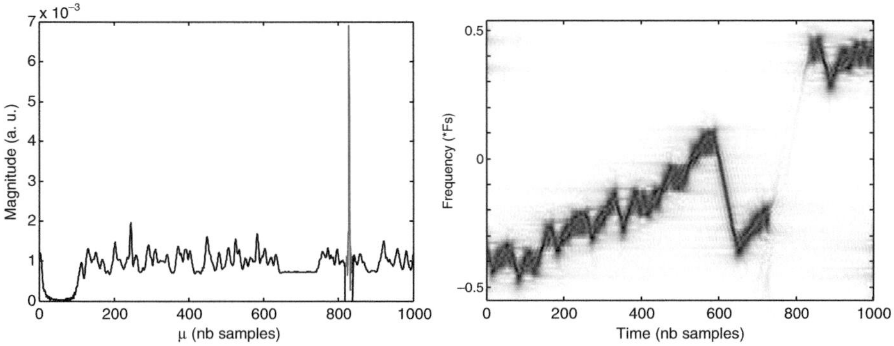

Fig. 12.12 *Left*: for θ_M, signal projections $|c_{\theta_M}(x_n)|$. *Right*: short-time Fourier transform of the signal S_1, partial reconstruction of S. The longest Lévy flight has been removed

is maximal for $\theta_M = \arctan(\frac{1}{b})$ and $x_M = c \sin \theta_m$.

The time-frequency tomogram will then be used to detect linear part in the phase derivative of the signal $S(t) = e^{i\varphi(t)}$ (Eq. 12.56), where the phase derivative is the arclength $s(t)$ of the particle (Eq. 12.55).

In order to detect the different slopes of the Lévy flights it is necessary to apply the time-frequency tomogram for different θ_k regularly spaced and search the maxima in the projections $c^S_{x_n,\theta_k}$. The number of selected θ_k is fixed by the user depending on how accurate he wants to be and is independent of the length of the signal N. The fast implementation of the time-frequency tomogram is of complexity $\mathcal{O}(N \log N)$, hence the overall complexity is of the same order.

The time-frequency tomogram can be reversed and it is possible to detect a linear part with slope $1/\tan\theta$ inside the signal then erase it in the (θ, μ) space and to re-synthesize the signal without this linear part by applying a time-frequency tomogram of angle $(-\theta)$.

12.4.4 Detection and Characterization of Lévy Flights

12.4.4.1 Method

On the signal shown in Fig. 12.11, one can see several Lévy flights (left) which have been turned into linear chirps in the frequency-time plane (right). For a specific angle θ_{M1}, the time-frequency tomogram defined (Eq. 12.59) will produce one sharp peak corresponding to the presence of a chirp as it is illustrated in Fig. 12.12 (left), where $|c_{\theta_{M1}}(x_n)|$ is plotted. For $x_{M1}{\sim}830$, the sharp peak $|c_{\theta_{M1}}(x_{M1})|$ gives evidence that there is a Lévy flight with a particular slope related to θ_{M1} and length related to amplitude of the peak. This search for maxima is the process that detects linear parts in the time-frequency plane.

Since the time-frequency tomogram is invertible (see Sect. 12.3.5.2), we can resynthesize the signal back to the initial representation after setting the values of the transform in red region of Fig. 12.12 (left) to zero. This result is illustrated in Fig. 12.12 (right), which represents the short-time Fourier transform of the newly recreated signal S_1. The largest frequency slope of S has been completely removed, the rest remaining untouched. This shows that indeed the peaks in the FRFT correspond to Lévy flights.

The method to detect linear parts of the phase derivative of the signal $S(t)$ is described by the following steps:

- compute the time-frequency tomogram of the signal $S(t)$, $|c^S_{\theta_k}(x_n)|$ for K values θ_k and N samples x_n:

$$c_{\theta_k}(x_n) =< S, \Psi^{\theta_k,T}_{x_n} >= \frac{1}{\sqrt{T}} \int_0^T S(t) e^{-i\left(\frac{-\cos\theta}{2\sin\theta} t^2 + \frac{x_n}{\sin\theta} t\right)} dt, \tag{12.60}$$

- extract the maximum from the $N \times K$ projections $|c^S_{\theta_k}(x_n)|$, θ_M, x_M will give the slope and the position of the first detection.
- reconstruction of the signal $S_1(t)$, where the linear part of the phase derivative is removed. A set $c^{S_1}_{\theta_M}(x_n)$ is obtained with the projections at the angle θ_M, where $|c^S_{\theta_M}(x_M)|$ and some coefficients of a small neighborhood are put to zero.

$$S_1(t) = \sum_{x_n} c^{S_1}_{\theta_M}(x_n) \Psi^{\theta_M,T}_{x_n}(t), \tag{12.61}$$

- repeat the process with $S_1(t)$ for other detections.

When the signal $S_p(t) = e^{i\phi_P(t)}$ is obtained, after p detections of Lévy flights in the phase derivative, we will estimate the arclength of trajectory $s_p(t)$ where the Lévy flights are removed:

$$s_P(t) = -i \frac{\partial S_p(t)/\partial t}{S_p(t)}. \tag{12.62}$$

Then $s_p(t)$ will be compared to $s(t)$ and the Lévy flights will be characterized by their length in time, Δl and their velocity $v_s = \Delta h/\delta l$.

This process is applied to the tracer trajectory s plotted in Fig. 12.11 (left). After two iterations, the linear part of the phase derivative of $S(t)$ is removed, as it can be seen in Fig. 12.13 (left). The tracer trajectory $s_2(t)$ where the Lévy flights have been removed is compared to the original $s(t)$ in Fig. 12.13 (right). Then, the flights are characterized by their length and velocity.

12.4.4.2 Results

We now consider blindly data obtained from the advection of 250 tracers in the point vector flow described in the previous subsection. That is to say, we analyze

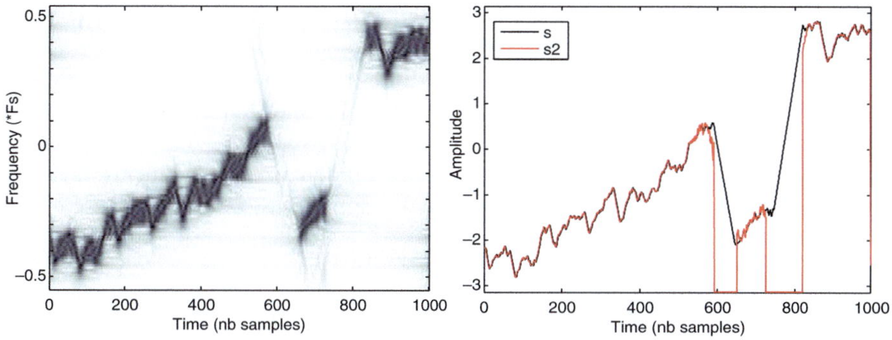

Fig. 12.13 *Left*: short-time Fourier transform of the signal S_2 where two Lévy flights have been removed. *Right*: original signal s (*black*) and partial reconstruction $s2$, without Lévy flights (*red*)

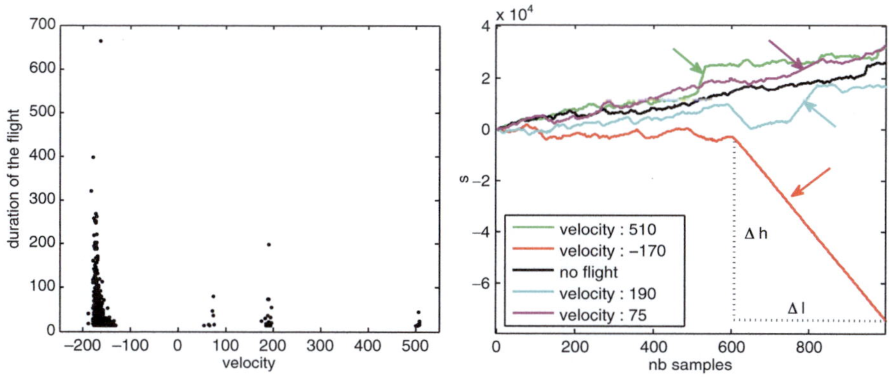

Fig. 12.14 *Left*: duration of the Lévy flights as a function of the velocity. *Right*: the velocity of the main Lévy flight is plotted for each trajectory

with our method 250 signals. We set up a threshold on the modulus of the projection coefficients in order to select only the most relevant Lévy flights. Similar transport data was analyzed in [54], with traditional tools and found to be anomalous and super diffusive. As mentioned, the starting point of the anomaly was traced back to a multi-fractal nature of transport linked to stickiness on four different regular regions. One would thus expect four different types of Lévy flights in the data (see Fig. 12.9).

In the present case, the method described above has been applied to the data set. Our goal is to detect the multi-fractal nature of the transport resulting from the sticky islands, which would serve as a proof of concept and pave the way for applying the method to numerical and experimental data. The results are presented in Fig. 12.14.

For each trajectory, Lévy flights have been detected and characterized by their length in time, Δl, and velocity, $\Delta h/\Delta l = s$. The process described in detail above gives, for each flight, its slope (related to the velocity) and length.

The Fig. 12.14 (left) is an illustration of the duration of the flights as a function of the velocity: four different values have been estimated (~ -170, ~ 75, ~ 190

and ~510), which means that there are four different types of Lévy flights, as anticipated. We mention as well that for some trajectories no Lévy flights have been detected. A few typical trajectories with Lévy flights have been plotted in Fig. 12.14 (right). The color coding corresponds to the one already used in Fig. 12.9, so that each specific detected flight can be easily associated with its originating sticky region. The agreement with the results found in [54] confirms that our method is successful and is thus ready to be applied to various numerical and experimental data.

12.5 Perspectives

After a first time frequency transformation, where the signal s is transformed as the phase derivative of a new signal S, the time-frequency tomogram is used to detect Lévy flights which are transformed as a linear phase derivative of S. This transformation makes use of the uncertainty principle: there is a "dilution effect" on the rapidly varying chaotic parts of the signal s while coherent patterns (Lévy flights) are only slightly affected. This part is critical for the robustness of the detection. Numerical simulations show that our technique is indeed extremely robust.

The time-frequency tomogram will give a sparse representation of the data of interest: Lévy flights become sharp peaks in the set of projections $c_{\theta k}(x_n)$. The key point is that we knew the pattern we want to detect and chose the transformation in consequence.

The door is open to further extension and generalization of our method, providing that one knows a priori the patterns to detect which may not be linear but curved or some other slowly varying shape (slowly varying with respect of the chaotic fluctuations). A different representation from the tomogram should be used based on the shape information. One may use a basis or a set of vectors different from the set of linear chirps. Possible alternatives may be found in, e.g., [16, 57] where what they call "tomograms" are bases of bended chirps and other more general time-frequency forms, associated with one or more parameters (equivalent of θ in the time-frequency tomogram case). One may also think of Gabor frames made of chirped windows [8]. Once the representation in which the relevant information is sparse has been found, the peak detection process remains the same.

12.6 Conclusion

This chapter is an attempt to show how non-commutative tomography can be used as an efficient and powerful signal processing tool. The approach is based on the physical analogies with the non-commutative nature between time and frequency, and actually use this to our advantage in order to "clean" signals from undesirable noise. For such purposes we started this chapter with a slow step by step introduction of the mathematical notions behind non-commutative tomography. We tried to emphasize on some simple specific examples in order to give the non-mathematical reader some possible intuitions on the nature of the considered transforms.

From then on we considered data originating from magnetized fusion plasmas, namely reflectometry data of the plasma. We first briefly introduce the field of tokamak plasmas and then discussed the ideas behind reflectometry and how reflectometers work, as well as how data is acquired and processed. We then showed how using tomogram techniques allowed to clearly separate relevant data from unnecessary reflections on the tokamak walls or on the initial porthole. In this context, the fact that the original signal sent into the plasma is a chirp, meaning a signal whose frequency varies linearly in time allowed us to select a specific family reminiscent of fractional Fourier transform, which are particularly adapted for such signals. The actual experimental data was then analyzed and for some specific value of frequency-time mixture, we were able to clearly distinguish between the different reflection of the original signal. Hence using the reconstructing technique we were able to filter out on the fly the data, in order to recover only the useful reflection on the plasma which is useful, for instance, in order to reconstruct time-dependent density profiles.

As a second application we considered data corresponding to the advection of the so-called passive tracers in the flow generated by three point vortices. The dynamics of these tracers is Hamiltonian but due to the time-dependent nature of the two-dimensional flow, their trajectories are chaotic. Actually the phase space of passive tracers corresponds to the so-called mixed phase spaces, meaning that there are regions where regular non-erratic motion is possible called regular islands, while there is a so-called stochastic sea, where the motion is chaotic. In these mixed phase space the phenomenon of stickiness is able to generate long memory effects which affects transport properties, generating anomalous diffusion of tracers and the existence of long-lasting Lévy flights. Using the analogy of considering a flight similar to the chirp signal used in the reflectometer, we performed a first simple transform of the signal in order to detect the chirp in the modified signal, which actually are flights in the original data. The method was shown to be successful in detecting the different Lévy flights present in the data, which were of different nature, as different sticking regions existed in the phase space.

In summary, we have showed in different contexts the efficiency of the signal processing method in two different cases, namely the case of reflectometry data and Lévy flights in advected data. Since in the context of magnetized fusion plasma there are some strong indications that transport is as well anomalous in the sense that it could be super-diffusive. It could be interesting to perform the Lévy analysis on reflectometer data, after the chirp flight trick has been performed. Should we detect as well some flights, it could be probably interesting to hard-code such signal processing treatment in a reflectometer to allow for fast plasma monitoring.

Acknowledgments Most of the works presented in this chapter are shared with our co-authors: Vladimir Man'ko, Margarita Man'ko, Rui Vilela Mendes, Ricardo Lima, Benjamin Ricaud,

Frederic Clairet, Christine Bottereau. We would like to take this opportunity to thank them and show our deep appreciation of our collaborations. We would like to thank as well Alberto Verga for a careful reading and suggestions which improved the manuscript.

References

1. Allen JB, Rabiner LR (1977) A Unified Approach to STFT Analysis and Synthesis. Proc IEEE 65:1558
2. Annibaldi SV, Manfredi G, Dendy RO, Drury LO (2000) Evidence for strange kinetics in Hasegawa-Mima turbulent transport. Plasma Phys Control Fusion 42:L13
3. Aref H (1984) Stirring by chaotic advection. J Fluid Mech 143:1
4. Aref H (1990) Chaotic advection of fluid particles. Phil Trans R Soc Lond A 333:273
5. Aref H, Pomphrey N (1980) Integrable and chaotic motion of four vortices. Phys Lett A 78:297
6. Bachelard R, Benzekri T, Chandre C, Leoncini X, Vittot M (2007) Targeted mixing in an array of alternating vortices. Phys Rev E 76(4):046217
7. Barbarossa S (1995) Analysis of multicomponent LFM signals by a combined Wigner-Hough transform. IEEE Trans Signal Process 43:1511
8. Baraniuk RG, Jones DL (1993) Shear madness: new orthonormal bases and frames using chirp functions. IEEE Trans Sig Proc 41(12):3543
9. Benzekri T, Chandre C, Leoncini X, Lima R, Vittot M (2006) Chaotic advection and targeted mixing. Phys Rev Lett 96(12):124503
10. Behringer R, Meyers S, Swinney H (1991) Chaos and mixing in geostrophic flow. Phys Fluids A 3:1243
11. Bottereau C, Briolle F, Clairet F, Giacalone J, Goniche M, Molina D, Poli S, Ricaud B, Sabot R (2011) New reflectometer in a Lower Hybrid Current Drive on Tore Supra. 10th international reflectometry workshop
12. Bottollier-Curtet H, Ichtchenko G (1987) Microwave reflectometry with the extraordinary mode on tokamaks: Determination of the electron density profile of PetuláB. Rev Sci Instrum 58(539)
13. Briolle F, Leoncini X, Ricaud R (2013) Fractional Fourier detection of Lévy Flights: application to Hamiltonian chaotic trajectories. Discontinuity, Nonlinearity, and Complexity, 2, 2,103–114.
14. Briolle F, Lima R, Vilela Mendes R (2009) Tomographic analysis of reflectometry data II: the phase derivative. Meas Sci Technol 20(10):105502
15. Briolle F, Lima R, Man'ko VI, Vilela Mendes R (2009) A tomographic analysis of reflectometry data I: Component factorization. Meas Sci Technol 20(10):105501
16. Briolle F, Man'ko VI, Ricaud B, Vilela Mendes R (2012) Noncommutative tomography: A tool for data analysis and signal processing. J Russian Laser Res 336(2):103
17. Brown M, Smith K (1991) Ocean stirring and chaotic low-order dynamics. Phys Fluids 3:1186
18. Carreras BA, Lynch VE, Garcia L, Edelman M, Zaslavsky GM (2003) Topological instability along filamented invariant surfaces. Chaos 13(4):1175
19. Chernikov AA, Petrovichev BA, Rogal'sky AV, Sagdeev RZ, Zaslavsky GM (1990) Anomalous Transport of Streamlines Due to their Chaos and their Spatial Topology. Phys Lett A 144:127
20. Chui CK (ed) (1992) Wavelets: a tutorial. Theory and applications, vol 2. Academic, New York
21. Clairet F, Bottereau C, Chareau JM, Paume M, Sabot R (2001) Edge density profile measurements by X-mode reflectometry on Tore Supra. Plasma Phys Control Fusion 43:429
22. Clairet F, Sabot R, Bottereau RC, Chareau JM, Paume M, Heuraux S, Colin M, Hacquin, Leclert G (2001) X-mode heterodyne reflectometer for edge density profile measurements on Tore Supra. Rev Sci Instrum 72:340
23. Clairet F, Bottereau C, Chareau JM, Sabot R (2003) Advances of the density profile reflectometry on TORE SUPRA. Rev Sci Instrum 74:1481

24. Clairet F, Ricaud B, Briolle F, Heuraux S, Bottereau C (2011) New signal processing technique for density profile reconstruction using reflectometry. Rev Sci Instrum 82(8)
25. Cohen L (1966) Generalized phase-space distribution functions. J Math Phys 7:781
26. Cohen L (1989) Time–frequency distributions. A review. Proc IEEE 77:941
27. Colchin RJ (1973) ORNL Technical Memo 93
28. Crisanti A, Falcioni M, Paladin G, Vulpiani A (1991) Lagrangian Chaos: Transport, Mixing and Diffusion in Fluids. La Rivista del Nuovo Cimento 14:1
29. Crisanti A, Falcioni M, Provenzale A, Tanga P, Vulpiani A, Phys. Fluids A (1992) Dynamics of passively advected impurities in simple two-dimensional flow models. 4:1805
30. Combes JM, Grossmann A, Tchamitchian P (eds) (1990) Wavelets, 2nd edn. Springer, Berlin
31. Daubechies I (1990) The wavelet transform: time–frequency localization and signal analysis. IEEE Trans Inform Theory 36(5):961
32. Deans SR (1983) The Radon transform and some of its applications. Wiley, New York
33. De Nicola S, Fedele R, Manko MA, Man'ko VI (2005) Fresnel Tomography: A Novel Approach to Wave-Function Reconstruction Based on the Fresnel Representation of Tomograms. Theor Math Phys 144:1206
34. del Castillo-Negrete D, Carreras BA, Lynch VE (2004) Fractional diffusion in plasma turbulence. Phys Plasmas 11(8):3584
35. Dupont F, McLachlan RI, Zeitlin V (1998) On possible mechanism of anomalous diffusion by Rossby waves. Phys Fluids 10(12):3185
36. Fourier JBJ (1888) Théorie Analytique de la Chaleur. Oeuvres de Fourier, vol. Tome premier. Gauthiers-Villars
37. Gelfand IM, Graev IM, Vilenkin NY (1966) Generalized functions, integral geometry and representation theory, vol 5. Academic, New York
38. Ginzburg VL (1964) The propagation of electro-magnetic waves in plasmas. Pergamon, New York
39. Grossmann A, Morlet J (1984) Decomposition of Hardy functions into square integrable wavelets of constant shape. SIAM J Math Anal 15:723
40. Horton W (1999) Drift waves and transport. Rev Mod Phys 71:735
41. Kuznetsov L, Zaslavsky GM (1998) Regular and Chaotic advection in the flow field of a three-vortex system. Phys Rev E 58:7330
42. Kuznetsov L, Zaslavsky GM (2000) Chaos, Fractional Kinetics, and Anomalous Transport. Phys Rev E 61:3777
43. Laviron C, Donne AJH, Manso ME, Sanchez J (1999) Reflectometry techniques for density profile measurements on fusion plasmas. Plasma Phys Control Fusion 38:905
44. Laforgia A, Leoncini X, Kuznetsov L, Zaslavsky GM (2001) Passive tracer dynamics in 4 point-vortex-flow. Eur Phys J B 20:427
45. Leoncini X (2011) Hamiltonian Chaos and Anomalous Transport in Two Dimensional Flows. In: Hamiltonian chaos beyond the KAM theory. Luo AC, Afraimovich V (eds) Nonlinear physical science. Springer, Berlin, pp 143–192
46. Leoncini X, Kuznetsov L, Zaslavsky GM (2000) Motion of Three Vortices near Collapse. Phys Fluids 12:1911
47. Leoncini X, Zaslavsky GM (2002) Jets, Stickiness and anomalous transport. Phys Rev E 65(4):046216
48. Leoncini X, Agullo O, Benkadda S, Zaslavsky GM (2005) Anomalous transport in Charney-Hasegawa-Mima flows. Phys Rev E 72(2):026218
49. Leoncini X, Verga A (2013) Dynamics of vortices and drift waves: a point vortex model. Eur Phys J B 86(3):95
50. Leoncini X, Barrat A, Josserand C, Villain-Guillot S (2011) Offsprings of a point vortex. Eur Phys J B 82:173
51. Leoncini X, Chandre C, Ourrad O (2008) Ergodicité, collage et transport anomal. C. R. Mecanique 336:530
52. Leoncini X, Kuznetsov L, Zaslavsky GM (2004) Evidence of fractional transport in point vortex flow. Chaos, Solitons and Fractals 19:259

53. Leoncini X, Neishtadt A, Vasiliev A (2009) Directed transport in a spatially periodic harmonic potential under periodic nonbiased forcing. Phys Rev E 79(2):026213
54. Leoncini X, Kuznetsov L, Zaslavsky GM (2001) Chaotic advection near a 3-vortex Collapse. Phys Rev E 63(3):036224
55. Leoncini X, Agullo O, Muraglia M, Chandre C (2006) From chaos of lines to lagrangian structures in flux conservative fields. Eur Phys J B 53(3):351
56. Man'ko VI, Vilela Mendes R (1999) Noncommutative time–frequency tomography. Phys Lett A 263:53
57. Man'ko MA, Man'ko VI, Vilela Mendes R (2001) Tomograms and other transforms: A unified view. J Phys A: Math Gen 34:8321
58. Marchioro C, Pulvirenti M (1994) Mathematical theory of uncompressible nonviscous fluids. Applied mathematical science, vol 96. Springer, New York
59. Meziani B, Ourrad O, Leoncini X (2012) Anomalous Transport and Phase Space Structures. In: Chaos, complexity and transport X. Leoncini, M. Leonetti Eds.. World Scientific, Singapore
60. Morlet J (1982) Sampling Theory and Wave propagation. 10, 233
61. Nawwab SH, Quatieri TF (1988) Short-time Fourier transform. Prentice Hall, Englewood, Cliffs, pp 289–337
62. Nazikian R, Mazzucato E (1995) Reflectometer measurements of density fluctuations in tokamak plasmas. Rev Sci Instrum 66:392
63. Novikov EA, Sedov YB (1978) Stochastic properties of a four-vortex system. Sov Phys JETP 48:440
64. Ottino J (1990) Mixing, Chaotic advection and turbulence. Ann Rev Fluid Mech 22:207
65. Ottino J (1989) The kinematics of mixing: streching, chaos, and transport. Cambridge U.P., Cambridge
66. Ozaktas HM, Zalevsky Z, Alper Kutay M (2001) The fractional Fourier transform with applications in optics and signal processing. Wiley, New York
67. Portonoff MR (1980) Time-frequency representations of Digitals signals and systems based on short-time Fourier analysis. IEEE Trans Acoust, Speech and Signal Proc ASSP-28(1):55
68. Poularikas AD (ed) (1996) The transforms and applications handbook. CRC/IEEE, New York
69. Qian S, Chen D (1995) Joint time–frequency analysis. Prentice-Hall, Englewood, Cliffs
70. Rax JM (2006) Physique des plasmas. Dunod, Paris
71. Stix TH (1962) The theory of plasma waves. McGraw-Hill, New York
72. Torresani B (1995) Analyse continue par ondelettes. InterEditions / CNRS Editions
73. Ricaud B., Briolle F., Clairet F., (2010) Traitement de données de réflectométrie pour la mise en évidence de phénomènes turbulents (Conférences URSI "Propagation et Plasma")
74. Briolle F, Clairet F (2009) Tomogram analysis and reflectometry 9nd international workshop on reflectometry
75. Ville J (1948) Théorie et applications de la notion de signal analytique. Cables et Transmission 2(A):61
76. Wigner E (1932) On the quantum correction for thermodynamic equilibrium. Phys Rev 40:749
77. Wolf KB (1979) Integral Transforms in Science and Engineering. Integral transforms in science and engineering. Plenum, New York
78. Woods JC, Barry DT (1994) Linear signal synthesis using the Radon–Wigner transform. IEEE Trans Signal Process 42:2105
79. Wood J, Barry DT (1994) Radon transformation of time-frequency distributions for analysis of multicomponent signals. IEEE Trans Signal Process 42:3166
80. Zaslavsky GM (2002) Chaos, Fractional Kinetics, and Anomalous Transport. Phys Rep 371:641
81. Zaslavsky GM, Sagdeev RZ, Usikov DA, Chernikov AA (1991) Weak chaos and quasiregular patterns. Cambridge University Press, Cambridge

Chapter 13
Projective Synchronization of Two Gyroscope Systems with Different Motions

Fuhong Min and Albert C. J. Luo

Abstract In this chapter, a simple nonlinear controller is applied to investigate the generalized projective synchronization for two gyroscopes with different dynamical behaviors. The projective synchronization conditions are developed through the theory of discontinuous dynamical systems. The synchronization invariant domain from the synchronization conditions is presented. The parameter maps are obtained for a better understanding of the synchronicity of two gyroscopes. Finally, the partial and full generalized projective synchronizations of two nonlinear coupled gyroscope systems are carried out to verify the effectiveness of the scheme. The scaling factors in such synchronization are observed through numerical simulations.

Keywords Projective synchronization • Gyroscope system • Discontinuous dynamical system

13.1 Introduction

Since Pecora and Carroll [1] investigated the synchronization between the dynamical systems, chaos synchronization has become a interesting topic due to its potential applications. The synchronization of many chaotic attractors was studied through different methods. Recently, chaos synchronization of gyroscopes with nonlinear damping has been studied extensively [2, 3]. In 2005, Lei et al. [4] discussed the global synchronization of two chaotic gyroscope systems through an

F. Min (✉)
Nanjing Normal University, Nanjing, Jiangsu 210042, China
e-mail: minfuhong@njnu.edu.cn

A.C.J. Luo
Southern Illinois University Edwardsville, Edwardsville, IL 62026-1805, USA
e-mail: aluo@siue.edu

J.A.T. Machado et al. (eds.), *Discontinuity and Complexity in Nonlinear Physical Systems*, Nonlinear Systems and Complexity 6, DOI 10.1007/978-3-319-01411-1_13,

active control method, and the sufficient conditions for the chaos synchronization were achieved. In 2006, Yan et al. [5] investigated the adaptive synchronization control for chaotic symmetric gyroscope systems through the adaptive sliding controller. In 2007, Yau et al. [6] investigated the complete synchronization of two chaotic nonlinear gyroscope systems through fuzzy logic control scheme. In 2008, Yau [7] adopted a fuzzy sliding mode control to synchronize two chaotic gyroscope systems with uncertainties and external disturbances. Hung et al. [8] used a sliding mode control technique to study the generalized projective synchronization of two chaotic gyroscope systems coupled with dead-zone nonlinear input. Salarieh and Alasty [9] used the modified sliding mode control to investigate the synchronization of two stochastic gyroscope systems with different parameters. From the above literature survey, the adopted techniques cannot present the necessary and sufficient conditions for synchronization, and the Lyapunov method was employed to determine the stability for such an error system. The control laws designed are often complicated, and the implementation becomes much difficult in practice.

In 2009, Luo [10] developed a theory for synchronization of dynamical systems with specific constraints via the theory of discontinuous dynamical systems. Such a theory for discontinuous dynamical systems can be found from [11–13]. In such a theory, the G-functions were introduced to determine the switchability of a flow from one domain to another in discontinuous dynamical systems. In Min and Luo [14], the complete synchronization of two chaotic gyroscope systems was investigated through the theory of discontinuous dynamical system. The parameter characteristics of chaotic synchronization were discussed from the analytical conditions of synchronization. In Min [15], the generalized projective synchronization of a noised chaotic gyroscope with a periodic gyroscope system was carried out initially. The partial and full projective synchronizations of two coupled chaotic gyros were observed. In Min and Luo [16], a comprehensive analytical mechanism of such synchronization will be discussed. The necessary and sufficient conditions for such synchronization will be derived from the theory of discontinuous dynamical systems in Luo [10–13].

In this chapter, the parameter characteristics for the generalized projective synchronization for two gyroscopes with different behaviors will be investigated. The synchronization will be presented in the theory of discontinuous dynamical systems. Numerical results for the partial and full generalized projective synchronizations for two dynamical systems with different behaviors are illustrated to demonstrate the usefulness and efficiency of the scheme.

13.2 Problem Statement

A periodically forced, symmetric gyroscope with linear-plus-cubic damping [5, 6] is considered as

$$\ddot{\theta} + c_1\dot{\theta} + c_2\dot{\theta}^3 + \alpha^2\frac{(1-\cos\theta)^2}{\sin^3\theta} - \beta\sin\theta = f\sin\omega t\sin\theta, \tag{13.1}$$

where θ is the angular displacement, $f \sin \omega t \sin \theta$ is parametric excitation, $c_1\dot{\theta}$ and $c_2\dot{\theta}^3$ are linear and nonlinear damping terms. The term $\alpha^2(1 - \cos\theta)^2/\sin^3\theta - \beta \sin\theta$ is a nonlinear force.

Let $x_1 = \theta$, $x_2 = \dot{\theta}$, and $h(\theta) = \alpha^2(1 - \cos\theta)^2/\sin^3\theta$, then the gyroscope system of Eq. (13.1) becomes

$$\begin{cases} \dot{x}_1 = x_2, \\ \dot{x}_2 = h(x_1) - c_1 x_2 - c_2 x_2^3 + (\beta + f_1 \sin \omega t) \sin x_1. \end{cases} \tag{13.2}$$

Consider Eq. (13.2) as a master system, a second controlled gyroscope system with different behavior is regarded as a slave system

$$\begin{cases} \dot{y}_1 = y_2, \\ \dot{y}_2 = h(y_1) - c_1 y_2 - c_2 y_2^3 + (\beta + f_1 \sin \omega t) \sin y_1, \end{cases} \tag{13.3}$$

where

$$h(y_1) = -\alpha^2(1 - \cos y_1)^2/\sin^3 y_1 \tag{13.4}$$

and the nonlinear control law $\mathbf{u}(t) = (u_1(t), u_2(t))^T$ is given by

$$u_1 = k_1 \operatorname{sgn}(y_1 - p_{1'} x_1) \text{ and } u_2 = k_2 \operatorname{sgn}(y_2 - p_2 x_2), \tag{13.5}$$

with p_1 and p_2 for the scaling factors, k_1 and k_2 for the controller parameters.

For simplicity, the state variables are introduced as

$$\mathbf{x} = (x_1, x_2)^{\mathrm{T}} \text{ and } \mathbf{y} = (y_1, y_2)^{\mathrm{T}} \tag{13.6}$$

and the corresponding vector fields are defined as

$$\bar{\mathbf{F}}(\mathbf{x}, t) = \left(\bar{F}_1(\mathbf{x}, t), \bar{F}_2(\mathbf{x}, t)\right)^{\mathrm{T}} \text{ and } \mathbf{F}(\mathbf{y}, t) = \left(F_1(\mathbf{y}, t), F_2(\mathbf{y}, t)\right)^{\mathrm{T}} \tag{13.7}$$

where

$$\begin{aligned} \bar{F}_1(\mathbf{x}, t) &= x_2, \\ \bar{F}_2(\mathbf{x}, t) &= h(x_1) - c_1 x_2 - c_2 x_2^3 + (\beta + f_1 \sin \omega t) \sin x_1, \\ F_1(\mathbf{y}, t) &= y_2 - u_1(t), \\ F_2(\mathbf{y}, t) &= h(y_1) - c_1 y_2 - c_2 y_2^3 + (\beta + f_2 \sin \omega t) \sin y_1 - u_2(t). \end{aligned} \tag{13.8}$$

Under the controlled law in Eq. (13.5), the controlled gyroscope system becomes discontinuous, and the corresponding vector fields are shown as follows:

1. For $y_1 > p_1 x_1$ and $y_2 > p_2 x_2$,

$$\begin{aligned} F_1(\mathbf{y},t) &= y_2 - k_1, \\ F_2(\mathbf{y},t) &= h(y_1) - c_1 y_2 - c_2 y_2^3 + (\beta + f_2 \sin \omega t) \sin y_1 - k_2; \end{aligned} \tag{13.9}$$

2. For $y_1 > p_1 x_1$ and $y_2 < p_2 x_2$,

$$\begin{aligned} F_1(\mathbf{y},t) &= y_2 - k_1, \\ F_2(\mathbf{y},t) &= h(y_1) - c_1 y_2 - c_2 y_2^3 + (\beta + f_2 \sin \omega t) \sin y_1 + k_2; \end{aligned} \tag{13.10}$$

3. For $y_1 < p_1 x_1$ and $y_2 < p_2 x_2$,

$$\begin{aligned} F_1(\mathbf{y},t) &= y_2 + k_1, \\ F_2(\mathbf{y},t) &= h(y_1) - c_1 y_2 - c_2 y_2^3 + (\beta + f_2 \sin \omega t) \sin y_1 + k_2; \end{aligned} \tag{13.11}$$

4. For $y_1 < p_1 x_1$ and $y_2 > p_2 x_2$,

$$\begin{aligned} F_1(\mathbf{y},t) &= y_2 + k_1 \\ F_2(\mathbf{y},t) &= h(y_1) - c_1 y_2 - c_2 y_2^3 + (\beta + f_2 \sin \omega t) \sin y_1 - k_2. \end{aligned} \tag{13.12}$$

With the above equations, there are four domains and four boundaries with different vector fields. As in Min and Luo [16], four domains Ω_α $(\alpha = 1,2,3,4)$ of the controlled slave systems in phase space are defined as

$$\begin{aligned} \Omega_1 &= \{(y_1, y_2) |\, y_1 - p_1 x_1(t) > 0,\, y_2 - p_2 x_2(t) > 0\}, \\ \Omega_2 &= \{(y_1, y_2) |\, y_1 - p_1 x_1(t) > 0,\, y_2 - p_2 x_2(t) < 0\}, \\ \Omega_3 &= \{(y_1, y_2) |\, y_1 - p_1 x_1(t) < 0,\, y_2 - p_2 x_2(t) < 0\}, \\ \Omega_4 &= \{(y_1, y_2) |\, y_1 - p_1 x_1(t) < 0,\, y_2 - p_2 x_2(t) > 0\}. \end{aligned} \tag{13.13}$$

and the boundaries $\partial \Omega_{\alpha\beta}$ $(\alpha, \beta = 1,2,3,4; \alpha \neq \beta)$ of the four domains are

$$\begin{aligned} \partial\Omega_{12} &= \{(y_1, y_2) |\, y_1 - p_1 x_1(t) > 0,\, y_2 - p_2 x_2(t) = 0\}, \\ \partial\Omega_{23} &= \{(y_1, y_2) |\, y_1 - p_1 x_1(t) = 0,\, y_2 - p_2 x_2(t) < 0\}, \\ \partial\Omega_{34} &= \{(y_1, y_2) |\, y_1 - p_1 x_1(t) < 0,\, y_2 - p_2 x_2(t) = 0\}, \\ \partial\Omega_{14} &= \{(y_1, y_2) |\, y_1 - p_1 x_1(t) = 0,\, y_2 - p_2 x_2(t) > 0\}. \end{aligned} \tag{13.14}$$

where the subscript $(\cdot)_{\alpha\beta}$ denotes the boundary from Ω_α to Ω_β.

From Eqs. (13.9) through (13.12), the controlled slave system is in a vector form of

$$\dot{\mathbf{y}}^{(\alpha)} = \mathbf{F}^{(\alpha)}(\mathbf{y}^{(\alpha)}, t), \tag{13.15}$$

where

$$\begin{aligned}
&\mathbf{F}^{(\alpha)}(\mathbf{y}^{(\alpha)},t)=(F_1^{(\alpha)},F_2^{(\alpha)})^{\mathrm{T}},\\
&F_1^{(\alpha)}(\mathbf{y}^{(\alpha)},t)=y_2^{(\alpha)}-k_1\ \ \text{for}\ \alpha=1,2;\\
&F_1^{(\alpha)}(\mathbf{y}^{(\alpha)},t)=y_2^{(\alpha)}+k_1\ \ \text{for}\ \alpha=3,4;\\
&F_2^{(\alpha)}(\mathbf{y}^{(\alpha)},t)=h(y_1^{(\alpha)})-c_1y_2^{(\alpha)}-c_2(y_2^{(\alpha)})^3\\
&\qquad\qquad+(\beta+f_2\sin\omega t)\sin y_1^{(\alpha)}-k_2\ \text{for}\ \alpha=1,4;\\
&F_2^{(\alpha)}(\mathbf{y}^{(\alpha)},t)=h(y_1^{(\alpha)})-c_1y_2^{(\alpha)}-c_2(y_2^{(\alpha)})^3\\
&\qquad\qquad+(\beta+f_2\sin\omega t)\sin y_1^{(\alpha)}+k_2\ \ \text{for}\ \alpha=2,3
\end{aligned}\tag{13.16}$$

The dynamical systems on the boundaries $\partial\,\Omega_{\alpha\beta}$ are presented by

$$\begin{aligned}
&\dot{\mathbf{y}}^{(\alpha\beta)}=\mathbf{F}^{(\alpha\beta)}\left(\mathbf{y}^{(\alpha\beta)},\mathbf{x}(t),t\right);\\
&\dot{\mathbf{x}}=\bar{\mathbf{F}}\left(\mathbf{x},t\right)
\end{aligned}\tag{13.17}$$

where

$$F_1^{(\alpha\beta)}\left(\mathbf{y}^{(\alpha\beta)},t\right)=y_2(t)=p_1x_2(t)\ \text{and}\ F_2^{(\alpha\beta)}(\mathbf{y}^{(\alpha\beta)},t)=p_2\dot{x}_2(t)\tag{13.18}$$

with

$$\begin{aligned}
&y_1^{(\alpha\beta)}=p_1x_1\ \text{and}\ y_2^{(\alpha\beta)}=p_2x_2\ \ \text{on}\partial\Omega_{\alpha\beta}\ \ \text{for}\ \ (\alpha,\beta)=\{(2,3),(1,4)\};\\
&y_1^{(\alpha\beta)}=p_1x_1+C\ \text{and}\ y_2^{(\alpha\beta)}=p_2x_2\ \ \text{on}\partial\Omega_{\alpha\beta}\ \ \text{for}\ \ (\alpha,\beta)=\{(1,2),(3,4)\}.
\end{aligned}\tag{13.19}$$

From the above equations, the boundary flows vary with time in the absolute coordinate, and it is difficult to develop the synchronization conditions. Then, introduce the relative coordinates

$$z_1=y_1-p_1x_1\ \text{and}\ z_2=y_2-p_2x_2.\tag{13.20}$$

The corresponding domains, boundaries, and the intersection point in the relative coordinates are presented as

$$\begin{aligned}
&\Omega_1=\{(z_1,z_2)|\,z_1>0,z_2>0\},\\
&\Omega_2=\{(z_1,z_2)|\,z_1>0,z_2<0\},\\
&\Omega_3=\{(z_1,z_2)|\,z_1<0,z_2<0\},\\
&\Omega_4=\{(z_1,z_2)|\,z_1<0,z_2>0\}.
\end{aligned}\tag{13.21}$$

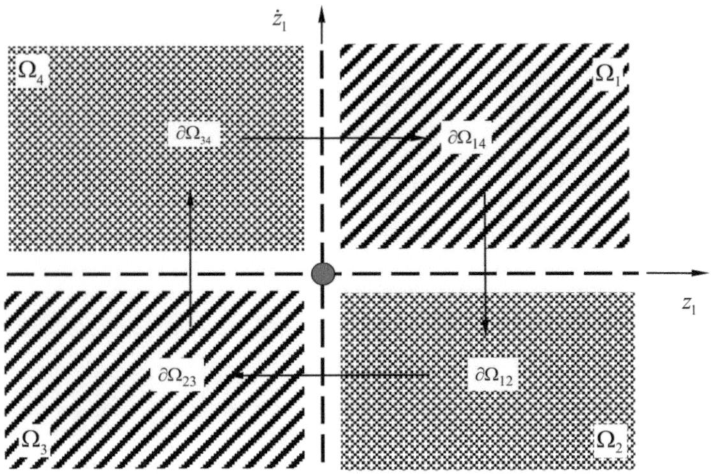

Fig. 13.1 Phase plane partitions and boundaries in the relative coordinates

$$
\begin{aligned}
\partial\Omega_{12} &= \{(z_1, z_2)|\, z_2 = 0, z_1 > 0\},\\
\partial\Omega_{23} &= \{(z_1, z_2)|\, z_1 = 0, z_2 < 0\},\\
\partial\Omega_{34} &= \{(z_1, z_2)|\, z_2 = 0, z_1 < 0\},\\
\partial\Omega_{14} &= \{(z_1, z_2)|\, z_1 = 0, z_2 > 0\}.
\end{aligned}
\tag{13.22}
$$

and

$$
\angle\Omega_{\alpha\beta} = \cap_{\alpha=1}^{4}\cap_{\alpha=1}^{4}\partial\Omega_{\alpha\beta} = \{(y_1, y_2)|\, z_1 = 0, z_2 = 0\}
\tag{13.23}
$$

From the above illustrations, the velocity and displacement boundaries in the relative frame are constant. Then, the partition of phase plane is sketched in Fig. 13.1. The intersection point is where the generalized projective synchronization of two gyroscopes with different motions. For this case, the analytical conditions for such synchronization can be developed easily through the theory of discontinuous dynamical systems. The controlled slave system in the relative coordinates becomes

$$
\begin{aligned}
&\dot{\mathbf{z}}^{(\alpha)} = \mathbf{g}^{(\alpha)}\left(\mathbf{z}^{(\alpha)}, \mathbf{x}, t\right)\\
&\text{with } \dot{\mathbf{x}} = \bar{\mathbf{F}}(\mathbf{x}, t)
\end{aligned}
\tag{13.24}
$$

where

$$
\begin{aligned}
&g_1^{(\alpha)}(\mathbf{z}^{(\alpha)}, \mathbf{x}, t) = z_2^{(\alpha)} - k_1 \quad \text{for } \alpha=1,2\\
&g_1^{(\alpha)}(\mathbf{z}^{(\alpha)}, \mathbf{x}, t) = z_2^{(\alpha)} + k_1 \quad \text{for } \alpha=3,4\\
&g_2^{(\alpha)}(\mathbf{z}^{(\alpha)}, \mathbf{x}, t) = \mathscr{G}(\mathbf{z}^{(\alpha)}, \mathbf{x}, t) - k_2 \quad \text{for } \alpha=1,4\\
&g_2^{(\alpha)}(\mathbf{z}^{(\alpha)}, \mathbf{x}, t) = \mathscr{G}(\mathbf{z}^{(\alpha)}, \mathbf{x}, t) + k_2 \quad \text{for } \alpha=2,3
\end{aligned}
\tag{13.25}
$$

with

$$\begin{aligned}\mathscr{G}(\mathbf{z}^{(\alpha)},\mathbf{x},t) = &-\alpha^2(1-\cos(p_1x_1+z_1^{(\alpha)}))^2/\sin^3(p_1x_1+z_1^{(\alpha)})-c_1z_2^{(\alpha)}\\ &-c_2(p_2x_2+z_2^{(\alpha)})^3+(\beta+f_2\sin\omega t)\sin(p_1x_1+z_1^{(\alpha)})\\ &-p_2h(x_1)+c_2x_2^3-(\beta+f_1\sin\omega t)\sin x_1,\end{aligned} \tag{13.26}$$

The dynamics on the boundary in the relative coordinates is also determined by

$$\dot{\mathbf{z}}^{(\alpha\beta)} = \mathbf{g}^{(\alpha\beta)}(\mathbf{z}^{(\alpha\beta)},\mathbf{x},t) \text{ with } \dot{\mathbf{x}} = \bar{\mathbf{F}}(\mathbf{x},t) \tag{13.27}$$

where

$$g_1^{(\alpha\beta)}(\mathbf{z}^{(\alpha\beta)},\mathbf{x},t) = z_2 = 0 \text{ and } g_2^{(\alpha\beta)}(\mathbf{z}^{(\alpha\beta)},t) = 0 \tag{13.28}$$

with

$$\begin{aligned} z_1^{(\alpha\beta)} &= 0 \text{ and } z_2^{(\alpha\beta)} = 0 \ \text{ on} \partial\Omega_{\alpha\beta} \text{ for } (\alpha,\beta) = (2,3),(1,4);\\ z_1^{(\alpha\beta)} &= C \text{ and } z_2^{(\alpha\beta)} = 0 \ \text{ on} \partial\Omega_{\alpha\beta} \text{ for } (\alpha,\beta) = (1,2),(3,4).\end{aligned} \tag{13.29}$$

13.3 Analytical conditions

Before discussing the synchronization conditions, the G-functions are introduced in the relative coordinates for $z_m \in \partial\,\Omega_{ij}$ at $t = t_m$, as in Luo [10–13]

$$G_{\partial\Omega_{ij}}^{(\alpha)}(\mathbf{z}_m,\mathbf{x},t_{m\pm}) = \mathbf{n}_{\partial\Omega_{ij}}^{\mathrm{T}}\cdot\left[\mathbf{g}^{(\alpha)}(\mathbf{z}_m,\mathbf{x},t_{m\pm}) - \mathbf{g}^{(ij)}(\mathbf{z}_m,\mathbf{x},t_{m\pm})\right] \tag{13.30}$$

$$G_{\partial\Omega_{ij}}^{(1,\alpha)}(\mathbf{z}_m,\mathbf{x},t_{m\pm}) = \mathbf{n}_{\partial\Omega_{ij}}^{\mathrm{T}}\cdot\left[D\mathbf{g}^{(\alpha)}(\mathbf{z}_m,\mathbf{x},t_{m\pm}) - D\mathbf{g}^{(ij)}(\mathbf{z}_m,\mathbf{x},t_{m\pm})\right] \tag{13.31}$$

From Eq. (13.24), the normal vectors of the relative boundaries are

$$\mathbf{n}_{\partial\Omega_{12}} = \mathbf{n}_{\partial\Omega_{34}} = (0,1)^{\mathrm{T}} \text{ and } \mathbf{n}_{\partial\Omega_{23}} = \mathbf{n}_{\partial\Omega_{14}} = (1,0)^{\mathrm{T}}. \tag{13.32}$$

From Eqs. (13.24) to (13.29), the corresponding G-functions in Eqs. (13.30) and (13.31) for a flow at the boundary are

$$\begin{aligned} G_{\partial\Omega_{12}}^{(\alpha)}(\mathbf{z}_m,\mathbf{x},t_{m\pm}) &= G_{\partial\Omega_{34}}^{(\alpha)}(\mathbf{z}_m,\mathbf{x},t_{m\pm}) = g_2^{(\alpha)}(\mathbf{z}_m,\mathbf{x},t_{m\pm}),\\ G_{\partial\Omega_{23}}^{(\alpha)}(\mathbf{z}_m,\mathbf{x},t_{m\pm}) &= G_{\partial\Omega_{14}}^{(\alpha)}(\mathbf{z}_m,\mathbf{x},t_{m\pm}) = g_1^{(\alpha)}(\mathbf{z}_m,\mathbf{x},t_{m\pm});\end{aligned} \tag{13.33}$$

$$\begin{aligned} G_{\partial\Omega_{12}}^{(1,\alpha)}(\mathbf{z}_m,\mathbf{x},t_{m\pm}) &= G_{\partial\Omega_{34}}^{(1,\alpha)}(\mathbf{z}_m,\mathbf{x},t_{m\pm}) = Dg_2^{(\alpha)}(\mathbf{z}_m,\mathbf{x},t_{m\pm}),\\ G_{\partial\Omega_{23}}^{(1,\alpha)}(\mathbf{z}_m,\mathbf{x},t_{m\pm}) &= G_{\partial\Omega_{14}}^{(1,\alpha)}(\mathbf{z}_m,\mathbf{x},t_{m\pm}) = Dg_1^{(\alpha)}(\mathbf{z}_m,\mathbf{x},t_{m\pm});\end{aligned} \tag{13.34}$$

To illustrate the flow switchability, the G-functions of a flow in domains with respect to the boundary are defined as

$$\begin{aligned}
G^{(\alpha)}_{\partial\Omega_{12}}(\mathbf{z}^{(\alpha)},\mathbf{x},t) &= G^{(\alpha)}_{\partial\Omega_{34}}(\mathbf{z}^{(\alpha)},\mathbf{x},t) = g^{(\alpha)}_{2}(\mathbf{z}^{(\alpha)},\mathbf{x},t),\\
G^{(\alpha)}_{\partial\Omega_{23}}(\mathbf{z}^{(\alpha)},\mathbf{x},t) &= G^{(\alpha)}_{\partial\Omega_{14}}(\mathbf{z}^{(\alpha)},\mathbf{x},t) = g^{(\alpha)}_{1}(\mathbf{z}^{(\alpha)},\mathbf{x},t);\\
G^{(1,\alpha)}_{\partial\Omega_{12}}(\mathbf{z}^{(\alpha)},\mathbf{x},t) &= G^{(1,\alpha)}_{\partial\Omega_{34}}(\mathbf{z}^{(\alpha)},\mathbf{x},t) = Dg^{(\alpha)}_{2}(\mathbf{z}^{(\alpha)},\mathbf{x},t),\\
G^{(1,\alpha)}_{\partial\Omega_{23}}(\mathbf{z}^{(\alpha)},\mathbf{x},t) &= G^{(1,\alpha)}_{\partial\Omega_{14}}(\mathbf{z}^{(\alpha)},\mathbf{x},t) = Dg^{(\alpha)}_{1}(\mathbf{z}^{(\alpha)},\mathbf{x},t).
\end{aligned} \tag{13.35}$$

where the total derivative functions are given by

$$\begin{aligned}
Dg^{(\alpha)}_{1}(\mathbf{z}^{(\alpha)},\mathbf{x},t) &= h_1(z^{(\alpha)}_1 + p_1x_1) - p_1h(x_1)\\
&\quad + (\beta + f_2\sin\omega t)\sin(z^{(\alpha)}_1 + p_1x_1) - c_1[z^{(\alpha)}_2 + p_2x_2 - p_1x_2]\\
&\quad - c_2[(z^{(\alpha)}_2 + p_2x_2)^3 - p_1x_2^3] - p_1(\beta + f_1\sin\omega t)\sin(x_1),\\
Dg^{(\alpha)}_{2}(\mathbf{z}^{(\alpha)},\mathbf{x},t) &= [h_1(z^{(\alpha)}_1 + p_1x_1) + (\beta + f_2\sin\omega t)\cos(z^{(\alpha)}_1 + p_1x_1)]F_1(\mathbf{z}^{(\alpha)} + \mathbf{x},t)\\
&\quad - [c_1 + 3c_2(z^{(\alpha)}_2 + p_2x_2)^2]F_2(\mathbf{z}^{(\alpha)} + \mathbf{x},t)\\
&\quad - p_2[h_1(x_1) + (\beta + f_1\sin\omega t)\cos x_1]F_1(\mathbf{x},t)\\
&\quad + p_2(c_1 + 3c_2x_2^2)F_2(\mathbf{x},t) + [f_2\sin(z^{(\alpha)}_1 + p_1x_1) - p_2f_1\sin x_1]\omega\cos\omega t;
\end{aligned} \tag{13.36}$$

As in Luo [10–13], the generalized projective synchronization state of the controlled slave system with the master system requires a sliding flow on the boundary. Similarly, the non-synchronization state at the boundary is a passable flow. The de-synchronization requires a source flow state to the boundary. Then the analytical conditions for the synchronization of two gyroscopes with different behaviors will be shown. Then, the synchronization conditions of two gyroscopes at the intersection point are

$$\begin{aligned}
&\left.\begin{aligned}
G^{(1)}_{\partial\Omega_{14}}(\mathbf{z}_m,\mathbf{x},t_{m-}) &= g^{(1)}_{1}(\mathbf{z}_m,\mathbf{x},t_{m-}) < 0,\\
G^{(1)}_{\partial\Omega_{12}}(\mathbf{z}_m,\mathbf{x},t_{m-}) &= g^{(1)}_{2}(\mathbf{z}_m,\mathbf{x},t_{m-}) < 0
\end{aligned}\right\} \text{for } \mathbf{z}_m \in \partial\Omega_{12}\cap\partial\Omega_{14} \text{ on } \Omega_1;\\
&\left.\begin{aligned}
G^{(2)}_{\partial\Omega_{12}}(\mathbf{z}_m,\mathbf{x},t_{m-}) &= g^{(2)}_{2}(\mathbf{z}_m,\mathbf{x},t_{m-}) > 0,\\
G^{(2)}_{\partial\Omega_{23}}(\mathbf{z}_m,\mathbf{x},t_{m-}) &= g^{(2)}_{1}(\mathbf{z}_m,\mathbf{x},t_{m-}) < 0
\end{aligned}\right\} \text{for } \mathbf{z}_m \in \partial\Omega_{12}\cap\partial\Omega_{23} \text{ on } \Omega_2;\\
&\left.\begin{aligned}
G^{(3)}_{\partial\Omega_{23}}(\mathbf{z}_m,\mathbf{x},t_{m-}) &= g^{(3)}_{1}(\mathbf{z}_m,\mathbf{x},t_{m-}) > 0,\\
G^{(3)}_{\partial\Omega_{34}}(\mathbf{z}_m,\mathbf{x},t_{m-}) &= g^{(3)}_{2}(\mathbf{z}_m,\mathbf{x},t_{m-}) > 0
\end{aligned}\right\} \text{for } \mathbf{z}_m \in \partial\Omega_{23}\cap\partial\Omega_{34} \text{ on } \Omega_3;\\
&\left.\begin{aligned}
G^{(4)}_{\partial\Omega_{34}}(\mathbf{z}_m,\mathbf{x},t_{m-}) &= g^{(4)}_{2}(\mathbf{z}_m,\mathbf{x},t_{m-}) < 0,\\
G^{(4)}_{\partial\Omega_{14}}(\mathbf{z}_m,\mathbf{x},t_{m-}) &= g^{(4)}_{1}(\mathbf{z}_m,\mathbf{x},t_{m-}) > 0
\end{aligned}\right\} \text{for } \mathbf{z}_m \in \partial\Omega_{34}\cap\partial\Omega_{14} \text{ on } \Omega_4.
\end{aligned} \tag{13.37}$$

From simplicity, four basic functions are introduced as

$$\begin{aligned} g_1(\mathbf{z}^{(\alpha)},\mathbf{x},t) &\equiv g_1^{(\alpha)}(\mathbf{z}^{(\alpha)},\mathbf{x},t) = z_2^{(\alpha)} - k_1 \text{ in } \Omega_\alpha \text{ for } \alpha = 1,2; \\ g_2(\mathbf{z}^{(\alpha)},\mathbf{x},t) &\equiv g_1^{(\alpha)}(\mathbf{z}^{(\alpha)},\mathbf{x},t) = z_2^{(\alpha)} + k_1 \text{ in } \Omega_\alpha \text{ for } \alpha = 3,4; \\ g_3(\mathbf{z}^{(\alpha)},\mathbf{x},t) &\equiv g_2^{(\alpha)}(\mathbf{z}^{(\alpha)},\mathbf{x},t) = \mathcal{G}(\mathbf{z}^{(\alpha)},\mathbf{x},t) - k_2 \text{ in } \Omega_\alpha \text{ for } \alpha = 1,4; \\ g_4(\mathbf{z}^{(\alpha)},\mathbf{x},t) &\equiv g_2^{(\alpha)}(\mathbf{z}^{(\alpha)},\mathbf{x},t) = \mathcal{G}(\mathbf{z}^{(\alpha)},\mathbf{x},t) + k_2 \text{ in } \Omega_\alpha \text{ for } \alpha = 2,3. \end{aligned} \tag{13.38}$$

The synchronization conditions in Eq. (13.37) become

$$\begin{aligned} g_1(\mathbf{z}_m,\mathbf{x},t_{m-}) &= z_{2m} + (p_2 - p_1)x_2 - k_1 < 0, \\ g_2(\mathbf{z}_m,\mathbf{x},t_{m-}) &= z_{2m} + (p_2 - p_1)x_2 + k_1 > 0, \\ g_3(\mathbf{z}_m,\mathbf{x},t_{m-}) &= \mathcal{G}(\mathbf{z}_m,\mathbf{x},t_{m-}) - k_2 < 0, \\ g_4(\mathbf{z}_m,\mathbf{x},t_{m-}) &= \mathcal{G}(\mathbf{z}_m,\mathbf{x},t_{m-}) + k_2 > 0. \end{aligned} \tag{13.39}$$

Let $\mathbf{z}_m = \mathbf{0}$, then the synchronization conditions of generalized projective synchronization for two gyroscopes are

$$\begin{aligned} g_1(\mathbf{z}_m,\mathbf{x},t_{m-}) &= (p_2 - p_1)x_2 - k_1 < 0, \\ g_2(\mathbf{z}_m,\mathbf{x},t_{m-}) &= (p_2 - p_1)x_2 + k_1 > 0, \\ g_3(\mathbf{z}_m,\mathbf{x},t_{m-}) &= \mathcal{G}(\mathbf{x},t_{m-}) - k_2 < 0, \\ g_4(\mathbf{z}_m,\mathbf{x},t_{m-}) &= \mathcal{G}(\mathbf{x},t_{m-}) + k_2 > 0. \end{aligned} \tag{13.40}$$

where

$$\begin{aligned} \mathcal{G}(\mathbf{x},t_{m-}) = & -\alpha^2(1-\cos(p_1x_1))^2/\sin^3(p_1x_1) - c_2[(p_2x_2)^3 - p_2x_2^3] \\ & + (\beta + f_2\sin\omega t_{m-})\sin(p_1x_1) - p_2h(x_1) - (\beta + f_1\sin\omega t_{m-})\sin x_1, \end{aligned} \tag{13.41}$$

If the control parameters k_1 and k_2 satisfy the conditions in Eq. (13.40), the projective synchronization of two coupled gyroscopes will be observed through numerical simulations. From the above conditions, the synchronization invariant set can be given by

$$-k_1 < (p_2 - p_1)x_2 < k_1 \text{ and } -k_2 < \mathcal{G}(\mathbf{x},t_{m\text{-}}) < k_2, \tag{13.42}$$

In a small neighborhood of $\mathbf{z}_m = \mathbf{0}$, the attractive conditions for $|\mathbf{z} - \mathbf{z}_m| < \varepsilon$ are

$$-k_1 < (p_2 - p_1)x_2 < k_1 \text{ and } -k_2 < \mathcal{G}(\mathbf{x},t_{m-}) < k_2, \tag{13.43}$$

From the foregoing equation, $z_1{}^*$ and $z_2{}^*$ are computed and the initial condition for the controlled slave system is computed by

$$y_1 = z_1^* + p_1x_1 \text{ and } y_2 = z_2^* + p_2x_2. \tag{13.44}$$

Once the generalized projective synchronization of two coupled gyroscopes disappears, the conditions of generalized projective synchronization vanishing for the controlled slave system on $\partial\Omega_{\alpha\beta}$ for $(\alpha,\beta)=\{(1,4),(2,3)\}$ are

$$\begin{aligned}
&0 \le z_2 < k_1 \text{ and } \mathscr{G}(\mathbf{z},\mathbf{x},t) < k_2 \text{ for } z_1 \in [0,\infty) \text{ in } \Omega_1,\\
&0 \le z_2 < k_1 \text{ and } -k_2 < \mathscr{G}(\mathbf{z},\mathbf{x},t) \text{ for } z_1 \in [0,\infty) \text{ in } \Omega_2,\\
&-k_1 < z_2 \le 0 \text{ and } -k_2 < \mathscr{G}(\mathbf{z},\mathbf{x},t) \text{ for } z_1 \in (-\infty,0] \text{ in } \Omega_3,\\
&-k_1 < z_2 \le 0 \text{ and } \mathscr{G}(\mathbf{z},\mathbf{x},t) < k_2 \text{ for } z_1 \in (-\infty,0] \text{ in } \Omega_4.
\end{aligned} \tag{13.45}$$

Let

$$y_1 = z_1^* + p_1 x_1 \text{ and } y_2 = z_2^* + p_2 x_2. \tag{13.46}$$

The vanishing conditions of generalized projective synchronization for the controlled slave systems on $\partial\Omega_{\alpha\beta}$ for $(\alpha,\beta)=\{(1,2),(4,3)\}$ are given by

$$\left.\begin{aligned}
g_1(\mathbf{z}_m^{(\alpha)},\mathbf{x},t_{m\mp}) &= z_{2m}^{(\alpha)} - k_1 = 0,\\
Dg_1(\mathbf{z}_m^{(\alpha)},\mathbf{x},t_{m\mp}) &= \mathscr{G}(\mathbf{z}_m^{(\alpha)},\mathbf{x},t_{m\mp}) > 0,\\
g_2(\mathbf{z}_m^{(\beta)},\mathbf{x},t_{m-}) &= z_{2m}^{(\beta)} + k_1 > 0,
\end{aligned}\right\} \text{ for } z_{m+\varepsilon} = y_1 - p_1 x_1 > 0, \tag{13.47}$$

and

$$\left.\begin{aligned}
g_1(\mathbf{z}_m^{(\alpha)},\mathbf{x},t_{m-}) &= z_{2m}^{(\alpha)} - k_1 < 0,\\
g_2(\mathbf{z}_m^{(\beta)},\mathbf{x},t_{m\mp}) &= z_{2m}^{(\beta)} + k_1 = 0,\\
Dg_2(\mathbf{z}_m^{(\beta)},\mathbf{x},t_{m\mp}) &= \mathscr{G}(\mathbf{z}_m^{(\beta)},\mathbf{x},t_{m\mp}) < 0,
\end{aligned}\right\} \text{ for } z_{m+\varepsilon} = y_1 - p_1 x_1 < 0. \tag{13.48}$$

The onset conditions of generalized projective synchronization for the controlled slave systems with $\mathbf{z}^{(\alpha)}(t_{m\mp}) = \mathbf{z}_m{}^{(\alpha)} = \mathbf{z}_m$ are

$$\left.\begin{aligned}
g_3(\mathbf{z}_m^{(\alpha)},\mathbf{x},t_{m\mp}) &= \mathscr{G}(\mathbf{z}_m^{(\alpha)},\mathbf{x},t_{m\mp}) - k_2 = 0,\\
Dg_3(\mathbf{z}_m^{(\alpha)},\mathbf{x},t_{m\mp}) &= D\mathscr{G}(\mathbf{z}_m^{(\alpha)},\mathbf{x},t_{m\mp}) > 0,\\
g_4(\mathbf{z}_m^{(\beta)},\mathbf{x},t_{m-}) &= \mathscr{G}(\mathbf{z}_m^{(\beta)},\mathbf{x},t_{m-}) + k_2 > 0,
\end{aligned}\right\} \text{for } \dot{z}_{m+\varepsilon} = y_2 - p_2 x_2 > 0, \tag{13.49}$$

from $z_{m-\varepsilon} = y_1 - p_1 x_1 > 0$, and

$$\left.\begin{aligned}
g_3(\mathbf{z}_m^{(\alpha)},\mathbf{x},t_{m-}) &= \mathscr{G}(\mathbf{z}_m^{(\alpha)},\mathbf{x},t_{m-}) - k_2 < 0,\\
g_4(\mathbf{z}_m^{(\beta)},\mathbf{x},t_{m\mp}) &= \mathscr{G}(\mathbf{z}_m^{(\beta)},\mathbf{x},t_{m\mp}) + k_2 = 0,\\
g_4(\mathbf{z}_m^{(\beta)},\mathbf{x},t_{m\mp}) &= D\mathscr{G}(\mathbf{z}_m^{(\beta)},\mathbf{x},t_{m\mp}) < 0,
\end{aligned}\right\} \text{for } \dot{z}_{m+\varepsilon} = y_2 - p_2 x_2 < 0. \tag{13.50}$$

from $z_{m+\varepsilon} = y_1 - p_1 x_1 < 0$.

The onset conditions of generalized projective synchronization for the controlled slave systems with $\mathbf{z}^{(\alpha)}(t_{m\pm}) = \mathbf{z}_m{}^{(\alpha)} = \mathbf{z}_m$ are

$$\left.\begin{aligned} g_1(\mathbf{z}_m^{(\alpha)}, \mathbf{x}, t_{m\pm}) &= z_{2m}^{(\alpha)} - k_1 = 0, \\ Dg_1(\mathbf{z}_m^{(\alpha)}, \mathbf{x}, t_{m\pm}) &= \mathscr{G}(\mathbf{z}_m^{(\alpha)}, \mathbf{x}, t_{m\pm}) > 0, \\ g_2(\mathbf{z}_m^{(\beta)}, \mathbf{x}, t_{m-}) &= z_{2m}^{(\beta)} + k_1 > 0; \end{aligned}\right\} \text{ for } (\alpha, \beta) = \{(1,4),(2,3)\} \tag{13.51}$$

from $\dot{z}_{m-\epsilon} = y_2 - p_2 x_2 > 0$ and

$$\left.\begin{aligned} g_1(\mathbf{z}_m^{(\alpha)}, \mathbf{x}, t_{m-}) &= z_{2m}^{(\alpha)} - k_1 < 0; \\ g_2(\mathbf{z}_m^{(\beta)}, \mathbf{x}, t_{m\pm}) &= z_{2m}^{(\beta)} + k_1 = 0, \\ Dg_2(\mathbf{z}_m^{(\beta)}, \mathbf{x}, t_{m\pm}) &= \mathscr{G}(\mathbf{z}_m^{(\beta)}, \mathbf{x}, t_{m\pm}) < 0; \end{aligned}\right\} \text{ for } (\alpha, \beta) = \{(1,4),(2,3)\} \tag{13.52}$$

from $\dot{z}_{m-\epsilon} = y_2 - p_2 x_2 < 0$.

13.4 Parameter Studies

From the previous analytical conditions, parameter studies will be carried out for a better understanding of the synchronization of two dynamical gyroscopes with different behaviors. The parameters of nonlinear gyroscope systems are first given as $\alpha^2 = 100$, $\beta = 1$, $c_1 = 0.5$, $c_2 = 0.05$, $\omega = 2$. As varying parameter f, the gyroscope system can exhibit different behaviors, including chaotic attractors and different periodic motions.

First, if $f_1 = 35.0$ and $f_2 = 35.7$, the master system is chaotic attractor, and the slave system is a period-4 motion. The initial conditions are given by $(x_1,x_2) = (0.17391, 1.36406)$ and $(y_1,y_2) = (0.32878, 0.67165)$. The scaling factors are given by $p_1 = -0.6$ and $p_2 = -1.0$. For a global view of the generalized projective synchronization of two coupled gyroscopes, the parameter map (k_1,k_2) is depicted in Fig. 13.2a. Acronyms "FS," "PS," and "NS" represent *full*, *partial*, and *non*-generalized projective synchronization. The partial generalized projective synchronization regions are shaded. For $k_1 > 0.59$ and $k_2 > 9.75$, the full generalized projective synchronization of the two coupled systems occurs. For $0 < k_1 < 0.59$ and $k_2 > 9.75$, only the partial generalized projective synchronization is obtained. For small k_2, the non-synchronization area is presented. Besides, varying the control parameter k_2, the switching phase for the generalized projective synchronization with $k_1 = 3$ is shown in Fig. 13.2b. The switching points of synchronization are satisfied $y_{1k} = p_1 x_{1k}$ and $y_{2k} = p_2 x_{2k}$. From the switching scenario, if $k_2 \in (0, 0.03)$, no synchronization is observed. If $k_2 \in (0.03, 9.74)$, the partial generalized projective synchronization of two gyroscopes occurs. If $k_2 \in (9.74, \infty)$, full generalized projective synchronization of the two systems exists. The switching scenarios are chaotic because the master system experiences the chaotic attractor.

Similarly, the master system is a period-4 motion and the slave system is chaotic with $f_1 = 35.0$ and $f_2 = 35.7$. The initial conditions are given by

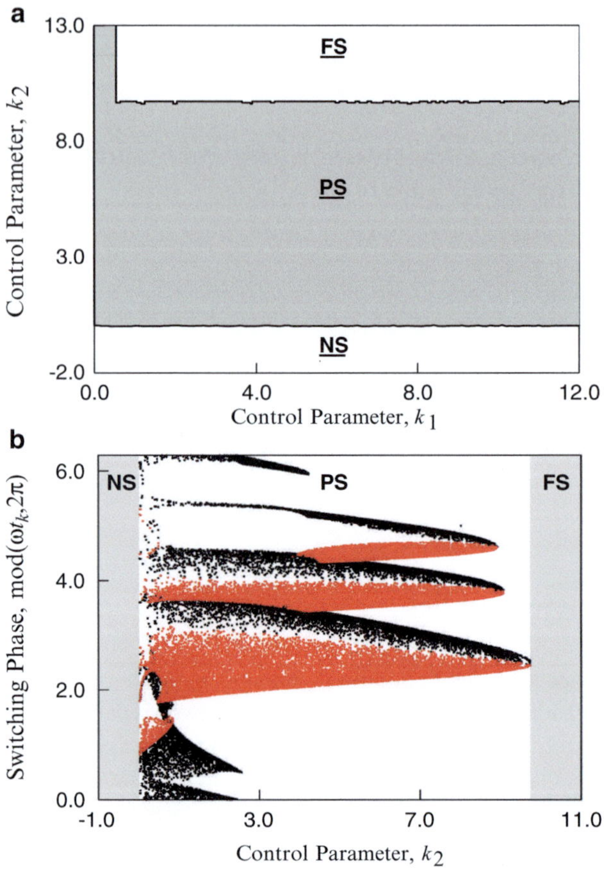

Fig. 13.2 Generalized projective synchronicity of two chaotic gyroscope dynamical systems. (**a**) Parameter maps for (k_1,k_2). (**b**) Switching phase with $k_1=3.0$. (The scaling factors: $p_1=-0.6$ and $p_2=-1.0$. System parameters: $c_1=0.5$, $c_2=0.05$, $\alpha=10$, $\beta=1.0$, $\omega=2.0$, $f_1=35.0$, $f_2=35.7$. *FS* full synchronization, *PS* partial synchronization, *NS* non-synchronization.)

$(x_1,x_2)=(0.08328,1.31797)$ and $(y_1,y_2)=(0.3,1.7)$. The scaling factors are given by $p_1=0.6$ and $p_2=1.2$. As varying the control parameters k_1 and k_2, the parameter map (k_1,k_2) is presented in Fig. 13.3a. Compared to Fig. 13.2a, the boundaries of the partial synchronization are much smooth because the master system is periodic. For $k_1>0.9$ and $k_2>14.05$, the full generalized projective synchronization of the two coupled gyroscopes yields. However, if $k_1<0.9$ and $k_2>14.05$, it is difficult to guarantee the synchronization conditions of two coupled gyroscopes, so only the partial synchronization exists. For small k_2, the non-synchronization area is presented. The switching phase for the generalized projective synchronization is illustrated in Fig. 13.3b with $k_1=3$. The switching scenario is regular because the master system is periodic. From the switching phase, no generalized projective

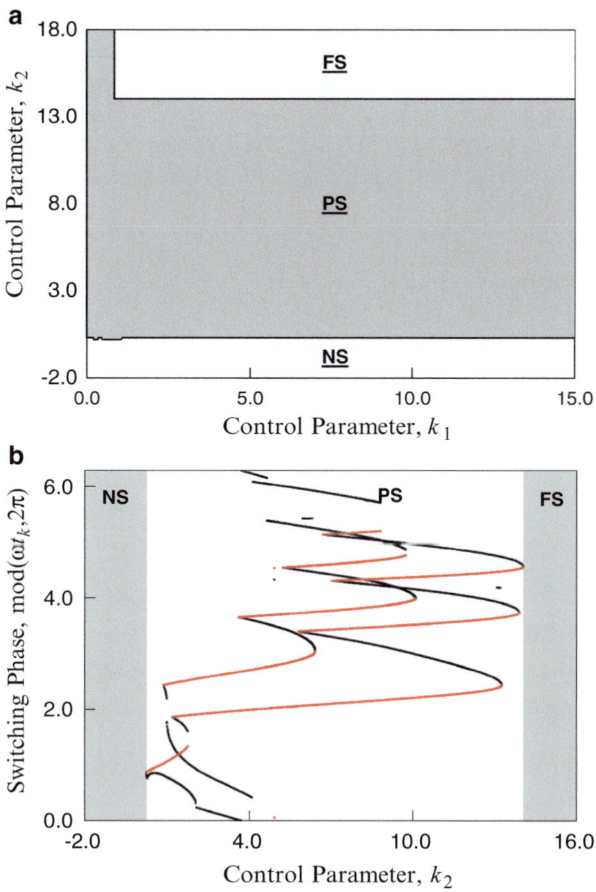

Fig. 13.3 Generalized projective synchronicity of two periodic gyroscopes. (**a**) Parameter maps for (k_1,k_2). (**b**) Switching phase with $k_1 = 3.0$. (The scaling factors: $p_1 = 0.6$ and $p_2 = 1.2$. System parameters: $c_1 = 0.5$, $c_2 = 0.05$, $\alpha = 10$, $\beta = 1.0$, $\omega = 2.0$, $f_2 = 35.0$, $f_1 = 35.7$. *FS* full synchronization, *PS* partial synchronization, *NS* non-synchronization.)

synchronization appears for $k_2 \in (0,0.3)$, and the partial generalized projective synchronization is in the range of $k_2 \in (0.3,14.05)$. The full synchronization of the two gyroscope systems can obtain for $k_2 \in (14.05,\infty)$.

13.5 Numerical Simulations

From the above parameter maps, the control parameters k_1 and k_2 can be chosen to do numerical simulations for the generalized projective synchronization of two coupled gyroscopes. According to the parameter maps in Fig. 13.2, the partial

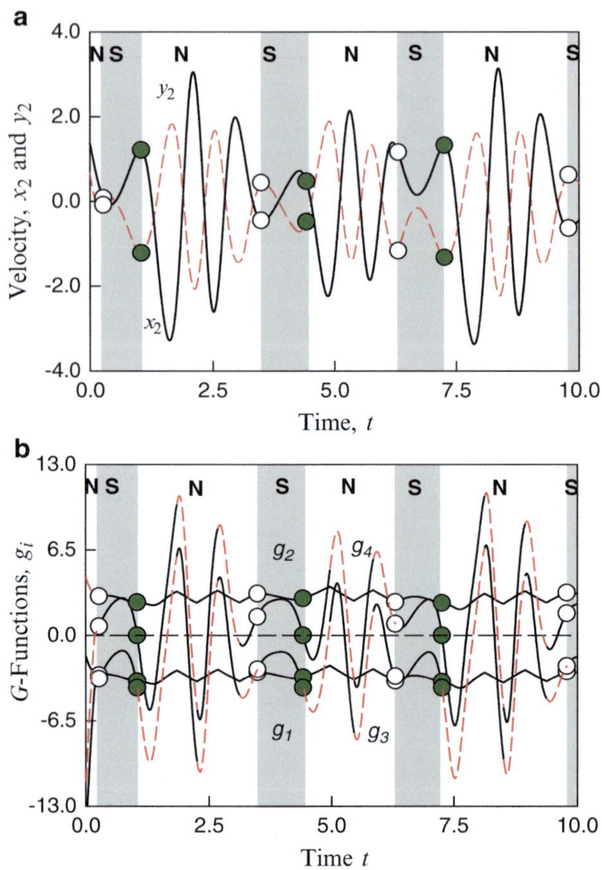

Fig. 13.4 The partial generalized synchronization of the periodic gyroscope with chaotic gyroscope system: (**a**) velocity responses, (**b**) G-function responses, (**c**) switching points of master system with the invariant domain, (**d**) switching points of slave system with the invariant domain. (The scaling factors: $p_1 = -0.6$ and $p_2 = -1.0$. System parameters: $c_1 = 0.5$, $c_2 = 0.05$, $\alpha = 10$, $\beta = 1.0$, $\omega = 2.0$, $f_1 = 35.0$, $f_2 = 35.7$. Initial condition: $(x_1, x_2) = (0.1739142, 1.3640564)$, and $(y_1, y_2) = (0.32877629, 0.67165378)$. *FS* full synchronization, *PS* partial synchronization, *NS* non-synchronization. *Hollow* and *filled circular symbols* are synchronization appearance and vanishing, respectively.)

synchronization of the coupled periodic gyroscope with the chaotic gyroscope system for $k_1 = 3.0$ and $k_2 = 2.0$ can be observed in Fig. 13.4. The symbols "S" and "N" represent "Synchronization" and "Non-synchronization." Hollow circulars represent the switching points for appearance, and filled circular symbols denote the synchronization disappearance. In Fig. 13.4a, the time-histories of velocities of two coupled gyroscopes are plotted. The trajectories for master system are depicted by solid curves and the trajectories for slave system are presented by dashed curves. In Fig. 13.4b, the corresponding G-functions are shown. The synchronization

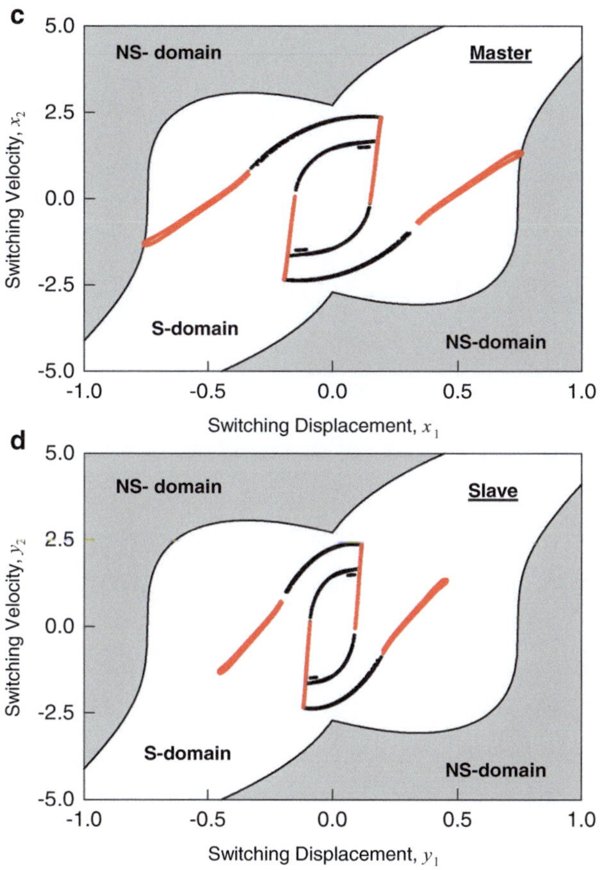

Fig. 13.4 (continued)

regions are shaded and the non-synchronization areas are not shaded. At all the switching points of appearance, the G-functions satisfy the onset conditions of projective synchronization in Eqs. (13.49)–(13.52). At all the switching points of disappearance, the G-functions satisfy the vanishing conditions of projective synchronization in Eqs. (13.45)–(13.48). The G-functions for non-synchronization are presented by dashed curve, which means imaginary flow. For instance, if the G-function of $g_3(t)$ is plotted by the dashed curve, the controlled slave system lies in domain $\Omega_\alpha(\alpha = 1, 4)$ and $y_2 < p_2x_2$ if the G-function of $g_4(t)$ is the dashed curve, the controlled slave system lies in domain Ω_α $(\alpha = 2, 3)$ and $y_2 > p_2x_2$. To observe the existence of the partial synchronization for a long time, the switching points of two coupled gyroscopes for 10,000 periods are shown in Fig. 13.4c, d, respectively. The black and red points are for the appearance and disappearance of projective synchronization, respectively. The invariant domain of synchronization is also inserted in Fig. 13.4c, d. "S-domain" denotes the synchronization invariant domain.

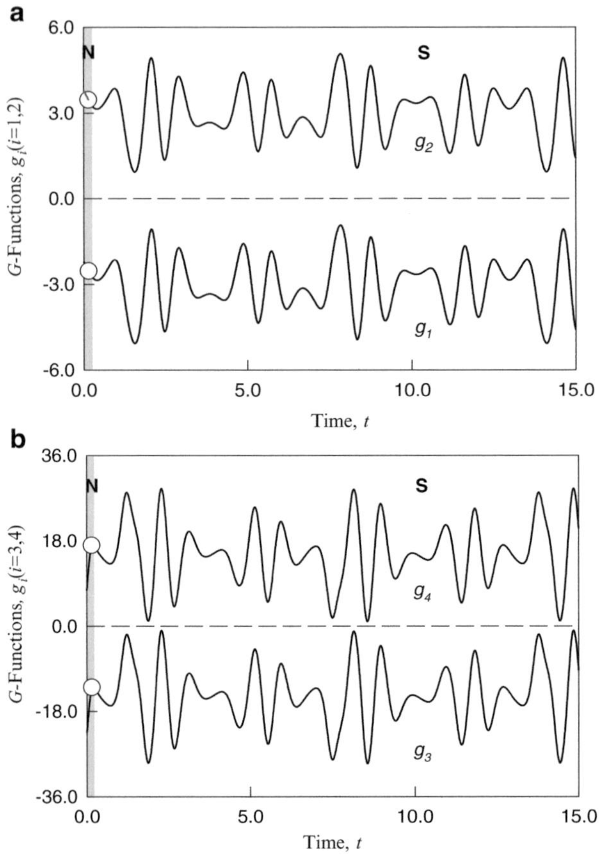

Fig. 13.5 Full generalized synchronization of the chaotic gyroscope with periodic gyroscope system: (**a** and **b**) time-histories of G-functions responses, (**c**) master system with the invariant domain, (**d**) slave system with the invariant domain. (The scaling factors: $p_1 = 0.6$ and $p_2 = 1.2$. System parameters: $c_1 = 0.5$, $c_2 = 0.05$, $\alpha = 10$, $\beta = 1.0$, $\omega = 2.0$, $f_1 = 35.7$, $f_2 = 35.0$. Initial condition: $(x_1, x_2) = (0.083285, 1.317967)$, and $(y_1, y_2) = (0.4, 1.6)$. *S* synchronization, *NS* non-synchronization. *Hollow circular symbols* are synchronization appearance.)

"NS-domain" represents the regions of non-synchronization. All the switching points lie in the invariant domain of projective synchronization.

From the parameter maps in Fig. 13.3, the full projective synchronization of the controlled chaotic gyroscope with a periodic gyroscope system is illustrated in Fig. 13.5 with control parameter $k_1 = 3$ and $k_2 = 15$. In Fig. 13.5a, b, the time-velocity history of the corresponding G-functions is shown. The shaded regions are for non-synchronization. Hollow symbols stand for the full generalized projective synchronization appearance. For time $t \in (0.14808, \infty)$, the G-function responses satisfy the conditions of full synchronization in Eq. (13.39), i.e., $g_1 < 0$, $g_2 > 0$, and $g_3 < 0$ and $g_4 > 0$, then the full synchronization with the scaling factors occurs

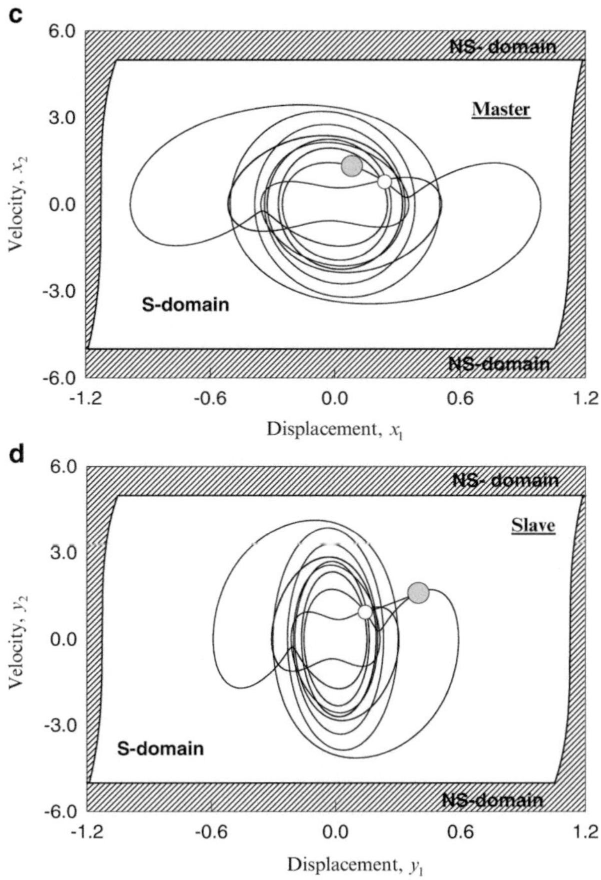

Fig. 13.5 (continued)

completely, $y_1(t) = 0.6x_1(t)$ and $y_2(t) = 1.2x_2(t)$. To make a better understanding the generalized projective synchronization for two coupled gyroscopes, the phase trajectories of master and slave systems are depicted in Fig. 13.5c, d, respectively. The invariant domain for full synchronization is also embedded. All the trajectories lie in the invariant domain of such synchronization.

13.6 Conclusion

In this chapter, the generalized projective synchronizations for two coupled gyroscopes with different motions were investigated through the theory of discontinuous dynamical systems. The analytical conditions for such synchronizations are shown.

The parameter studies for slave system with master system were carried out. Finally, the partial synchronization for a controlled periodic gyroscope with a chaotic gyroscope was illustrated, and the full synchronization of a controlled chaotic gyroscope with periodic gyroscope was developed. The scaling factors in such synchronization are observed through numerical simulations.

References

1. Pecora LM, Carroll TL (1990) Synchronization in chaotic systems. Phys Rev Lett 64(8): 821–824
2. Chen HK (2002) Chaos and chaos synchronization of symmetric gyro with linear-plus-cubic damping. J Sound Vib 255:719–740
3. Dooren RV (2003) Comments on chaos and chaos synchronization of symmetric gyro with linear-plus-cubic damping. J Sound Vib 268:632–635
4. Lei YM, Xu W, Zheng HC (2005) Synchronization of two chaotic nonlinear gyros using active control. Phys Lett A 343:153–158
5. Yan JJ, Hung ML, Liao TL (2006) Adaptive sliding mode control for synchronization of chaotic gyros with fully unknown parameters. J Sound Vib 298:298–306
6. Yau HT (2007) Nonlinear rule-based controller for chaos synchronization of two gyros with linear-plus-cubic damping. Chaos Solitons Fract 34:1357–1365
7. Yau HT (2008) Chaos synchronization of two uncertain chaotic nonlinear gyros using fuzzy sliding mode control. Mech Syst Signal Process 22:408–418
8. Hung MT, Yan JJ, Liao TL (2008) Generalized projective synchronization of chaotic nonlinear gyros coupled with dead-zone input. Chaos Solitons Fract 35:181–187
9. Salarieh H, Alasty A (2008) Chaos synchronization of nonlinear gyros in presence of stochastic excitation via sliding mode control. J Sound Vib 313:760–771
10. Luo ACJ (2009) A theory for synchronization of dynamical systems. Commun Nonlinear Sci Numer Simul 14:1901–1951
11. Luo ACJ (2008) A theory for flow switchability in discontinuous dynamical systems. Nonlinear Anal Hybrid Syst 2(4):1030–1061
12. Luo ACJ (2008) Global transversality, resonance and chaotic dynamics. World Scientific, Hackensack
13. Luo ACJ (2009) Discontinuous dynamical systems on time-varying domains. Higher Education Press, Beijing
14. Min FH, Luo ACJ (2012) On parameter characteristics of chaotic synchronization in two nonlinear gyroscope systems. Nonlinear Dyn 69:1203–1223
15. Min FH (2012) Analysis of generalized projective synchronization for a chaotic gyroscope with a periodic gyroscope. Commun Nonlinear Sci Numer Simul 17:4917–4929
16. Min FH, Luo ACJ (2013) Parameter characteristics of projective synchronization of two gyroscope systems with different dynamical behaviors. Discontinuity Nonlinearity Complex 3:1–16

Chapter 14
Measuring and Analysing Nonlinearities in the Lung Tissue

Clara M. Ionescu

Abstract This paper introduces the concept of fractional order models for characterizing viscoelasticity in the lungs. A technique to detect and analyse these nonlinear, low-frequency contributions in the lung tissue is presented, along with some experimental data. The measurements are performed using the forced oscillation technique and a non-invasive lung function testing procedure which takes only 40 s, while the patient is breathing at rest. The index introduced to quantify the nonlinear contributions in the lungs in healthy is then employed in a theoretical analysis to show that the values are changing in case of disease. The results indicate that the proposed method and index are useful for clinical classification of viscoelastic properties in the lungs.

Keywords Nonlinear distortion • Respiratory system • Viscoelasticity • Frequency response

14.1 Introduction

An optimal lung function parameter follow-up is a key element in allowing early detection of respiratory disorders and managing treatment strategies to maximize their positive effect on the patient. Since most lung diseases affect the viscoelastic properties of the respiratory tissue [1,6], it is optimal that lung function tests provide information upon the low-frequency dynamics in the airways and tissue [2, 14, 19].

The viscoelastic properties characterize materials such as polymers, found to be very similar to lung tissue [10]. In the human lung, these properties are changing with diseases and they may be detected at early stages by evaluating the respiratory

C.M. Ionescu (✉)
Faculty of Engineering and Architecture, Ghent University, Technologiepark 913, 9052 Gent-Zwijnaarde, Belgium
e-mail: ClaraMihaela.Ionescu@UGent.be

J.A.T. Machado et al. (eds.), *Discontinuity and Complexity in Nonlinear Physical Systems*, Nonlinear Systems and Complexity 6, DOI 10.1007/978-3-319-01411-1_14,

impedance at low frequencies, i.e. closer to the breathing frequency of the patient [2, 11, 14, 18]. However, the lower the frequency one wants to investigate in terms of signal processing methods, the more difficult the filtering problem between the breathing (which acts as a disturbance) and the effect of the excitation signal in the lungs (which is in fact the useful information one wants to extract). In particular, this paper addresses the detection of nonlinear contributions at frequencies below 10 Hz.

There is one single group of healthy young adult volunteers measured for the purpose of this paper. These are male and female, with age between 24–33 years, height between 164–180 cm and weight between 49–78 kg.

The paper is organized as follows: the forced oscillations technique is presented in the next section, along with some of the advantages over spirometry, in order to motivate our choice of method. The third section presents the theoretical background for signal processing and the algorithms employed in order to obtain the respiratory impedance in its best linear approximation, while minimizing the biasing effects introduced by the breathing of the patient. The fourth section delivers the results and the discussion of these results obtained for the prototype device, calibration tube and a group of 11 volunteers. A conclusion section summarizes the main outcome of this work and offers some perspectives.

14.2 FOT: Applications, Devices and Impedance Measurement

The forced oscillation technique (FOT) consists of superimposing external pressure signals on spontaneous breathing (tidal breathing) [3, 5, 12, 17]. It provides an effort independent assessment of respiratory mechanics [4].

The measurements of the signals analysed in this paper have been performed using the device depicted in Fig. 14.1, assessing respiratory mechanics in the range from 0.1 Hz to 5 Hz. The low-frequency multisine in the prototype allows excitation

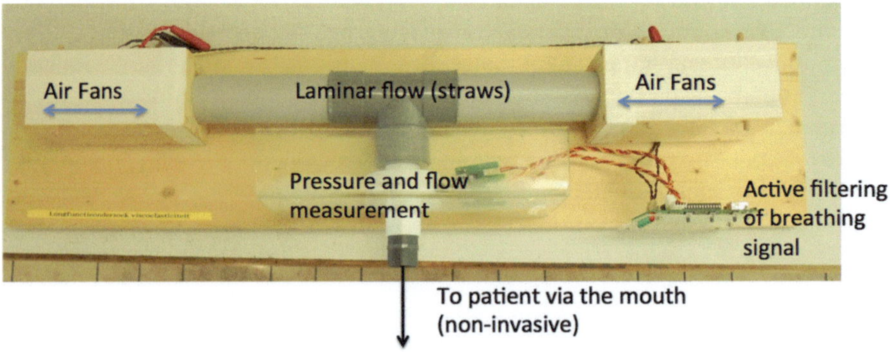

Fig. 14.1 A photo of the prototype device

of respiratory tissue at frequencies where viscoelastic properties become important and relevant for clinical insight. The measurement of air-pressure P (kPa) and air-flow $Q = dV/dt$ (liter/s) (with V as the air volume) during the FOT lung function test is done at the mouth of the patient. The FOT lung function tests were performed according to the recommendations described in [13]. The FOT excitation signal was kept within a range of a peak-to-peak size of 0.1–0.3 kPa, in order to ensure patient comfort and safety. The test has a maximum duration of 40 s and the sampling frequency is 1 kHz.

Although the prototype device introduced in this paper can only produce excitation signals as low as 0.1 Hz, it serves to perform a feasibility study upon detecting the nonlinear distortions and evaluating respiratory mechanics at frequencies closer to the breathing of the patient. Later on, the prototype can be further used to gather data from a larger number of volunteers to obtain reference intervals.

Viscoelasticity is well defined in materials exhibiting nonlinear dynamics, e.g. polymers. The properties of lungs are fairly similar to those of polymers [2] and viscoelastic effects in the human tissue take place at low frequencies [10]. When the respiratory mechanics are characterized at frequencies below the resonant frequency, one investigates the balance between elastic and viscous properties of the lung tissue and parenchyma [11, 14, 19].

14.3 Signal Processing Methods

This section addresses two problems: (1) the problem of breathing interference with the excitation signal and (2) the detection of nonlinear contributions in the measured signals. The common solution to these problems is the optimization of the excitation signal, detailed in [7, 15]. In short, the optimized excitation signal is an odd random phase multisine defined as:

$$U_{FOT} = \sum_{k=0}^{109} A_k \, sin(2\pi(2k+1) f_0 \, t \; + \phi_k) \tag{14.1}$$

with:

- frequency interval from 0.1 to 10.9 Hz
- frequency resolution f_0 of 0.1 Hz
- only odd harmonics
- only harmonics which are not overlapping with the first 5 breathing harmonics are used
- equal amplitude A_k for all excited harmonics
- the phase ϕ_k uniformly distributed between $[0, 2\pi]$
- 1 so-called *detection line* for each group of 4 excited odd harmonics is not excited in order to check for odd nonlinear distortion.

The standard procedure to obtain the impulse response $g(t)$ of a linear system is based on the correlation analysis [15]:

$$R_{yu}(t) = g(t) * R_{uu}(t) \tag{14.2}$$

with $u(t)$ the input signal, $y(t)$ the output signal and $*$ denoting the convolution product. $R_{yu}(t)$ and $R_{uu}(t)$ are the cross- and auto-correlations, respectively:

$$\begin{aligned} R_{yu}(\tau) &= E\{y(t)u(t-\tau)\} \\ R_{uu}(\tau) &= E\{u(t)u(t-\tau)\} \end{aligned} \tag{14.3}$$

with τ the shift interval. Applying Fourier-transform to (14.2) results in

$$G(j\omega) = \frac{S_{YU}(j\omega)}{S_{UU}(j\omega)} \tag{14.4}$$

where the cross-spectrum $S_{YU}(j\omega)$, the auto-spectrum $S_{UU}(j\omega)$, and the frequency response function (FRF) $G(j\omega)$ are the Fourier transforms of $R_{YU}(t)$, $R_{UU}(t)$ and $g(t)$, respectively.

The Best Linear Approximation (BLA) [15, 16] of a nonlinear system $g_{BLA}(t)$ minimizes the mean squared error (MSE) between the real output of a nonlinear system $y(t) - E\{y(t)\}$ and the output of a linear model approximation $g_{BLA}(t) * (u(t) - E\{u(t)\})$:

$$E\{\|(y(t) - E\{y(t)\}) - g_{BLA}(t) * (u(t) - E\{u(t)\})\|^2\} \tag{14.5}$$

where E denotes the expected value with respect to realizations of the input. In the frequency domain, the solution to the optimization problem from (14.5) is given by:

$$\hat{G}_{BLA}(j\omega) = \frac{\hat{S}_{YU}(j\omega)}{\hat{S}_{UU}(j\omega)} \tag{14.6}$$

where the cross-spectrum $\hat{S}_{YU}(j\omega)$, the auto-spectrum $\hat{S}_{UU}(j\omega)$, and the FRF $\hat{G}_{BLA}(j\omega)$ are the Fourier transforms of $R_{YU}(t)$, $R_{UU}(t)$ and $g_{BLA}(t)$, respectively. In practice, this relation is simplified for periodical signals as:

$$\hat{G}_{BLA}(j\omega_k) = \frac{1}{M}\sum_{m=1}^{M}\frac{Y^{[m]}(k)}{U^{[m]}(k)} \tag{14.7}$$

where the notation $X^{[m]}(k)$ has been used to describe the DFT-spectrum of the mth multisine realization. The estimation of BLA, $\hat{G}_{BLA}$, described by relation (14.7) can be re-written as:

$$\hat{G}_{BLA}(j\omega_k) = G_{BLA}(j\omega_k) + G_S(j\omega_k) + N_G(j\omega_k) \tag{14.8}$$

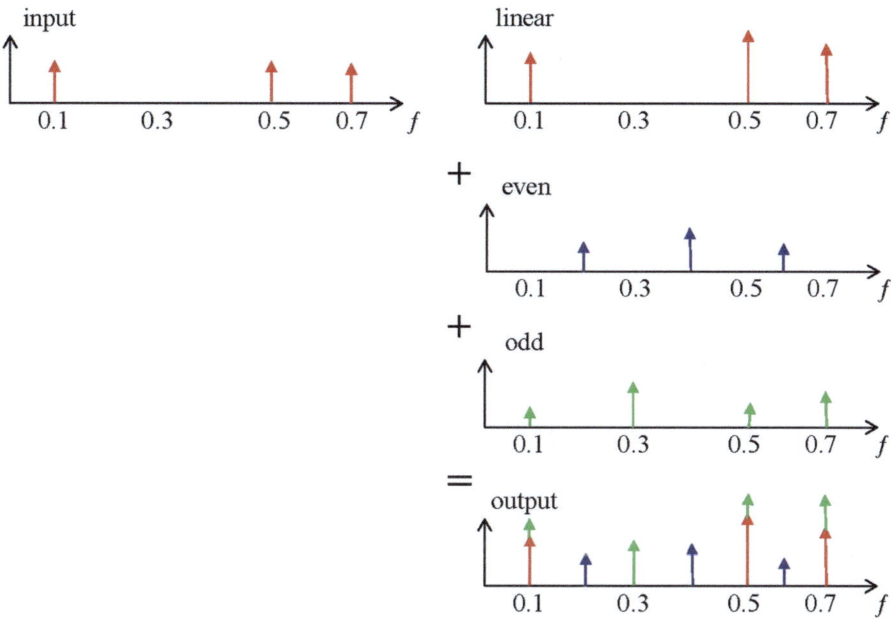

Fig. 14.2 A schematic representation of the input–output contributions. It can be observed that the proposed algorithm allows separation of the even, odd contributions which come from non-excited bins of frequency. In this way, the response to the excited bins (i.e. linear and nonlinear contributions) and the response from the non-excited bins (i.e. nonlinear contributions) can be separated and used for analysis

where G_S is the nonlinear noise term ($E\{G_S\} = 0$) and N_G is the measurement noise. The nonlinear stochastic contribution G_S depends on the power spectrum and the power distribution of the input signal, as well as on the even and odd nonlinear contributions. The effect of G_S can be reduced by averaging the measurements over several multisine realizations (i.e. multiple measurements m of the same system, with different inputs of same amplitude distribution, but different random phase realization in (14.1)). The effect of N_G can be reduced by measuring longer records (i.e. larger number of periods p during each measurement).

In Fig. 14.2, the underpinning principle of detecting these nonlinearities is represented.

In time domain, the output $y(t)$ of a nonlinear system can be written as:

$$y(t) = g_{BLA}(t) * u(t) + y_s(t) \tag{14.9}$$

where $g_{BLA}(t)$ is the impulse response of the linear BLA, and $y_s(t)$ is the term in the output signal as a result of the stochastic nonlinear distortion.

Given that $n_y(t)$ is a stochastic process and $y_s(t)$ is a periodical signal dependent on the realization $r(t)$, the FRF of the mth realization and p^{th} period, $G^{[m,p]}(j\omega_k)$ can be described as [15]:

$$G^{[m,p]}(j\omega_k) = \frac{Y^{[m,p]}(k)}{U_0^{[m]}(k)} = G_{BLA}(j\omega_k) + \frac{Y_S^{[m]}(k)}{U_0^{[m]}(k)} + \frac{N_Y^{[m,p]}(k)}{U_0^{[m]}(k)} \tag{14.10}$$

where $X^{[m,p]}(k)$ is the DFT spectrum of the pth period of the mth multisine realization and $\hat{X}^{[m]}$ is the estimated spectrum of the mth multisine realization.

Consequently, one can estimate the BLA, the variance of the stochastic nonlinear distortions and the noise variance using:

$$\hat{G}^{[m]}(j\omega_k) = \frac{1}{P}\sum_{p=1}^{P} G^{[m,p]}(j\omega_k)$$

$$\hat{G}_{BLA}(j\omega_k) = \frac{1}{M}\sum_{m=1}^{M} \hat{G}^{[m]}(j\omega_k) \tag{14.11}$$

$$\hat{\sigma}^2_{\hat{G}^{[m]}}(k) = \sum_{p=1}^{P} \frac{|G^{[m,p]}(j\omega_k) - \hat{G}^{[m]}(j\omega_k)|^2}{P(P-1)}$$

$$\hat{\sigma}^2_{\hat{G}_{BLA}}(k) = \sum_{m=1}^{M} \frac{|G^{[m]}(j\omega_k) - \hat{G}_{BLA}(j\omega_k)|^2}{M(M-1)} \tag{14.12}$$

$$\hat{\sigma}^2_{\hat{G}_{BLA,n}}(k) = \frac{1}{M^2}\sum_{m=1}^{M} \hat{\sigma}^2_{\hat{G}^{[m]}}(k) \tag{14.13}$$

$$var(G_S(j\omega_k)) \approx M(\hat{\sigma}^2_{\hat{G}_{BLA}}(k) - \hat{\sigma}^2_{\hat{G}_{BLA,n}}(k)) \tag{14.14}$$

where $\hat{G}_{BLA}(j\omega_k)$ is the estimated BLA, $\hat{\sigma}^2_{\hat{G}_{BLA}}(k)$ is the estimated total variance (stochastic nonlinear variance + noise variance) averaged over the m realizations, $\hat{\sigma}^2_{\hat{G}_{BLA,n}}(k)$ is the estimated noise variance averaged over the m experiments and $var(G_S(j\omega_k))$ the variance of the stochastic nonlinear distortion with respect to one multisine realization. This estimations can be done for odd and even frequencies separately, depending on the selection of ω_k.

The total variance and noise variance are averaged over the m experiments and provide insight into the reliability of the FRF measurements over m different multisine realizations. The variance of the stochastic nonlinear distortion with respect to one realization provides insight into the amount of nonlinear distortion in the system. A comprehensive description of these methods and a manifold of illustrative examples are given in [15].

In order to obtain a quantification of these nonlinear contributions, we introduce the following index:

$$T = \frac{P_{even} + P_{odd}}{P_{exc}} \cdot \frac{U_{exc}}{U_{even} + U_{odd}} \tag{14.15}$$

where each variable is the sum of the absolute values of all the contributions in pressure signal and input flow signal, respectively, at the even non-excited frequencies, the odd non-excited frequencies and the excited odd frequencies. Only the corrected output pressure (i.e. corrected from the bias coming from the device itself) has been taken into account when calculating (14.15).

14.4 Results and Validation

14.4.1 Device and Calibration Tube

In order to validate the correctness of the measurement, a known impedance is required. A calibration tube with the characteristics shown in Fig. 14.3 has been measured by means of the prototype device. The corresponding BLA of the calibration tube is given in Fig. 14.4, which corresponds to a theoretical value for the reference tube impedance.

14.4.2 Volunteers

The nonlinear distortions introduced in the input signal due to the device itself are corrected in the measured pressure before calculating the BLA or the respiratory impedance of the volunteers. This is done using the BLA of the device itself and (14.10). For the signal processing part, we used m=6 realizations, p=3 intervals and n=5000 samples.

Figure 14.5 shows the results obtained for a healthy volunteer. In this figure, one may observe the excited and non-excited frequency contributions in the pressure signal.

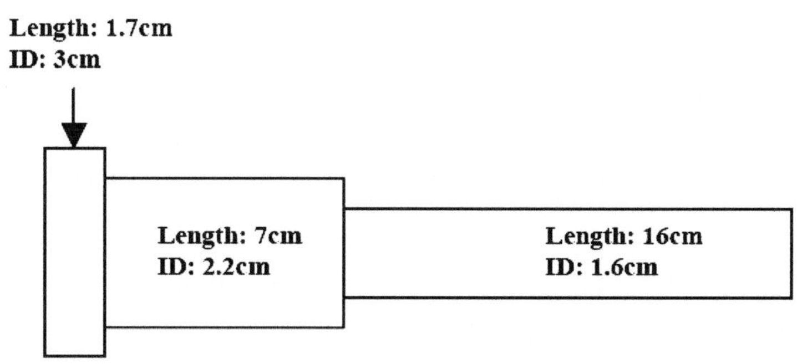

Fig. 14.3 Schematic representation of the calibration tube. ID: inner diameter

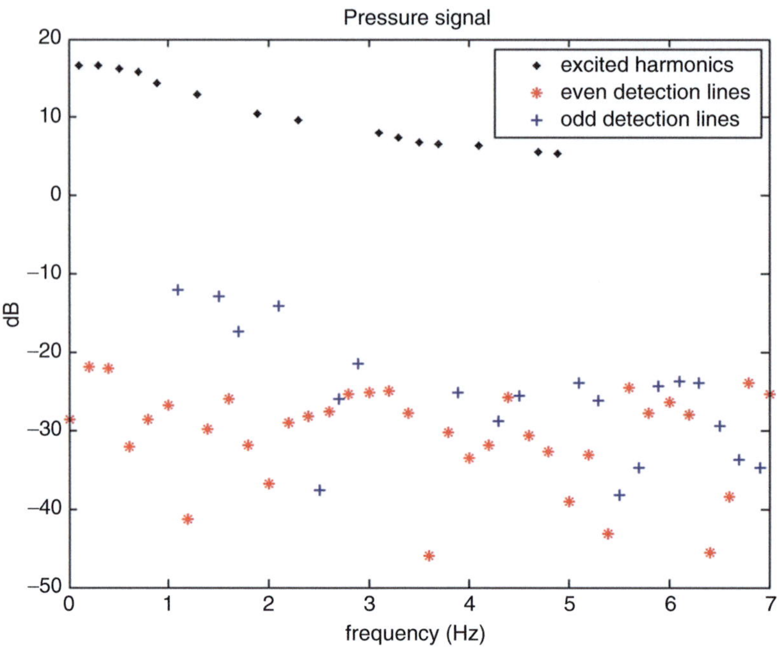

Fig. 14.4 BLA of the calibration tube, for the pressure signal. *Bold black line*: BLA; *blue dashed line*: total variance (noise + stochastic nonlinear distortion); *red dotted line*: noise variance; *green dash-dot line*: variance of the stochastic nonlinear distortion with respect to one multisine realization

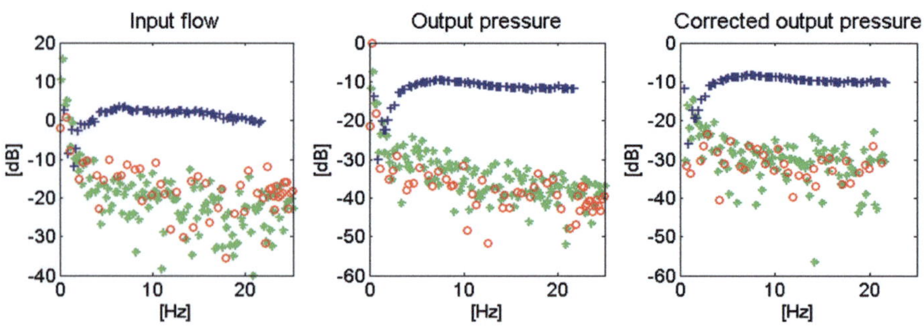

Fig. 14.5 Input (*left column*), output (*middle column*) and corrected DFT spectrum for nonlinear contributions in the device (*right column*) of a *healthy* volunteer. *Blue* '+': excited odd harmonics; *red* 'o': non-excited odd harmonics; *green* '*': non-excited even harmonics

Figure 14.6 shows the results obtained for all the volunteers, in terms of the new index from (14.15). Statistical analysis has been performed using standard *t-tests* from statistical toolbox of Matlab. The 5 % confidence intervals are 0.1586 and 0.1887, respectively, with a mean of 0.1736, median of 0.1715 and standard deviation of only 0.0224. The fact that the standard deviation is rather small

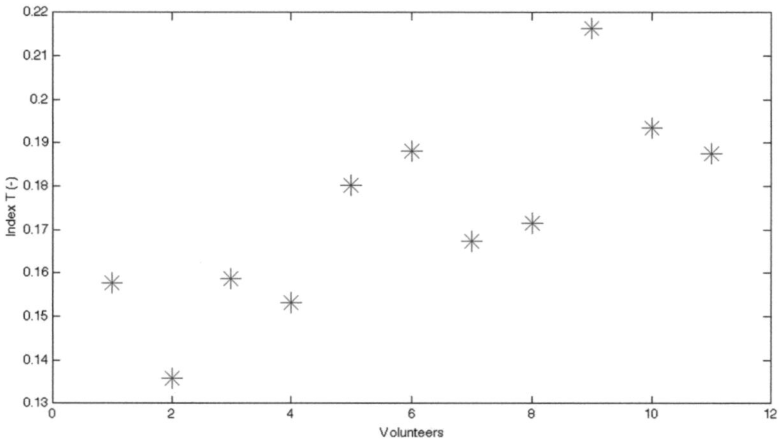

Fig. 14.6 The values of the novel index from (14.15) plotted for the group of volunteers

compared to the mean value is an indication that the group of volunteers have been chosen close to each other in terms of pulmonary characteristics (this was also prerequisite to be included in the test group). This is further supported by the fact that the mean and median values are also close together.

One may expect that the nonlinear distortions tend to be significantly increased in patients diagnosed with respiratory disease than in healthy subjects. From clinical insight, this indeed makes sense. The respiratory system affected by chronic obstructive pulmonary disease contains broken alveolar sacs which will change the heterogeneous appearance of the tissue and introduce nonlinear effects originated by turbulent flow, stiffness, with additional inflammation and clogging of the airways. The respiratory system affected by asthma is subject to airway hyperresponsiveness leading to airway chronic inflammation. This affects the airway remodelling, changing airflow dynamics and hence introducing nonlinear effects from turbulent flow, airway obstruction, airway muscle fibrosis, etc. In both cases, changes in structure and morphology will change the nonlinear response of the respiratory system, hence the values of the proposed index will change as well. This assumption is also supported by earlier works on the dynamic analysis of the respiratory system in healthy volunteers and diagnosed patients [8, 9].

Although this preliminary evaluation was performed on a limited number of volunteers, it suggests that measuring nonlinear contributions may hold significant information upon the evolution of respiratory diseases. Respiratory mechanics at low frequencies have inherent information on the viscoelastic properties of airways and tissue. The challenge is that the amplitude and frequency of the breathing signal may vary within the measurement and from one measurement to another, making the detection lines prone to biased values. The results obtained in these initial steps are a proof of concept which motivates further development of the technique.

14.5 Conclusions and Perspectives

This contribution introduced a novel device and an improved method which allow detection and quantification of the nonlinear effects in the measurement instrumentation and in the respiratory system. The proposed algorithm extracts the best linear approximation of the nonlinear dynamics present in the respiratory system. An optimized multisine excitation signal has been applied on a group of healthy volunteers. A novel index has been proposed to quantify these nonlinear contributions in the signals measured from the volunteers.

Acknowledgements The author gratefully acknowledge Hannes Maes, Stig Dooms and Dana Copot for their technical assistance. This work has been financially supported by the Flanders Research Foundation (FWO) grant nr: 3E009811.

References

1. Barnas GM, Yoshino K, Loring H, Mead J (1987) Impedance and relative displacements of relaxed chest wall up to 4 Hz. J Appl Physiol 62:71–81
2. Bates J (2009) Lung mechanics, an inverse modelling approach, Chap. 9–11. Cambridge University Press, Cambridge
3. Bates J, Irvin C, Farre R, Hantos Z (2011) Oscillation mechanics of the respiratory system. Am Physiol Soc Compr Physiol 3(1):1233–1272
4. Birch M, MacLeod D, Levine M (2001) An analogue instrument for the measurement of respiratory impedance using for the forced oscillation technique. Phys Meas 22:323–339
5. DuBois AB, Brody AW, Lewis DH, Burgess BF Jr (1956) Oscillation mechanics of lungs and chest in man. J Appl Physiol 8:587–594
6. Guyton A, Hall J (2005) Textbook of medical physiology, 11th edn. W.B. Sanders, Philadelphia, PA; London: Saunders
7. Ionescu CM, Schoukens J, De Keyser R (2011a) Detecting and analyzing non-linear effects in respiratory impedance measurements. American control conference, 29 June-01 July, San Francisco, USA, pp 978–1-4577–0079-8, 5412–5417
8. Ionescu C, Machado JT, De Keyser R (2011b) Fractional-order impulse response of the respiratory system. In: Yong Z (ed) Computers and mathematics with application, special issue on advances in fractional differential equations II, vol 62, pp 845–854
9. Ionescu C, Machado JT, De Keyser R (2011c) Is multidimensional scaling suitable for mapping the input respiratory impedance in subjects and patients ?. Comput Methods Prog Biomed 104:e189–e200, DOI information: 10.1016/j.cmpb.2011.02.009
10. Lakes RS (2009) Viscoelastic materials. Cambridge University Press, Cambridge
11. Lande B, Mitzner W (2006) Analysis of lung parenchyma as a parametric porous medium. J Appl Physiol 101:926–933
12. Michaelson ED, Grassman ED, Peters W (1975) Pulmonary mechanics by spectral analysis of forced random noise. J Clin Invest 56:1210–1230
13. Oostveen E, Macleod D, Lorino H, Farré R, Hantos Z, Desager K, Marchal F (2003) The forced oscillation technique in clinical practice: methodology, recommendations and future developments, Eur Respir J 22(6):1026–1041
14. Pelin R, Duvivier C, Bekkari H, Gallina C (1990) Stress adaptation and low frequency impedance of rat lungs. J Appl Physiol 69:1080–1086

15. Schoukens J, Pintelon R (2012) System identification. A frequency domain approach, 2nd edn. IEEE, New Jersey
16. Schoukens J, Pintelon R, Dobrowiecki T, Rolain Y (2005) Identification of linear systems with nonlinear distorsions. Automatica 41(3):491–504
17. Smith HJ, Reinhold P, Goldman MD (2005) Forced oscillation technique and impulse oscillometry. Eur Respir Mon 31:72–105
18. Suki B, Barabasi AL, Lutchen K (1994) Lung tissue viscoelasticity: a mathematical framework and its molecular basis. J Appl Physiol 76(6):2749–2759
19. Zhang Q, Lutchen K, Suki B (1999) A frequency domain approach to nonlinear and structure identification for long memory systems: application to lung mechanics. Ann Biomed Eng 27:1–13

Part III
Discontinuous Dynamics

Chapter 15
Drilling Systems: Stability and Hidden Oscillations

M.A. Kiseleva, N.V. Kuznetsov, G.A. Leonov, and P. Neittaanmäki

Abstract There are many mathematical models of drilling systems Despite, huge efforts in constructing models that would allow for precise analysis, drilling systems, still experience breakdowns. Due to complexity of systems, engineers mostly use numerical analysis, which may lead to unreliable results.

Nowadays, advances in computer engineering allow for simulations of complex dynamical systems in order to obtain information on the behavior of their trajectories. However, this simple approach based on construction of trajectories using numerical integration of differential equations describing dynamical systems turned out to be quite limited for investigation of stability and oscillations of these systems. This issue is very crucial in applied research; for example, as stated in Lauvdal et al. (Proceedings of the IEEE control and decision conference, 1997) the following phrase: "Since stability in simulations does not imply stability of the physical control system (an example is the crash of the YF22) stronger theoretical understanding is required".

In this work, firstly a mathematical model of a drilling system developed by a group of scientists from the University of Eindhoven will be considered. Then a mathematical model of a drilling system with perfectly rigid drill-string actuated by induction motor will be analytically and numerically studied. A modification of the first two models will be considered and it will be shown that even in such simple models of drilling systems complex effects such as hidden oscillations may appear, which are hard to find by standard computational procedures.

Keywords Drilling system • Induction motor • Hidden oscillation • Simulation

M.A. Kiseleva • N.V. Kuznetsov (✉) • G.A. Leonov • P. Neittaanmäki
University of Jyväskylä, Finland, Saint Petersburg State University, Russia
e-mail: maria.kiseleva.87@gmail.com; nkuznetsov239@gmail.com; leonov@math.spbu.ru; pekka.neittaanmaki@jyu.fi

J.A.T. Machado et al. (eds.), *Discontinuity and Complexity in Nonlinear Physical Systems*, Nonlinear Systems and Complexity 6, DOI 10.1007/978-3-319-01411-1_15,

Among the problems the drilling industry faces, drill string failure is of particular interest because of its frequency of occurrence. This issue suggests that the drill string was under a load which led to its break or operation cutoff. The costs derived from these failures are of such importance that finding solutions for reducing them has been a concern for industrial research for many years [8, 39]. For example, in 1985, among all the deep well drilling problems, 45% were related to drill string failures. In 1991, Shell Expro suffered exceptionally high losses due to drill string failures. These losses were estimated to be more than US $2MM in a two-month span and were caused by a specific drilling unit experiencing five failures within this period. To reduce these failures, a Drill String Prevention Quality Improvement Project was implemented, and it succeeded in reducing nonproductive time costs associated with drill string failures from US $6.5MM in 1992 to less than US $1MM in 1994. Also, by the end of 1994 total failures had been reduced by 55%, which translated into about US $8.5MM in savings associated with these costs. The cost associated with each failure averages roughly US $106000, and drill string failure occurs at some point in 1 out of 7 drill rigs, thus research in drilling systems to reduce both the rate and costs of these failures continues being crucial.

The drill string undergoes various types of vibrations during drilling [11, 19, 35, 40]: axial (longitudinal), lateral (bending), hydraulic and torsional (rotational) vibrations. Axial vibrations are compression alternations and are due to the rebound (bouncing) of the drill against the formation during the rotation. Lateral vibrations are also called transversal of whirling vibrations. This type of vibration is caused by the eccentricity of the strings which leads to centripetal forces during rotation. Hydraulic vibrations appear in circulation system stemming from pulp pulsation.

Torsional vibrations are caused by the nonlinear interaction between the bit and the rock. In [3, 14, 36] it was concluded that the negative damping in the friction force that appears due to the contact of the bit and the borehole is the reason for torsional vibrations. Negative damping in the friction force may lead to stick-slip phenomenon [4, 9, 20, 37, 38], when drill string and borehole wall alternate between sticking to each other and sliding over each other. The consequence of stick-slip vibration may be severe enough to provoke a sudden stop of the drill rotation.

There are many mathematical models of drilling systems [5, 12, 22]. Despite, huge efforts in constructing models that would allow for precise analysis, drilling systems still experience breakdowns. Due to complexity of systems, engineers mostly use numerical analysis, which may lead to unreliable results.

Nowadays, advances in computer engineering allow for simulations of complex dynamical systems in order to obtain information on the behavior of their trajectories. However, this simple approach based on construction of trajectories using numerical integration of differential equations describing dynamical systems turned out to be quite limited for investigation of stability and oscillations of these systems. This issue is very crucial in applied research; for example, as stated in [18] the following phrase: *"Since stability in simulations does not imply stability of the physical control system (an example is the crash of the YF22) stronger theoretical understanding is required"*.

In this work, firstly a mathematical model of a drilling system developed by a group of scientists from the University of Eindhoven will be considered. Then a mathematical model of a drilling system with perfectly rigid drill-string actuated by induction motor will be analytically and numerically studied. A modification of the first two models will be considered and it will be shown that even in such simple models of drilling systems complex effects such as hidden oscillations may appear, which are hard to find by standard computational procedures.

15.1 Two-Mass Mathematical Model of a Drilling System

Let us first consider the "two-mass" mathematical model of a drilling system studied in [5, 36]. This model consists of an upper disc actuated by a drive part (consisting of a power amplifier, DC-motor, and a gear box), a no-mass string, and a lower disc (see Fig. 15.1). The upper disc is connected to the lower disc by the string, which is a low stiffness connection between the discs. There are two friction torques acting on the upper and the lower discs. The upper friction torque is mainly caused by the electromagnetic field in the drive part of the model. The lower friction torque is a result of the friction against the workpiece which the drill bit cuts. This model is described by equations of motion for the upper and lower discs:

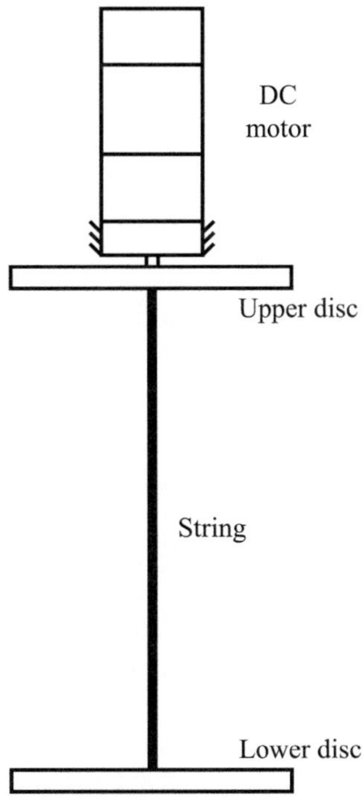

Fig. 15.1 Two-mass mathematical model of a drilling system

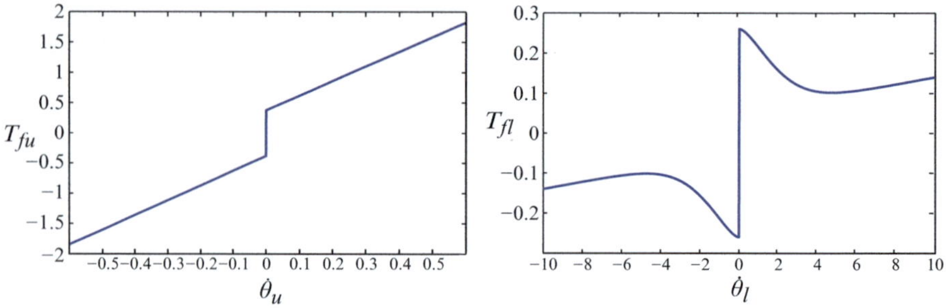

Fig. 15.2 Upper and lower friction models

$$
\begin{aligned}
J_u\ddot{\theta}_u + k_\theta(\theta_u - \theta_l) + b(\dot{\theta}_u - \dot{\theta}_l) + T_{fu}(\dot{\theta}_u) - k_m u = 0,\\
J_l\ddot{\theta}_l - k_\theta(\theta_u - \theta_l) - b(\dot{\theta}_u - \dot{\theta}_l) + T_{fl}(\dot{\theta}_l) = 0.
\end{aligned}
\tag{15.1}
$$

Here θ_u and θ_l—angular displacements of the upper and lower discs, respectively, ($\dot{\theta}_u$, $\dot{\theta}_l$—derivatives of θ_u, θ_l with respect to time t), J_u and J_l—inertia torques, k_θ, k_m, b—nonnegative coefficients, u—constant input voltage. T_{fu} and T_{fl}—friction torques acting on the upper and the lower discs. In order to model both frictions set-valued force laws are needed. The friction torque acting on the upper disc is described as follows:

$$
T_{fu}(\dot{\theta}_u) \in \begin{cases} T_{cu}(\dot{\theta}_u)\text{sign}(\dot{\theta}_u), & \text{for } \dot{\theta}_u \neq 0 \\ [-T_{su} + \Delta T_{su}, T_{su} + \Delta T_{su}], & \text{for } \dot{\theta}_u = 0, \end{cases}
\tag{15.2}
$$

$$
T_{cu}(\dot{\theta}_u) = T_{su} + \Delta T_{su}\text{sign}(\dot{\theta}_u) + b_u|\dot{\theta}_u| + \Delta b_u\dot{\theta}_u,
\tag{15.3}
$$

where T_{su}, ΔT_{su}, b_u, Δb_u—nonnegative coefficients.

The model of the friction torque acting on the lower disc is:

$$
T_{fl}(\dot{\theta}_l) \in \begin{cases} T_{cl}(\dot{\theta}_l)\text{sign}(\dot{\theta}_l), & \text{for } \dot{\theta}_l \neq 0 \\ [-T_{sl}, T_{sl}], & \text{for } \dot{\theta}_l = 0, \end{cases}
\tag{15.4}
$$

$$
T_{cl}(\dot{\theta}_l) = T_{fm} + (T_{sl} - T_{fm})e^{-|\frac{\dot{\theta}_l}{\omega_{sl}}|^{\delta_{sl}}} + b_l|\dot{\theta}_l|,
\tag{15.5}
$$

where T_{sl}, T_{fm}, ω_{sl}, δ_{sl} and b_l—nonnegative coefficients.

Both friction models are depicted in Fig. 15.2. The usage of discontinuous friction models allows to properly describe stick-slip effect and most of the important friction phenomena. In the discontinuous region the solution of system (15.1) is understood in the sense of [6, 41].

Performing nonsingular change of variables $\omega_u = \dot{\theta}_u$, $\omega_l = \dot{\theta}_l$, $\alpha = \theta_u - \theta_l$ we obtain the system

$$\begin{aligned}\dot{\omega}_u &= -k_\theta\alpha - b(\omega_u - \omega_l) - T_{fu}(\omega_u) + k_m u,\\ \dot{\omega}_l &= k_\theta\alpha + b(\omega_u - \omega_l) - T_{fl}(\omega_l),\\ \dot{\alpha} &= \omega_u - \omega_l.\end{aligned} \tag{15.6}$$

Upper and lower friction torques transform to

$$T_{fu}(\omega_u) \in \begin{cases} T_{cu}(\omega_u)\text{sign}(\omega_u), & \text{for } \omega_u \neq 0\\ [-T_{su} + \Delta T_{su}, T_{su} + \Delta T_{su}], & \text{for } \omega_u = 0,\end{cases} \tag{15.7}$$

$$T_{fl}(\omega_l) \in \begin{cases} T_{cl}(\omega_l)\text{sign}(\omega_l), & \text{for } \omega_l \neq 0\\ [-T_{sl}, T_{sl}], & \text{for } \omega_l = 0.\end{cases} \tag{15.8}$$

Here

$$T_{cu}(\omega_u) = T_{su} + \Delta T_{su}\text{sign}(\omega_u) + b_u|\omega_u| + \Delta b_u\omega_u \tag{15.9}$$

and

$$T_{cl}(\omega_l) = T_{fm} + (T_{sl} - T_{fm})e^{-|\frac{\omega_l}{\omega_{sl}}|^{\delta_{sl}}} + b_l|\omega_l|. \tag{15.10}$$

Due to the complexity of the friction models only numerical analysis of system (15.6) is possible. During the local analysis it was found that the system has either a stable or an unstable equilibrium state. Then the global analysis of the system was done in order to check whether there were any oscillations in the system.

Since, for computing an oscillation in nonlinear dynamical system, one of the key factors is its basin of attraction, the attractors can be regarded [2, 21, 24, 30, 33] as *self-exciting* or *hidden attractors*, depending on simplicity of finding its basin of attraction in the phase space. *Self-exciting attractors* can be numerically localized by *standard computational procedure, in which after a transient process a trajectory, started from a point of unstable manifold in a neighborhood of equilibrium, reaches a state of oscillation therefore one can easily identify it*. In contrast, for a *hidden attractor, its basin of attraction does not intersect with small neighborhoods of equilibria* [1].

[1]In the 1950–1960s of last century the investigations of widely known Markus-Yamabe's, Aizerman's conjecture (Aizerman problem), and Kalman's conjecture (Kalman problem) on absolute stability led to the finding of hidden oscillations in automatic control systems with nonlinearity, which belongs to the sector of linear stability (see, e.g., [2, 17, 23, 27, 29] and others). In 1961,

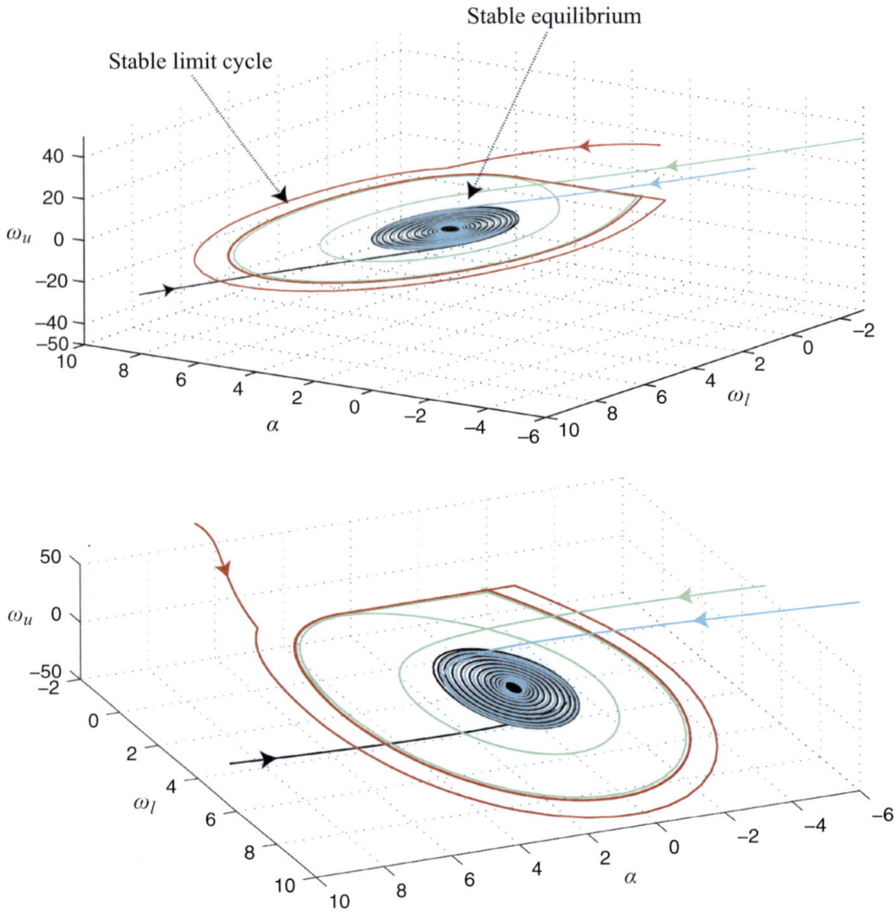

Fig. 15.3 Hidden oscillations and stable equilibrium

During the analysis of the model hidden oscillations were found (Fig. 15.3). One can see both the stable equilibrium state and the stable limit cycle. Note that here the mentioned above stick-slip vibrations appear. In Fig. 15.4 limit cycle for the same data is depicted. Here sections with $\omega_l = 0$ correspond to moments when the drill got stuck against borehole. Finding such hidden oscillations is a quite complicated problem due to the fact that standard computation (in which a trajectory from a neighborhood of an unstable equilibrium reaches and identifies an attractor) does not

---

Gubar' [7] showed analytically the possibility of hidden oscillations existence in two-dimensional system of phase locked-loop [32, 33]. In 2010 chaotic hidden oscillations (hidden attractors) were discovered for the first time [15, 16, 28, 30, 31] in Chua's circuit.

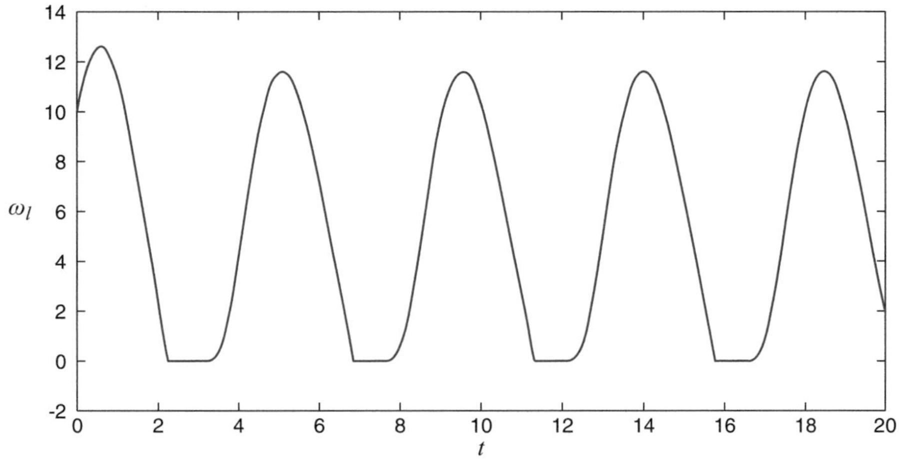

Fig. 15.4 Stick-slip vibrations phenomenon

work here, since the only equilibrium is stable. Integration with random initial data is a challenging task due to the fact that the system is of 3rd order and the basin of attraction is not large.

This model considers only the dynamics of the drill since it uses DC motor. In real systems, the DC motor may heat to the point where the windings of the rotor burn out (for example, if the load is too high). Using an induction motor in a drive part of a drilling system allows to avoid this problem. In the next two sections models of a drilling systems actuated by induction motor will be introduced.

15.2 Mathematical Model of Drilling System with Perfectly Rigid Drill-String

Assume that the drill-string is a perfectly rigid body, stiffly connected with the rotor which rotates due to magnetic field created by the stator of the induction motor. During the operation this system experiences sudden load changes provoked by the interaction of the drill bit with the rock, so the problem of drilling system behavior understanding arises.

There are different mathematical models of induction motor [25, 26, 34]. Here, in order to provide qualitative analysis of the system, we use one of the low-dimensional models proposed in [13, 22] supplemented by friction torque appearing when the drill bit cuts the rock.

Induction motor consists of fixed stator and rotating rotor stiffly connected to the drill-string. Alternating current in stator windings creates alternating magnetic field. In order to simplify the model several assumptions are made: stator's windings are made in such a way that intensity vector of magnetic field and angular velocity of

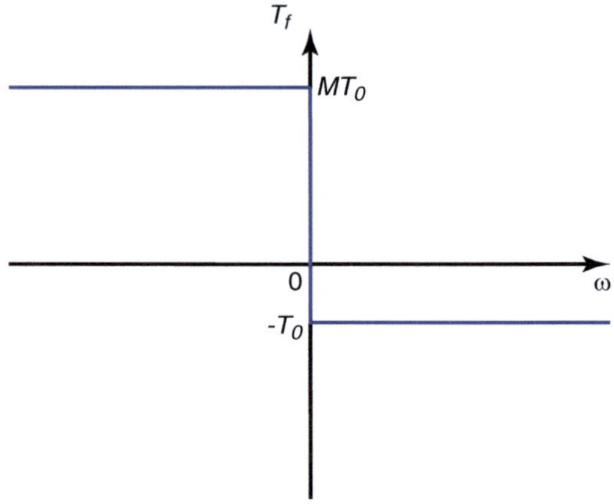

Fig. 15.5 Friction torque T_f

magnetic field are constant (Tesla-Ferraris effect, see [10]), these electromagnetic processes in rotor windings don't influence the currents in the stator windings and the rotor contains only two orthogonal windings with currents i_1 and i_2 passing through them. Consider rotating coordinate system stiffly connected with the intensity vector of magnetic field. In this case we define currents with the help of Faraday's law and Ohm's law (the reciprocal inductance is not taken into account):

$$\begin{aligned} L\dot{i}_1(t) + Ri_1(t) &= \Phi_B(sin\theta(t))\dot{\theta}(t), \\ L\dot{i}_2(t) + Ri_2(t) &= \Phi_B(cos\theta(t))\dot{\theta}(t). \end{aligned} \tag{15.11}$$

Here θ—angular displacement of the drill with respect to the magnetic field created by the stator, which rotates with constant velocity ω_{mf}; $i_1(t)$, $i_2(t)$—currents in rotor windings; R—resistance of the windings; L—inductance of the windings; Φ_B magnetic flux through the windings.

The equations of the motion of the drill-string connected to the rotor with respect to the rotating magnetic field takes the following form:

$$J\ddot{\theta}(t) = -\beta\Phi_B(i_1(t)\sin\theta(t) + i_2(t)\cos\theta(t)) + T_f(\omega_{mf} + \dot{\theta}(t)). \tag{15.12}$$

where J—inertia torque of the drill, β—proportionality coefficient; $\omega = \dot{\theta} + \omega_{mf}$—angular speed of the drill-string rotation with respect to the motionless coordinates system. Equations (15.11) and (15.12) are equations of the mathematical model of drilling the system perfectly rigid drill-string actuated by induction motor.

Let us assume that T_f is of Coulomb type [41]. Here in contrast to the classical Coulomb friction law with symmetrical discontinuous characteristics we consider T_f with asymmetrical characteristics shown in Fig. 15.5:

$$T_f = \begin{cases} -T_0 & \text{if } \omega > 0 \\ MT_0 & \text{if } \omega < 0, \end{cases}$$

where $M, T_0 > 0$, M is sufficiently large number. This corresponds to the fact that the drilling process happens only when $\omega > 0$. In real systems during transient processes such characteristics don't allow change from positive to negative ω. In such cases the system may only get stuck at $\omega = 0$ for some period of time. Such effects will be shown during the analysis of system (15.11)–(15.12) and happen quite often [1] during the drilling process.

Performing nonsingular change of variables

$$
\begin{aligned}
s &= -\dot{\theta}, \\
x &= \frac{L}{\Phi_B}(i_1\cos\theta - i_2\sin\theta), \\
y &= \frac{L}{\Phi_B}(i_1\sin\theta + i_2\cos\theta),
\end{aligned}
$$

from (15.11)–(15.12) we obtain the following system:

$$
\begin{aligned}
\dot{s} &= ay + \xi(s, y), \\
\dot{y} &= -cy - s - xs, \\
\dot{x} &= -cx + ys,
\end{aligned}
\tag{15.13}
$$

where $a = \frac{\beta {\Phi_B}^2}{IL}$, $c = \frac{R}{L}$. Here x, y determine electrical variables in rotor windings, and s defines the slip. In the discontinuous region, $\xi(s, y)$ should be defined as:

$$
\xi(s, y) = \begin{cases} \gamma, & \text{if } s = \omega_{mf}, y < -\frac{\gamma}{a} \text{ or } s < \omega_{mf}; \\ -\gamma M, & \text{if } s = \omega_{mf}, y > \frac{M\gamma}{a} \text{ or } s > \omega_{mf}; \\ -ay, & \text{if } s = \omega_{mf}, -\frac{\gamma}{a} \le y \le \frac{M\gamma}{a}, \end{cases}
$$

where $\gamma = \frac{T_0}{I}$.

Limit load problem. Let there be a sudden load change from γ_0 to γ_1 at the moment $t = \tau$, where $0 < \gamma_0 < \gamma_1 < \gamma_{max} = \frac{a}{2}$. Such situation happens during the transition to a harder medium. For $\gamma = \gamma_0$ the system has one stable equilibrium state $s_0 = \frac{c(a - \sqrt{a^2 - 4{\gamma_0}^2})}{2\gamma_0}$, $y_0 = -\frac{\gamma_0}{a}$, $x_0 = -\frac{\gamma_0 s_0}{ac}$. It is important that the solution $s(t)$, $x(t)$, $y(t)$ of system (15.13) in a new transient mode with $\gamma = \gamma_1$ and initial data $s(\tau) = \frac{c(a - \sqrt{a^2 - 4{\gamma_0}^2})}{2\gamma_0}$, $y(\tau) = -\frac{\gamma_0}{a}$, $x(\tau) = -\frac{\gamma_0 s_0}{ac}$ tends the equilibrium state $s_1 = \frac{c(a - \sqrt{a^2 - 4{\gamma_1}^2})}{2\gamma_1}$, $y_1 = -\frac{\gamma_1}{a}$, $x_1 = -\frac{\gamma_1 s_1}{ac}$ for $t \to +\infty$.

Using the results obtained in [22] we obtain the following theorem.

Theorem 1. *Let the following conditions be fulfilled:*

$$\gamma_0 < \gamma_{max}, \tag{15.14}$$

$$\gamma_1 < min\left\{\gamma_{max}, \quad 2c^2\right\}, \tag{15.15}$$

$$\frac{(\gamma_1-\gamma_0)^2}{2c^2}{s_0}^2 + \frac{(\gamma_1-\gamma_0)^2}{2} \leq \int\limits_{s_0}^{\omega_{mf}} \phi(s)ds + \frac{(1+M)^2}{2}{\gamma_1}^2. \tag{15.16}$$

Then the solution of system (15.13) with $\gamma = \gamma_1$ *and initial data* $s(\tau) = \frac{c(a-\sqrt{a^2-4{\gamma_0}^2})}{2\gamma_0}$, $y(\tau) = -\frac{\gamma_0}{a}$, $x(\tau) = -\frac{\gamma_0 s_0}{ac}$ *tends to the equilibrium state of this system for* $t \to +\infty$.

In this theorem, condition on permissible parameters γ_0 and γ_1 corresponding to two different medium types is given, such that the transient process during sudden medium change is stable.

Proof. Let us present the scheme of the proof of the theorem. Consider the region $\left\{s(t) < \omega_{mf}\right\}$ of the phase space of system (15.13).

Performing nonsingular change of variables

$$\eta = ay + \gamma_1, \qquad z = -x - \frac{\gamma_1}{ac}s,$$

from (15.13) we obtain

$$\begin{aligned} \dot{s} &= \eta, \\ \dot{\eta} &= -c\eta + azs - \phi(s), \\ \dot{z} &= -cz - \frac{1}{a}s\eta - \frac{\gamma_1}{ac}\eta. \end{aligned} \tag{15.17}$$

Here $\phi(s) = -\frac{\gamma_1}{c}s^2 + as - c\gamma_1$.

Let's introduce the function

$$V(s,\eta,z) = \frac{a^2}{2}z^2 + \frac{1}{2}\eta^2 + \int\limits_{s_1}^{s} \phi(s)ds.$$

For any solution of system (15.17) from region $s(t) < \omega_{mf}$ the following relation is satisfied:

$$\dot{V}(s(t),\eta(t),z(t)) = -a^2cz(t)^2 - \frac{a\gamma_1}{c}\eta(t)z(t) - c\eta(t)^2 \leq 0. \tag{15.18}$$

The quadratic form in the right-hand side of (15.18) is negative definite taking into account (15.15).

Introduce the set

$$\Omega_{mf} = \left\{ V(s, \eta, z) \le \int\limits_{s_1}^{\omega_{mf}} \phi(s)ds + \frac{(1+M)^2}{2}\gamma_1^2, s \in [s_2, \omega_{mf}] \right\},$$

where the point $s_2 < \omega_{mf}$ is such that

$$\int\limits_{s_2}^{\omega_{mf}} \phi(s)ds + \frac{(1+M)^2}{2}\gamma_1^2 = 0.$$

Set Ω_{mf} is limited, and for $s(t) = \omega_{mf}$ it takes the form:

$$\frac{a^2}{2}z^2 + \frac{1}{2}\eta^2 \le \frac{(1+M)^2}{2}\gamma_1^2.$$

Going back to initial coordinates(x, y, s) we obtain:

$$(x + \frac{\gamma_1}{a})^2 + (y + \frac{\gamma_1}{a})^2 \le \frac{(1+M)^2}{a^2}\gamma_1^2.$$

Note that this circle lies below the lower boundary $y = \frac{M\gamma_1}{a}$ of the slip region $\Delta = \left\{ s = \omega_{mf}, -\frac{\gamma_1}{a} \le y \le \frac{M\gamma_1}{a} \right\}$ of system (15.13).

In the slip region Δ of system (15.13) can be transformed to the following form:

$$\begin{aligned} \dot{y} &= -cy - \omega_{mf} - \omega_{mf}x, \\ \dot{x} &= -cx + \omega_{mf}y. \end{aligned}$$

There are no equilibrium states in the slip region if the condition (15.15) is valid. The solution which falls into the slip region necessarily goes out through the lower boundary $y = -\frac{\gamma_1}{a}$ into the region $s < \omega_{mf}$ ($\dot{s} < 0$ if $s = \omega_{mf}, y < -\frac{\gamma_1}{a}$). Condition (15.18) implies that this solution is found to be inside the region $\left\{ V(s, \eta, z) \le \int_{s_1}^{\omega_{mf}} \phi(s)ds \right\}$, it doesn't fall further into the slip region and tends to the equilibrium state (s_1, y_1, x_1) of the system due to the boundedness of Ω_{mf}. Obviously, other trajectories which fall into Ω_{mf} but don't pass through the slip region will also tend to the equilibrium state.

Thus the system is dichotomic (i.e., every solution bounded on $[t_0, \infty)$, where $t_0 \in \mathbb{R}$ tends to a stationary set, see [41]), if condition (15.15) is fulfilled.

The set Ω_{mf} contains the point $s = s_0$, $\eta = \gamma_1 - \gamma_0$, $z = \dfrac{\gamma_0 - \gamma_1}{ac} s_0$, if

$$\frac{(\gamma_1 - \gamma_0)^2}{2c^2} {s_0}^2 + \frac{(\gamma_1 - \gamma_0)^2}{2} \leq \int\limits_{s_0}^{\omega_{mf}} \phi(s) ds + \frac{(1+M)^2}{2} {\gamma_1}^2. \tag{15.19}$$

Due to $\gamma_0 < \gamma_1$ and condition (15.16)

$$\frac{(\gamma_1 - \gamma_0)^2}{2} \leq \int\limits_{0}^{\omega_{mf}} \phi(s) ds + \frac{(1+M)^2}{2} {\gamma_1}^2. \tag{15.20}$$

Let's show that

$$\frac{(\gamma_1 - \gamma_0)^2}{2c^2} {s_0}^2 \leq \int\limits_{s_0}^{0} \phi(s) ds. \tag{15.21}$$

Indeed, taking into account $\gamma_0 \leq 2\gamma_1$, we get: $\frac{\gamma_1}{3c} {s_0}^2 - \frac{a}{2} s_0 - \frac{(\gamma_1 - \gamma_0)^2}{2c^2} s_0 + c\gamma_1 = \frac{1}{12c^2{\gamma_0}^2}(c^2(a - \sqrt{a^2 - 4{\gamma_0}^2})^2\gamma_1 - 3a^2c^2\gamma_0 + 3a^2c\sqrt{a^2 - 4{\gamma_0}^2}\gamma_0 - 3(\gamma_1 - \gamma_0)^2(a - \sqrt{a^2 - 4{\gamma_0}^2})\gamma_0 + 12c^2\gamma_1{\gamma_0}^2) \geq \frac{1}{12c^2{\gamma_0}^2}(2a^2c^2 - 2ac^2\sqrt{a^2 - 4{\gamma_0}^2}\gamma_1 + 3ac^2\sqrt{a^2 - 4{\gamma_0}^2} - 3a^2c^2\gamma_0 + 3\sqrt{a^2 - 4{\gamma_0}^2}\gamma_0{\gamma_1}^2 - 3a{\gamma_1}^2\gamma_0 + 8c^2{\gamma_0}^2\gamma_1) \geq 0$. Hence, from inequalities (15.20) and (15.21) we obtain the condition (15.19).

Thus the solution $s(t)$, $\eta(t)$, $z(t)$ with initial data

$$s(\tau) = s_0, \quad \eta(\tau) = \gamma_1 - \gamma_0, \quad z(\tau) = \frac{\gamma_0 - \gamma_1}{ac} s_0$$

tends to the equilibrium state of the system. □

The following corollary is formulated for the case when the rotating speed of the magnetic field is equal to the maximum value of the static characteristics of the induction motor.

Corollary 1. *Let the following conditions be fulfilled*

$$\gamma_0 < \gamma_{max},$$

$$\gamma_1 < min\left\{\gamma_{max}, \quad 2c^2\right\},$$

$$3(M^2 + 2M)\gamma_1^2 - 8c^2\gamma_1 + 3ac^2 \geq 0.$$

Then the solution of system (15.13) with $\omega_{mf} = c$, $\gamma = \gamma_1$ *and initial data*

$$s(\tau) = \frac{c(a - \sqrt{a^2 - 4{\gamma_0}^2})}{2\gamma_0}, \quad y(\tau) = -\frac{\gamma_0}{a}, \quad x(\tau) = -\frac{\gamma_0 s_0}{ac}$$

tends to the equilibrium state of this system for $t \to +\infty$.

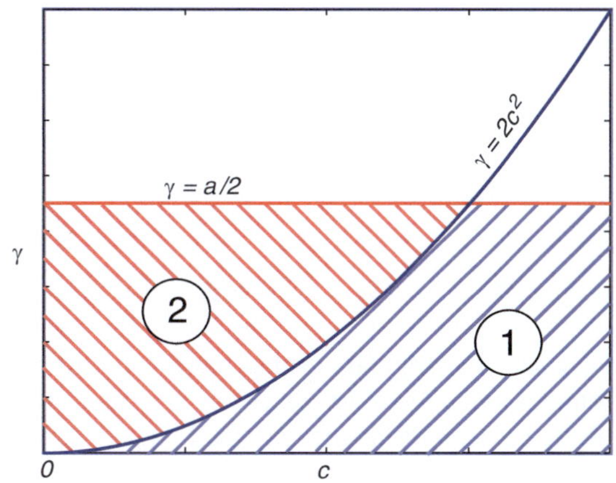

Fig. 15.6 1 permissible load region due to the theorem, 2 permissible load region due to the computer modeling of the system

The next corollary states the stability of transient process for drilling system when drill may rotate only in one direction.

Corollary 2. *Let M be a sufficiently large number,* $\omega_{mf} = c$, $\gamma_0 = 0$ *and*

$$\gamma_1 < min\left\{\frac{a}{2}, \quad 2c^2\right\}. \tag{15.22}$$

Then the solution of system (15.13) with $\gamma = \gamma_1$ *and initial data* $s(\tau) = 0$, $y(\tau) = 0$, $x(\tau) = 0$ *tends to the equilibrium state of this system when* $t \to +\infty$.

For $\gamma_1 \in \left\{2c^2, \frac{a}{2}\right\}$ (i.e., the condition (15.22) is not valid) computer modeling of system (15.13) was done (region 2 in Fig. 15.6), which showed that the statement of the corollary 2 is valid.

Thus, with the help of analytical methods and computer simulation, it was shown that the limit permissible discontinuous load depends on the maximum value of the constant load under which the system has the steady mode. As opposed to the previous model, no oscillations were found in this model.

15.3 Mathematical Model of the Drilling System Actuated by Induction Motor

In order to take into account the dynamics of the motor let us consider the modification of the first model supplemented by the equations of the induction motor used in the second model. The rotation of the discs will be considered with respect to the rotating magnetic field created by the stator of the induction motor, but we will use the same notation as before.

The equations of the new two-mass model of the drilling system actuated by induction motor are as follows:

$$
\begin{aligned}
&L\dot{i}_1 + Ri_1 = \Phi_B(\sin\theta_u)\dot{\theta}_u,\\
&L\dot{i}_2 + Ri_2 = \Phi_B(\cos\theta_u)\dot{\theta}_u,\\
&J_u\ddot{\theta}_u + k_\theta(\theta_u - \theta_l) + b(\dot{\theta}_u - \dot{\theta}_l) + \beta\Phi_B(i_1\sin\theta_u + i_2\cos\theta_u) = 0,\\
&J_l\ddot{\theta}_l - k_\theta(\theta_u - \theta_l) - b(\dot{\theta}_u - \dot{\theta}_l) + T_{fl}(\omega_{mf} + \dot{\theta}_l) = 0.
\end{aligned}
\tag{15.23}
$$

Here θ_u, θ_l—angular displacements of rotor and the lower disc relative to the rotating magnetic field created by the stator of the induction motor, $\omega_m f$—speed of the rotation of magnetic field, $T_{fl}(\omega_{mf} + \dot{\theta}_l)$— friction torque (same as in the first model). Here the first two equations are the equations of the induction motor from the second model. Third and fourth equations are taken from the first model, but in the third equation expression $T_{fu}(\dot{\theta}_u) - k_m u$ from the first model is replaced by the expression $\beta\Phi_B(i_1(t)\sin\theta_u(t) + i_2(t)\cos\theta_u(t))$ which represents the effect of the induction motor on the upper disc. Only $T_{fl}(\dot{\theta})$ changed to $T_{fl}(\omega_{mf} + \dot{\theta}_l)$ due to the fact that $\theta_u - \theta_l$ is same in both systems and, obviously, the derivatives of $\omega_{mf} + \dot{\theta}_u$ and $(\omega_{mf} + \dot{\theta}_l)$ are equal to $\dot{\theta}_u$ and $\dot{\theta}_l$, respectively.

Performing nonsingular change of variables

$$
\begin{aligned}
\omega_u &= -\dot{\theta}_u,\\
x &= \frac{L}{\Phi_B}(i_1\cos\theta_u - i_2\sin\theta_u),\\
y &= \frac{L}{\Phi_B}(i_1\sin\theta_u + i_2\cos\theta_u),\\
\omega_l &= -\dot{\theta}_l,\\
\theta &= \theta_u - \theta_l,
\end{aligned}
$$

we obtain the system of 5th order

$$
\begin{aligned}
\dot{y} &= -cy - \omega_u - x\omega_u,\\
\dot{x} &= -cx + y\omega_u,\\
\dot{\theta} &= \omega_l - \omega_u,\\
\dot{\omega}_u &= \frac{k_\theta}{J_u}\theta + \frac{b}{J_u}(\omega_l - \omega_u) + \frac{a}{J_u}y,\\
\dot{\omega}_l &= -\frac{k_\theta}{J_l} - \frac{b}{J_l}(\omega_l - \omega_u) + \frac{1}{J_l}T_{fl}(\omega_{mf} - \omega_l),
\end{aligned}
\tag{15.24}
$$

Here $a = \dfrac{\beta{\Phi_B}^2}{L}$, $c = \dfrac{R}{L}$.

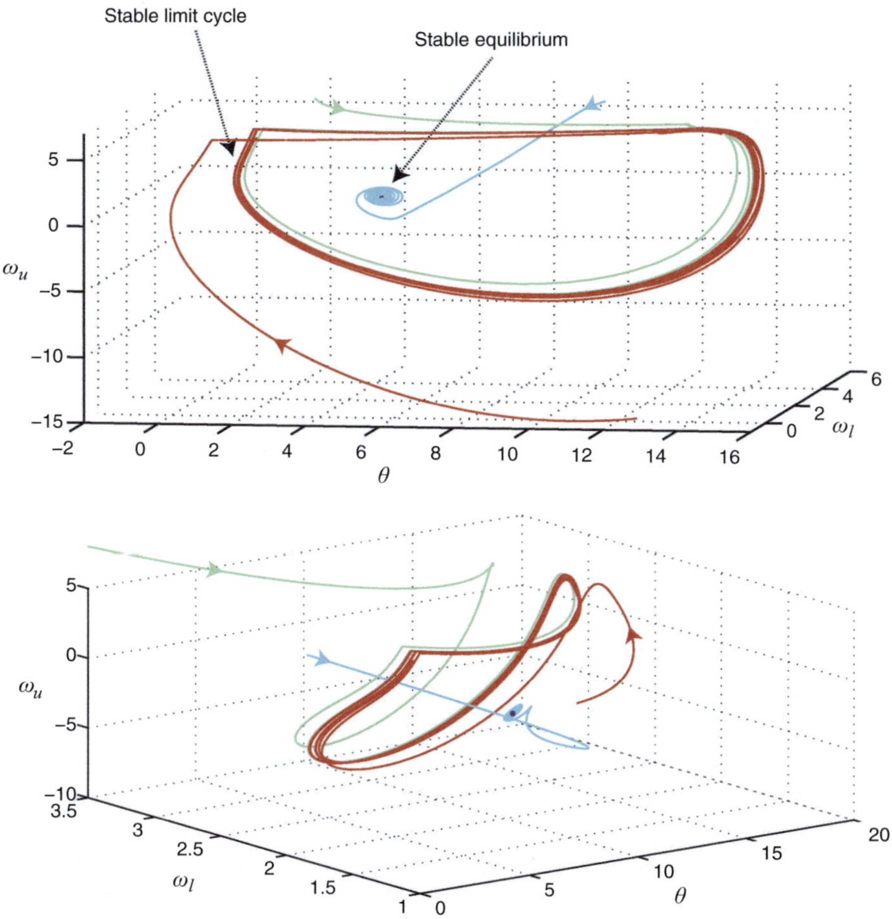

Fig. 15.7 Hidden oscillations and stable equilibrium in the mathematical model of a drill actuated by induction motor—projection onto $(\theta, \omega_u, \omega_l)$

Due to the complexity of $T_{fl}(\omega_{mf} - \omega_l)$ and high order of the system it is hard to provide in-depth qualitative analysis for system (15.24). Using computer modeling it is shown that under certain parameters the system has unique stable equilibrium state and hidden oscillations represented by stable limit cycle (See Fig. 15.7). Here the modeling is done in the system of 5th order so the chance of finding hidden oscillations was much lower than in the first model described above. In Fig. 15.8 it can be seen that these oscillations are also of a stick-slip type. Here $\omega_{mf} - \omega_u$, $\omega_{mf} - \omega_l$ are speeds of the upper and lower discs relatively to the fixed coordinate system.

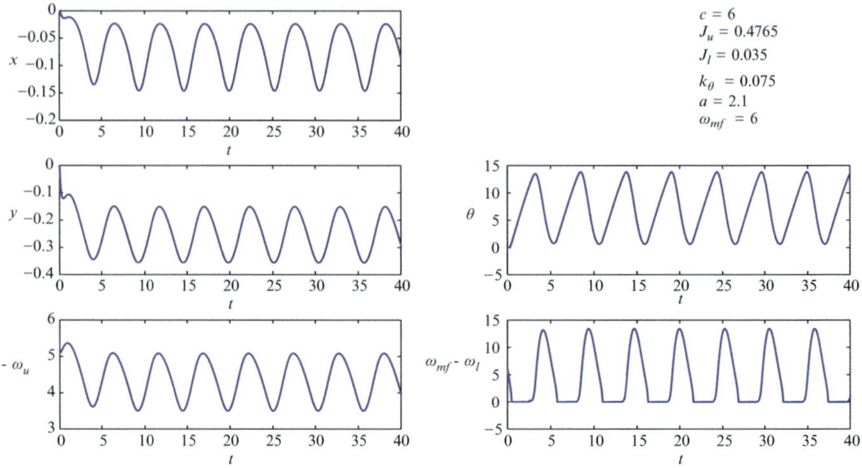

Fig. 15.8 Stick-slip vibrations phenomenon in the mathematical model of a drill actuated by induction motor

15.4 Conclusion

Despite the numerous models describing drilling systems, drill-string failures still occur, which cause enormous cost losses for drilling industry. In this work, a two-mass model of a drilling system, a mathematical model of drilling system with perfectly rigid drill-string, actuated by induction motor, and a modified version of these two models are considered. For the second model both analytical and numerical methods are used and it is shown that the limit permissible discontinuous load is equal to the value of the highest constant load at which the system has the steady mode. For the other two models it is demonstrated that along with the stable equilibrium, a stable limit cycle was found in both cases. This result shows that such complex effects such as hidden oscillations appear even in rather simple models. It is possible that the breakdowns in real drilling systems happen due to the existence of hidden oscillations which were not found because of difficulties during numerical analysis of those systems.

References

1. Al-Bender F, Lampaert V, Swevers J (2004) Modeling of dry sliding friction dynamics: From heuristic models to physically motivated models and back. Chaos 14(2):446–460
2. Bragin VO, Vagaitsev VI, Kuznetsov NV, Leonov GA (2011) Algorithms for finding hidden oscillations in nonlinear systems. The Aizerman and Kalman conjectures and Chua's circuits. J Comput Syst Sci Int 50(4):511–543, DOI 10.1134/S106423071104006X
3. Brett J (1992) Genesis of torsional drillstring vibrations. SPE Drilling Eng 7(3):168–174

4. Brockley C, Cameron R, Potter A (1967) Friction-induced vibrations. ASME J Lubricat Technol 89:101–108
5. de Bruin J, Doris A, van de Wouw N, Heemels W, Nijmeijer H (2009) Control of mechanical motion systems with non-collocation of actuation and friction: a Popov criterion approach for input-to-state stability and set-valued nonlinearities. Automatica 45(2):405–415
6. Filippov AF (1988) Differential equations with discontinuous right-hand sides. Kluwer, Dordrecht
7. Gubar' NA (1961) Investigation of a piecewise linear dynamical system with three parameters. J Appl Math Mech 25(6):1011–1023
8. Horbeek J, Birch W (1995) In: Proceedings of the society of petroleum engineers offshore, Europe, pp 43–51
9. Ibrahim R (1994) Friction-induced vibration, chatter, squeal, and chaos: dynamics and modeling. Appl Mech Rev: ASME 47(7):227–253
10. Ivanov-Smolensky A (1980) Electrical machines. Energiya, Moscow
11. Jansen J (1991) Non-linear rotor dynamics as applied to oilwell drillstring vibrations. J Sound Vibration 147(1):115–135
12. Kiseleva MA, Kuznetsov NV, Leonov GA, Neittaanmäki P (2012) Drilling systems failures and hidden oscillations. In: IEEE 4th international conference on nonlinear science and complexity, NSC 2012 – Proceedings, pp 109–112, DOI 10.1109/NSC.2012.6304736 http://ieeexplore.ieee.org/xpl/articleDetails.jsp?arnumber=6304736
13. Kondrat'eva N, Leonov G, Rodjukov F, Shepeljavyj A (2001) Nonlocal analysis of differential equation of induction motors. Technische Mechanik 21(1):75–86
14. Kreuzer E, Kust O (1996) Analyse selbsterregter drehschwingugnen in torsionsstäben. ZAMM – J Appl Math Mech 76(10):547–557
15. Kuznetsov N, Kuznetsova O, Leonov G, Vagaitsev V (2013) Informatics in control, automation and robotics. Lecture notes in electrical engineering, vol 174, Part 4, Chap. Analytical-numerical localization of hidden attractor in electrical Chua's circuit. Springer, Berlin, pp 149–158. DOI 10.1007/978-3-642-31353-0_11
16. Kuznetsov NV, Leonov GA, Vagaitsev VI (2010) Analytical-numerical method for attractor localization of generalized Chua's system. IFAC Proc Vol (IFAC-PapersOnline) 4(1):29–33, DOI 10.3182/20100826-3-TR-4016.00009
17. Kuznetsov NV, Leonov GA, Seledzhi SM (2011) Hidden oscillations in nonlinear control systems. IFAC Proc Vol (IFAC-PapersOnline) 18(1):2506–2510, DOI 10.3182/20110828-6-IT-1002.03316
18. Lauvdal T, Murray R, Fossen T (1997) Stabilization of integrator chains in the presence of magnitude and rate saturations: a gain scheduling approach. In: Proceedings of the 36th IEEE Conference on Decision and Control, Vol. 4, pp 4404–4005, DOI 10.1109/CDC.1997.652491
19. Leine R (2000) Bifurcations in discontinuous mechanical systems of filippov-type. Ph.D. thesis, Eindhoven University of Technology, The Netherlands
20. Leine R, Campen DV, Keultjes W (2003) Stick-slip whirl interraction in drillstring dynamics. ASME J Vibrat Acoustics 124
21. Leonov G, Kuznetsov N (2013) IWCFTA2012 Keynote Speech I - Hidden attractors in dynamical systems: From hidden oscillation in Hilbert-Kolmogorov, Aizerman and Kalman problems to hidden chaotic attractor in Chua circuits. In: 2012 Fifth International Workshop on Chaos-Fractals theories and applications (IWCFTA), pp XV–XVII, DOI 10.1109/IWCFTA.2012.8
22. Leonov GA, Kiseleva MA (2012) Analysis of friction-induced limit cycling in an experimental drill-string system. Doklady Phys 57(5):206–209
23. Leonov GA, Kuznetsov NV (2011) Algorithms for searching for hidden oscillations in the Aizerman and Kalman problems. Doklady Math 84(1):475–481, DOI 10.1134/S1064562411040120
24. Leonov GA, Kuznetsov NV (2011) Analytical-numerical methods for investigation of hidden oscillations in nonlinear control systems. IFAC Proc Vol (IFAC-PapersOnline) 18(1): 2494–2505, DOI 10.3182/20110828-6-IT-1002.03315

25. Leonov GA, SolovTeva EP (2012) The nonlocal reduction method in analyzing the stability of differential equations of induction machines. Doklady Math 85(3):375–379
26. Leonov GA, Solov'eva EP (2012) On a special type of stability of differential equations for induction machines with double squirrel -cage rotor. Vestnik St Petersburg Univ Math 45(3):128–135
27. Leonov GA, Bragin VO, Kuznetsov NV (2010) Algorithm for constructing counterexamples to the Kalman problem. Doklady Math 82(1):540–542, DOI 10.1134/S1064562410040101
28. Leonov GA, Vagaitsev VI, Kuznetsov NV (2010) Algorithm for localizing Chua attractors based on the harmonic linearization method. Doklady Math 82(1):693–696, DOI 10.1134/S1064562410040411
29. Leonov GA, Kuznetsov NV, Kuznetsova OA, Seledzhi SM, Vagaitsev VI (2011) Hidden oscillations in dynamical systems. Trans Syst Contl 6(2):54–67
30. Leonov GA, Kuznetsov NV, Vagaitsev VI (2011) Localization of hidden Chua's attractors. Phys Lett A 375(23):2230–2233, DOI 10.1016/j.physleta.2011.04.037
31. Leonov GA, Kuznetsov NV, Vagaitsev VI (2012) Hidden attractor in smooth Chua systems. Physica D 241(18):1482–1486, DOI 10.1016/j.physd.2012.05.016
32. Leonov GA, Kuznetsov NV, Yuldahsev MV, Yuldashev RV (2012) Analytical method for computation of phase-detector characteristic. IEEE Trans Circ Syst – II: Express Briefs 59(10):633–647, DOI 10.1109/TCSII.2012.2213362
33. Leonov GA, Kuznetsov GV (2013) Hidden attractors in dynamical systems. From hidden oscillations in Hilbert-Kolmogorov, Aizerman, and Kalman problems to hidden chaotic attractors in Chua circuits. Int J Bifurcat Chaos 23(1):1–69, DOI 10.1142/S0218127413300024
34. Marino R, Tomei P, Verrelli C (2010) Induction motor control design. Springer, The Netherlands
35. Mihajlović N (2005) Torsional and lateral vibrations in flexible rotor systems with friction. Ph.D. dissertation, Eindhoven University of Technology, Eindhoven, Netherlands
36. Mihajlovic N, van Veggel A, van de Wouw N, Nijmeijer H (2004) Analysis of friction-induced limit cycling in an experimental drill-string system. J Dyn Syst Meas Control 126(4):709–720
37. Olsson H (1996) Control systems with friction. Ph.D. thesis, Lund Institute of Technology, Sweden
38. Popp K, Stelter P (1990) Stick-slip vibrations and chaos. Philosoph Trans R Soc Lond 332: 89–105
39. Shokir E (2004) A novel pc program for drill string failure detection and prevention before and while drilling specially in new areas. J Oil Gas Bus (1)
40. den Steen LV (2005) Suppressing stick-slip-induced drill-string oscillations: a hyper stability approach. Ph.D. dissertation, University of Twente
41. Yakobovich VA, Leonov GA, Gelig AK (2004) Stability of Stationary Sets in Control Systems with Discontinuous Nonlinearities. World Scientific, Singapore

Chapter 16
Chaos in a Piecewise Linear System with Periodic Oscillations

Chunqing Lu

Abstract The paper studies a second order nonlinear differential equation whose right hand side is a piecewise linear function. It shows the coexistence of a countable set of periodic solutions and an uncountable set of bounded non-periodic solutions. The result can be also used to explain the chaos on some smooth nonlinear dynamical systems.

Keywords Chaos • Piecewise linearity • Weierstrass theorem

16.1 Introduction

Consider a second order nonlinear differential equation

$$x'' + p(x) = \gamma \cos \epsilon t \tag{16.1}$$

where $p(x)$ is an S-shaped polynomial and γ and ϵ are positive constants. One of such equations is the well-known Duffing equation without damping in which $p(x) = x^3 - x$ or a third order polynomial. Many researchers have investigated the rich phenomenon of its solutions numerically. There are also some analytical results based on the perturbation method and Poincare maps [1]. However, there are still more questions to be investigated. For example, if the solutions are chaotic, can we find the pattern of such solutions? This paper uses a piecewise linear function to approximate the polynomial $p(x)$ to explore certain types of chaotic solutions of (16.1). As an example, we study the equation

---

C. Lu (✉)
Southern Illinois University Edwardsville, Edwardsville, IL62026-1805, USA
e-mail: clu@siue.edu

J.A.T. Machado et al. (eds.), *Discontinuity and Complexity in Nonlinear Physical Systems*, Nonlinear Systems and Complexity 6, DOI 10.1007/978-3-319-01411-1_16,

$$x'' - f(x) = \gamma \cos \epsilon t \tag{16.2}$$

where $f(x)$ is a piecewise linear function

$$f(x) = \begin{cases} x \text{ for } |x| \leq 1, \\ 2 - x \text{ if } x > 1, \\ -2 - x \text{ if } x < -1. \end{cases} \tag{16.3}$$

By the Weierstrass approximation theorem, for any $\Delta > 0$, there exists a polynomial $p(x)$ such that $|p(x) - f(x)| < \Delta$ for all x on a finite interval $[\alpha, \beta]$. This means also that $f(x)$ can be used to approximate the polynomial $p(x)$ for all x on any given finite interval.

Note that the piecewise linear function $f(x)$ is Lipschitz conditioned and therefore, the existence and uniqueness theorem of solutions to (16.2) can be applied. In addition, solutions of (16.2) continuously depend on its initial values, parameters, and the function $f(x)$ on any finite interval of t. Precisely, we can take Δ small enough so that the solutions of equations (16.2) and (16.1) and their first order derivatives can be sufficiently close over any fixed finite interval. In this way, the behavior of solutions of (16.1) is determined by the solutions of (16.2) on the finite interval. The analysis in this paper shows that the behavior of the solutions of (16.2) on the finite interval will determine the behavior of the solutions for $-\infty < t < \infty$, which becomes chaotic in the sense that there exist a countable set of periodic solutions and a non-countable set of bounded non-periodic solutions. The N-shaped function $(-f(x))$ is used to approximate the S-shaped polynomial. We then will observe the existence of the chaotic solutions of a generalized Duffing equation (16.1).

Equation (16.2) is not a smooth nonlinear dynamical system. It may be used as a real mathematical model for some dynamical systems. The perturbation of Poincare maps and the related theory including the so-called Melnikov's method may not be applied. However, (16.2) can be solved explicitly in different intervals. Thus, the direct classical analysis can be implemented, which gives us a clear insight about how chaos happens in this system. This paper modifies the analyses in [6] and gives clearer proofs. But, some approximation analyses are still given by their outlines.

Using approximation theory and the direct classical analysis to study Chaos was first accomplished by N. Levinson [3] in 1949. Levinson first proved the existence of chaotic solutions of the generalized van der Pol equation

$$x'' + p(x)x' + x = c \sin t, \tag{16.4}$$

where $p(x)$ is a polynomial and c is a constant. He used a piecewise constant function to approximate the polynomial $p(x)$ and obtained some chaotic solutions of the equation. Levinson's work also showed the existence of the strange attractors, the Levinson ring, which is a ring-shaped closed connected set in the phase plan. It was Levinson's analysis that led to Smale's introduction of the horseshoe map.

The relation between Levinson's ring and Smale's horseshoe was explained by Levi [1,4].

16.2 Asymptotic Solutions for $|x| \leq 1$

We assume that $\epsilon > 0$ is sufficiently small in this paper. We first make a change of variables: $\tau = \epsilon t, u(\tau) = x(t) = x(\tau/\epsilon)$. For convenience, we replace τ by t. Thus, (16.2) takes the following form

$$\epsilon^2 u'' = f(u) + \gamma \cos t, \tag{16.5}$$

where

$$f(u) = \begin{cases} u \text{ for } |u| \leq 1 \\ 2 - u \text{ if } u > 1 \\ -2 - u \text{ if } u < -1 \end{cases} \tag{16.6}$$

The solutions of (16.5) can be given explicitly. For $|u| < 1$, the equation is linear which takes the form

$$\epsilon^2 u'' = u + \gamma \cos t. \tag{16.7}$$

Then its solutions are given by

$$u = A_1 e^{(t-t_0)/\epsilon} + A_2 e^{-(t-t_0)/\epsilon} - \frac{\gamma}{1+\epsilon^2} \cos t. \tag{16.8}$$

Thus,

$$u' = \frac{1}{\epsilon} \left[A_1 e^{(t-t_0)/\epsilon} - A_2 e^{-(t-t_0)/\epsilon} \right] + \frac{\gamma}{1+\epsilon^2} \sin t. \tag{16.9}$$

In these two expressions, A_1, A_2, and t_0 are constants.

If $u > 1$, the equation is another linear equation,

$$\epsilon^2 u'' = 2 - u + \gamma \cos t, \tag{16.10}$$

which has the general solution

$$u = B_1 \cos\left(\frac{t - t_1}{\epsilon} + \delta_1 \right) + 2 + \frac{\gamma}{1-\epsilon^2} \cos t, \tag{16.11}$$

where t_1 is an initial time, and B_1 and B_2 are constants determined by initial conditions.

Similarly, for $u < -1$,

$$u = B_2 \cos\left(\frac{t - t_0}{\epsilon} + \delta_2\right) - 2 + \frac{\gamma}{1 - \epsilon^2} \cos t. \tag{16.12}$$

Since the right-hand side of (16.5) satisfies the Lipschitz condition, its solutions are in C^1 over any finite interval. This enables us to determine the above coefficients A_i, B_i, and δ_i for $i = 1, 2$.

We begin with a continuous family of solutions of (16.5) with initial conditions

$$u_0 = -1, u_0' > 0 \tag{16.13}$$

where $u(t_0) = u_0, u'(t_0) = u_0'$, and $t_0 = \tau$. From the solution form (16.8), we see

$$A_1 + A_2 - \frac{\gamma}{1 + \epsilon^2} \cos\tau = -1 \tag{16.14}$$

and

$$\frac{1}{\epsilon} A_1 - \frac{1}{\epsilon} A_2 + \frac{\gamma}{1 + \epsilon^2} \sin\tau = u_0'. \tag{16.15}$$

Thus,

$$A_1 = \left(u_0 + \epsilon u_0' + \frac{\gamma}{1 + \epsilon^2} \cos\tau - \frac{\epsilon\gamma}{1 + \epsilon^2} \sin\tau\right) / 2 \tag{16.16}$$

and

$$A_2 = \left(u_0 - \epsilon u_0' + \frac{\gamma}{1 + \epsilon^2} \cos\tau + \frac{\epsilon\gamma}{1 + \epsilon^2} \sin\tau\right) / 2 \tag{16.17}$$

To make these solutions increasingly cross the line $u = 1$ we must require $A_1 \geq 0$, $A_2 < 0$. Since

$$\epsilon u_0' = 2A_1 + 1 - \frac{\gamma}{1 + \epsilon^2} \cos\tau + \frac{\epsilon\gamma}{1 + \epsilon^2} \sin\tau, \tag{16.18}$$

we set

$$u_0' = \frac{1}{\epsilon}\left(1 - \gamma \frac{\cos\tau}{\epsilon^2 + 1} + \gamma\epsilon \frac{\sin\tau}{\epsilon^2 + 1} + 2\rho\right) \tag{16.19}$$

It follows that $A_1 = \rho$ and $A_2 = -1 - \rho + \frac{\gamma}{1+\epsilon^2} \cos\tau$, where $\rho > 0$ is sufficiently small. It then follows that $u_0' = 1/\epsilon(1 - \gamma + o(\gamma))$. Hence the solution is given as

$$u = \rho e^{(t-\tau)/\epsilon} - \left(1 + \rho - \frac{\gamma}{1+\epsilon^2}\cos\tau\right)e^{-(t-\tau)/\epsilon} - \frac{\gamma}{1+\epsilon^2}\cos t, \tag{16.20}$$

and

$$u' = \frac{1}{\epsilon}\left[\rho e^{(t-\tau)/\epsilon} + \left(1 + \rho - \frac{\gamma}{1+\epsilon^2}\cos\tau\right)e^{-(t-\tau)/\epsilon}\right] + \frac{\gamma}{1+\epsilon^2}\sin t. \tag{16.21}$$

It then follows

$$u' = \frac{1}{\epsilon}\left[u + 2\left(1 + \rho - \frac{\gamma}{1+\epsilon^2}\cos\tau\right)e^{-(t-\tau)/\epsilon} + \frac{\gamma}{1+\epsilon^2}\cos t\right] + \frac{\gamma}{1+\epsilon^2}\sin t \tag{16.22}$$

Note that $u' > 0$ as long as $t \in [\tau, \pi]$ for any $\rho \geq 0$. This is true if $\tau \geq 0$. If $\tau < 0$ and $|\tau|$ is sufficiently small, $u'(t) > e^{-(t-\tau)/\epsilon}/2\epsilon + \frac{\gamma}{1+\epsilon^2}\sin t$ for $t > \tau$ if $\gamma < 1/4$. Then, $u'(t) > \frac{1}{\epsilon}e^{-|\tau|/\epsilon} - \frac{1}{2}\sin|\tau| > 0$ for $0 > t > \tau$ if $\epsilon > 0$ is sufficiently small. Notice that τ can be negative and independent of ρ. Let $E(t)$ be the sum of the first two terms in (16.20), i.c.,

$$E(t) = \rho e^{(t-\tau)/\epsilon} - \left(1 + \rho - \frac{\gamma}{1+\epsilon^2}\cos\tau\right)e^{-(t-\tau)/\epsilon}.$$

It then follows that $E(\tau) \geq -1 + \frac{\gamma}{1+\epsilon^2}\cos\tau$, and $E' > 0$ for all $t > \tau$ and for sufficiently small $|\tau|$. This implies that $E(t) > E(\pi) = \rho e^{(\pi-\tau)/\epsilon} - (1 + \rho - \frac{\gamma}{1+\epsilon^2}\cos\tau)e^{-(\pi-\tau)/\epsilon}$ as long as $u(t) < 1$ for $t > \pi$. Thus, $u(t) \geq E(\pi) - \frac{\gamma}{1+\epsilon^2}$ for all $t \geq \tau$ as long as $u \leq 1$, if $\gamma < 1/4$. It is seen that $u(t) > E(\pi) - \frac{\gamma}{1+\epsilon^2} > -\frac{1}{2}e^{-2\pi/\epsilon} - \frac{\gamma}{1+\epsilon^2} > -\frac{1}{2} > -1$ for all $t > \tau$, since $E(\pi) > -\frac{1}{2}e^{-(\pi-\tau)/\epsilon}$. It is impossible to have $u = -1$ for some $t > \pi$, for otherwise, u must be equal to $-1/2$ for some $t > \pi$ before it reaches -1, which is a contradiction. Thus, we conclude that $u > -1$ for all $t > \tau$ as long as $u < 1$ (since the expression (16.20) is valid).

One of sufficient conditions for $u \leq 1$ is the inequality, $\rho e^{t-\tau/\epsilon} < 1 - \gamma$, where $t \in [\tau, \pi + \tau]$, which comes from the expression (16.20). We first let $\rho_1 = f(\gamma, \epsilon)e^{-\pi+\delta/\epsilon}$. In this case, it is seen that at $t_1 = \pi + \tau - \delta$, $u = f(\gamma, \epsilon) - (1 - \gamma + \rho_1)e^{(-\pi+\delta)/\epsilon} + \frac{\gamma}{1+\epsilon^2}\cos(\tau - \delta) + o(\delta) = 1$ for some continuous function $f(\gamma, \epsilon) \approx 1 - \gamma$ if $|\delta|$ is sufficiently small. This comes from the fact $u(\pi, \rho_1) = f(\gamma, \epsilon) + \gamma + o(\delta)$, and hence, $f(\gamma, \epsilon) = 1 - \gamma + o(\delta)$. Similarly, we may choose another $\rho_2 = g(\gamma, \epsilon)$ such that $u(3\pi + \tau - \delta, \rho_2) = 1$. For simplicity, we denote $u(t, \rho_1) = u_1(t), t_1 = \pi + \tau - \delta, t_2 = 3\pi + \tau - \delta$ and $u(t, \rho_2) = u_2(t)$. In addition, we may assume that t_i is the first time for u_i to reach the line $u = 1$ from the region $u < 1$. From (16.22)

$$u_1'(t_1) = \frac{1}{\epsilon}\left[1 + 2\left(1 + \rho - \frac{\gamma}{1+\epsilon^2}\cos\tau\right)e^{-(t_1-\tau)/\epsilon} + \frac{\gamma}{1+\epsilon^2}\cos t_1\right]$$
$$+ \frac{\gamma}{1+\epsilon^2}\sin t_1$$

$$= \frac{1}{\epsilon}\left[1 - \frac{\gamma}{1+\epsilon^2}\cos(\tau-\delta)\right] + \frac{\gamma}{1+\epsilon^2}\sin t_1 + O(e^{-(\pi-\delta)/\epsilon}/\epsilon) \qquad (16.23)$$

Note that $E' > 0$ for $t > \tau$ and $E \approx \rho e^{t-\tau/\epsilon}$ for $t > \pi$ and for sufficiently small $|\epsilon|$. Thus, the first time for u to reach the line $u = 1$ can be estimated by solving

$$\rho e^{(t-\tau)/\epsilon} - \frac{\gamma}{1+\epsilon^2}\cos t = 1. \qquad (16.24)$$

Letting $\rho = (1-\gamma)e^{-(2n-1)\pi+\delta/\epsilon}$, we see the ascending time for $u = 1$ is around $t = \tau + (2n-1)\pi$ for $n = 1, 2$ by choosing τ and δ such that $e^{-(\pi-\delta)/\epsilon}/\epsilon$ is sufficiently small .

16.3 Solutions for $x > 1$

Consider the solution (16.11) with the initial condition $u_1(t_1) = 1$. We see

$$B_1 \cos\delta_1 + 2 + \frac{\gamma}{1-\epsilon^2}\cos t_1 = 1. \qquad (16.25)$$

Thus,

$$B_1 \cos\delta_1 = -1 - \frac{\gamma}{1+\epsilon^2}\cos t_1 \qquad (16.26)$$

Again, from (16.11),

$$u' = -\frac{B_1}{\epsilon}\sin\left(\frac{t-t_1}{\epsilon} + \delta_1\right) - \frac{\gamma}{1-\epsilon^2}\sin t, \qquad (16.27)$$

and

$$u'(t_1) = -\frac{B_1}{\epsilon}\sin\delta_1 - \frac{\gamma}{1-\epsilon^2}\sin t_1. \qquad (16.28)$$

Therefore, from (16.23), B_1 and δ_1 satisfy (16.26) and

$$-\frac{B_1}{\epsilon}\sin\delta_1 - \frac{\gamma}{1-\epsilon^2}\sin t_1 = \frac{1}{\epsilon}\left[1 - \frac{\gamma}{1+\epsilon^2}\cos(\tau-\delta)\right] + \frac{\gamma}{1+\epsilon^2}\sin t_1 + O(e^{-(\pi-\delta)/\epsilon}/\epsilon). \qquad (16.29)$$

From (16.26) and (16.29), we see that

$$B_1 \sin\delta_1 = -\left[1 + \frac{\gamma}{1+\epsilon^2}\cos t_1)\right] - \frac{2\gamma\epsilon}{1-\epsilon^4}\sin t_1$$

It then turns out that

$$u(t) = B_1 \cos\left(\frac{t-t_1}{\epsilon} + \delta_1\right) + \frac{\gamma}{1-\epsilon^2}\cos t$$

where $\delta_1 = 5\pi/4 + o(\epsilon)$. We then see that $u(t)$ increases to its maximum and then gets back to the line $u = 1$ around $\frac{t-t_1}{\epsilon} + \delta_1 = 2\pi + 2\pi - \delta_1$, or $t - t_1 = \frac{3\epsilon\pi}{2}$. This can be seen from the fact $t_1 \approx \pi$ and that if $t = \pi + o(\epsilon)$, then $\cos t = -1 + o(\epsilon^2)$ and the value of $\cos(\frac{t-t_1}{\epsilon} + \delta_1)$ can run over the interval $[-1, 1]$ because this cosine function has the period $2\pi\epsilon$. If $\cos(\frac{t-t_1}{\epsilon} + \delta_1) = (-1+\gamma+o((\tau-\delta))/B_1$ while $\cos t$ would be still close to -1, then Therefore, $u(t)$ returns to $u = 1$ around $t = t_1 + \frac{3\epsilon\pi}{2}$ where $t_1 = \pi + \tau - \delta$.

Similarly, if we choose $t_1 = \pi + \tau + \delta$, then the corresponding solution $u(t)$ would return to $u = 1$ before $t_1 + \epsilon(2\pi - \delta_0)$ for $t_1 = \pi + \delta$. By continuity of solutions we see that there is a sufficiently small $\rho \approx (1 - \gamma)e^{-\pi/\epsilon}$ such that $u(t, \rho)$ satisfies the following properties: $u(0) = -1$, and $u(t)$ crosses the line $u = 1$ and reaches its maximum. Then, $u(t)$ descends to the line $u = 1$ at $t = \tau_1$ where $\tau_1 \approx t_1 + \frac{3\epsilon\pi}{2}$.

Once $u(t)$ reaches the line $u = 1$ at $t = \tau_1$ around π, the solution shall take the form (16.8), i.e.,

$$u = A_1 e^{(t-\tau_1)/\epsilon} + A_2 e^{-(t-\tau_1)/\epsilon} - \frac{\gamma}{1+\epsilon^2} \cos t.$$

Using the information about $u'(\tau_1)$, we can adjust the value $\delta - \tau$ (or u_0') to make $A_1 \approx -(1 - \gamma)e^{-(2n-1)\pi/\epsilon}$ for $n = 1, 2$. Similarly, we can analyze the case when $t_2 = 3\pi + \tau - \delta$.

16.4 Chaotic Solutions

If we set the initial value $u(\tau) = -1$ and $u'(\tau)$ as in (16.19) where ρ is in the neighborhood of $(1 - \gamma)e^{-\pi/\epsilon}$ and follow the argument above, we will get the solution that starts at the line $u = -1$ and then crosses the line $u = 1$ and returns to $u = 1$ around the time $t = \pi + \tau$. In what follows, the returning time to line $u = 1$ means the first time it gets back to the line $u = 1$ after it crosses the line $u = 1$. From the continuity of solutions in initial conditions, we may extend the value of $u'(\tau)$ to an interval $(\alpha_1, \beta_1) \subset (0, 2/\epsilon)$ so that the above properties hold for $u'(\tau) \in (\alpha_1, \beta_1)$. This can be done by adjusting the value δ. Similarly, we extend the value of ρ around $(1 - \gamma)e^{-3\pi/\epsilon}$ so that for $u'(\tau) \in (\alpha_2, \beta_2)$ around $1/\epsilon$. Of course, $(\alpha_1, \beta_1) \cap (\alpha_2, \beta_2) = \emptyset$. We can prove similarly that there exists an interval (α_n, β_n) such that if $\alpha \in (\alpha_n, \beta_n)$ then the solution $u(t, \alpha)$ crosses the line $u = 1$ at the first time $t_1 = (2n - 1)\pi$ for $n = 1, 2$. Thus, we just proved the following theorem.

Theorem 1. *Let $\epsilon, \gamma > 0$ be sufficiently small. There exist at least two disjoint subintervals (α_1, β_1) and (α_2, β_2) of $(0, 2/\epsilon)$ such that the descending time T_1 of $u(t, \alpha)$ is in the interval $(\pi - \lambda, \pi + \lambda)$ and the ascending time S_1 of $u(t, \beta)$ is in $(3\pi - \lambda, 3\pi + \lambda)$, where $\alpha \in (\alpha_1, \beta_1)$ and $\beta \in (\alpha_2, \beta_2)$.*

Similarly, we can prove Theorem 2 as follows.

Theorem 2. *Let $\epsilon, \gamma > 0$ be sufficiently small. There exists at least two disjoint subinterval (μ_1, ν_1) and (μ_2, ν_2) of (α_1, β_1) such that if $\alpha \in (\mu_1, \nu_1)$ then $u(t, \alpha)$ descends to the line $u = -1$ and then ascends back to the line $u = -1$ around $t = 2\pi$ and if $\alpha \in (\mu_2, \nu_2)$ then $u(t, \alpha)$ descends to the line $u = -1$ and then ascends back to the line $u = -1$ around $t = 4\pi$.*

Let $I_1 = (\alpha_1, \beta_1)$ and $I_2 = (\alpha_2, \beta_2)$. Denote $I_{11} = (\mu_1, \nu_1)$ and $I_{12} = (\mu_2, \nu_2)$. Then, $I_{11} \cap I_{12} = \emptyset$ and $I_{11}, I_{12} \subset I_1$. If $\alpha \in I_{11}$, then the corresponding solution has the first descending time $T_1 = \pi$ and the first ascending time at 2π. If $\alpha \in I_{12}$, then the corresponding solution has the first descending time $T_1 = \pi$ and the first ascending time at 4π. Similarly, we can define I_{21} and I_{22} such that $I_{21} \cap I_{22} = \emptyset$ and $I_{21}, I_{22} \subset I_2$ and if $\alpha \in I_{21}$ the corresponding solution has the first descending time $T_1 = 3\pi$ and the first ascending time at 4π. If $\alpha \in I_{22}$, then the corresponding solution has the first descending time $T_1 = 3\pi$ and the first ascending time at 6π. We now can see that the descending time is always at $(2n - 1)\pi$ and ascending time is always at $2m\pi$. The difference between the two successive ascending time and descending time is either π or 2π. Assume the solution starts from $u = -1$ at $t = 0$. Then the first descending time is τ_1 which is either $T_1 = \pi$ or $S_1 = 3\pi$, and the next ascending time would be θ_1 which is either $\pi + \tau_1$ or $3\pi + \tau_1$. We then obtain a sequence $\{\tau_k, \theta_k\}$ for $k = 1, 2, \ldots$. Let $d_k = \theta_k - \tau_k$. We then can prove the following theorems.

Theorem 3. *Assume that $\epsilon > 0$ and $\gamma > 0$ are sufficiently small. For any sequence $\{d_k\}$ for $k = 1, 2, \ldots$, where d_k is either π or 3π, there exist at least two solutions such that the difference between the k_{th} successive descending time and k_{th} ascending time is d_k of the solutions.*

Proof. The infinite sequence $\{d_k\}, k = 1, 2, \ldots$ corresponds to a nested sequence of open intervals $I_1 \supset I_{1j_1} \supset I_{1j_1j_2} \supset \ldots \supset I_{1j_1j_2\cdots j_k} \supset \ldots$. Since each open interval is a proper subset of the former interval in the nested sequence, we may choose a closed interval from each open interval, say $J_{1j_1j_2\cdots j_k} \subset I_{1j_1j_2\cdots j_k}$. This implies that we can make a nested closed intervals $J_1 \supset J_{1j_1} \supset J_{1j_1j_2} \supset \ldots \supset J_{1j_1j_2\cdots j_k} \supset \ldots$. Therefore there exists at least one $\alpha \in I_{1j_1j_2\cdots j_k}$ for $k = 1, 2, \ldots$ such that the solution $u(t, \alpha)$ has the successive spacing $\{d_k\}, k = 1, 2, \ldots$. Of course, this solution has the first descending time at $T_1 = \pi$. Similarly, we can start with the interval I_2, which gives another solution having the first descending time $S_1 = 3\pi$.

Notice that the descending time is always odd and ascending time is always even. With the similar arguments, we can generalize Theorem 3 as the following theorem.

Theorem 4. *Let $S_1 = 2m + 1$ and $T_1 = 2n + 1$ and $m > n \geq 0$ are integers. There exists γ_0 and $\epsilon_0 > 0$ such that for any $\epsilon \in (0, \epsilon_0)$ and $\gamma \in (0, \gamma_0)$ and for any sequence $\{d_k\}$ there exist at least two solutions of the equation with the successive time spacing $d_k, k = 1, 2, \ldots$ where $d_k = (2m + 1)\pi$ or $(2n + 1)\pi$.*

Theorem 3 is the case $m = 0$ and $n = 1$ of Theorem 4. The proof of Theorem 4 is almost the same as that of Theorem 3.

16.5 Conclusion

The cardinality of the set of all sequences $\{d_k\}, k = 1, 2, \ldots$, where d_k is either 3π or π, is the continuum. Among such sequences, there is a countable set of periodic sequences. We may prove that to each periodic sequence $\{d_k\}$ there is at least one periodic solution of the equation. Therefore, this dynamical system admits at least a countable set of periodic solutions. Of course, the other sequences correspond to bounded non-periodic solutions, which form an uncountable set of solutions. This coexistence of periodic and non-periodic solutions shows the chaos of the system, (16.2) and therefore (16.1). Note that some part of this paper, the asymptotic analyses are given only approximately. The more delicate analyses and the proof of existence of periodic solutions will be given in another paper. The technique used in this paper is a classical shooting method. The detailed explanation of this method can be found in [2]. The method was also applied by this author to get the chaos for the pendulum equations [5]. This paper only studies the equation with piecewise linearity. However, further investigation can show that Theorems 1–4 in the paper also hold for (16.1) as well as the equation

$$x'' + cx' = p(x) + \gamma \cos \epsilon t \tag{16.30}$$

provided the values $|c|$, $|\gamma|$, and $|\epsilon|$ are sufficiently small. This and the existence of the strange attractors for the nonlinear dynamic systems will be reported in another paper.

References

1. Guckenheimer J, Holmes P (1983) Nonlinear oscillations, dynamical systems, and bifurcations of vector fields. Spring-Verlag, New York
2. Hastings SP, McLeod JB (2012) Classical methods in ordinary differential equations with applications to boundary value problems. AMS, Providence, vol. 129
3. Levinson N (1949) A second order differential equation with singular solutions. Ann Math. 50(1):127–153
4. Levi M (1981) Qualitative analysis of the periodically forced relaxation oscillations. Mem Ams 214:1–147
5. Lu C (2007) Chaos of a parametrically excited undamped pendulum. Comm Nonlin Sci Num Sim 12:45–57
6. Lu C (2012) Chaotic solutions of a differential equation with piecewise linearity, NSC 2012, IEEXplore, pp 117–120

Chapter 17
Basins of Attraction in a Simple Harvesting System with a Stopper

Marek Borowiec, Grzegorz Litak, and Stefano Lenci

Abstract We examine the dynamical response and the power output of a vibration energy harvesting electromechanical system with kinematic ambient excitation and impact. Due to the stopper nonlinearities the examined system exhibits multiple solutions. We characterize their properties and stability by the voltage output and the corresponding basins of attraction.

Keywords Piecewise linear system • Energy harvesting • Basins of attractions

17.1 Introduction

Many mechanical systems with nonlinearities show complex responses characterized by multiple solutions with different amplitudes of vibrations and specific basins of attraction. Their existence make unrivalled opportunity to improve the effectiveness of kinetic energy harvesters through the so-called broadband frequency effect [1, 3, 9, 11]. Energy harvesting devices based on the existence of multiple attractors are equipped with mechanical nonlinear resonators and appropriate energy transducers, transforming ambient mechanical energy into electric form.

Recently, kinetic energy harvesters based on mechanical resonator and electromagnetic transducers [8, 10, 12] were explored extensively. See also the interesting

M. Borowiec (✉) • G. Litak
Faculty of Mechanical Engineering, Lublin University of Technology, Nadbystrzycka 36, PL-20-618 Lublin, Poland
e-mail: m.borowiec@pollub.pl; g.litak@pollub.pl

S. Lenci
Department of Civil and Building Engineering, and Architecture, Polytechnic University of Marche, 60131 Ancona, Italy
e-mail: lenci@univpm.it

J.A.T. Machado et al. (eds.), *Discontinuity and Complexity in Nonlinear Physical Systems*, Nonlinear Systems and Complexity 6, DOI 10.1007/978-3-319-01411-1_17,

work by Blystad and Halvorsen [2]. Micro-electromechanical systems (MEMS) electrostatic devices were proposed and studied by Gu [4] and Le et al. [5, 6].

In this chapter we continue to study on the nonlinear impacting electro-magnetic harvesters with stoppers, derived from the idea of Soliman et al. [13, 14]. For the considered system, at least two different solutions appear due to the presence of a stopper of the moving structure, if the amplitude of mechanical resonator is large enough. The impact with the stopper both limits vertical displacements and simultaneously changes the elastic characteristics of the system.

17.2 The Model

The model of energy harvester is made up of a main body frame, which contains both the electrical harvester and the internal mechanical system (see Fig. 17.1a). The subsystem within the frame includes the effective magnet mass m which is linked to the frame through the springs and dampers. The frame system is moving vertically due to a ground harmonic excitation $y = A\cos(\omega_e t)$. The transducer on the frame harvests the kinetic energy, converting into the electric power output. This energy transformation causes the electromagnetic damper be via the moving magnet inside the coil located appropriately on the frame.

When a given distance z_d is reached by the mass m an impact occurs and a second spring k_2 activates, the effective spring force is changing as shown in Fig. 17.1b. Due to impacts both the stiffness and the mechanical damping take two differently values ($i = 1$ and $i = 2$), from k_1 and b_{m1} when impacts do not take place, to k_2 and b_{m2}, while contacting. Then the mechanical restoring force Fr is simultaneously modified according to:

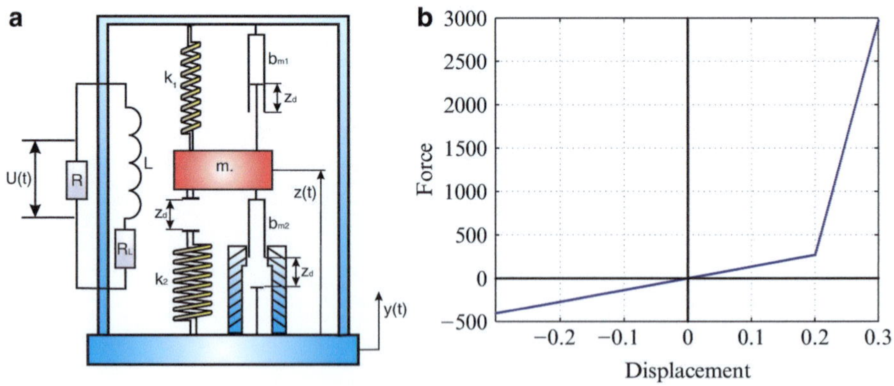

Fig. 17.1 Schematics of the mechanical resonator of energy harvesting system (**a**). The additional electrical circuit is powered by Faraday electromotive force via the moving coil across the magnetic field. In the calculations we neglect self-induction of the coil L_c. The stiffness characteristics of the effective model (**b**)

$$F_r = \begin{cases} k_1 z & \text{for} \quad z < z_d \ (i = 1) \\ k_2 z + (k_1 - k_2) z_d & \text{for} \quad z \geq z_d \ (i = 2) \end{cases} \tag{17.1}$$

and the damping restoring force F_d is modified according to:

$$F_d = \begin{cases} b_1 \dot{z}, & \text{for} \quad z < z_d \ (i = 1) \\ b_2 \dot{z}, & \text{for} \quad z \geq z_d \ (i = 2) \end{cases} \tag{17.2}$$

The equation of motion of the system reads

$$m\ddot{z} + b_i \dot{z} + k_i z = -m\ddot{y} + (k_2 - k_1) z_d \Theta(z - z_d), \tag{17.3}$$

where $\Theta(.)$ is a s the Heaviside step function.

Finally, voltage induced across the load resistor R can be estimated as

$$U = \frac{RB\ell}{R + R_c} \dot{z}. \tag{17.4}$$

where R and R_c denote the load and the coil resistances, B is the magnetic induction and ℓ is the coil effective length.

Using the dimensionless variables:

$$\tau = \omega_1 t, \ \ \Omega = \frac{\omega_e}{\omega_1}, \ \ \mathcal{Z} = \frac{z}{z_d}, \ \ \mathcal{Y} = \frac{y}{z_d}, \tag{17.5}$$

where the natural frequency used for introducing dimensionless time τ is $\omega_1 = \sqrt{k_1/m}$, the equation of motion in dimensionless form becomes:

$$\ddot{\mathcal{Z}} + 2\eta_i \dot{\mathcal{Z}} + r_i^2 \mathcal{Z} = -\ddot{\mathcal{Y}} + (\rho^2 - 1)\Theta(\mathcal{Z} - 1). \tag{17.6}$$

The function $\Theta(\mathcal{Z} - 1)$ is the Heaviside function, switching the system whether the mass m is in contact with the spring of stiffness k_2 or not.

The parameters r and η depend on conditions of Eqs. 17.1, 17.2 and for two cases $(i = 1, 2)$, $r_i = \sqrt{k_i/k_1}$ and $\eta_i = (b_e + b_{mi})/(2\sqrt{k_1 m})$ (see Table 17.1).

$$\begin{cases} r_1 = 1, & \text{and} \quad \eta_1 = 0.0074, \quad \text{for} \quad z < z_d \ (i = 1) \\ r_2 = \rho = \sqrt{20} & \text{and} \quad \eta_2 = 0.45, \quad\ \ \text{for} \quad z \geq z_d \ (i = 2) \end{cases} \tag{17.7}$$

The excitation frequency range used in simulation is $f_e = \frac{\omega_e}{2\pi} = (90 - 110)$Hz for crossing the resonant area, which was found at $f_e = f_n = \frac{\omega_1}{2\pi} = 94.8$Hz [14] ($f_n$—natural frequency), (see Figs. 17.2).

The other parameters used in simulations are listed in table 17.1.

Table 17.1 System parameters

Symbol and value	Description
$m = 0.0038\,\text{kg}$	The effective mass of the magnet
$k_1 = 1348\,\text{N/m}$	The stiffness of the upper spring 1
$k_2 = 26960\,\text{N/m}$	The stiffness of the lower spring 2
$b_{m1} = 0.0175\,\text{Ns/m}$	The mechanical damping coefficient of the upper damper 1
$b_{m2} = 2.0208\,\text{Ns/m}$	The mechanical damping coefficient of the lower damper 2
$B = 0.57\,\text{T}$	The magnetic induction
$\ell = 0.44\,\text{m}$	The effective length of an electric coil
$R = 2.7\Omega$	The load resistance
$R_c = 1.2\Omega$	The internal resistance of an electric coil
$b_e = \frac{(B\ell)^2}{R+Rc}$	The electric damping coefficient
$\eta_i = \frac{b_e+b_{mi}}{2\sqrt{k_1 m}}$	The dimensionless damping coefficient of the system

17.3 The Results of Simulations

The main feature of our nonlinear systems is the appearance of two solutions. In Fig. 17.2a we show the resonance curve of the voltage output U versus excitation frequency. Note that the black one shows the results for the system without a stopper impacts at an enough large gap distance z_d. After shifting the gap to an appropriate smaller value, the stopper hits and the situation changes drastically. First of all the resonance region amplitude is limited to some value, but on the right-hand side of the black curve we observe a substantial increase of the voltage output due to continuation of the impacting solutions (red curve) with increasing the excitation frequency. Simultaneously the second non-impacting solution exists (blue curve and points) in the same region of frequency competing with the impacting one. This solution coincides with the black curve solution without a stopper. In Fig. 17.2b we show additionally a stroboscopic bifurcation diagram versus excitation frequency. It is possible to see that the impacting solution disappears entirely at the frequency f_e at about 106 Hz. Finally in Fig. 17.3a, b we show the corresponding time series and phase portraits for impacting and non-impacting solutions for chosen frequency at $f_e = 100$Hz. It confirms that for different initial conditions the mechanical resonator vibrates at different amplitudes and velocities, respectively, and so it leads to the larger or smaller voltage output. For distinguishing the different behaviour of the system in the case presented in Fig. 17.3, the dimensionless initial conditions $(\mathcal{Z}(\tau = 0), \dot{\mathcal{Z}}(\tau = 0)) = (z_0, \dot{z}_0)$ were chosen in accordance with Fig. 17.4f as $z_0 = 0, \dot{z}_0 = 0$ (no impacts) and $z_0 = 1, \dot{z}_0 = 0$ (with impacts).

Having two competing solutions a new question arises. What are the basins of attraction of corresponding solutions and how they evolve with increasing frequency? The answer to this question is the focus the next part of our discussion.

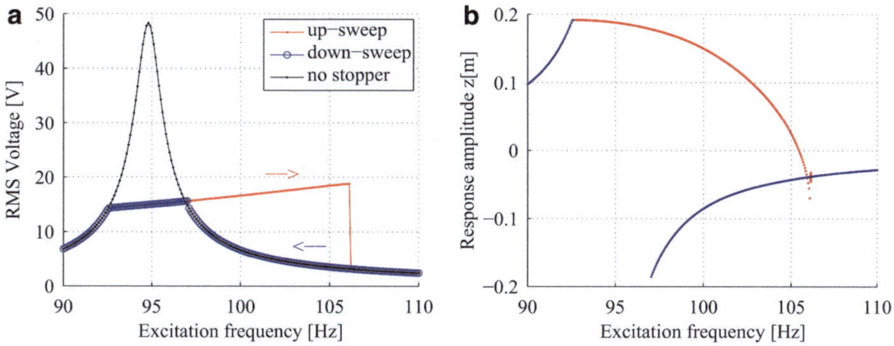

Fig. 17.2 RMS of voltage output versus frequency (**a**), *blue* colour denotes the solution with impacts for $f_e = (92.6 - 97.1)$Hz (swept by quasi-static decreasing of frequency) while *red* one corresponds to the impacting solution for $f_e = (92.6 - 106.1)$Hz (swept with increasing frequency), additionally *black* curve illustrates the solution without stopper. Bifurcation diagram (**b**)

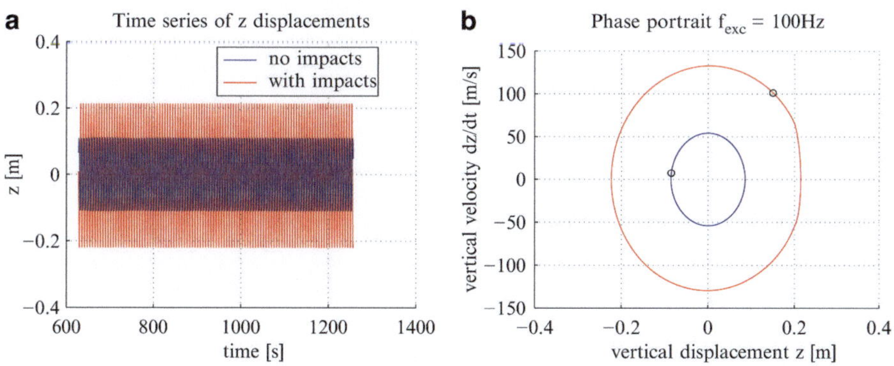

Fig. 17.3 Time series (**a**) and corresponding phase portraits with Poincare points (**b**) for two solutions. *Blue* colour denotes the solution without impacts while *red* one corresponds to the impacting solution

For better clarity, simulations were done (Fig. 17.4) for increasing frequency. Obviously the impacting solution basin (red colour) is fairly reduced by increasing excitation frequency and about $f_e = 106$ Hz it almost disappears. To follow the quantitative changes of the basin size, we defined the ratio between the area of the basin of the impacting solution and the area on the considered rectangular window of the phase space. These results are plotted in Fig. 17.5. Note, Figs. 17.4 and 17.5 show erosion in the basin of attraction with increasing frequency. One can clearly observe in Fig. 17.5 the erosion increasing from the impacting solutions at f_e about 97 Hz (red background in Fig. 17.4a) to the solutions nearly without impacting (white background in Fig. 17.4i).

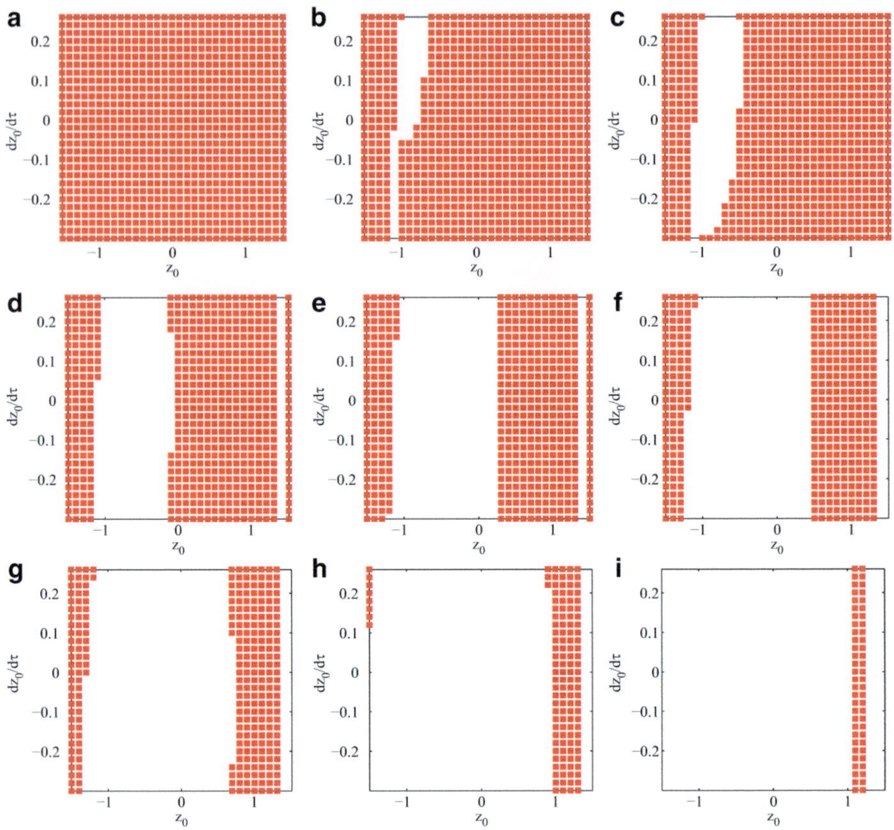

Fig. 17.4 Basins of attraction for impacting solution for increasing frequency: (**a**) $f_e = 97\,\text{Hz}$, (**b**) $f_e = 97.2\,\text{Hz}$, (**c**) $f_e = 97.4\,\text{Hz}$, (**d**) $f_e = 98\,\text{Hz}$, (**e**) $f_e = 99\,\text{Hz}$, (**f**) $f_e = 100\,\text{Hz}$, (**g**) $f_e = 102\,\text{Hz}$, (**h**) $f_e = 105\,\text{Hz}$, (**i**) $f_e = 106\,\text{Hz}$. Note that in this figure, the dimensionless variables were used. Following Eq. 15.5 $(\mathcal{Z}(\tau = 0), \dot{\mathcal{Z}}(\tau = 0)) = (z_0, \dot{z}_0)$

Fig. 17.5 Fraction of the impacting solution (basin of attraction) versus frequency

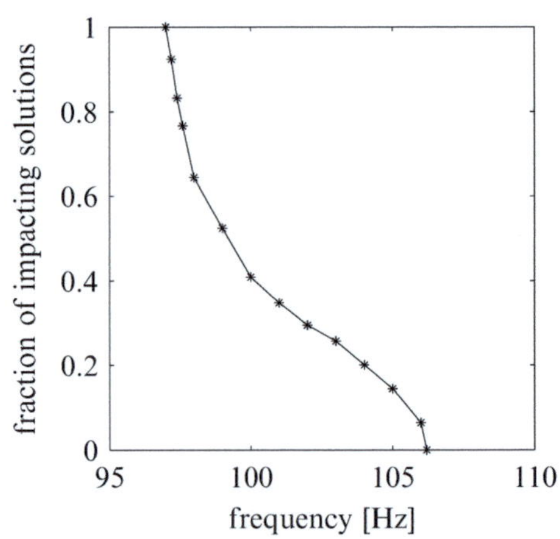

17.4 Conclusions

In summary we note that the nonlinear characteristics of the mechanical resonator with impacts provide a much broader frequency range for the power (RMS voltage in Fig. 17.2a). Two existing solutions (Fig. 17.2: with and without impacts) are characterized by different resonator amplitudes. The results show that the basin of attraction for the impacting solution erodes strongly with the increasing frequency (Figs. 17.4 and 17.5). The influence of initial conditions on output energy is significant within the broaden band resonance curve, effecting multi-solution phenomenon.

A possible development of the proposed analysis consists in applying more detailed dynamical integrity arguments [7] to the basins of attraction reported in Fig. 17.4. This will allow us to better detect the robustness of the two competing solutions with respect to changes in initial conditions, and thus will permit to judge on the reliability of the proposed system in harvesting energy.

Acknowledgements The authors gratefully acknowledge the support of the 7th Framework Programme FP7-REGPOT-2009-1, under Grant Agreement No. 245479. MB and GL were partially supported by the Polish National Science Center under the grant No. 2012/05/B/ST8/00080.

References

1. Beeby SP, Tudor MJ, White NM (2006) Measurem Sci Technol 17:R175–R195
2. Blystad L-CJ, Halvorsen E (2011) Microsyst Technol 17:505–511
3. Erturk A, Inman D (2011) Piezoelectric energy harvesting. Wiley, Chichester
4. Gu L, Livermore C (2011) J Smart Mater Struct 20:045004
5. Le CP, Halvorsen E, Sorasen O, Yeatman EM (2012) J Intellig Math Syst Struct 23:1409
6. Le CP, Halvorsen E (2012) J Micromech Microeng 22:074006
7. Lenci S, Rega G (2011) Physica D 240:814–824
8. Mann BP, Barton DAW, Owens BAM (2012) J Intellig Math Syst Struct 23:1451
9. Mitcheson PD, Yeatman EM, Rao GK, Holmes AS, Green TC (2008) Proc. IEEE 96: 1457–1486
10. Owens BAM, Mann BP (2012) J Sound Vibr 331:922
11. Pellegrini SP, Tolu N, Schenk M, Herder JL (2012) J Intellig Math Syst Struct doi: 10.1177/1045389X12444940
12. Spreemann D, Manoli Y (2012) Electromagnetic vibration energy harvesting devices. Springer, Berlin
13. Soliman MSM, Abdel-Rahman EM, El-Saadany EF, Mansour RR (2008) J Micromech Microeng 18:115021 11
14. Soliman MSM, Abdel-Rahman EM, El-Saadany EF, Mansour RR (2009) J Mircoelectromech Syst 18:1288–1299

Chapter 18
Analytical Dynamics of a Mass–Damper–Spring Constrained System

Albert C. J. Luo and Richard George

Abstract This chapter discusses the dynamics of a mass–damper–spring system with two rigid constraints and impact interactions. Impacting chatter and stuck phenomena are investigated for the mass with constraints and the corresponding conditions for such phenomena are determined. Analytical predictions are presented for the system to give a more precise and complete demonstration of the phenomena in the system. Finally, an analytical parameter map is given to show how the system changes for varying parameters. From these conditions, numerical simulations are performed to demonstrate these phenomena in the system.

Keywords Discontinuous dynamical systems • Constrained mass-damper-Spring system • G-Functions • Grazing motion • Chatter motion • Stuck motion

18.1 Equations of Motion

A sinusoidal forcing function is applied to a mass–damper–spring system, as shown in Fig. 18.1. The system is further constrained, both above and below, by two rigid walls, which the mass will interact with when the displacement of the mass from equilibrium is equal to the position of either wall. There are different types of interactions that can occur between the mass and either wall, which are discussed in this chapter. The possible interactions include impact, chatter, and stuck motions. Impact occurs when the mass comes to the wall with some velocity, hits the wall, and then leaves the wall. When multiple impacts occur with one wall, impacting chatter occurs. This phenomenon is explained as follows. The impacts continue to cause the mass to bounce back from the wall and the system is still trying to continue through

A.C.J. Luo (✉) • R. George
Southern Illinois University Edwardsville, Edwardsville, IL 62026-1805, USA
e-mail: aluo@siue.edu; rgeorge@siue.edu

J.A.T. Machado et al. (eds.), *Discontinuity and Complexity in Nonlinear Physical Systems*, Nonlinear Systems and Complexity 6, DOI 10.1007/978-3-319-01411-1_18,

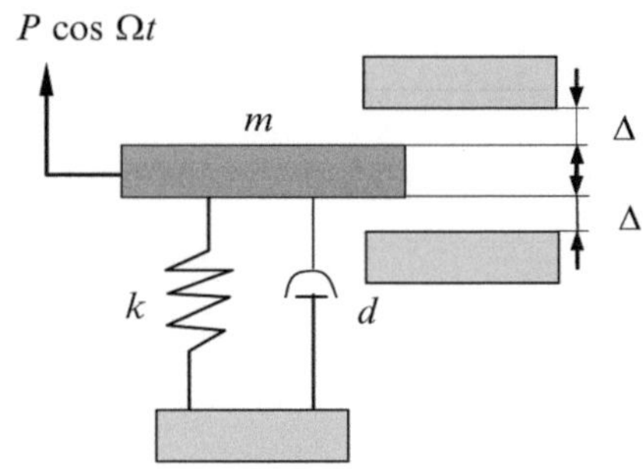

Fig. 18.1 Mechanical model

the wall, when the mass comes to the wall with zero velocity and continues to try and force its way into the wall, the mass will appear to stuck to the wall. There are two states of motion that can exist in this system and they are both defined here with their respective equations of motion. The first is free flight motion, which exists when the mass is moving in between the two walls. The second is stuck motion, which occurs when the mass appears to be stuck to one of the walls. Free flight motion includes the impact and chattering motions with the wall. The previous studies on this topic can be referred to [1–12]. However, in this chapter, the analytical condition will be developed for a better understanding of complex motions.

The system's motion is described by the ordinary differential equation for the forced vibration of a mass–spring–damper system

$$m\ddot{x} + d\dot{x} + kx = P\cos\Omega t \tag{18.1}$$

A coefficient of restitution impact model is used in this system. The coefficient of restitution, e, relates the velocity of the mass before and after impact

$$\dot{x}_{m+} = -e\dot{x}_{m-}. \tag{18.2}$$

Stuck motion occurs during the intervals where the mass appears to be stuck to the wall. This motion begins with stuck initiation, which is when the mass comes to one of the walls with zero velocity and continues to try and force its way through the wall. The motion continues until stuck vanishing, where the system leaves the wall and enters back into free flight motion. Stuck motion is defined as the point where the mass is either at the top or bottom wall and its velocity is equal to zero

$$\begin{aligned} &x = \pm\Delta \text{ and } \dot{x} = 0 \\ &F = P\cos\Omega t - k\Delta \geq 0 \text{ at } x = +\Delta \\ &F = P\cos\Omega t + k\Delta \leq 0 \text{ at } x = -\Delta \end{aligned} \tag{18.3}$$

18.2 Domains and Boundaries of Motion

Before determining how the system will interact with the wall when it comes to it, and what conditions need to be satisfied to determine the state of motion the mass is in, the domains and boundaries of motion must be defined. The domains and boundaries are all defined on the phase plane in Fig. 18.2. For this system, there are three domains of motion. The first is the free flight domain, defined as Ω_1, which is when the position of the mass is between the two walls, with any velocity. The remaining two are stuck domains, defined as ${\Omega_0}^{(\pm)}$, where the mass is "stuck" to either the top or bottom wall. The domains are defined as,

$$\begin{aligned} \Omega_1 &= \left\{(x,\dot{x})\,\middle|\, x\in(-\Delta,\Delta),\dot{x}\in(-\infty,\infty)\right\} \\ \Omega_0^{(\pm)} &= \left\{(x,\dot{x})\,\middle|\, x=\pm\Delta,\dot{x}=0\right\} \end{aligned} \tag{18.4}$$

There are four boundaries in this system, two are impact boundaries, defined as $\partial\Omega_{1\infty}^{(\pm)}$, and two are stuck boundaries, defined as $\partial\Omega_{10}^{(\pm)}$. There is one impact boundary at each wall, both defined as when the mass comes to the wall with a nonzero velocity. There is also one stuck boundary at the each wall, but these are defined at the wall where the velocity of the mass is equal to zero. The boundaries are defined as,

$$\begin{aligned} \partial\Omega_{1\infty}^{(\pm)} &= \left\{(x,\dot{x})\,\middle|\, x=\pm\Delta,\dot{x}\neq 0\right\} \\ \partial\Omega_{10}^{(\pm)} &= \left\{(x,\dot{x})\,\middle|\, x=\pm\Delta,\dot{x}=0\right\} \end{aligned} \tag{18.5}$$

The free flight domain is represented by the subscript 1 and the stuck domains by 0. The top and bottom walls are represented by + and –, respectfully. The impact boundary is represented by ∞, which is used to show that the boundary cannot be passed through. The stuck domains and stuck boundaries are both given as the same

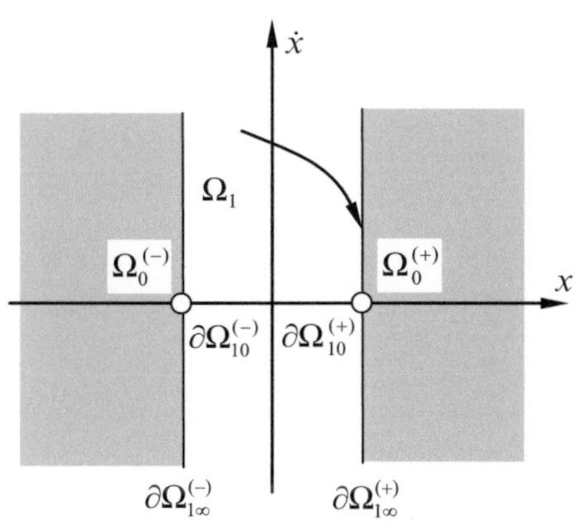

Fig. 18.2 Domains and boundaries

two points for the top and bottom walls in phase plane. For the stuck boundary, the mass is still not able to move through the wall, but can still apply a force into the wall while remaining at this point.

18.3 G-Functions

When the system is in one of the domains, the motion is determined by the respective equation of motion previously presented. In order to determine the necessary conditions for motion switchability at the boundaries, the G-Function will be introduced. To begin, introduce vectors as,

$$\mathbf{x} = (x, \dot{x})^{\mathrm{T}} = (x, y)^{\mathrm{T}} \text{ and } \mathbf{F} = (y, F)^{\mathrm{T}} \tag{18.6}$$

From the definitions of the domains, the equation of motion can be expressed in the vector form of,

$$\dot{\mathbf{x}}^{(i)} = \mathbf{F}^{(i)}\left(\mathbf{x}^{(i)}, t, \mathbf{p}^{(i)}\right), (i = 1, 2) \tag{18.7}$$

Where $\mathbf{p}$ are system parameters and,

$$\begin{aligned} F^{(1)} &= \tfrac{1}{m}\left(P\cos\Omega t - dy^{(1)} - kx^{(1)}\right); \\ F^{(0)} &= \tfrac{1}{m}\left(P\cos\Omega t - k\Delta\right) \text{ at } x^{(0)} = +\Delta \text{ and } y^{(0)} = 0, \\ F^{(0)} &= \tfrac{1}{m}\left(P\cos\Omega t + k\Delta\right) \text{ at } x^{(0)} = -\Delta \text{ and } y^{(0)} = 0. \end{aligned} \tag{18.8}$$

The lower and higher order G-Functions can then be introduced as,

$$\begin{aligned} G^{(0,\alpha)}_{\partial\Omega_{\alpha\beta}}(\mathbf{x}, t_{m\pm}) &= \mathbf{n}^{\mathrm{T}}_{\partial\Omega_{\alpha\beta}} \cdot \mathbf{F}^{(\alpha)}(\mathbf{x}, t_{m\pm}), \\ G^{(1,\alpha)}_{\partial\Omega_{\alpha\beta}}(\mathbf{x}, t_{m\pm}) &= \mathbf{n}^{\mathrm{T}}_{\partial\Omega_{\alpha\beta}} \cdot D\mathbf{F}^{(\alpha)}(\mathbf{x}, t_{m\pm}), \end{aligned} \tag{18.9}$$

where $D\mathbf{F}$ is the total derivative of $\mathbf{F}$,

$$D\mathbf{F} = \frac{\partial \mathbf{F}}{\partial \mathbf{x}}\dot{\mathbf{x}} + \frac{\partial \mathbf{F}}{\partial t} \tag{18.10}$$

and $\mathbf{n}_{\partial\Omega_{\alpha\beta}}$ represents the vector normal to the boundary in phase plane, given by

$$\mathbf{n}_{\partial\Omega_{\alpha\beta}} = \nabla\varphi_{\alpha\beta} = \left(\frac{\partial\varphi_{\alpha\beta}}{\partial x}, \frac{\partial\varphi_{\alpha\beta}}{\partial y}\right)^{\mathrm{T}}. \tag{18.11}$$

For the impact boundaries, $\partial\Omega^{(\pm)}_{1\infty}$, and the stuck boundaries, $\partial\Omega^{(\pm)}_{10}$, the corresponding normal vectors are,

$$\mathbf{n}_{\partial\Omega_{1\infty}^{(\pm)}} = (1,0)^{\mathrm{T}} \text{ and } \mathbf{n}_{\partial\Omega_{10}^{(\pm)}} = (0,1)^{\mathrm{T}} \tag{18.12}$$

The corresponding G-Functions at the impact boundaries, $\partial\Omega_{1\infty}^{(\pm)}$, can then be calculated as,

$$\begin{aligned} G_{1\infty}^{(0,1)}(\mathbf{x}, t_{m\pm}) &= y^{(1)} \\ G_{1\infty}^{(1,1)}(\mathbf{x}, t_{m\pm}) &= F^{(1)}\left(\mathbf{x}^{(1)}, t_{m\pm}\right). \end{aligned} \tag{18.13}$$

The corresponding G-Functions at the stuck boundaries, $\partial\Omega_{10}^{(\pm)}$, can similarly be calculated as,

$$\begin{aligned} G_{\partial\Omega_{10}^{(\pm)}}^{(0,\alpha)}(\mathbf{x}, t_{m\pm}) &= F^{(\alpha)}\left(\mathbf{x}^{(\alpha)}, t_{m\pm}\right), \\ G_{\partial\Omega_{10}^{(\pm)}}^{(1,\alpha)}(\mathbf{x}, t_{m\pm}) &= DF^{(\alpha)}\left(\mathbf{x}^{(\alpha)}, t_{m\pm}\right); \end{aligned} \tag{18.14}$$

Where

$$\begin{aligned} DF^{(1)} &= \tfrac{1}{m}\left(-P\Omega\sin\Omega t - F^{(1)}d - ky^{(1)}\right); \\ DF^{(0)} &= -\tfrac{1}{m}P\Omega\sin\Omega t \ \text{ at } x^{(0)} = +\Delta \text{ and } y^{(0)} = 0, \\ DF^{(0)} &= -\tfrac{1}{m}P\Omega\sin\Omega t \ \text{ at } x^{(0)} = -\Delta \text{ and } y^{(0)} = 0. \end{aligned} \tag{18.15}$$

18.4 Analytical Conditions

Using the G-Function, the conditions necessary to determine the motion switchabilities can be defined. In other words, the G-Function is used to determine what type of interaction when the system comes to a boundary and how the system is able to move relative to the boundary.

From Luo [10], the conditions for impact to occur in the system are given by the lower order G-Function

$$\begin{aligned} &G_{\partial\Omega_{1\infty}^{(+)}}^{(0,1)}(\mathbf{x}_m, t_{m-}) > 0 \text{ and } G_{\partial\Omega_{1\infty}^{(+)}}^{(0,1)}(\mathbf{x}_m, t_{m+}) < 0 \text{ on } \partial\Omega_{1\infty}^{(+)} \\ &G_{\partial\Omega_{1\infty}^{(-)}}^{(0,1)}(\mathbf{x}_m, t_{m-}) < 0 \text{ and } G_{\partial\Omega_{1\infty}^{(-)}}^{(0,1)}(\mathbf{x}_m, t_{m+}) > 0 \text{ on } \partial\Omega_{1\infty}^{(-)} \end{aligned} \tag{18.16}$$

In other words, based on the calculated G-Functions,

$$\begin{aligned} &y_{m-} > 0 \text{ and } y_{m+} < 0 \text{ on } \partial\Omega_{1\infty}^{(+)} \\ &y_{m-} < 0 \text{ and } y_{m+} > 0 \text{ on } \partial\Omega_{1\infty}^{(-)} \end{aligned} \tag{18.17}$$

For impact to occur at the top wall, the mass will come to the wall with a positive velocity. Using the coefficient of restitution impact model given in Eq. (18.2), the

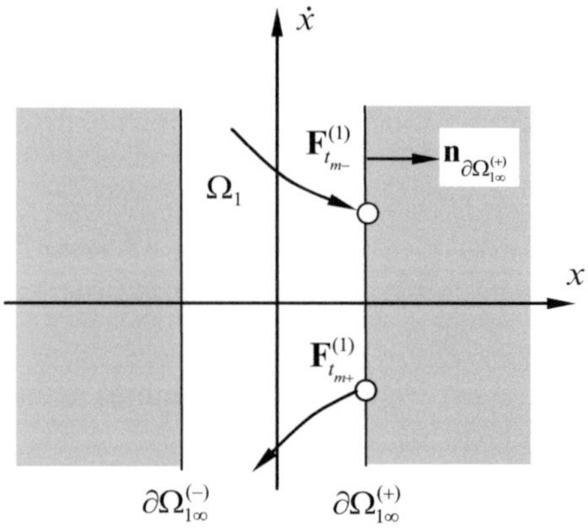

Fig. 18.3 Impact with the top wall

mass will instantaneously bounce back from the top wall with a negative velocity, completing an impact interaction with the top wall, as shown in Fig. 18.3. Similarly for the bottom wall, the mass will come to the wall with a negative velocity and bounce back into the free flight domain with a positive velocity.

Stuck motion is initiated in the system when the mass comes to one of the walls and the system stays at that position with zero velocity. The conditions for stuck onset are given by the lower order G-Function at the stuck boundary.

$$\begin{aligned}
&G^{(0,1)}_{\partial\Omega^{(+)}_{10}}\left(\mathbf{x}^{(1)}_m, t_{m-}\right) > 0 \text{ and } G^{(0,0)}_{\partial\Omega^{(+)}_{10}}\left(\mathbf{x}^{(0)}_m, t_{m-}\right) > 0\\
&\text{at } x_m = \Delta \text{ and } y_m = 0 \text{ for } \Omega_1 \to \Omega^{(+)}_0;\\
&G^{(0,1)}_{\partial\Omega^{(-)}_{10}}\left(\mathbf{x}^{(1)}_m, t_{m-}\right) < 0 \text{ and } G^{(0,0)}_{\partial\Omega^{(-)}_{10}}\left(\mathbf{x}^{(0)}_m, t_{m-}\right) < 0\\
&\text{at } x_m = -\Delta \text{ and } y_m = 0 \text{ for } \Omega_1 \to \Omega^{(-)}_0.
\end{aligned} \tag{18.18}$$

From the previous G-Function, these can also be represented as,

$$\begin{aligned}
&F^{(1)}\left(\mathbf{x}^{(1)}_m, t_{m-}\right) > 0 \text{ and } F^{(0)}\left(\mathbf{x}^{(0)}_m, t_{m+}\right) > 0\\
&\text{at } \partial\Omega^{(+)}_{10} \text{ with } x_m = \Delta \text{ and } y_m = 0\\
&\text{for } \Omega_1 \to \Omega^{(+)}_0;\\
&F^{(1)}\left(\mathbf{x}^{(1)}_m, t_{m-}\right) < 0 \text{ and } F^{(0)}\left(\mathbf{x}^{(0)}_m, t_{m+}\right) < 0\\
&\text{at } \partial\Omega^{(+)}_{10} \text{ with } x_m = -\Delta \text{ and } y_m = 0\\
&\text{for } \Omega_1 \to \Omega^{(-)}_0.
\end{aligned} \tag{18.19}$$

The system becomes stuck when it remains at the wall with a velocity of zero. It is unable to pass through the rigid wall, but continues to force its way into the wall, as shown in Fig. 18.4 on the higher order phase plane. For the duration of the stuck motion, there is an equal and opposite reaction force of the wall supporting the mass that keeps the system at equilibrium for that time, as shown in Fig. 18.5. Since the wall is considered to be rigid, this force will always oppose the mass, preventing motion into the wall.

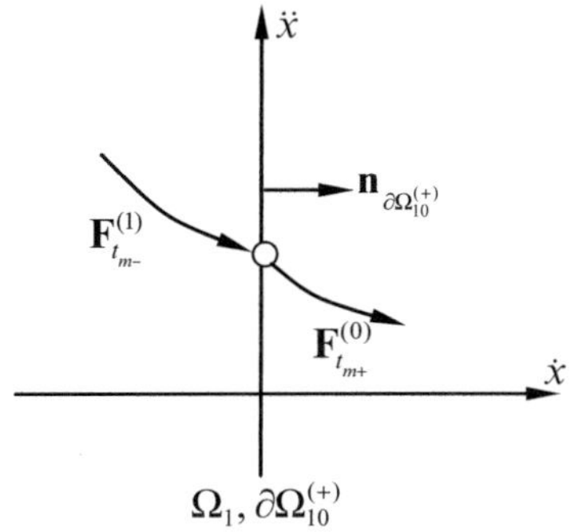

Fig. 18.4 Stuck conditions for the top wall

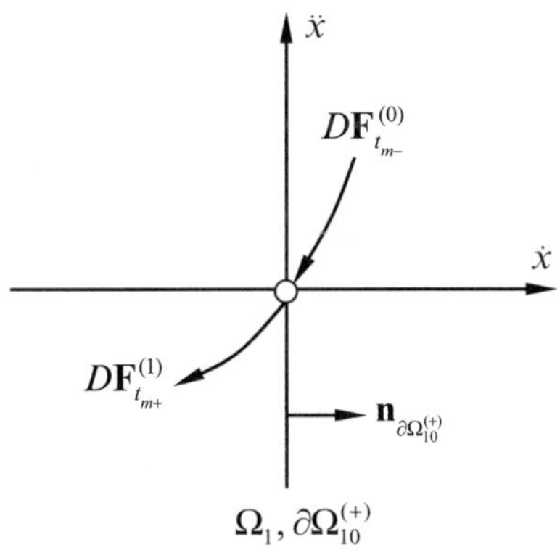

Fig. 18.5 Stuck vanishing conditions for the top wall

Once stuck initiation occurs in the system, the G-Functions need to be used to determine when stuck vanishing will occur. The higher order G-Function for the stuck boundary is used to determine when the system will move back into the free flight domain. After some amount of time the system will return to the stuck boundary where the lower order G-Function will be equal to zero. At this point, the stuck motion will vanish if the conditions given by the higher order G-Function are met. If these conditions are not satisfied, the system will remain in the stuck domain. The conditions for stuck vanishing are,

$$\left.\begin{aligned}
&G^{(0,0)}_{\partial\Omega_{10}^{(+)}}\left(\mathbf{x}_m^{(0)}, t_{m-}\right) = 0 \text{ and } G^{(1,0)}_{\partial\Omega_{10}^{(+)}}\left(\mathbf{x}_m^{(0)}, t_{m-}\right) < 0,\\
&G^{(1,1)}_{\delta\Omega_{10}^{(+)}}\left(\mathbf{x}_m^{(1)}, t_{m+}\right) < 0
\end{aligned}\right\}$$
$$\text{at } x_m = \Delta \text{ and } y_m = 0 \text{ for } \Omega_0^{(+)} \to \Omega_1;$$
$$\left.\begin{aligned}
&G^{(0,0)}_{\partial\Omega_{10}^{(-)}}\left(\mathbf{x}_m^{(0)}, t_{m-}\right) = 0 \text{ and } G^{(1,0)}_{\partial\Omega_{10}^{(-)}}\left(\mathbf{x}_m^{(0)}, t_{m-}\right) > 0,\\
&G^{(1,1)}_{\partial\Omega_{10}^{(-)}}\left(\mathbf{x}_m^{(1)}, t_{m+}\right) > 0
\end{aligned}\right\}$$
$$\text{at } x_m = -\Delta \text{ and } y_m = 0 \text{ for } \Omega_0^{(-)} \to \Omega_1. \tag{18.20}$$

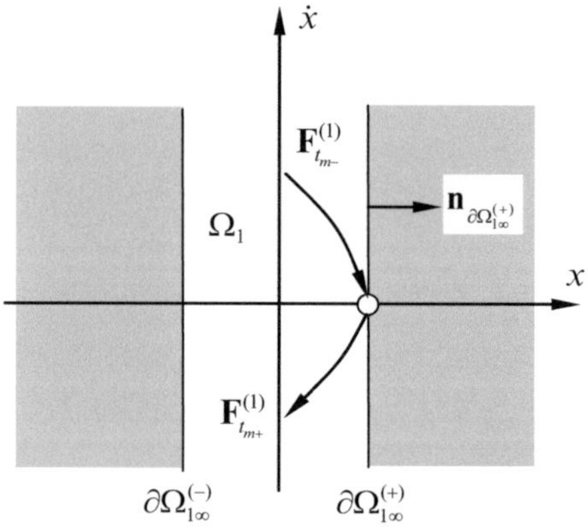

Fig. 18.6 Grazing conditions for the top wall

From the G-Functions, the conditions that must be satisfied for stuck vanishing can also be given as,

$$\left.\begin{array}{l} F^{(0)}(\mathbf{x}_m^{(0)}, t_{m-}) = 0 \text{ and } DF^{(0)}(\mathbf{x}_m^{(0)}, t_{m-}) < 0 \\ DF^{(1)}(\mathbf{x}_m^{(1)}, t_{m+}) < 0 \end{array}\right\} \\ \text{at } \partial\Omega_{10}^{(+)} \text{ with } x_m = \Delta \text{ and } y_m = 0 \text{ for } \Omega_0^{(+)} \to \Omega_1; \\ \left.\begin{array}{l} F^{(0)}\left(\mathbf{x}_m^{(0)}, t_{m-}\right) = 0 \text{ and } DF^{(0)}(\mathbf{x}_m^{(0)}, t_{m-}) > 0 \\ DF^{(1)}(\mathbf{x}_m^{(1)}, t_{m+}) > 0 \end{array}\right\} \\ \text{at } \partial\Omega_{10}^{(-)} \text{ with } x_m = -\Delta \text{ and } y_m = 0 \text{ for } \Omega_0^{(-)} \to \Omega_1. \tag{18.21}$$

The lower order G-Function is the same for both stuck vanishing and stuck initiation. Therefore, the first condition that must be satisfied is that the system stops trying force its way through the wall. The higher order G-Function is basically a jerk term. Once the zero-order term comes to zero, the direction of the first-order term will determine whether the system will accelerate away from the wall, back into the free flight domain, or return to attempting to force its way back into the wall and remain stuck.

The phenomenon of grazing occurs in the system when the mass comes to one of the walls from the free flight domain, just touches the wall with velocity of zero, then moves back into the free flight domain, as shown in Fig. 18.6. The conditions for grazing to occur at one of the walls are determined using both the lower and higher order G-Functions. The conditions are

$$G^{(0,1)}_{\partial\Omega^{(+)}_{1\infty}}(\mathbf{x}_m, t_{m\pm}) = 0 \text{ and } G^{(1,1)}_{\partial\Omega^{(+)}_{1\infty}}(\mathbf{x}_m, t_{m\pm}) < 0 \text{ on } \partial\Omega^{(+)}_{1\infty}, \\ G^{(0,1)}_{\partial\Omega^{(-)}_{1\infty}}(\mathbf{x}_m, t_{m\pm}) = 0 \text{ and } G^{(1,1)}_{\partial\Omega^{(-)}_{1\infty}}(\mathbf{x}_m, t_{m\pm}) > 0 \text{ on } \partial\Omega^{(-)}_{1\infty}. \tag{18.22}$$

or

$$
\begin{aligned}
&G^{(0,1)}_{\partial\Omega^{(+)}_{10}}(\mathbf{x}_m, t_{m\pm}) = 0 \text{ and } G^{(1,1)}_{\partial\Omega^{(+)}_{10}}(\mathbf{x}_m, t_{m\pm}) < 0 \text{ on } \partial\Omega^{(+)}_{10},\\
&G^{(0,1)}_{\partial\Omega^{(-)}_{10}}(\mathbf{x}_m, t_{m\pm}) = 0 \text{ and } G^{(1,1)}_{\partial\Omega^{(-)}_{10}}(\mathbf{x}_m, t_{m\pm}) > 0 \text{ on } \partial\Omega^{(-)}_{10}.
\end{aligned}
\tag{18.23}
$$

Calculating out these conditions,

$$
\begin{aligned}
&y^{(1)}_{m\pm} = 0 \text{ and } F^{(1)}\left(\mathbf{x}^{(1)}_m, t_{m\pm}\right) < 0 \text{ on } \partial\Omega^{(+)}_{1\infty},\\
&y^{(1)}_{m\pm} = 0 \text{ and } F^{(1)}\left(\mathbf{x}^{(1)}_m, t_{m\pm}\right) > 0 \text{ on } \partial\Omega^{(-)}_{1\infty}.
\end{aligned}
\tag{18.24}
$$

or

$$
\begin{aligned}
&F^{(1)}\left(\mathbf{x}^{(1)}_m, t_{m\pm}\right) = 0 \text{ and } DF^{(1)}\left(\mathbf{x}^{(1)}_m, t_{m\pm}\right) < 0 \text{ on } \partial\Omega^{(+)}_{10},\\
&F^{(1)}\left(\mathbf{x}^{(1)}_m, t_{m\pm}\right) = 0 \text{ and } DF^{(1)}\left(\mathbf{x}^{(1)}_m, t_{m\pm}\right) > 0 \text{ on } \partial\Omega^{(-)}_{10}.
\end{aligned}
\tag{18.25}
$$

The first set is for grazing against the impact boundary. The zero-order G-Function is the velocity of the mass in the free flight domain. The first order G-Function is essentially the acceleration of the mass as calculated from the equation of motion. When the system comes to one of the boundaries with a velocity of zero and is accelerating back towards the free flight domain, the mass will graze the wall. The second set is for grazing against the stuck boundary. The zero order G-Function is basically an acceleration term and the first-order term is essentially a jerk term. For this case, the velocity of the mass at the boundary will still be zero, but if the acceleration term is also zero, then the jerk term is needed to determine where the system will go. If the mass comes to the wall with zero velocity and zero acceleration, and the jerk is directed back towards the free flight domain, the mass will graze the wall. Higher order G-Functions would need to be calculated and considered as much lower order G-Functions are equal to zero at the boundary. These can be found by taking more total derivatives of the G-Functions already presented.

18.5 Generic Mappings

In order for periodic and chaotic motions to be described and determined, switching sets should be introduced. From the switching boundary, the six switching sets for this system are,

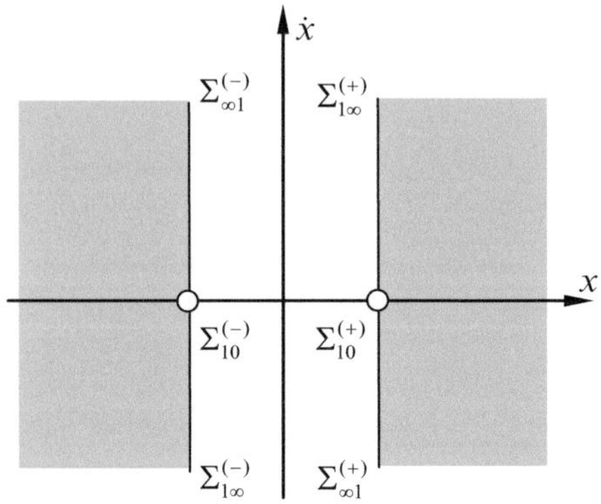

Fig. 18.7 Switching sets

$$
\begin{aligned}
\Sigma_{1\infty}^{(+)} &= \left\{(t_k, \dot{x}_k) \middle| x_k = \Delta, \dot{x}_k > 0\right\},\\
\Sigma_{\infty 1}^{(+)} &= \left\{(t_k, \dot{x}_k) \middle| x_k = \Delta, \dot{x}_k < 0\right\},\\
\Sigma_{1\infty}^{(-)} &= \left\{(t_k, \dot{x}_k) \middle| x_k = -\Delta, \dot{x}_k < 0\right\},\\
\Sigma_{\infty 1}^{(-)} &= \left\{(t_k, \dot{x}_k) \middle| x_k = -\Delta, \dot{x}_k > 0\right\},\\
\Sigma_{10}^{(+)} &= \left\{(t_k, \dot{x}_k) \middle| x_k = \Delta, \dot{x}_k = 0, F^{(0)} > 0\right\},\\
\Sigma_{01}^{(+)} &= \left\{(t_k, \dot{x}_k) \middle| x_k = \Delta, \dot{x}_k = 0, F^{(0)} = 0\right\},\\
\Sigma_{10}^{(-)} &= \left\{(t_k, \dot{x}_k) \middle| x_k = -\Delta, \dot{x}_k = 0, F^{(0)} < 0\right\},\\
\Sigma_{01}^{(-)} &= \left\{(t_k, \dot{x}_k) \middle| x_k = -\Delta, \dot{x}_k = 0, F^{(0)} = 0\right\}.
\end{aligned}
\tag{18.26}
$$

The switching sets defined in Eq. (18.26) are used for the purpose of mapping. ${\Sigma_{10}}^{(+)}$ and ${\Sigma_{01}}^{(+)}$ are the same switching plane with different conditions for stuck motion initiation and vanishing at the boundary $\partial{\Omega_{10}}^{(+)}$. Similarly, ${\Sigma_{10}}^{(-)}$ and ${\Sigma_{01}}^{(-)}$ are the same switching plane at the boundary $\partial{\Omega_{10}}^{(-)}$. ${\Sigma_{1\infty}}^{(\pm)}$ and ${\Sigma_{\infty 1}}^{(\pm)}$ are the same switching planes at the boundaries $\partial\,{\Omega_{1\infty}}^{(\pm)}$. The switching sets are shown in Fig. 18.7.

Generic mappings can be defined using these switching sets in order to describe the free flight and stuck motions that exist in the system. In all, there are six generic mappings in this system. Two of these mappings are stuck mappings and the remaining four mappings pass through the free flight domain. The basic six mappings are defined as,

$$
\begin{aligned}
&P_1 : \Sigma_{\infty 1}^{(-)} \to \Sigma_{1\infty}^{(+)}, \quad P_2 : \Sigma_{\infty 1}^{(+)} \to \Sigma_{1\infty}^{(+)}\\
&P_3 : \Sigma_{\infty 1}^{(+)} \to \Sigma_{1\infty}^{(-)}, \quad P_4 : \Sigma_{\infty 1}^{(-)} \to \Sigma_{1\infty}^{(-)}\\
&P_5 : \Sigma_{10}^{(+)} \to \Sigma_{01}^{(+)}, \quad P_6 : \Sigma_{10}^{(-)} \to \Sigma_{01}^{(-)}.
\end{aligned}
\tag{18.27}
$$

Fig. 18.8 Generic mappings

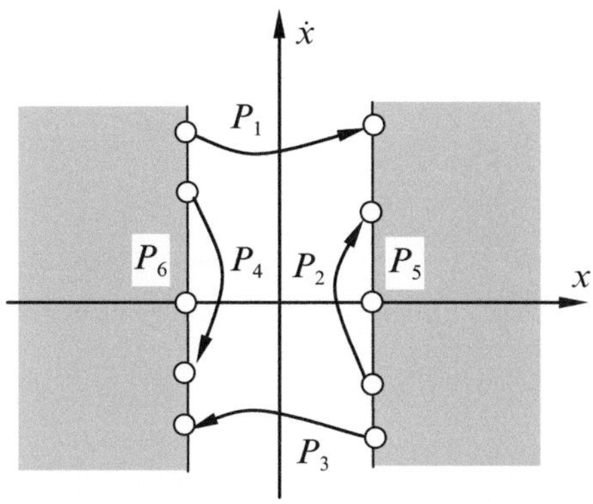

Mappings P_1 and P_3 are global mappings that map the system from one switching plane to another switching plane. The initial impact with one wall after interacting with the other wall is represented by these mappings. Mappings P_2 and P_4 are local mappings, which map from a switching plane back to itself. These will be used to describe the impacting chatter phenomena. Based on the defined switching sets, the generic mappings can be expanded to show all possibilities of mappings

$$\begin{array}{lll} P_1 : \Sigma_{\infty 1}^{(-)} \rightarrow \Sigma_{10}^{(+)}, & P_1 : \Sigma_{01}^{(-)} \rightarrow \Sigma_{1\infty}^{(+)}, & P_1 : \Sigma_{\infty 1}^{(-)} \rightarrow \Sigma_{1\infty}^{(+)}; \\ P_2 : \Sigma_{\infty 1}^{(+)} \rightarrow \Sigma_{10}^{(+)}, & P_2 : \Sigma_{01}^{(+)} \rightarrow \Sigma_{1\infty}^{(+)}, & P_2 : \Sigma_{\infty 1}^{(+)} \rightarrow \Sigma_{1\infty}^{(+)}; \\ P_3 : \Sigma_{\infty 1}^{(+)} \rightarrow \Sigma_{10}^{(-)}, & P_3 : \Sigma_{01}^{(+)} \rightarrow \Sigma_{1\infty}^{(-)}, & P_3 : \Sigma_{\infty 1}^{(+)} \rightarrow \Sigma_{1\infty}^{(-)}; \\ P_4 : \Sigma_{\infty 1}^{(-)} \rightarrow \Sigma_{10}^{(-)}, & P_4 : \Sigma_{01}^{(-)} \rightarrow \Sigma_{1\infty}^{(-)}, & P_4 : \Sigma_{\infty 1}^{(-)} \rightarrow \Sigma_{1\infty}^{(-)}. \end{array} \quad (18.28)$$

Mappings P_5 and P_6 are local at stuck boundaries and are used to describe the stuck motions. These stuck mappings map the system from stuck initiation to stuck vanishing. Grazing is a singular point that can be shown in different circumstances with any of the mapping structures that move through the free flight domain. All of the generic mapping structures are shown in Fig. 18.8.

18.6 Analytical Predictions and Stability

Based on the equations and switchability conditions developed thus far, a rough view of the system can be generated in the form of numerical simulations. In order to get an initial idea of how the system changes for as a parameter changes, a numerical simulation of a bifurcation scenario is given in Fig. 18.9. The scenario is based on the parameters,

$$m = 5, \ \ d = 3, \ \ k = 10, \ \ P = 15, \ \ \Delta = .01, \ \ e = .3 \quad (18.29)$$

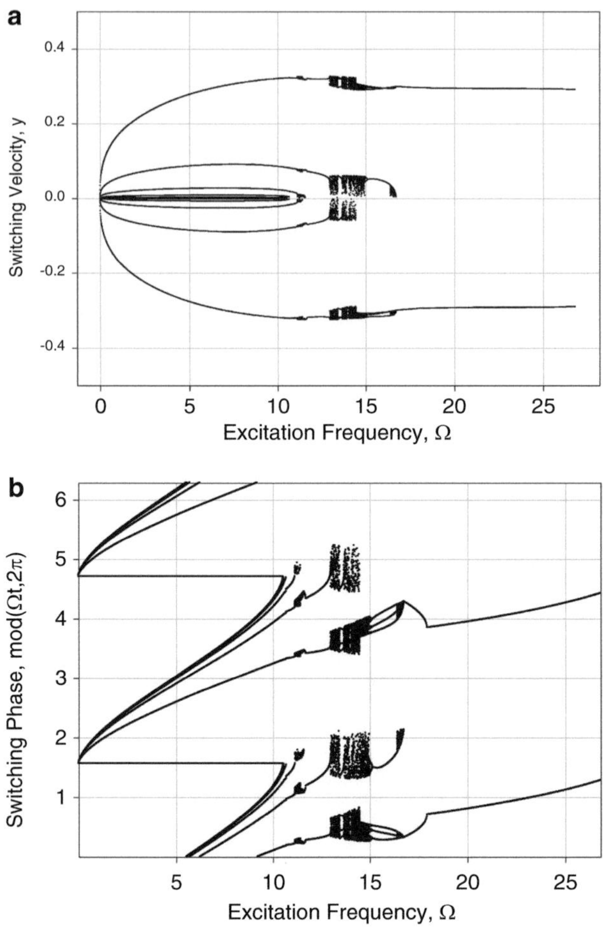

Fig. 18.9 A bifurcation scenario. (**a**) Switching velocity, (**b**) switching phase

While these numerical simulations do offer some insight to the system, they also suffer from numerical error and fail to show the stability of the system. In order to improve on these simulations, the system must be predicted analytically based on the equations of motion. Analytical predictions of the periodic motions, using the mapping structures, improve on the results of the numerical simulations and allow the stabilities of the motions to be directly calculated. Using the switching sets and mapping structures, any given periodic motion can be predicted analytically. A single mapping structure is a collection of functions of velocity and time. Displacements from the switching set are fixed because the boundaries are rigid walls, but velocity and time will change with the parameters. A mapping can be written mathematically as a collection of two functions of the initial and final switching set points

$$P_j : \begin{cases} f_1\left(y_{k+1}, y_k, t_{k+1}, t_k\right) = 0 \\ f_2\left(y_{k+1}, y_k, t_{k+1}, t_k\right) = 0 \end{cases} \tag{18.30}$$

where the index j is used to represent any of generic mappings already discussed. In order to simplify the analysis, introduce the vectors,

$$\mathbf{f} = (f_1, f_2)^{\mathrm{T}} \text{ and } \mathbf{z} = (y, t)^{\mathrm{T}} \tag{18.31}$$

Then the mapping structure can be written as,

$$P_j : \ \mathbf{f}\left(\mathbf{z}_{k+1}, \mathbf{z}_k\right) = \mathbf{0} \tag{18.32}$$

Using these notations, a periodic motion can be defined as a collection of mapping structures. For example, the simplest motion occurring in this system has the mass impacting on one wall and then impacting the other wall and repeat. This periodic motion can be described as,

$$P_{31} = P_3 \circ P_1 \tag{18.33}$$

Using an initial guess of the switching points for some periodic motion, the mapping equations, Eqs. (18.32) and (18.33), can be solved for switching set points of the same periodic motion with varying parameters.

To determine the stability of some periodic motion, the periodicity condition needs to be considered. For the simplest mapping of one iteration, the stability can be found by perturbing the switching sets from the fixed points

$$\mathbf{z}^*_{k+1} = \mathbf{z}^*_k \tag{18.34}$$

$$\mathbf{z}_{k+1} = \mathbf{z}^*_{k+1} + \Delta\mathbf{z}_{k+1}, \quad \mathbf{z}_k = \mathbf{z}^*_k + \Delta\mathbf{z}_k \tag{18.35}$$

After inserting these perturbed values back into the mapping equations, the equations can be expanded around the fixed point using Taylor's series

$$\mathbf{f}\left(\mathbf{z}^*_{k+1}, \mathbf{z}^*_k\right) + \left.\frac{\partial\mathbf{f}}{\partial\mathbf{z}_{k+1}}\frac{\partial\mathbf{z}_{k+1}}{\partial\mathbf{z}_k}\right|_{\mathbf{z}^*_k} \Delta\mathbf{z}_k + \left.\frac{\partial\mathbf{f}}{\partial\mathbf{z}_k}\right|_{\mathbf{z}^*_k} \Delta\mathbf{z}_k + o\|\Delta\mathbf{z}_k\| = 0 \tag{18.36}$$

Neglecting the higher order terms and reducing yields,

$$\left.\frac{\partial\mathbf{f}}{\partial\mathbf{z}_{k+1}}\frac{\partial\mathbf{z}_{k+1}}{\partial\mathbf{z}_k}\right|_{\mathbf{z}^*_k} + \left.\frac{\partial\mathbf{f}}{\partial\mathbf{z}_k}\right|_{\mathbf{z}^*_k} = 0 \tag{18.37}$$

The Jacobian matrix of the mapping structure of periodic motion can then be solved for

$$DP_j = \frac{\partial \mathbf{z}_{k+1}}{\partial \mathbf{z}_k} = -\left[\frac{\partial \mathbf{f}}{\partial \mathbf{z}_{k+1}}\right]^{-1}\left[\frac{\partial \mathbf{f}}{\partial \mathbf{z}_k}\right] \tag{18.38}$$

The Jacobian matrix can be found in a similar fashion as the periodic motions become more complex. Starting with an initial guess for all of the switching points of a periodic motion, each point will map to the next. The Jacobian matrix for the system can then be found

$$DP = \underbrace{DP_{j_n} \cdot \cdots \cdot DP_{j_2} \cdot DP_{j_1}}_{n-term} \tag{18.39}$$

In which leads to the equation for an n-iterative periodic motion and $j_l \in \{1, 2 \cdots, 6\}$ with $l = 1, 2, \cdots, n$

$$\Delta \mathbf{z}_{k+n} = DP \Delta \mathbf{z}_k \tag{18.40}$$

To determine the stability of a periodic motion in the system, the periodicity condition needs to be considered at the perturbed fixed points. These perturbations need to be scaled to ensure the periodic motion

$$\Delta \mathbf{z}_{k+n} = \lambda \Delta \mathbf{z}_k \tag{18.41}$$

The eigenvalues, which determine the stability of the system, can then be calculated from Eqs. (18.40) and (18.41), i.e.,

$$(DP - \lambda I)\, \Delta \mathbf{z}_k = 0 \tag{18.42}$$

with

$$|DP - \lambda I| = 0 \tag{18.43}$$

For any periodic motion that exists in the system, the proceeding determinate will yield two eigenvalues from which the stability of the periodic motion can be determined. To show the preceding analysis, the analytical bifurcation scenario corresponding to the previous numerical scenario is presented in Fig. 18.10. The analytical predictions show good correlation with the numerical simulations, but are able to improve on the defined ends of the existing periodic motions.

The dashed lines mark the points where the mapping structure of the periodic motion changes. The empty areas are areas of more complex motions that exist in the system. The locations of these lines are found by tracking the stability of the periodic motions. As an example, the stabilities of the simplest periodic motions that exist in this system, P_{31}, are presented by their eigenvalues in Fig. 18.11. This simplest mapping structure has areas of symmetric and asymmetric motion, which are separated by a saddle node at $\Omega \approx 17.928$. As the excitation frequency is increased,

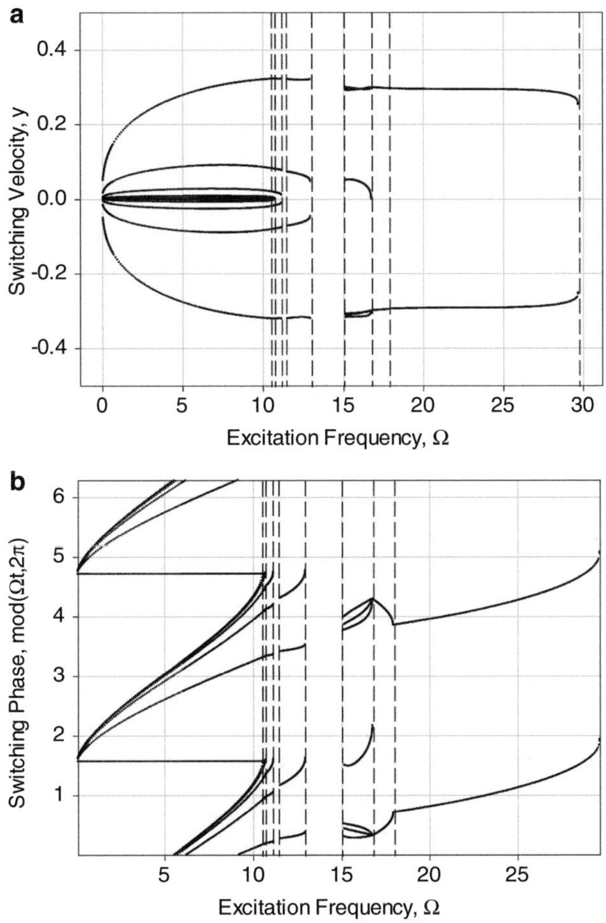

Fig. 18.10 Analytical prediction of the bifurcation scenario. (**a**) Switching velocity, (**b**) switching phase

the symmetric motion comes to another saddle node, after which the mass no longer interacts with the walls, at $\Omega \approx 29.6998$. As the excitation frequency is decreased, the asymmetric motion is found to be bound by a grazing bifurcation that occurs in the system, and the mapping structure of the motion changes despite having a stable motion according to the eigenvalues, at $\Omega \approx 16.7422$. The boundaries of all other periodic motions can be found in a similar method.

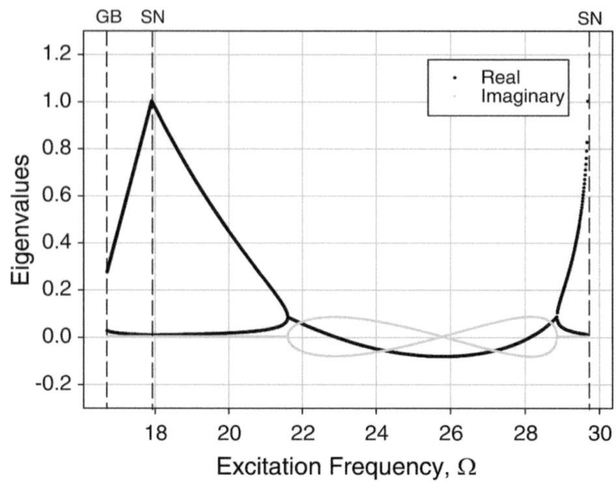

Fig. 18.11 Stability of the P_{31} periodic motions

18.7 Numerical Simulations

The locations of the various types of periodic motion in the system can be found for different parameters using the completed predictions of the bifurcation scenarios. To better show the periodic motions that exist in the system, numerical simulations can then be performed for specific parameters and initial conditions. To demonstrate the impact, impacting chatter, and stuck motion phenomena previously discussed, a periodic motion involving all of these is presented here. Using the mapping structure notation, the periodic motion considered is

$$P_{64^4352^41} = P_6 \circ P_{4^4} \circ P_3 \circ P_5 \circ P_{2^4} \circ P_1 \tag{18.44}$$

Using the parameters,

$$m = 5, d = 3, k = 10, \Omega = 7, P = 15, \Delta = .01, e = .3 \tag{18.45}$$

This periodic motion is presented in Fig. 18.12 for the initial conditions,

$$t_0 = 0.021443, \;\; x_0 = 0.01, \;\; y_0 = -0.02700 \tag{18.46}$$

In the numerical simulations, the four impacting chatter mappings, along with the stuck motion, can be observed on both the upper and lower boundaries. The discontinuity of the impact model is more clearly observed from the velocity responses and trajectory in phase plane. The stuck motion is clearly shown in both

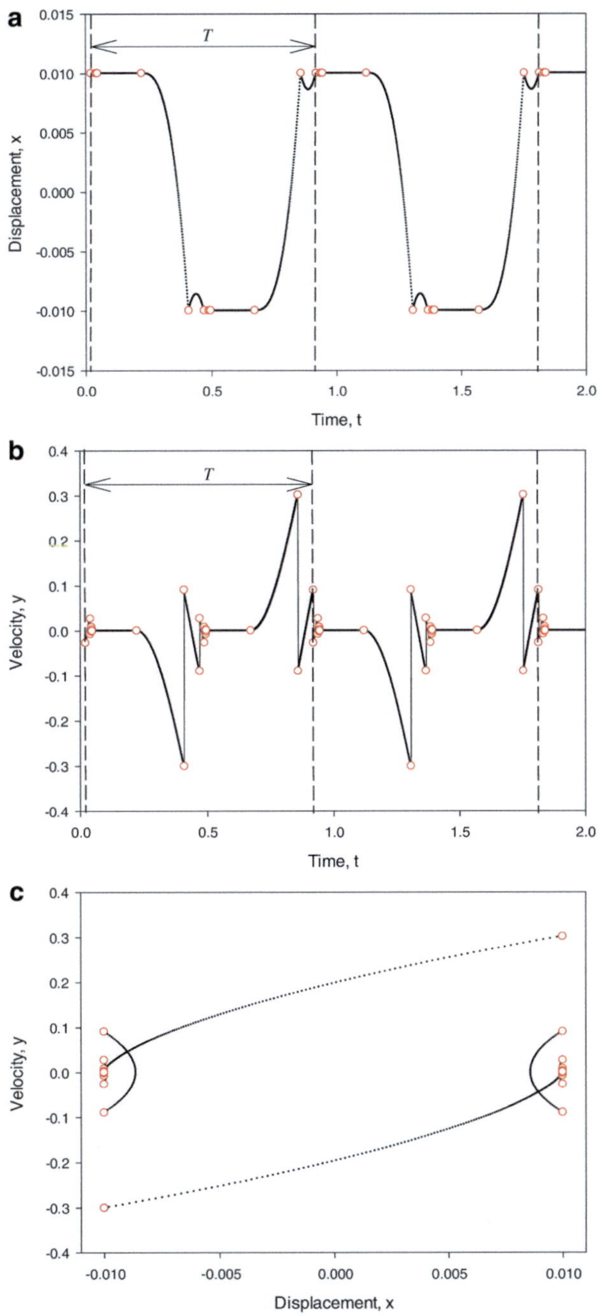

Fig. 18.12 Numerical simulation with impact, chatter, and stuck. (**a**) Displacement time history, (**b**) velocity time history, (**c**) phase plane

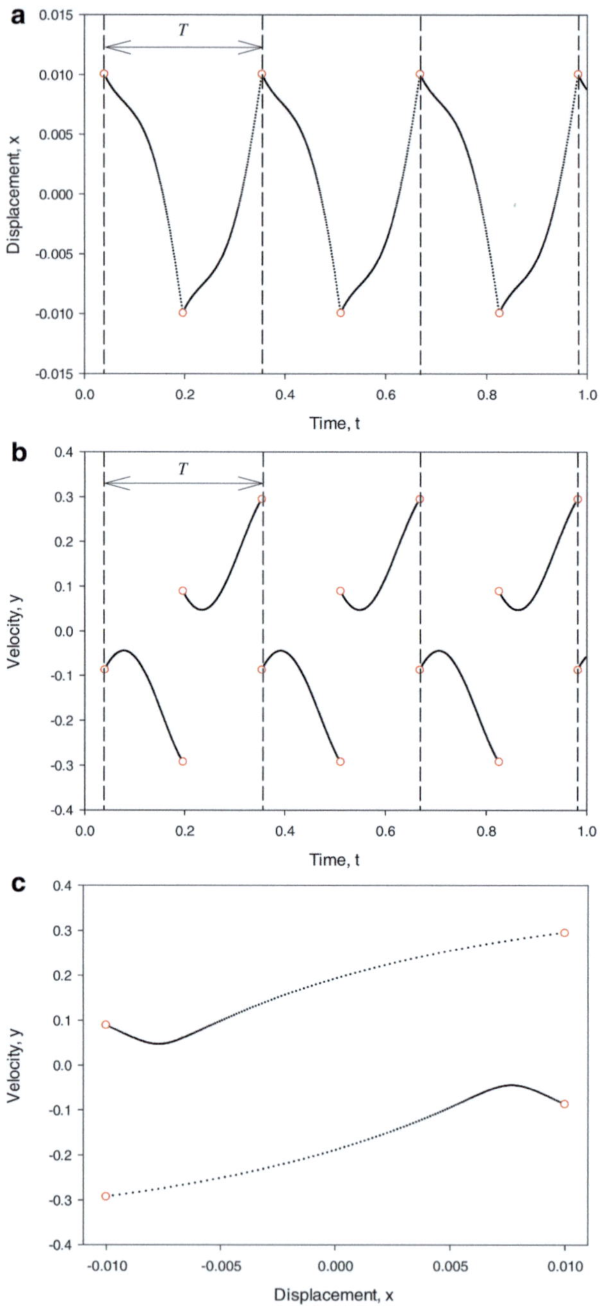

Fig. 18.13 Numerical simulation of symmetric P_{31} motion. (**a**) Displacement time history, (**b**) velocity time history, (**c**) phase plane

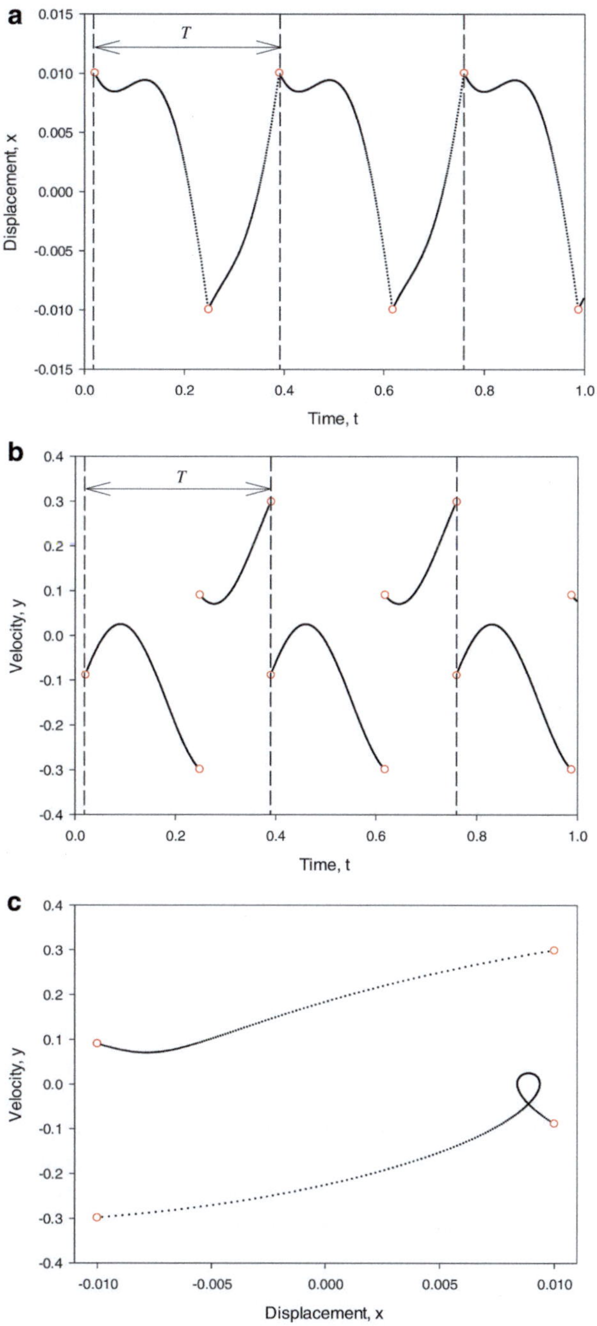

Fig. 18.14 Numerical simulation of asymmetric P_{31} motion. (**a**) Displacement time history, (**b**) velocity time history, (**c**) phase plane

the displacement and velocity responses, with the displacement remains constant and the velocity being zero for the duration of the stuck motion.

As was previously shown in the example of a stability analysis, the system contains two types of the simplest periodic motion. For the P_{31} periodic motions, there are both symmetric and asymmetric cases. In order to illustrate the difference between these two motions, numerical simulations are presented in Fig. 18.13 for the symmetric case and Fig. 18.14 for the asymmetric case. For the symmetric case the parameters and initial conditions are,

$$m = 5, d = 3, k = 10, \Omega = 20, P = 15, \Delta = .01, e = .3 \tag{18.47}$$

$$t_0 = 0.04033, \;\; x_0 = 0.01, \;\; y_0 = -0.08814 \tag{18.48}$$

For the asymmetric case the parameters and initial conditions are,

$$m = 5, d = 3, k = 10, \Omega = 17, P = 15, \Delta = .01, e = .3 \tag{18.49}$$

$$t_0 = 0.021341, \;\; x_0 = 0.01, \;\; y_0 = -0.08940 \tag{18.50}$$

From Fig. 18.14 for asymmetric motion it can also be seen how the motion disappears as the excitation frequency is lowered. From the stability analysis it was stated that as the excitation frequency is lowered, the periodic motion disappears due to a grazing bifurcation. The little loop back towards the wall that can be seen in the figure eventually comes back into contact with the wall for lower excitation frequencies and the motion switches to a more complex periodic motion.

18.8 Conclusion

The impact dynamics of a mass–damper–spring constrained system was considered herein. The analytical conditions governing the impact and stuck motions at the boundaries were developed and implemented. Periodic motion for the system was discussed based on the defined switching sets and mapping structures. To show the effectiveness of the analysis, numerical simulations were presented showing the discussed motions. Finally, in order to improve on the numerical simulations, analytical predictions were presented showing good correspondence with, and then extending farther than the numerical simulations.

References

1. Popplewell N, Bapat CN, McLachlan K (1983) Stable periodic vibroimpacts of an oscillator. J Sound Vib 87:41–59
2. Salapaka S, Dahleh M, Mezic I (2001) On the dynamics of a harmonic oscillator undergoing impacts with a vibrating platform. Nonlinear Dyn 24:333–358
3. Foale S, Bishop SR (1994) Bifurcations in impact oscillations. Nonlinear Dyn 6:285–299
4. Heiman MS, Sherman PJ, Bajaj AK (1987) On the dynamics and stability of an inclined impact pair. J Sound Vib 114:535–547
5. Lo CC (1980) A cantilever beam chattering against a stop. J Sound Vib 69:245–255
6. Dumont Y (2002) Vibrations of a beam between stops: numerical simulations and comparison of several numerical schemes. Math Comput Simul 60:45–83
7. Toulemonde C, Gontier C (1998) Sticking motions of impact oscillators. Eur J Mech Solids 17:339–366
8. Wagg DJ (2005) Periodic sticking motion in a two-degree-of-freedom impact oscillator. Int J Non Linear Mech 40:1076–1087
9. Luo ACJ (2006) Singularity and dynamics on discontinuous vector fields. Elsevier, Amsterdam
10. Luo ACJ (2009) Discontinuous dynamical systems on time-varying domains. Higher Education Press, Beijing
11. Luo ACJ, O'Connor D (2009) Impacting chatter and stick in a transmission system with two oscillators. J Multi Body Dyn 223:159–188
12. Luo ACJ, Guo Y (2009) Motion switching and chaos of a particle in a generalized Fermi-acceleration oscillator. Math Probl Eng 2009 Article ID: 298906, p 40

Part IV
Engineering and Financial Nonlinearity

Chapter 19
Formations of Transitional Zones in Shock Wave with Saddle-Node Bifurcations

Jia-Zhong Zhang, Yan Liu, Pei-Hua Feng, and Jia-Hui Chen

Abstract The formations of transitional zones in shock wave, governed by Burgers' equation, are studied from viewpoint of saddle-node bifurcations. First, the inviscid Burgers' equation is studied in detail, the solution of the system with a certain smooth initial condition is obtained, and the solution in vector form is reduced into a Map in order to investigate the stability and bifurcation in the system. It is proved that there exists a thin spatial zone where a saddle-node bifurcation occurs in finite time, and the velocity of the fluid behaves as jumping, namely, the characteristic of shock wave. Further, the period-doubling bifurcation is captured, that means there exist multiple states as time increases, and the complicated spatio-temporal pattern is formatted. In addition to above, the viscous Burgers' equation is further studied to extend to dissipative systems. By traveling wave transformation, the governing equation is reduced into an ordinary differential equation. More, the instability or bifurcation condition is obtained, and it is proved that there are three singular points in the system as the bifurcation condition is satisfied. The results show that the discontinuity resulting from saddle-node bifurcations is removed with the introduction of viscosity, and another kind of velocity change with strong gradient is obtained. However, the change of velocity is continuous with sharp slopes. As a conclusion, it can be drawn that all results can provide a fundamental understanding of the nonlinear phenomena relevant to shock wave and other complicated nonlinear phenomena, from viewpoint of nonlinear dynamics.

Keywords Burgers • Shock wave • Saddle-node bifurcation • Discontinuity

J.-Z. Zhang (✉) • P.-H. Feng • J.-H. Chen
School of Energy and Power Engineering, Xi'an Jiaotong University, Xi'an, P. R. China
e-mail: jzzhang@mail.xjtu.edu.cn; f-peihua@stu.xjtu.edu.cn; jiahui_chen@stu.xjtu.edu.cn

Y. Liu
School of Mechanical Engineering, Northwestern Polytechnical University, Xi'an, P. R. China
e-mail: liuyan@nwpu.edu.cn

J.A.T. Machado et al. (eds.), *Discontinuity and Complexity in Nonlinear Physical Systems*, Nonlinear Systems and Complexity 6, DOI 10.1007/978-3-319-01411-1_19,

19.1 Introduction

The motion of continuum media is often accompanied by the formation of transitional zones, where the parameters, including velocity, density, pressure, temperature, etc., vary rapidly [1]. In aerodynamics, there are a lot of circumstances where shock waves are present. For example, the aerodynamic performance in supersonic aircraft, turbo-machinery, helicopter are much affected by shock waves, which are of prime importance in air intakes, drag, lift, etc. Indeed, the interaction between convection and diffusion in many processes, such as fluid flow, chemical reaction, plays an important role in the dynamic behaviors, and can lead to some complicated nonlinear phenomena. Burgers' equation can be considered as a model equation or approach to the Navier–Stokes equation in fluid dynamics since the main terms are remained, and it can be used to study the turbulence, shock wave, soliton, etc. There are a very rich variety of nonlinear phenomena in it [2–5]. In particular, one of the phenomena is the sharp jumping or discontinuities as the Reynolds becomes higher or the viscosity coefficient is lower, and such phenomenon is relevant to the shock wave which is encountered frequently in the aircraft with supersonic speed. It is clear that the governing equation is a hyperbolic equation which is used to describe the traveling wave, and the singularities are the results from the counterbalance between the dispersive and convective effects. That is, as the hyperbolicity condition is violated, a qualitative change of the system and a bifurcation will occur. Because of the nonlinear phenomena mentioned above in the Burgers' equation, some special numerical methods are normally used to study the phenomena listed. Among them, the Discontinuous Finite Element Method and Spectral Method are the popular one. Indeed, Burgers' equation can be considered as a nonlinear dynamic system, and dynamic system ideas or theories have increasingly be applied to the analysis of fluid dynamics. For example, the Approximate Inertial manifolds, which is a global compact manifold and global attractor is included in, has been introduced to the analysis of Burgers' equation and Navier–Stokes equations, and the computing time will be saved as the numerical method is used to study the systems [6, 7].

Roughly speaking, the study of nonlinear dynamics is a fascinating question that is at the very heart of understanding of many important problems of natural science and engineering. In decades, the ideas from nonlinear dynamics are of interest in turbulent flows. For the fluid dynamics, some researchers are interested in numerical schemes that approximate the solutions of the Navier–Stokes equation for a long time, the connections between complex continuum mechanics (fluid dynamics, etc.), partial differential equations and nonlinear dynamical system, and the route to instability from viewpoint of bifurcation are the main focuses in this field. Because bifurcation theory and others in nonlinear science are deemed to be the fundamental nature of some nonlinear phenomena which a linear-world-view fails to capture and can give a deep insight into the mathematical nature [8–14]. As an example, the static stall of airfoil, which is typical discontinuity in lift as angle of attack is increased, is proved to be a result from saddle-node bifurcation, with introducing a map to study the nonlinear dynamics of the lift of the airfoil. More, the results show that static stall can be postponed by an external perturbation, and the ensuing lift could be enhanced significantly. Hence, the static stall of airfoil could be controlled feasibly by the external perturbation [15].

Under such background, this paper will focus on the nature of shock wave and other complex phenomena in Burgers' equation from viewpoint of bifurcations, and a fundamental analysis is carried out theoretically and numerically. Moreover, some comparisons between inviscid and viscous Burgers' equations are given.

19.2 Governing Equations and Analysis

19.2.1 Inviscid Burgers' Equation

The inviscid Burgers' equation in general form is

$$\begin{cases} \dfrac{\partial u}{\partial t} + u\dfrac{\partial u}{\partial x} = 0, & [0,1] \times [0,\mathrm{T}] \\ u\,(x,0) = u_0(x), & [0,\mathrm{T}] \end{cases} \tag{19.1}$$

where u is the velocity.

In this study, the smooth initial condition is

$$u\,(x,0) = 0.5\sin\left(2\pi x\right) \tag{19.2}$$

Then, the solution to Eqs. (19.1) and (19.2) can be obtained as

$$u\,(x,t) = 0.5\sin\left[2\pi\,(x - ut)\right] \quad t \geq 0 \tag{19.3}$$

It is clear that Eq. (19.3) in implicit form is a nonlinear equation, and iterative method can used to approach the solution or "the equilibrium position" from viewpoint of dynamic system. Hence, Eq. (19.3) is reduced into a Map g as following,

$$g: \quad u_{n+1} = 0.5\sin\left[2\pi\,(x - u_n t)\right] \tag{19.4}$$

The fixed point or P-1 solution to Map Eq. (19.4), u^*, is the solution to Eq. (19.3).

Considering x and t as bifurcation parameters, and keeping constant, the stability of the fixed point or the state of the system at certain position and time can be described by the Floquet multiplier,

$$Dg\,|_{u^*} = -\pi t\,\cos\left[2\pi\left(x - u^* t\right)\right] \tag{19.5}$$

For Map g, governed by Eq. (19.4), there exist following bifurcations,

1. As $Dg|_{u^*} = -1$, a period-doubling bifurcation can occur as parameters are varied.
2. $Dg|_{u^*} = 1$, a saddle-node bifurcation can occur as parameters are varied.

19.2.2 Viscous Burgers' Equation

The viscous Burgers' equation in general form is

$$\begin{cases} \dfrac{\partial u}{\partial t} + u\dfrac{\partial u}{\partial x} = \nu\dfrac{\partial^2 u}{\partial x^2}, & [0, 1] \times [0, \mathrm{T}] \\ u(x, 0) = u_0(x), & [0, \mathrm{T}] \end{cases} \tag{19.6}$$

where u is the velocity, ν the viscosity.

For Eq. (19.6), the traveling wave solution can be in the following form,

$$\xi = x - ct \tag{19.7}$$

where c is the wave speed.

Substitute Eq. (19.7) into Eq. (19.6), yields,

$$-c\dot{u} + u\dot{u} - \nu\ddot{u} = a \tag{19.8}$$

where a is a integral constant, relevant to initial condition.

Further, we have

$$\dot{u} = \frac{1}{2\nu}\left(u^2 - 2cu - 2a\right) \tag{19.9}$$

Then, Eq. (19.8) can be transformed into the form with first order in state space,

$$\begin{cases} \dot{u} = v \\ \dot{v} = \frac{1}{2\nu^2}(u - c)\left(u^2 - 2cu - 2a\right) \end{cases} \tag{19.10}$$

And, the instability or bifurcation condition can be obtained as

$$c \geq \sqrt{2\pi\nu} \tag{19.11}$$

As the condition, namely, Eq. (19.11), is satisfied, there are three steady states, A(c,0), B(u_1,0), C(u_2,0), and A(c,0) is a trivial solution to Eq. (19.10).

The Jacobean matrix for a steady state can be obtained as,

$$D = \begin{bmatrix} 0 & 1 \\ \frac{v^*}{\nu} & \frac{u^* - c}{\nu} \end{bmatrix} \tag{19.12}$$

For steady state A(c,0), the eigenvalue of its Jacobean matrix is 0, and it is a non-hyperbolic equilibrium.

For steady states B(u_1,0) and C(u_2,0), the eigenvalues of their Jacobean matrix are $\lambda_1 = 0, \lambda_2 = \frac{u^* - c}{\nu}$, and it is clear that both of them are non-hyperbolic equilibriums.

Normally, as the hyperbolicity condition is violated, a qualitative change of the system and a bifurcation occur. For a continuous-time dynamical system, the loss of hyperbolicity of an equilibrium generally happens, by the approach to zero of a simple real eigenvalue of the Jacobian (tangent or fold bifurcation) or by a pair of simple complex eigenvalues crossing the imaginary axis (Andronov–Hopf bifurcation). Indeed, an important and well-known aspect of nonlinear dynamics is the sensitive dependence of the solution on the perturbations, and such perturbation can come from the imperfection to the system. A slight perturbation to the system may produce very significant changes in the system's configuration after a long time.

It is clear that there exist many complex nonlinear phenomena in the system governed by Eq. (19.6), and the complex bifurcation will be investigated in the further work.

19.3 Stability and Bifurcation Analysis

19.3.1 Inviscid Burgers' Equation

For Eq. (19.3), as discussed above, it is time-dependent. Figure 19.1 shows the evolution of the system, it is clear that the slopes of curve will become sharp as time increases, leading to the appearance of discontinuities. At $t = 0.31$ s and location $x = 0.464$, the Floquet multiplier of Map Eq. (19.4) is -0.973, that means a period-doubling bifurcation may appear.

From Fig. 19.1, the discontinuity will appear as time is beyond 0.31, that is, there will be a jumping in the following time. In nonlinear dynamics, jumping is normally relevant to saddle-node bifurcation. In the system studied, namely, inviscid Burgers' equation, a saddle-node bifurcation will be induced as system is evolving, and this will be proved in the following.

Considering x as the function of u, and taking derivative of Eq. (19.3) with respect to u, yields,

$$1 - \frac{1}{2}\cos\left[2\pi\left(x - ut\right)\right] 2\pi\left(x_u - t\right) = 0 \tag{19.13}$$

At saddle-node points, following condition should satisfy,

$$x_u = 0 \tag{19.14}$$

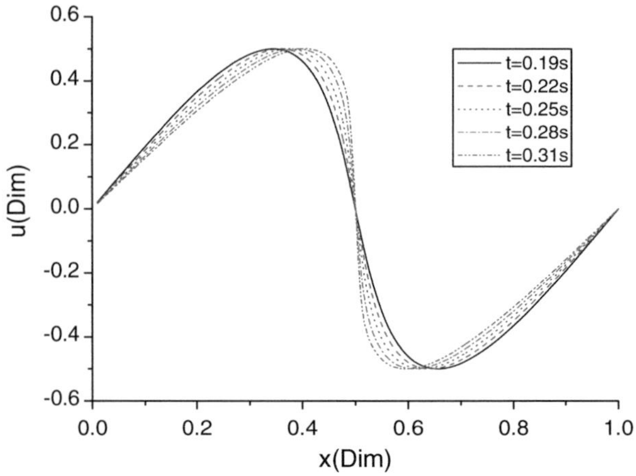

Fig. 19.1 The velocity distribution

Substitute Eq. (19.14) into Eq. (19.13), we have

$$\cos\left(2\pi\left(x - ut\right)\right) = -\frac{1}{\pi t} \tag{19.15}$$

Obviously, there exists saddle-node bifurcation only if $t \geq \frac{1}{\pi}$.
With Eqs. (19.3) and (19.15), the following can be obtained,

$$\left(\frac{1}{\pi t}\right)^2 + (2u)^2 = 1 \tag{19.16}$$

Hence, at saddle-node points, the velocities are

$$u_c = \pm\frac{1}{2}\sqrt{1 - \left(\frac{1}{\pi t}\right)^2} \tag{19.17}$$

By Eq. (19.15), the critical time is 0.318 s. As time increases beyond it, a saddle-node bifurcation occurs, the ensuing velocity behaves as jumping around $x = 0.5$. As $t = 0.32$ s, $x = 0.50006$, $u = 0.05132$, the Floquet multiplier governed by Eq. (19.5) is 0.99999, implying a saddle-node bifurcation is induced in this parameter family. Also, another saddle-node bifurcation appears at this moment with $x = 0.49994$, $u = -0.05132$. All the results mentioned above are shown in Fig. 19.2.

More, a period-doubling bifurcation spatially occurs around $x = 0.04600$ and 0.95200, as shown in Fig. 19.2. At this moment and location $x = 0.46500$, the Floquet multiplier of Map Eq. (19.4) is -1.0040, that means the state is critical and a period-doubling bifurcation appears.

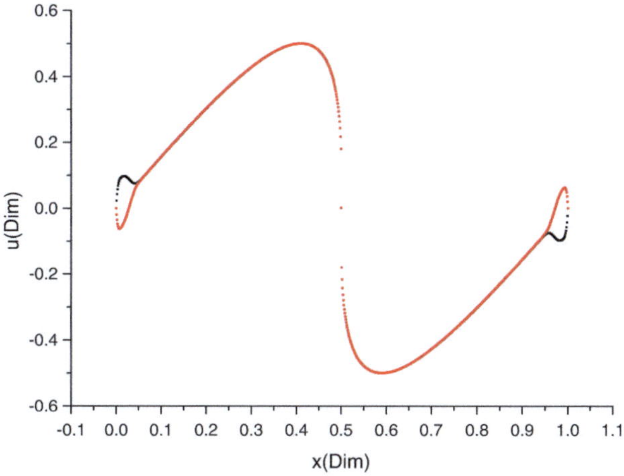

Fig. 19.2 Period-doubling bifurcation at t = 0.32 s

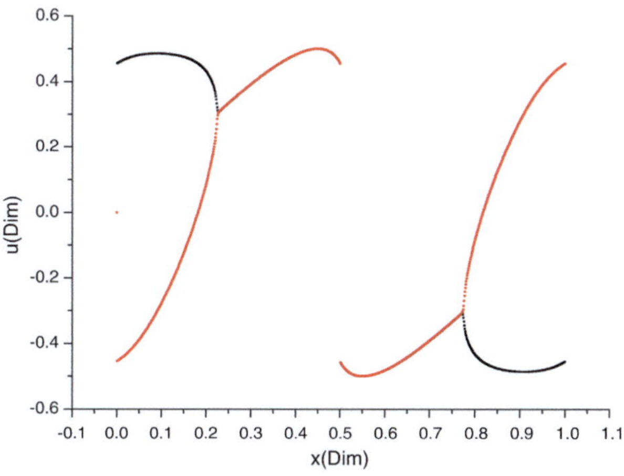

Fig. 19.3 Discontinuity or shock wave at t = 0.40 s

As time increases to 0.4 s, the jumping or shock wave appears obviously, and the velocity behaves as discontinuity, as shown in Fig. 19.3. As $t = 0.40$ s, $x = 0.51759$, $u = 0.30280$, the Floquet multiplier governed by Eq. (19.5) is 0.99999, implying a saddle-node bifurcation is induced in this parameter family. Also, another saddle-node bifurcation appears at this moment with $x = 0.48241$, $u = -0.30280$. All the results can be proved numerically in Fig. 19.3.

At this moment and location $x = 0.48700$, the Floquet multiplier of Map Eq. (19.4) is -1.25500, implying the state is unstable.

As time increases to 0.50 s, the velocity distribution is shown in Fig. 19.4. As $t = 0.50$ s, $x = 0.55263$, $u = 0.38559$, the Floquet multiplier is 0.99999, implying a

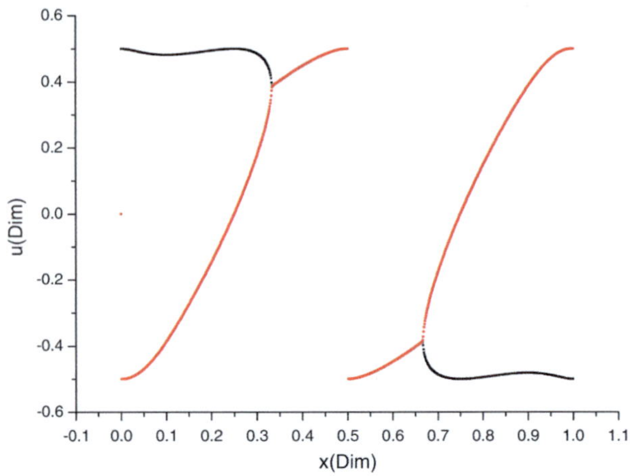

Fig. 19.4 Discontinuity or shock wave at t = 0.50 s

saddle-node bifurcation is induced in this parameter family. Also, another saddle-node bifurcation appears at this moment with $x = 0.44737$, $u = -0.38559$. In the next work, the unstable path between the two saddle-node points will be followed by numerical path-following method, and only the stable paths are shown in the following figures. Notice that the two saddle-node points are located around $x = 0.5$, and there is a thin zone where the shock wave appears. All the results mentioned above are shown in Fig. 19.4. Moreover, the discontinuity is located around $x = 0.5$, and the velocity is within [−0.5, 0.5] in this thin zone, as proved by Eq. (19.17).

At this moment and location $x = 0.50300$, the Floquet multiplier of Map Eq. (19.4) is −1.56900, that means the state is unstable.

As time increases to 0.60 s, the velocity distribution is shown in Fig. 19.5. As $t = 0.60$ s, $x = 0.59330$, $u = 0.42384$, the Floquet multiplier is 0.99999, implying a saddle-node bifurcation is induced in this parameter family. Also, another saddle-node bifurcation appears at this moment with $x = 0.40670$, $u = 0.42384$.

In contrast to above, at this moment, there is a sequence of period-doubling bifurcation, that is, the velocity will oscillate between some values. Note that the discontinuity is still located around $x = 0.5$, and the velocities is within [−0.5, 0.5] in a thin zone.

As time increases to 0.70 s, the velocity distribution is shown in Fig. 19.6. As $t = 0.70$ s, $x = 0.63685$, $u = 0.44532$, the Floquet multiplier is 0.99999, implying a saddle-node bifurcation is induced in this parameter family. Also, another saddle-node bifurcation appears at this moment with $x = 0.36315$, $u = -0.44532$. It is obvious that the thin zone where the shock wave appears is changing.

At this moment, the velocity will oscillate violently. The discontinuity is located around $x = 0.5$, and the velocity is within [−0.5, 0.5] in this thin zone.

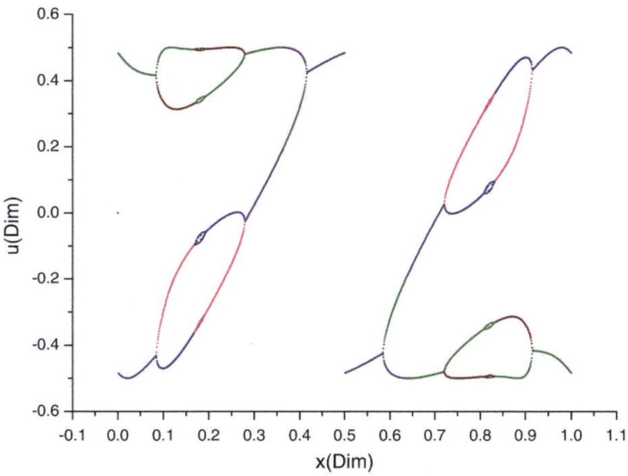

Fig. 19.5 Sequence of period-doubling bifurcation at t = 0.60 s

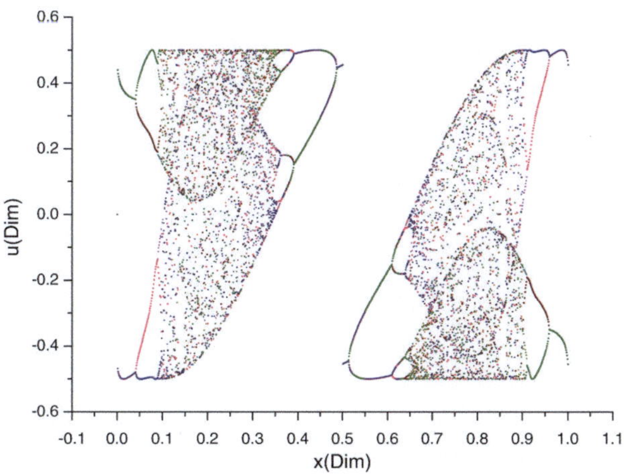

Fig. 19.6 Chaotic behaviors at t = 0.70 s

As time increases to 0.90 s, the velocity distribution is in chaotic state, as shown in Fig. 19.7. As $t = 0.90$ s, $x = 0.72845$, $u = 0.46768$, the Floquet multiplier is 0.99999, implying a saddle-node bifurcation is induced in this parameter family. Also, another saddle-node bifurcation appears at this moment with $x = 0.27155$, $u = -0.46768$. It is obvious that the thin zone where the shock wave appears is changing. All the results can be proved numerically in Fig. 19.7.

At this moment, the discontinuity is located around $x = 0.5$, and the velocities is within [−0.5, 0.5] in this thin zone. And there are period-1 windows in the chaotic state.

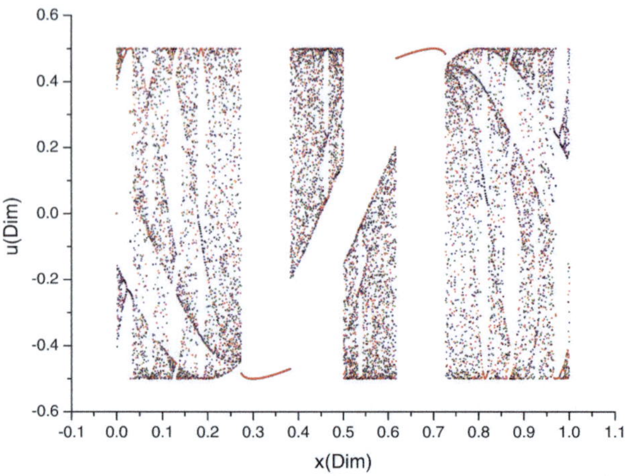

Fig. 19.7 Period-1 window in chaotic state at t = 0.90 s

Fig. 19.8 Velocity distribution versus ξ

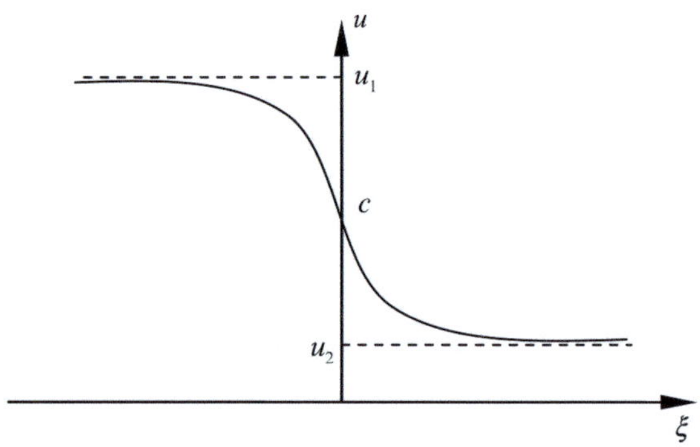

19.3.2 Viscous Burgers' Equation

As the viscosity is considered, the system becomes dissipative, and the dynamic behaviors will be changed at a certain degree. With a certain initial condition, the analytical solution to Eq. (19.9) can be obtained as,

$$u(x,t) = c - \frac{1}{2}(u_1 - u_2)\, th \frac{u_1 - u_2}{4\nu}\xi \tag{19.18}$$

The curve governed by Eq. (19.18) can be sketched in Fig. 19.8.

With the introduction of viscosity, the slope of the curve becomes smooth. As the viscosity decreases, the changing of velocity with respect to ξ becomes rapid, and the limit is the discontinuity captured in Inviscid Burgers' equation stated above.

As a conclusion in this section, it can be drawn that fluid flow is often accompanied by the formation of zones with strong gradients, and the flow parameters (velocity, density, pressure, temperature, etc.) vary rapidly in them. If dissipation is ignored or becomes slight, the discontinuity will appear in thin zones.

19.4 Conclusions

The nature of the shock wave is studied with inviscid Burgers' equation in detail, and it has been proved that there exists a thin spatial zone where a saddle-node bifurcation occurs as time goes on, and the velocity of the fluid behaves as jumping, namely, the characteristic of shock wave. Further, the period-doubling bifurcation is captured, implying there exist multiple states as time increases, and the complicated spatio-temporal pattern can be formatted. With viscous Burgers' equation, the instability or bifurcation condition is obtained, and it is proved that there are three singular points in the system as the bifurcation condition is satisfied. With a comparison between them, it shows that the discontinuity resulting from saddle-node bifurcation is removed with the introduction of viscosity, and another kind of velocity change with strong gradient is obtained.

Acknowledgments This research is supported by National 973 Program in China, No. 2012CB026000, and the National High Technology Research Program of China (863 Program), No. SS2012AA052303. The authors would like to gratefully acknowledge their supports.

References

1. Blokhin AM, Tkachevb DL, Baldanb LO (2007) Well-posedness of a modified initial-boundary value problem on stability of shock waves in a viscous gas, Part I. J Math Anal Appl 331: 408–423
2. Sakaguchi H (1999) Chaotic dynamics of an unstable Burgers equation. Phys D 129:57–67
3. Basto M, Semiao V, Calheiros F (2006) Dynamics in spectral solutions of Burgers equation. J Comput Appl Math 205:296–304
4. Basto M, Semiao V, Calheiros F (2009) Dynamics and synchronization of numerical solutions of the Burgers equation. J Comput Appl Math 231:793–806
5. Dang-Vu H (1995) Hopf bifurcation and strange attractors in Chebyshev spectral solutions of the Burgers equation. Appl Math Comput 73:99–113
6. Zhang J-Z, Liu Y, Feng P-H (2011) Approximate inertial manifolds of Burgers equation approached by nonlinear Galerkin's procedure and its application. Commun Nonlinear Sci Numer Simulat 16:4666–4670
7. Zhang J-Z, Ren S, Mei G-H (2011) Model reduction on inertial manifolds for N–S equations approached by multilevel finite element method. Commun Nonlinear Sci Numer Simulat 16:195–205
8. Henderson RD (1997) Nonlinear dynamics and pattern formation in turbulent wake transition. J Fluid Mech 352:65–112

9. Barkley D, Tuckerman LS, Golubitsky M (2000) Bifurcation theory for three-dimensional flow in the wake of a circular cylinder. Phys Rev E 61:5247–5252
10. Haller G (2000) Finding finite-time invariant manifolds in two dimensional velocity fields. Chaos 10:9–108
11. Surana A, Haller G (2008) Ghost manifolds in slow-fast systems, with applications to unsteady fluid flow separation. Phys D 237:1507–1529
12. Bakker PG (1991) Bifurcations in flow patterns. Kluwer, Dordrecht
13. Brøns M, Jakobsen B, Niss K, Bisgaard AV, Voigt LK (2007) Streamline topology in the near wake of a circular cylinder at moderate Reynolds numbers. J Fluid Mech 584:23–43
14. Bisgaard AV (2005) Structures and bifurcation in fluid flows: with application to vortex breakdown and wakes. PhD thesis, Technical University of Denmark, Denmark
15. Liu Y, Li K, Wang H, Liu L, Zhang J (2012) Numerical bifurcation analysis of static stall of airfoil and dynamic stall under unsteady perturbation. Commun Nonlinear Sci Numer Simulat 17:3427–3434

Chapter 20
Dynamics of Composite Milling: Application of Recurrence Plots to Huang Experimental Modes

G. Litak, R. Rusinek, K. Kecik, A. Rysak, and A. Syta

Abstract We study the dynamics of a milling process of a composite material basing on the experimental time series of cutting force components measured in the feeding direction. By using the recurrence plots we observe the differences in the response of the system depending on the feeding direction with respect to composite fibers orientation. This effect has been found after decomposition on the Huang experimental modes. Showing the results of recurrences in particular experimental modes we advocate to use this quantity to analyze the stability of the cutting of composites. The difference between different cases was also noticed using Fourier transform and statistical parameters such as RMS and kurtosis, but for these methods the necessary time interval of the examined time series has to be much longer, while recurrence approach is designed for shorter time series.

Keywords Dynamics of milling process • Chatter identification • Recurrence plots • Experimental modes decomposition

20.1 Introduction

The cutting process is a basic technology to get the desired shape and surface parameters. Some conditions may be affected by vibration types of "chatter" as manifested in unexpected waves on the machined surface of the workpiece. This effect was noticed and described by Taylor in the early twentieth century [1]. However first attempts to explain this phenomenon took place 50 years after its discovery. The sources of these vibrations was seen in a number of nonlinear deterministic

G. Litak (✉) • R. Rusinek • K. Kecik • A. Rysak • A. Syta
Faculty of Mechanical Engineering, Lublin University of Technology, Nadbystrzycka 36, PL-20-618 Lublin, Poland
e-mail: g.litak@pollub.pl; r.rusinek@pollub.pl; k.kecik@pollub.pl; rysak.andrzej@gmail.com; a.syta@pollub.pl

J.A.T. Machado et al. (eds.), *Discontinuity and Complexity in Nonlinear Physical Systems*, Nonlinear Systems and Complexity 6, DOI 10.1007/978-3-319-01411-1_20,

effects, which include the mechanisms of self-excited vibration generation [2], the effects of regenerative cutting [3], the structural dynamics of the process [4, 5], and finally dry friction [6, 7]. It should also be noted that these effects are not mutually exclusive. As a result, the elimination of vibration and stabilize cutting accompanying met with great interest in science and technology [8–10]. Short time series studies have become important to understand the process and develop a better control strategy [11].

Recently, dynamics of milling process have been investigated intensively. The authors of papers [10, 12–16] focused on stability of milling process, bifurcations leading to chatter vibrations, and finally on identification of various types of system vibrations using nonlinear methods.

Resistance of fibers in the composites and possible damage mechanisms (such as fiber pullout, fiber fragmentation and delamination, matrix burning, and/or cracking) influence on the surface quality of a machining process [17–19].

20.2 Experimental Setup and Measured Time Series

Milling process of composite material photo and schematics are presented in Fig. 20.1. In climb thread cutting process a finger cutter (full circuit milling) without colling. The parameters of the investigated milling process are shown in Table 20.1. In the experiment, we sampled values feed component forces F_x (Fig. 20.1) with frequency of 10 kHz. For tests we used Carbon-fiber-reinforced polymer CFRP based on unidirectional Carbon-epoxyde prepreg (Hexcel) with carbon fiber -AS7J 12K and epoxide resin M12.

The measured time series for the force F_x for fiber orientation angle β are presented in Fig. 20.2. The angle β was chosen $\beta = 90°$ and $\beta = 75°$ for D1 and D2, respectively.

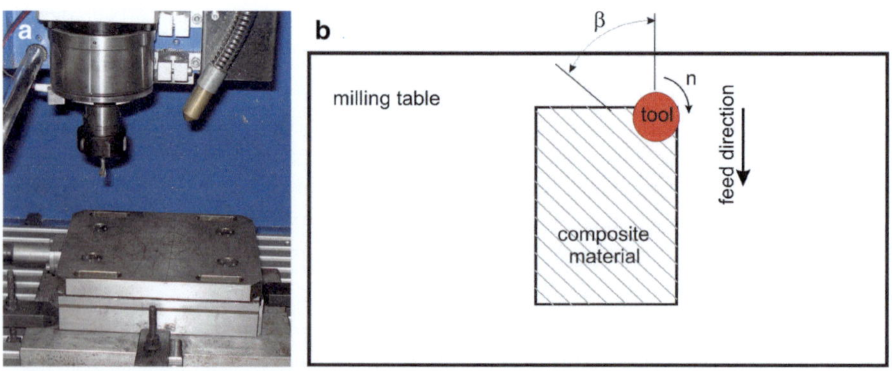

Fig. 20.1 Photo of an experimental stand (**a**) and a schematic plot (**b**) of milling process configuration. Note, the angle β denotes the milling direction with respect to composite fibers (Table 20.1)

Table 20.1 Milling process parameters

No. measur.	Cutting depth [mm]	Feed ratio [mm/rev.]	Feed ratio [mm/min]	Milling width [mm]	Rot. speed [rpm]	Cutting speed [m/min]	Angle β [deg]
(D1)	0.8	0.125	1,500	12	12,000	45.216	90
(D2)	0.8	0.0625	1,500	12	12,000	45,216	75

The angle β ($\beta = 90°$ for D1 and $\beta = 75°$ for D2) was denoted in Fig. 20.1b)

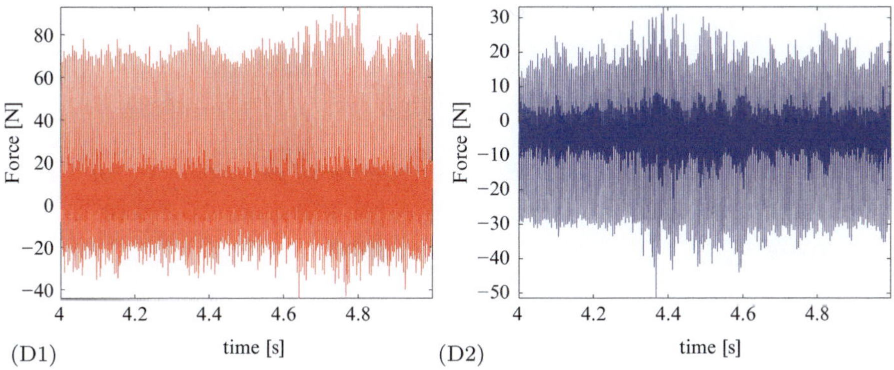

Fig. 20.2 Time series of measured force F_x (feed direction in Fig. 20.1). Note, the difference of scale in vertical axes

20.3 Analysis of the Experimental Response

The corresponding frequency spectrum is shown in Fig. 20.3. One can see the characteristic frequency of 200 Hz cutter rotation and its multiples. In these two cases, other components are dominant. If D2 is the frequency of 400 Hz, while in the case of D1 is 200 Hz. Note that the appearance multiples of 200 Hz implies nonlinear dynamics.

It also appears that in the case of D1 the spectrum is also showing additional structure of less than 200 Hz, which may be associated with the orientation of composite fibers. Namely, this is the most transparent difference in the spectra of these two cases.

20.4 Recurrence Plots for Huang Experimental Modes

In the analysis by Hilbert–Huang one performs the so-called signal decomposition into experimental modes (Huang decomposition): $F_x^1(t)$, $F_x^2(t)$, ..., $F_x^m(t)$ [20,21]:

$$F_x(t) = \sum_{j=1}^{m} F_x^j(t) + r_m, \tag{20.1}$$

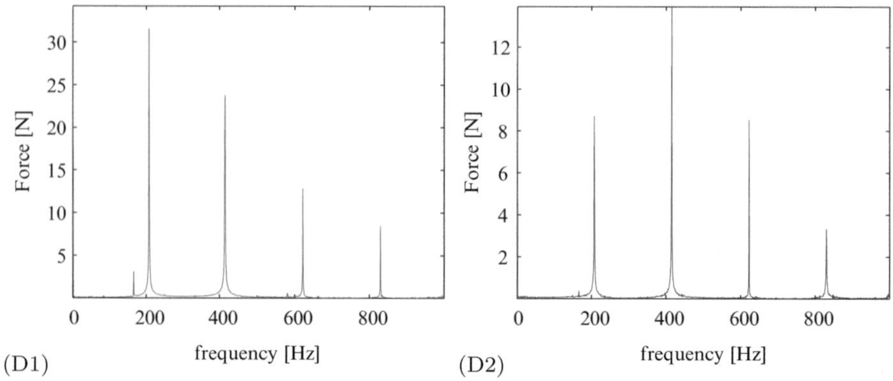

Fig. 20.3 Fourier analysis of measured F_x time series: frequency response

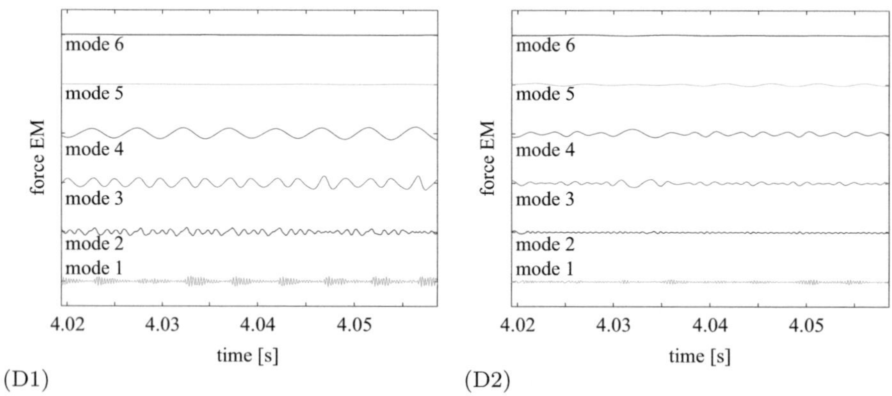

Fig. 20.4 Huang experimental mode (EM) decomposition (of measured F_x). Consecutive modes 1–6 from *bottom to top* of figures

where r_m is a truncation error. Each next experimental j mode is defined after subtracting average of maximum and minimum values interpolated by a cubic splines of the local envelope $F_x^{j-1}(t)$. Note that the first mode $F_x^1(t)$ is obtained from the original signal $F_x(t) = F_x^0(t)$ and the Huang decomposition procedure.

The first 6 Huang modes obtained using the above schema are plotted in Fig. 20.4. One can see that the amplitude reach maximum for mode 4, which could be the most important to distinguish the type of vibrations.

In the next step we provide the second coordinate as the numerical derivative $F_x^{\prime i}(t)$ for each mode $F_x^i(t)$. After normalization of each two variables $\tilde{\mathbf{F}}_x^i(t) = (\tilde{F}_x^{\prime i}(t), \tilde{F}_x^i(t))$ for given mode through the corresponding standard deviations we get phase vector representation

$$\tilde{\mathbf{F}}_x^1(t), \tilde{\mathbf{F}}_x^2(t), \ldots, \tilde{\mathbf{F}}_x^m(t), \tag{20.2}$$

where m is a natural number of the highest mode truncation (in our case $m = 6$).

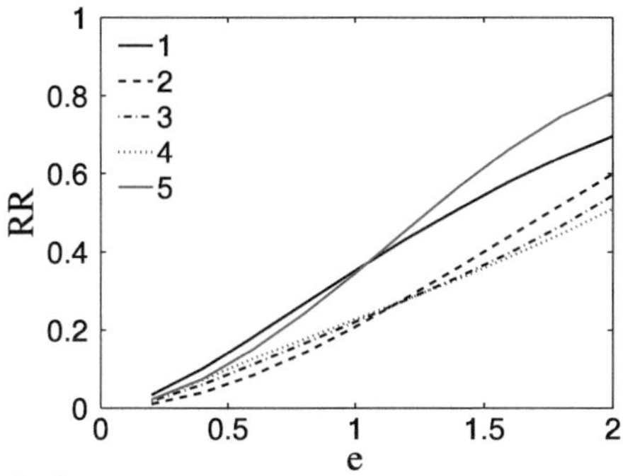

(D1)

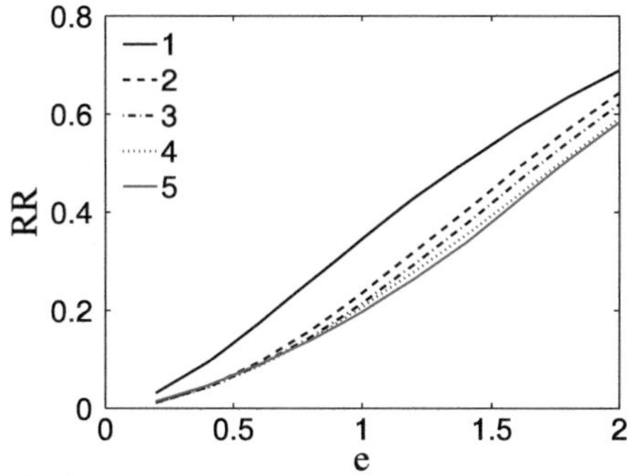

(D2)

Fig. 20.5 Recurrence rate *RR* versus the threshold value *e*

Using such formulation we performed recurrence analysis for each mode to compare D1 and D2 time series (Fig. 20.2). The recurrence rate (RR) parameter can be defined for each mode separately. It is defined as a fraction of off diagonal $i \neq j$ following inequality [22]

$$\|\tilde{\mathbf{F}}_x^n(t_i) - \tilde{\mathbf{F}}_x^n(t_j)\| < e, \tag{20.3}$$

where e is the given threshold number.

Namely, RR reads

$$RR = \frac{1}{N(N-1)} \sum_{ij} \theta(e - \|\tilde{\mathbf{F}}_x^n(t_i) - \tilde{\mathbf{F}}_x^n(t_j)\|) \quad (\text{for } i \neq j), \tag{20.4}$$

where $\theta(.)$ defines the Heaviside step function and N denotes the length of considered time series. while n indicates the corresponding mode. Note that RR has already been proposed as a good quantity to distinguish some different responses of dynamical systems [23–25]. In this paper we adopt this idea. The results for the first 5 corresponding modes (see Fig. 20.5) have been used for calculations of RR One can clearly see the difference in mode functions versus threshold value e. The most prominent difference is expressed in modes 4 and 5 and also in modes 1–3 intersection behavior in Fig. 20.5 D1 in contrary to the separate (no-crossing) grow tendency in Fig. 20.5 D2. These behaviors have the origin in different mode vibrations for D1 and D2 time series. For better clarity we have also plotted the corresponding recurrence plots (Fig. 20.6 a–j). One can clearly see different patterns in particular RP figures. The most regular is Fig. 20.6g (for D1) where the diagonal long lines are most repeatable. This opposes to Fig. 20.6h (for D2) where the lines have the fairly shorter lengths. These results, and also other figures (from the set of figures: Fig. 20.6a–j) in a smaller extent, imply that the time series D1 are more periodic than D2. Note that this conclusion can be drawn from fairly short time series. Interestingly in that case (D1) the cutter feed direction is oriented perpendicularly to the composite fibers.

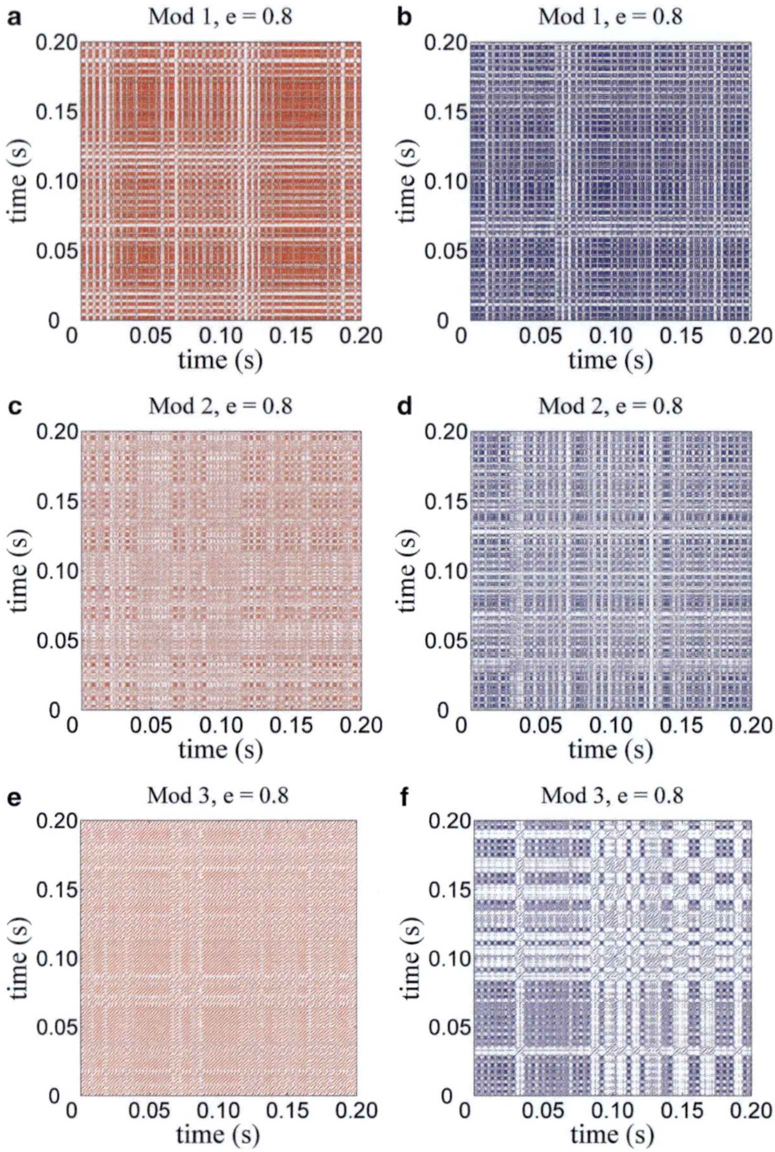

Fig. 20.6 (continued)

The modal results helped to capture visible changes in the statistical measures (Table 20.2). They can also be used to the design of improved control algorithm milling. Note, the nonmonotonic evolution of *RR* values for D1 series (Table 20.2). Fairly larger kurtosis in D2 case (see mode 3 in Table 20.2) implies intermittency

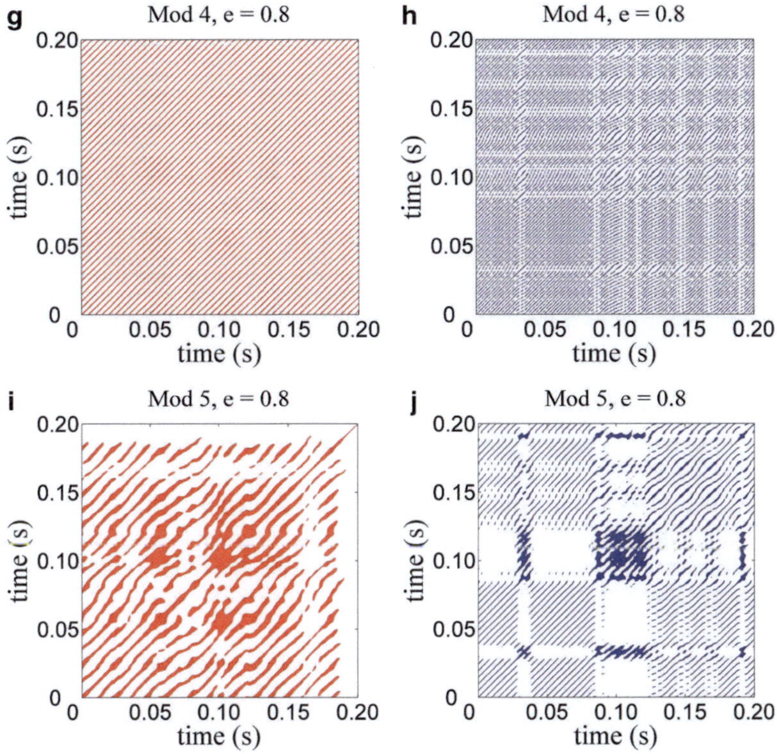

Fig. 20.6 Recurrence plots for $e = 0.8$ in two cases D1 (**a,c,e,g,i**) and D2 (**b,d,f,h,j**) for particular experimental modes. Note that, the plot is made using the 20.3

confirmed also by RP figures (see the difference between Fig. 20.6e and f, also Fig. 20.6g and h, and Fig. 20.6i and j). It seems that the configuration of the angle introduces an additional fairly low frequency modulation in the D1 case, which is clearly visible in the mode 5 recurrence plot (Fig. 20.6g). In contrast to it, the recurrence plots of the D2 case exhibits less regular behaviour. This has been confirmed by the results of the frequency spectrum (Fig. 20.3 (D2)), where the second harmonic of D2 series reaches the highest values.

20.5 Conclusions

The results of measurement and analysis of signals are based on multi-resolution method for experimental modes. Unlike the Fourier transform, it is applied to non-stationary signals as well as those that exhibit the phenomenon of intermittency. In our case, we have examined the process of composite milling tools with different orientations relative to the direction of the fibers.

Table 20.2 Summary of statistics and recurrences for the milling process F_x component: number of experimental modes (No. EM), root mean square (RMS), kurtosis, recurrence rate (RR)

No. EM	RMS (D1)[N]	RMS (D2) [N]	Kurtosis (D1)	Kurtosis (D2)	RR (D1) $e = 0.8$	RR (D2) $e = 0.8$
1	73.33	182.35	2.608	3.117	0.268	0.260
2	37.08	154.91	2.299	3.454	0.141	0.159
3	106.39	280.21	1.730	4.159	0.166	0.145
4	141.34	308.66	1.586	2.651	0.178	0.141
5	69.67	13.20	2.053	1.929	0.243	0.140

Note: RMS and kurtosis have been calculated for intervals of 1 s as shown in Fig. 20.2 while recurrence for the first 0.2 s of the corresponding time series (see Figs. 20.2 and 20.6)

Recurrences inform about a specific modulation and may also indicate a nonlinear nature of these oscillations. However, to provide specific guidance and make a more systematic study some more information about the nature of identified vibrations can be learned from other parameters which are in use in recurrence quantification analysis [22]. However repeating the procedure for other parameters in the adopted processing conditions goes beyond this paper and will be reported in a separate article.

Acknowledgements The financial support of Structural Funds in the Operational Programme–Innovative Economy (IE OP) financed from the European Regional Development Fund–Project "Modern material technologies in aerospace industry," No. POIG.01.01.02-00-015/08-00 is gratefully acknowledged.

References

1. Taylor F (1907) On the art of cutting metals. Trans ASME 28:310–350
2. Arnold RN (1946) The mechanism of tool vibration in the cutting of steel. Proc Inst Mech Eng 154:261–284
3. Tobias SA, Fishwick W (1958) A theory of regenerative chatter. The Engineer, London
4. Tlusty J, Polacek M (1963) The stability of machine tool against self-excited vibrations in machining. In: Proceedings of ASME international research in production engineering, Pittsburgh, Pa, USA pp 465–474
5. Merrit HE (1965) Theory of self-excited machine-tool chatter. ASME J Eng Ind 87:447–454
6. Wu DW, Liu CR (1985) An analytical model of cutting dynamics. Part 1: model building. ASME J Eng Ind 107:107–111
7. Wu DW, Liu CR (1985) An analytical model of cutting dynamics. Part 2: verification. ASME J Eng Ind 107:112–118
8. Altintas Y (2000) Manufacturing automation: metal cutting mechanics, machine tool vibrations, and CNC design. Cambridge University Press, Cambridge
9. Warminski J, Litak G, Cartmell MP, Khanin R, Wiercigroch W (2003) Approximate analytical solutions for primary chatter in the nonlinear metal cutting model. J Sound Vib 259:917–933
10. Insperger T, Gradisek J, Kalveram M, Stepan G, Winert K, Govekar E (2006) Machine tool chatter and surface location error in milling processes. J Manuf Sci Eng 128:913–920

11. Ganguli A, Deraemaeker A, Preumont A (2007) Regenerative chatter reduction by active damping control. J Sound Vib 300:847–862
12. Mann BP, Insperger T, Bayly PV, Stepan G (2003) Stability of up-milling and down-milling, part 2: experimental verification. Int J Mach Tools Manuf 43:35–40
13. Mann BP, Bayly PV, Davies MA, Halley JE (2004) Limit cycles, bifurcations, and accuracy of the milling process. J Sound Vib 277:31–48
14. Insperger T, Mann BP, Surmann T, Stepan G (2008) On the chatter frequencies of milling processes with runout. Int J Mach Tools Manuf 48:1081–1089
15. Litak G, Syta A, Rusinek R (2011) Dynamical changes during composite milling: recurrence and multiscale entropy analysis. Int J Adv Manuf Technol 56:445–453
16. Sen AK, Litak G, Syta A, Rusinek R (2013) Intermittency and multiscale dynamics in milling of fiber reinforced composites. Meccanica 48:738–789
17. Rao GVG, Mahajan P, Bhatnagar N (2007) Micromechanical modeling of machining of FRP composites cutting force analysis. Compos Sci Technol 67:579–593
18. Koplev A, Lystrup A, Vorm T (1983) The cutting process, chips, and cutting forces in machining CFRP. Composites 14:371–376
19. Rusinek R (2010) Cutting process of composite materials: an experimental study. Int J Non Linear Mech 45:458–462
20. Huang NE, Shen Z, Long SR, Wu MLC, Shih HH, Zheng QN, Yen NC, Tung CC, Liu HH (1998) The empirical mode decomposition and the Hilbert spectrum for nonlinear and non-stationary time series analysis. Proc R Soc Lond 454:903–993
21. Litak G, Kecik K, Rusinek R (2013) Cutting force response in milling of Inconel: analysis by wavelet and Hilbert-Huang transforms. Lat Am J Solids Struct 10:133–140
22. Marwan N, Romano MC, Thiel M, Kurths J (2007) Recurrence plots for the analysis of complex systems Phys Rep 438:237–329
23. Litak G, Sawicki JT, Kasperek R (2009) Cracked rotor detection by recurrence plots. Nondestruct Test Eval 24:347–351
24. Litak G, Wiercigroch M, Horton BW, Xu X (2010) Transient chaotic behaviour versus periodic motion of a parametric pendulum by recurrence plots. Z Angew Math Mech 90:33–41
25. Syta A, Jonak J, Jedlinski L, Litak G (2012) Failure diagnosis of a gear box by recurrences. J Vib Acoust 134:041006

Chapter 21
The Dynamics of Shear-Type Frames Equipped with Chain-Based Nonlinear Braces

Enrico Babilio

Abstract In recent years a number of bracing devices have been proposed, analyzed, and applied to real cases, since in engineering applications the construction of frames equipped with braces is a widespread practice. In the present contribution, a nonlinear bracing system is introduced and applied to the case of shear-type moment-resistant frames. The frame is considered here as the primary structure and is assumed to have linear elastic behavior and the bracing system is considered as a secondary, additional structure. The bracing system is made of two chains, each of them constructed as the assemblage of two axial elements (springs) undergoing axial force, only. The springs that are assumed to have linear elastic behavior are connected to each other in the chain and to the frame through hinges. The global behavior of the system is nonlinear, since the restoring force of the bracing system is a piecewise-defined function. In order to asses the performance of the whole nonlinear system, its behavior is compared with that of the linear primary structure alone, through a suitable concise descriptor.

Keywords Shear-type buildings • Lateral loads • Nonlinear braces

21.1 Introduction

In engineering applications, the construction of frames endowed with braces is a widespread practice, especially for, but not limited to, the case of steel structures. In recent years a number of bracing devices [5, 6, 13], including elastic, elastic-plastic

E. Babilio (✉)
Department of Structures for Engineering and Architecture (DiSt), University of Naples "Federico II", via Forno Vecchio 36, 80134 Naples, Italy
e-mail: enrico.babilio@unina.it

J.A.T. Machado et al. (eds.), *Discontinuity and Complexity in Nonlinear Physical Systems*, Nonlinear Systems and Complexity 6, DOI 10.1007/978-3-319-01411-1_21,

and viscous braces, have been proposed, analyzed, and applied to real cases. Braces allow structures, as buildings or bridges, for instance, to efficiently resist to lateral loads due to earthquakes or wind.

Large amounts of deformation energy can be dissipated under cyclic loads, if plastic behavior of the braces is taken into account [7–9], or if viscous or friction brace dampers are mounted in structures, finding the performance of such structures superior to that of conventional buildings: see [17, 18], where the performance of a framed building equipped with friction devices is numerically studied, and [19], where brace dampers, installed at the cross points in braced-frame structure systems, are introduced, and it is shown that the stiffness and damping in a structure can be altered by setting properly the stiffness and damping of the damper.

In the present contribution, a bracing system based on linear springs, connected to each other and to a shear-type moment-resistant frame through hinges, is considered. The idea comes from a previous work [1], with the main objective to simplify the device presented there, since, at a deeper insight, it resulted to have some practical difficulties, both in parameter tuning and, mainly, from a fabrication point of view. The device proposed here is still based on axial elements, that we call as springs, experiencing only axial force, but in the present work, in contrast to [1], the adopted links have elastic, instead of viscoelastic, behavior and no additional mass is considered. The global behavior of the assemblage of springs is nonlinear, although the single parts have linear elastic behavior. It is indeed possible for a system with only linear components to exhibit nonlinear characteristics, as highlighted in [10]. The restoring force is a piecewise-defined function with non-smooth corners and each branch of the restoring force-displacement curve is close to have linear behavior.

In engineering fields, there are many examples of dynamical systems modeled as multi-linear (bilinear or trilinear) oscillators or, more in general, oscillators whose restoring force is a piecewise-defined function, as in the present case. Such oscillators are indeed of great importance in the modeling of the nonlinear phenomena occurring in structures and machines and their knowledge is helpful in the design, control, and fault detection. A number of analytical and numerical studies on such oscillators have appeared in the literature. In what follows, we cite some of them.

In [20], an articulated mooring tower is modeled as a bilinear oscillator, with different stiffness for positive and negative deflections, due to the slackening of mooring lines, showing the model responses a good agreement with experimental results. The response of the same model under irregular seas is studied in [11]. In [12], the dynamic response of an offshore structure subjected to a nonzero mean, oscillatory fluid flow is studied. The interaction between the stiffness characteristic and the asymmetric hydrodynamic drag force is taken into account. In [4] to investigate the behavior of an articulated offshore platform, the structure is modeled as an upright pendulum with bilinear springs at the top, having different stiffness for positive and negative displacements. In [21] a multi-bay, multi-story scaffold with loose tube-in-tube connecting joints is modeled as a plane structure in sway under lateral base excitations. The loose restraining joint between adjacent stories is treated as a bilinear stiffness. In [2] the effects of a clearance on the normal mode frequencies of an N-DOF mechanical system, with bilinear stiffness and without damping, is investigated. In [15] the periodic motion and stability for a system with a symmetric, trilinear spring is numerically determined and in [16] an analytical procedure to determine the exact, single-crossing, periodic response of a class of harmonically excited piecewise linear oscillators is applied.

In the following sections of the present contribution, the model and the motion equations of the shear-type frame, equipped with nonlinear braces, will be introduced and some numerical applications will be presented and discussed.

21.2 The Model

Let us consider the structure depicted in Fig. 21.1, to which we refer for notations. Assume the system is made of a primary structure plus additional devices designed to improve the response of the system against lateral (i.e. horizontal or parallel to x-direction) loads.

Since we are mainly interested in civil and structural engineering applications, we assume the primary structure is a multi-story shear-type moment-resistant frame. We assume that the total mass of the system is concentrated in a number N of lumped masses, N equal to the number of the stories, and consider all the columns as massless and inextensible and the beams as rigid.

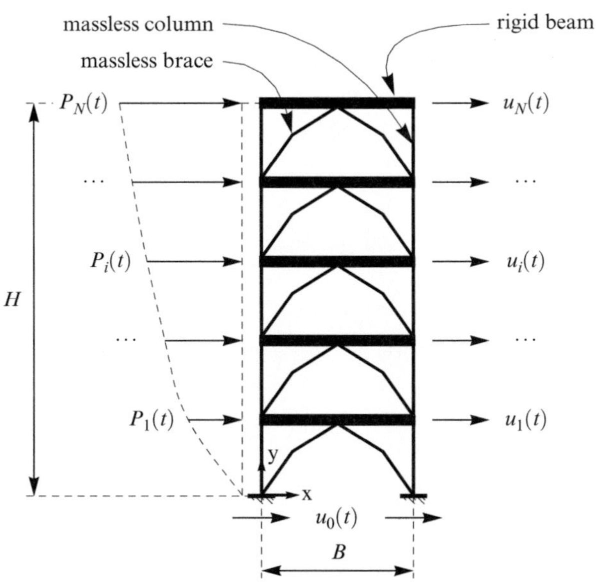

Fig. 21.1 The model of the shear-type moment-resistant building equipped with the braces

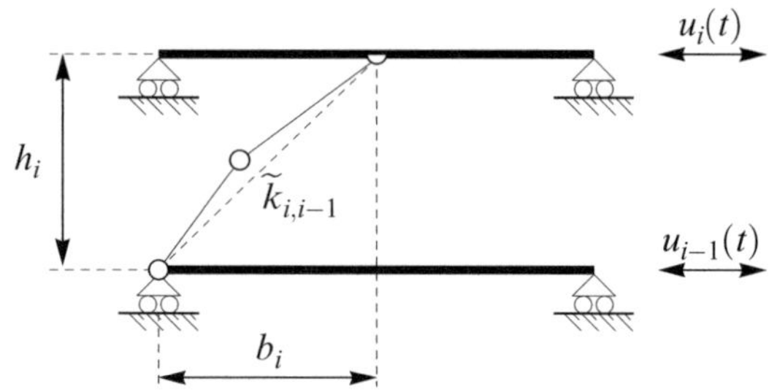

Fig. 21.2 The basic cell of the bracing device, in between two successive stories. $\widetilde{k}_{i,i-1}$ is the stiffness of each link

The single massless brace is designed as an assemblage of two equal springs, connected to each other and to the frame itself by means of hinges. We call it as a *basic cell* (see Fig. 21.2).

Assume that a couple of such basic cells is inserted in every inter-story height. In what follows we call it as a *bracing couple*.

With these ingredients, the dynamics of the building of total height H, made of N stories, endowed with N inter-story bracing couples, can be modeled by

$$\mathbf{M}\ddot{\mathbf{u}} + \mathbf{C}\dot{\mathbf{u}} + \mathbf{K}\mathbf{u} + \mathbf{f} = \mathbf{p}, \tag{21.1}$$

where $\mathbf{M},\ \mathbf{C},\ \mathbf{K}$ are $N \times N$ symmetric matrices (mass, damping and stiffness, respectively), $\mathbf{u}$, $\dot{\mathbf{u}}$, and $\ddot{\mathbf{u}}$ are the displacement vector and its derivatives (velocity and acceleration) w.r.t. the time t, $\mathbf{f}$ is the nonlinear restoration force vector, and $\mathbf{p}$ is the external load vector.

Being the columns of the primary structure inextensible, each displacement component of $\mathbf{u}$ is parallel to the x-direction (see Fig. 21.1).

The (i, j)th entry of the matrices $\mathbf{M},\ \mathbf{C},\ \mathbf{K}$ is given by

$$M_{i,j} = \begin{cases} m_i, & \text{if } i = j, \\ 0, & \text{otherwise;} \end{cases} \tag{21.2}$$

$$C_{i,j} = \begin{cases} c_{i,j-1} + c_{i,j+1}, & \text{if } i = j \text{ and } i < N, \\ c_{i,j-1}, & \text{if } i = j \text{ and } i = N, \\ -c_{i,j}, & \text{if } |i - j| = 1, \\ 0, & \text{otherwise;} \end{cases} \tag{21.3}$$

$$K_{i,j} = \begin{cases} k_{i,j-1} + k_{i,j+1}, & \text{if } i = j \text{ and } i < N, \\ k_{i,j-1}, & \text{if } i = j \text{ and } i = N, \\ -k_{i,j}, & \text{if } |i - j| = 1, \\ 0, & \text{otherwise,} \end{cases} \tag{21.4}$$

being m_i the mass of the ith story and $c_{i,j}$ and $k_{i,j}$ the (i, j)th inter-story total damping and stiffness, respectively.

The definition (21.3) is for the case of *relative* damping, that is

$$\mathbf{C} = \beta \mathbf{K}, \tag{21.5}$$

being β a positive constant. The relation (21.5) is a special kind of *proportional* damping (see [3]), given by

$$\mathbf{C} = \alpha \mathbf{M} + \beta \mathbf{K}, \tag{21.6}$$

with $\alpha = 0$, and $\beta \neq 0$. The case of *absolute* damping corresponds to $\alpha \neq 0$, and $\beta = 0$.

The ith component of the nonlinear force vector $\mathbf{f}$ in (21.1) is given by

$$f_i(t) = f(u_i(t) - u_{i-1}(t)) + \sigma_i \ f(u_i(t) - u_{i+1}(t)), \tag{21.7}$$

where σ_i is defined as

$$\sigma_i = \begin{cases} 1, & \text{if } i < N, \\ 0, & \text{if } i = N. \end{cases} \tag{21.8}$$

The terms appearing in (21.7) will be discussed in Sect. 21.2.1.

The components of the external load vector $\mathbf{p}$ in (21.1) are given here as sine functions, though in actual applications, in many engineering fields, it is quite rare to meet loads variable in time according to a sine function. Nevertheless sine loads have the advantage to allow us to construct response diagrams in the simplest way. The load is assumed variable along the height of the structure, with the ith component of $\mathbf{p}$ given by

$$p_i(t) = P_i \ \sin(\nu_1 \, t + \phi_i) - \delta_{i,1} U_g \left(k_{0,1} \ \sin \nu_2 \, t \, + \, \nu_2 \, c_{0,1} \ \cos \nu_2 \, t\right), \tag{21.9}$$

where $\delta_{i,j}$ is the Kronecker delta defined as

$$\delta_{i,j} = \begin{cases} 1, & \text{if } i = j, \\ 0, & \text{otherwise}. \end{cases} \tag{21.10}$$

In (21.9), P_i, $\phi_i, i = \{1, \cdots N\}$ and ν_1, are, respectively, amplitudes, phases, and frequency of the external loads applied to the floors and U_g and ν_2 are amplitude and frequency of the ground motion, given by

$$u_0(t) = u_g(t) = U_g \ \sin(\nu_2 \, t). \tag{21.11}$$

In case of seismic events, buildings are loaded by ground motion, while wind or gust loads are variable with the height. It is worth recalling that both seismic inputs and wind are random in nature and, as a consequence, in real applications they induce random and multi-frequency loads [14].

21.2.1 The Nonlinear Restoring Force

In order to explicitly write the components (21.7) of the nonlinear restoring force vector **f**, consider the system depicted in Fig. 21.2, showing a brace mounted in between two moving elements, representing two successive stories (numbered as $i-1$ and i) of the shear-type frame. The brace is built by assembling two links (springs) assumed equal in terms of stiffness $\tilde{k}_{i,i-1}$ and initial length l_i.

We define the relative displacement Δu_i between the two stories as

$$\Delta u_i = u_i(t) - u_{i-1}(t), \tag{21.12}$$

introduce two *limit* relative-displacements

$$\Delta u_i^m = \sqrt{4l_i^2 - h_i^2}, \tag{21.13}$$

$$\Delta u_i^p = -b_i + \sqrt{4l_i^2 - h_i^2}, \tag{21.14}$$

and assume that

$$l_i > \frac{\sqrt{b_i^2 + h_i^2}}{2}. \tag{21.15}$$

The stress experienced by the chain of the two springs is equal to zero if Δu_i satisfies the inequality

$$-\Delta u_i^p < \Delta u_i < \Delta u_i^p. \tag{21.16}$$

Otherwise, for

$$\Delta u_i^p \leq \Delta u_i < \Delta u_i^m, \tag{21.17}$$

the force exerted is given by

$$f^{(1)}(\Delta u_i) = \tilde{k}_{i,i-1}(\Delta u_i + b_i)\left(1 - \sqrt{\frac{(b_i + \Delta u_i^p)^2 + h_i^2}{(b_i - \Delta u_i)^2 + h_i^2}}\right), \tag{21.18}$$

and for

$$\Delta u_i \geq \Delta u_i^m, \tag{21.19}$$

Fig. 21.3 The restoring force plot of a bracing couple

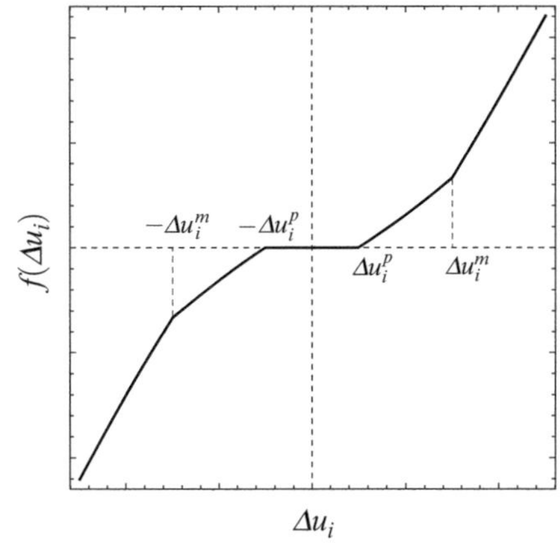

$$f^{(2)}(\Delta u_i) = \tilde{k}_{i,i-1}(\Delta u_i + b_i) \times \left(2 - \sqrt{\frac{(b_i + \Delta u_i^m)^2 + h_i^2}{(b_i + \Delta u_i)^2 + h_i^2}} - \sqrt{\frac{(b_i + \Delta u_i^p)^2 + h_i^2}{(b_i + \Delta u_i)^2 + h_i^2}}\right). \tag{21.20}$$

Since in each inter-story height the bracing couple is assumed composed by two braces symmetrically mounted (see Fig. 21.1, again), the nonlinear term is given by

$$f(\Delta u_i) = \begin{cases} -f^{(2)}(-\Delta u_i), & \text{if} \quad \Delta u_i < -\Delta u_i^m, \\ -f^{(1)}(-\Delta u_i), & \text{if} \; -\Delta u_i^m \leq \Delta u_i < -\Delta u_i^p, \\ 0, & \text{if} \; -\Delta u_i^p \leq \Delta u_i < \Delta u_i^p, \\ f^{(1)}(\Delta u_i), & \text{if} \quad \Delta u_i^p \leq \Delta u_i < \Delta u_i^m, \\ f^{(2)}(\Delta u_i), & \text{if} \quad \Delta u_i \geq \Delta u_i^m. \end{cases} \tag{21.21}$$

In Fig. 21.3 the restoring force (21.21) is shown.

It is worth noting that, because of the definition (21.12), if $U_g \neq 0$, the function (21.11), besides that in (21.9), appears in the left-hand side of (21.1), in the first component $f_1(t)$ of the nonlinear term $\mathbf{f}$.

21.2.2 Initial Conditions

The initial conditions, in terms of displacements and velocities of each of the stories, are

$$u_i(0) = 0, \quad \dot{u}_i(0) = 0, \quad i = \{1, 2, \ldots, N\}. \tag{21.22}$$

21.3 Dimensionless Equations

We rescale m_i, $c_{i,j}$, and $k_{i,j}$ w.r.t. the trace (that is equal to the sum of the eigenvalues) of the mass, damping, and stiffness matrices, respectively. To this end, we define

$$M = \mathrm{tr}\,\mathbf{M}, \qquad C = \mathrm{tr}\,\mathbf{C}, \qquad K = \mathrm{tr}\,\mathbf{K}. \tag{21.23}$$

Notice that M gives the total mass of the system.

By choosing the total height H of the building as the length scale, and

$$\sqrt{\frac{M}{K}} = \frac{1}{\omega}, \tag{21.24}$$

as the time scale, the following set of dimensionless variables can be defined:

$$w_i = \frac{u_i}{H}, \quad \tau = t\omega, \quad \mu_i = \frac{m_i}{M}, \quad \gamma_{i,j} = \frac{c_{i,j}}{C}, \quad \mathscr{K}_{i,j} = \frac{k_{i,j}}{K}. \tag{21.25}$$

Applying construction rules similar to those given by (21.2), (21.3), and (21.4) we get the dimensionless version of (21.1) as

$$\hat{\mathbf{M}}\ddot{\mathbf{w}} + \frac{C}{\sqrt{M\,K}}\hat{\mathbf{C}}\dot{\mathbf{w}} + \hat{\mathbf{K}}\mathbf{w} + \hat{\mathbf{f}} = \hat{\mathbf{p}}, \tag{21.26}$$

where all the symbols are the dimensionless counterpart of those in (21.1). In particular, $\dot{\mathbf{w}}$ and $\ddot{\mathbf{w}}$ are the derivatives of the dimensionless displacement vector $\mathbf{w}$ w.r.t. the dimensionless time τ (notice that in (21.1), the over-dot symbol stands for the derivative w.r.t the time t, instead).

The components of the dimensionless vector $\hat{\mathbf{f}}$ are given by

$$\hat{f}_i(\tau) = \hat{f}(w_i(\tau) - w_{i-1}(\tau)) + \sigma_i\, \hat{f}(w_i(\tau) - w_{i+1}(\tau)), \tag{21.27}$$

with σ_i defined in (21.8). The next Sect. 21.3.1 will be devoted to rewrite (21.27) explicitly. In order to do that, we need to consider the dimensionless counterparts of constants and functions appearing in Sect. 21.2.1.

The external force (21.9) is rewritten as

$$\hat{p}_i(\tau) = \hat{P}_i\, \sin(\Omega_1\tau + \phi_i) - \delta_{i,1}\hat{U}_g\left(\mathscr{K}_{0,1}\, \sin\Omega_2\tau + \frac{C}{\sqrt{M\,K}}\,\Omega_2\,\gamma_{0,1}\, \cos\Omega_2\tau\right), \tag{21.28}$$

where $\hat{P}_i$, $i = \{1, \cdots, N\}$, $\hat{U}_g$, Ω_1 and Ω_2, are given by

$$\hat{P}_i = \frac{P_i}{HK}, \quad \hat{U}_g = \frac{U_g}{H}, \quad \Omega_1 = \frac{\nu_1}{\omega}, \quad \Omega_2 = \frac{\nu_2}{\omega}, \tag{21.29}$$

and $\delta_{i,j}$ is the Kronecker delta defined in (21.10). Initial conditions are nondimensionalized accordingly.

21.3.1 The Dimensionless Restoring Force

Lengths and relative displacements are rescaled w.r.t. the length scale H, as in the first of the rules (21.25):

$$\hat{b}_i = \frac{b_i}{H}, \quad \hat{h}_i = \frac{h_i}{H}, \quad \Delta w_i^p = \frac{\Delta u_i^p}{H}, \quad \Delta w_i^m = \frac{\Delta u_i^m}{H}, \quad \Delta w_i = \frac{\Delta u_i}{H}. \tag{21.30}$$

In particular Δw_i, i.e. the dimensionless version of Δu_i, can be rewritten as

$$\Delta w_i = w_i(\tau) - w_{i-1}(\tau). \tag{21.31}$$

The forces (21.18) and (21.20) are rewritten as

$$\hat{f}^{(1)}(\Delta w_i) = \kappa_{i,i-1}(\hat{b}_i + \Delta w_i)\left(1 - \sqrt{\frac{(\hat{b}_i + \Delta w_i^p)^2 + \hat{h}_i^2}{(\hat{b}_i - \Delta w_i)^2 + \hat{h}_i^2}}\right), \tag{21.32}$$

$$\hat{f}^{(2)}(\Delta w_i) = \kappa_{i,i-1}(\hat{b}_i + \Delta w_i) \times \left(2 - \sqrt{\frac{(\hat{b}_i + \Delta w_i^m)^2 + \hat{h}_i^2}{(\hat{b}_i + \Delta w_i)^2 + \hat{h}_i^2}} - \sqrt{\frac{(\hat{b}_i + \Delta w_i^p)^2 + \hat{h}_i^2}{(\hat{b}_i + \Delta w_i)^2 + \hat{h}_i^2}}\right), \tag{21.33}$$

where

$$\kappa_{i,j} = \frac{\tilde{k}_{i,j}}{K}, \tag{21.34}$$

and the ith component of the nonlinear term $\hat{\mathbf{f}}$ is given by

$$\hat{f}(\Delta u_i) = \begin{cases} -\hat{f}^{(2)}(-\Delta w_i), & \text{if} \quad \Delta w_i < -\Delta w_i^m, \\ -\hat{f}^{(1)}(-\Delta w_i), & \text{if} \ -\Delta w_i^m \le \Delta w_i < -\Delta w_i^p, \\ 0, & \text{if} \ -\Delta w_i^p \le \Delta w_i < \Delta w_i^p, \\ \hat{f}^{(1)}(\Delta w_i), & \text{if} \quad \Delta w_i^p \le \Delta w_i < \Delta w_i^m, \\ \hat{f}^{(2)}(\Delta w_i), & \text{if} \quad \Delta w_i \ge \Delta w_i^m. \end{cases} \tag{21.35}$$

21.4 A Sample Problem

For the numerical applications we present in Sect. 21.5, the special case, in which all the stories have the same mass $m_i = m$, the inter-story massless columns have the same stiffness $k_{i,i-1} = k_{i,i+1} = k$ and damping property $c_{i,i-1} = c_{i,i+1} = c$ and all the nonlinear braces have the same stiffness $\tilde{k}_{i,i-1} = \tilde{k}_{i,i+1} = \tilde{k}$ and are mounted in the same way in the inter-story height, is considered. In particular, we assume that

$$h_i = h = \frac{H}{N}, \qquad b_i = b = \rho h = \rho \frac{H}{N}, \qquad l_i = l = \eta h = \eta \frac{H}{N}, \tag{21.36}$$

where ρ and η are positive real constants, and, as before, H is the total height of the frame and N is the number of the stories. These assumptions allow us to simplify relations (21.13)–(21.15) as

$$\Delta w_m = \frac{\sqrt{4\,\eta^2 - 1}}{N}, \tag{21.37}$$

$$\Delta w_p = \Delta w_m - \frac{\rho}{N}, \tag{21.38}$$

$$\eta > \frac{\sqrt{\rho^2 + 1}}{2}, \tag{21.39}$$

and the terms in (21.26) as follows

$$\hat{\mathbf{M}} = \frac{1}{N}\mathbf{I}, \qquad \hat{\mathbf{C}} = \hat{\mathbf{K}} = \frac{1}{2N-1}(\mathbf{I} + \mathbf{Q}), \qquad \frac{C}{\sqrt{M\,K}} = \frac{c}{\sqrt{m\,k}}\sqrt{\frac{2N-1}{N}}, \tag{21.40}$$

where $M, C,$ and K are defined in (21.23), $\mathbf{Q}$ is an $N \times N$ symmetric matrix whose entries are defined as

$$Q_{i,j} = \begin{cases} 1, & \text{if } i = j \text{ and } i < N, \\ -1, & \text{if } |i - j| = 1, \\ 0, & \text{otherwise,} \end{cases} \tag{21.41}$$

and $\mathbf{I}$ is an $N \times N$ identity matrix whose entries $I_{i,j} = \delta_{i,j}$ are given by the Kronecker delta, defined in (21.10).

The matrix equation (21.26) is rewritten as

$$\frac{1}{N}\mathbf{I}\ddot{\mathbf{w}} + \frac{1}{2N-1}(\mathbf{I} + \mathbf{Q})\left(\frac{2\zeta}{\sqrt{1 + \lambda_{\min}}}\sqrt{\frac{2N-1}{N}}\dot{\mathbf{w}} + \mathbf{w}\right) + \hat{\mathbf{f}} = \hat{\mathbf{p}}, \tag{21.42}$$

where $\lambda_{\min}$ is the smaller eigenvalue of the matrix $\mathbf{Q}$ (see Fig. 21.4, where the variation of $\lambda_{\min}$ w.r.t. N is depicted) and the dimensionless constant ζ is defined as follows

$$\zeta = \frac{c}{2\sqrt{m\,k}}\sqrt{1 + \lambda_{\min}}. \tag{21.43}$$

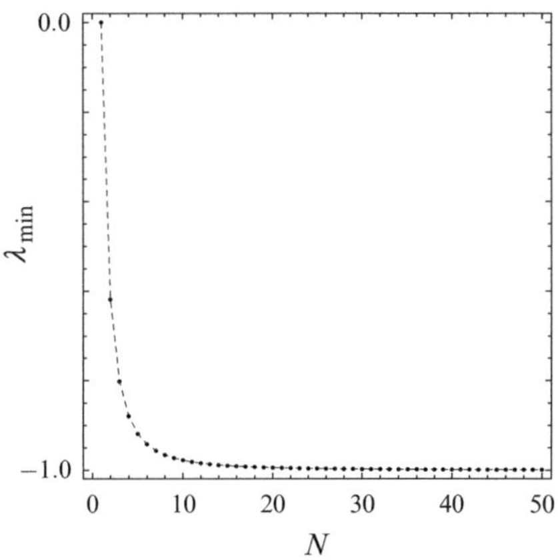

Fig. 21.4 The variation of $\lambda_{\min}$ with the respect to N

We call ζ as the *overall* damping ratio, since the value of ζ determines if all the displacement components w_i of the damped ($\zeta > 0$) linear N-DOF system associated with (21.42) (with $\hat{\mathbf{f}} = \hat{\mathbf{p}} = \mathbf{0}$) approaches a static equilibrium oscillating ($0 < \zeta < 1$) or decaying without oscillating ($\zeta \geq 1$).

Notice that for the single-DOF system ($N = 1$), $\lambda_{\min} = 0$ holds, and the definition of ζ, as given here, is coincident with the *standard* definition for the damping ratio.

The forces (21.32) and (21.33) are rewritten as

$$\hat{f}^{(1)}(\Delta w_i) = \frac{\kappa}{N}(\rho + N\Delta w_i)\left(1 - \sqrt{\frac{(\rho + N\Delta w^p)^2 + 1}{(\rho - N\Delta w_i)^2 + 1}}\right), \tag{21.44}$$

$$\begin{aligned}\hat{f}^{(2)}(\Delta w_i) &= \frac{\kappa}{N}(\rho + N\Delta w_i) \\ &\quad \times \left(2 - \sqrt{\frac{(\rho + N\Delta w^m)^2 + 1}{(\rho + N\Delta w_i)^2 + 1}} - \sqrt{\frac{(\rho + N\Delta w^p)^2 + 1}{(\rho + N\Delta w_i)^2 + 1}}\right),\end{aligned} \tag{21.45}$$

and the component of the nonlinear term $\hat{\mathbf{f}}$ is still given by (21.35), provided that (21.44) and (21.45) are taken into account.

The ith external load component (21.28) is rewritten as

$$\hat{p}_i(\tau) = \hat{P}_i \sin \Omega_1 \tau - \frac{\delta_{i,1}\hat{U}_g}{2N-1}\left(\sin \Omega_2 \tau + \frac{2\,\zeta\,\Omega_2}{\sqrt{1+\lambda_{\min}}}\sqrt{\frac{2N-1}{N}}\cos \Omega_2 \tau\right), \tag{21.46}$$

with $\phi_i = 0,\ i = \{1, \cdots, N\}$.

In what follows the ground motion $\hat{U}_g$ and the live loads $\hat{P}_i$ are assumed not contemporaneous events. In particular, if $\hat{U}_g = 0$, to evaluate the effect of the distribution of loads along the height, we consider N different external force vectors acting separately, each of them associated with a linear mode shape of the system

$$\frac{1}{N}\mathbf{I}\ddot{\mathbf{w}} + \frac{1}{2N-1}(\mathbf{I}+\mathbf{Q})\mathbf{w} = \mathbf{0}. \tag{21.47}$$

Finally, in order to evaluate the performance of the nonlinear system ($\kappa > 0$), in comparison with the linear system ($\kappa = 0$), the definition of a suitable concise descriptor is needed. To this end, we consider the Poincaré sections taken at

$$\tau_n = \tau_0 + \frac{2\,n\,\pi}{\Omega_j}, \qquad n = \{0, 1, \cdots\}, \tag{21.48}$$

where j takes the values 1 or 2 depending on the assigned load: for $\hat{P}_i \neq 0$, and $\hat{U}_g = 0$, $j = 1$; for $\hat{P}_i = 0$, and $\hat{U}_g \neq 0$, $j = 2$. Time τ_0 is taken large enough to have the system in its steady state. On each Poincaré section, we consider the vector

$$\mathbf{v}_i(\tau_n) = w_i(\tau_n)\,\mathbf{e}_1 + \dot{w}_i(\tau_n)\,\mathbf{e}_2, \tag{21.49}$$

$\mathbf{e}_1$, $\mathbf{e}_2$ being orthonormal vectors. The concise descriptor we choose to consider is $|\mathbf{v}_i|$, i.e. the norm of the vector $\mathbf{v}_i$.

21.5 Numerical Results

The simplified system described in Sect. 21.4 is numerically analyzed. The cases of 1 and 2-DOF systems are considered in Sects. 21.5.1 and 21.5.2, respectively.

21.5.1 *The Case of a Single Degree of Freedom*

The examples we consider first are focused on the single-DOF system ($N = 1$). The values of parameters adopted in the numerical simulations are reported in Table 21.1. Both the case of a live load (applied to the height of the story) and the case of ground motion are considered. The simulations are performed for different

Table 21.1 Simulation parameters for tests on 1-DOF system

N	η	ρ	ζ	$\lambda_{\min}$	$\hat{P}_1$ [a]	$\hat{U}_g$ [a]
1	0.848528	1.0	0.1	0.0	1.0	1.0

[a]Live load and ground motion do not act simultaneously

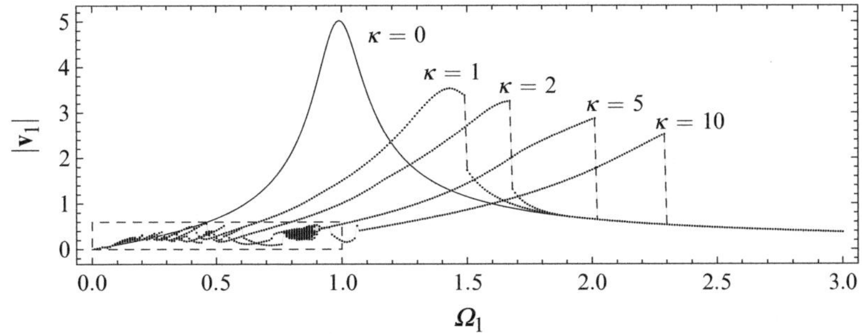

Fig. 21.5 The response of the 1-*DOF* system, for different values of κ

values of the dimensionless stiffness κ of the braces. The results are shown in Fig. 21.5 (live load) and in Fig. 21.9 (ground motion). All the reported curves are obtained performing a number of computations large enough to get diagrams as "smooth" and readable as possible. In both cases of loading, the frequency of the external load is taken in the interval $0 < \Omega_j \leq 3,\ j = \{1, 2\}$.

The *response diagrams* (or the *bifurcation diagrams*) for the nonlinear case show the expected huge richness of behaviors; in the nonlinear case, the norm $|\mathbf{v}_1|$ near the linear undamped frequency ($\Omega = 1.0$) is smaller than that in the linear case, and the (higher) peak of each diagram is shifted w.r.t. the peak in the linear case, due to the *foldover effect*. After the peak, the system jumps to the lower branch of each diagram and, although the typical hysteretic behavior in between the *jump-up* and the *jump-down* frequencies is expected, it is not documented here.

In Fig. 21.6, three enlargements in narrower and narrower frequency ranges are reported (for $\kappa = 10$). In particular, the first enlargement shows a number of superharmonic resonances responsible for the wiggled pattern in the frequency range $0 < \Omega_1 \leq 0.4$. The second enlargement, in the range $0.454 < \Omega_1 \leq 0.480$, shows that the system passes from periodic to chaotic responses through period doubling and again to a periodic solution through period halving (and similarly the system does in the frequency range $0.75 < \Omega_1 \leq 0.90$). A number of jump-down or jump-up phenomena, smaller in amplitude with respect to that detected at a frequency around $\Omega_1 = 2.3$ (see Fig. 21.5), are found.

Diagrams in Fig. 21.7 (again for $\kappa = 10$) summarize the results of a number of computations performed by setting Ω_1 at the values reported in each diagram and $\hat{P}_1$ variable in the interval $0 < \hat{P}_1 \leq 4.25$.

Beside the system again shows a richness of behaviors (period doubling, chaotic behavior, periodic windows, mainly for $\Omega_1 = 0.85$ or $\Omega_1 = 1.00$), these diagrams are interesting also from a system performance point of view.

For low values of $\hat{P}_1$, the growth of $|\mathbf{v}_1|$ in the nonlinear case is linear (as in the linear case, dashed line in Fig. 21.7) since the displacement w_1 satisfies the dimensionless counterpart of the inequality (21.16) and the bracing couple remains unstressed. As $\hat{P}_1$ increases, the bracing couple starts to exert force on the primary

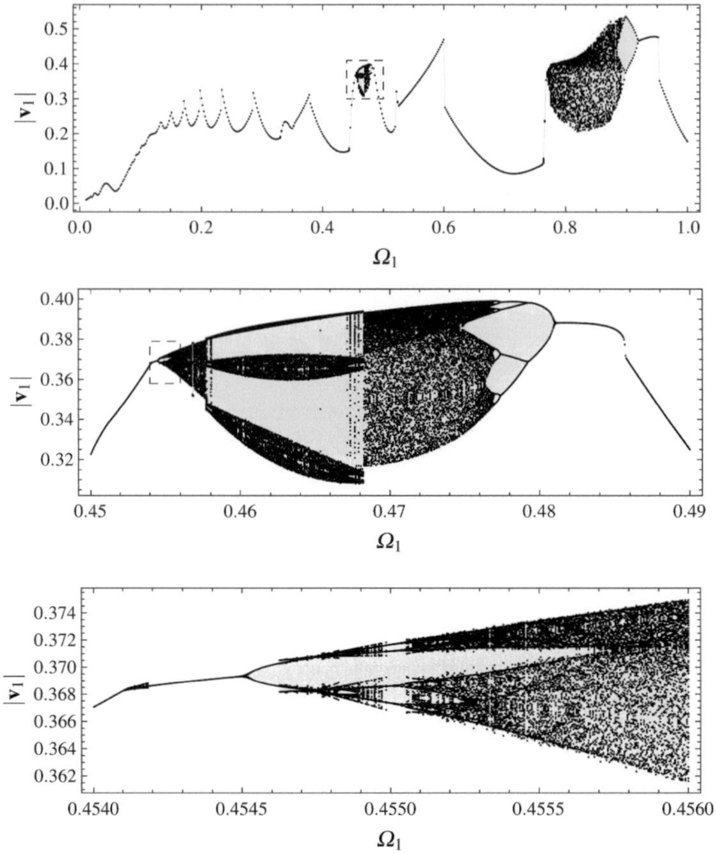

Fig. 21.6 The response of the 1-*DOF* system for $\kappa = 10$. Three successive enlargements in narrower and narrower frequency ranges, starting from $0.0 \leq \Omega_1 \leq 1.0$ (*dashed rectangle* in Fig. 21.5)

structure and the system response leaves the linear branch of the diagram. It is worth noting that the growth rate of $|\mathbf{v}_1|$ is slower than that in the linear case, and depending on the frequency of the loading, the quantity $|\mathbf{v}_1|$, in the nonlinear case, remains smaller or jumps to a higher value than that in the linear case. Combining the results shown in Figs. 21.5–21.9, the stiffness of the bracing system can be tuned in order to get the best performance in the frequency and amplitude ranges as large as possible. Indeed, $|\mathbf{v}_1|$ is related to the amplitude of the response, and the smaller it is, the smaller is the stress experienced by the primary structure.

Finally, for the record, four selected solutions, for different values of frequency Ω_1 and amplitude $\hat{P}_1$, are shown in Fig. 21.8. The reported plots show three periodic solutions of periods 1, 2, and 3, and a chaotic solution. Time plots (over twelve cycles of the forcing), phase portraits and corresponding Poincaré maps (dots inside the phase portraits) are reported.

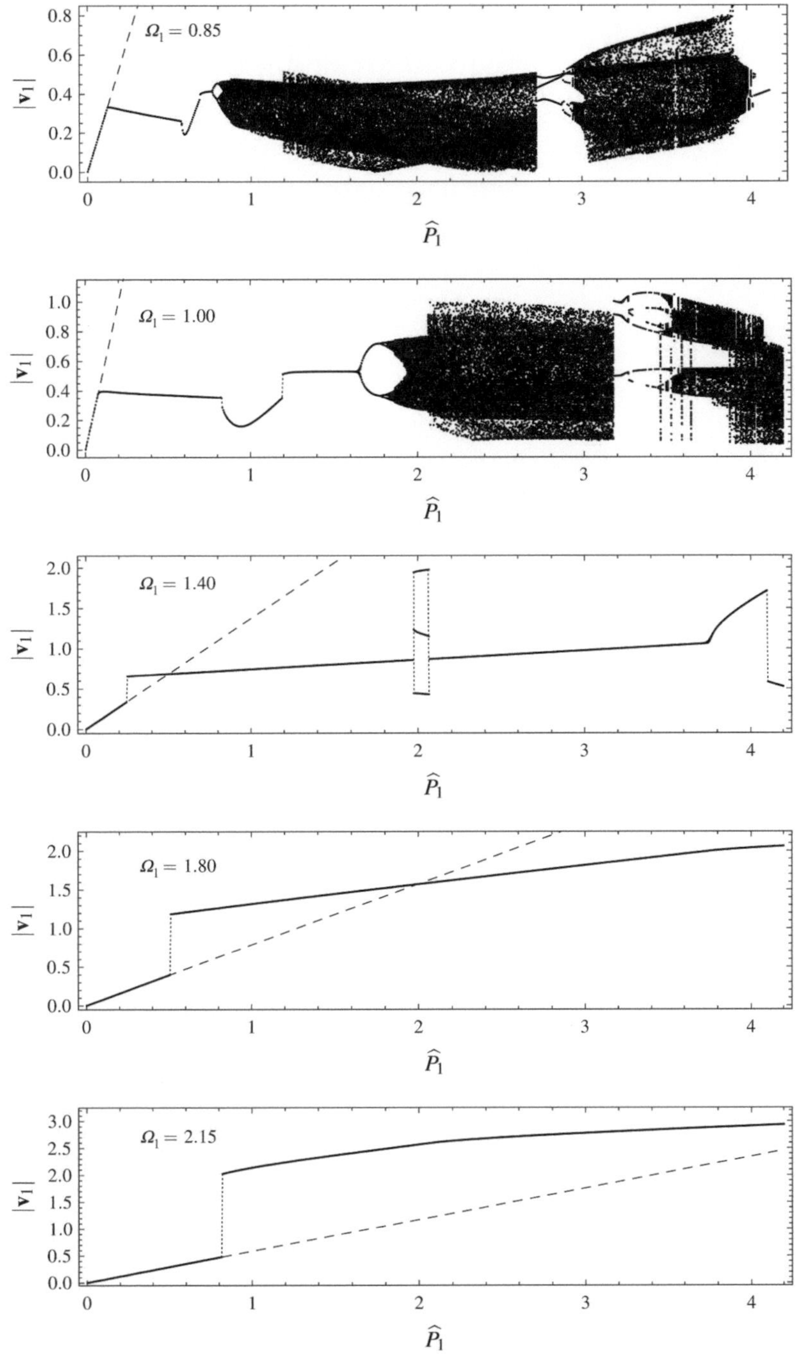

Fig. 21.7 The response of the 1-*DOF* system for different values of the frequency Ω_1 (reported into each graph) with respect to the value of the load amplitude $\hat{P}_1$. For each diagram $\kappa = 10$ is set, and the response of the linear system ($\kappa = 0$) is reported (*dashed line*)

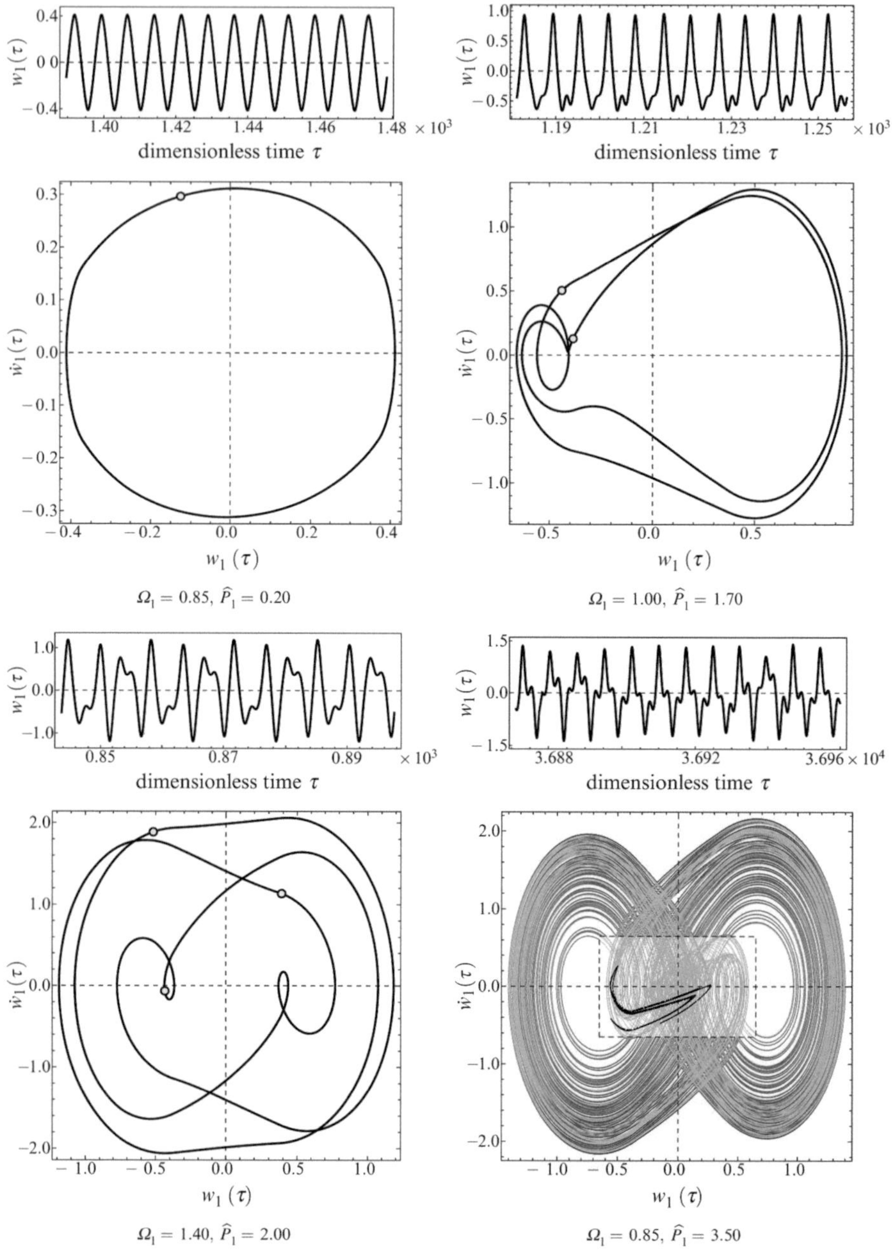

Fig. 21.8 Four selected solutions for different values of frequency Ω_1 and amplitude $\hat{P}_1$: period-1 (*top left*), period-2 (*top right*), period-3 (*bottom left*), and chaotic (*bottom right*) solutions. Time plots (over 12 cycles of the forcing), phase portraits, and corresponding Poincaré maps (*dots* inside the phase portraits) are reported

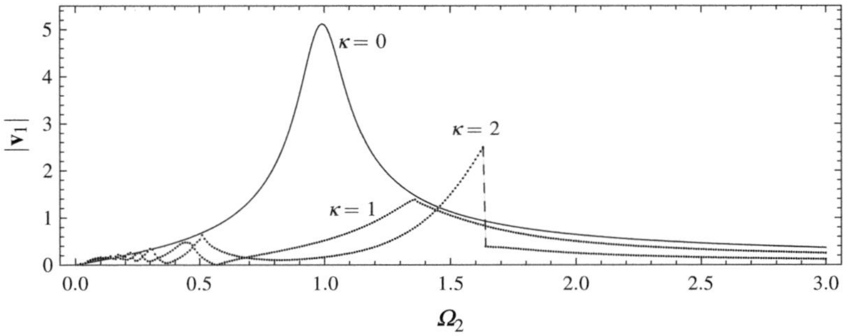

Fig. 21.9 The response of the 1-*DOF* system driven by the ground motion, for different values of κ

Table 21.2 Simulation parameters for tests on 2-*DOF* system

N	η	ρ	ζ	$\lambda_{\min}$	$\hat{P}_1$[a]	$\hat{P}_2$[a]	$\hat{U}_g$[a]
2	0.848528	1.0	0.1	−0.618034	0.525731	0.850651	1.0
					0.850651	−0.525731	

[a]Live loads and ground motion do not act simultaneously

21.5.2 The Case of Two Degrees of Freedom

The present subsection is devoted to the case of 2-*DOF* system ($N = 2$). The values of parameters adopted in the numerical simulations are reported in Table 21.2.

Both the case of live loads (applied to the height of each of the stories) and the case of ground motion are considered and, again, the simulations are performed for different values of the parameter κ. The results are shown in Figs. 21.10 and 21.11 (live loads) and in Fig. 21.12 (ground motion) and, in both cases of loading, the frequency of the external load is taken in the interval $0 < \Omega_i \leq 3,\ i = \{1, 2\}$.

Frequencies and damping ratio of the linear 2-*DOF* system associated with (21.42) (with $\hat{\mathbf{f}} = \hat{\mathbf{p}} = \mathbf{0}$) are reported in Table 21.3.

Remarks made for the previously considered case ($N = 1$) are essentially still valid in the present case. Moreover the response of the linear system is amplified around the first and the second natural frequencies, while the nonlinear system may present frequency peaks not only due to harmonic and superharmonic resonances but also due to internal coupling between the two degrees of freedom.

21.6 Conclusions

The present contribution is focused on the study of nonlinear dynamics of a shear-type moment-resistant frame equipped with nonlinear braces, improving the response of the system against lateral loads. The bracing system is based on linear springs connected to each other and to the frame through hinges.

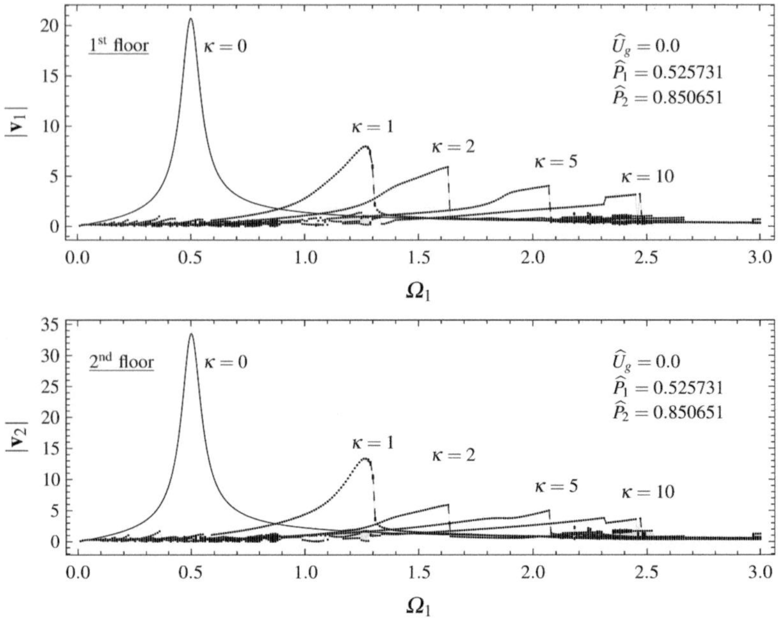

Fig. 21.10 The response of the 2-*DOF* system under loads applied along the height similarly to the first linear eigenmode, for different values of κ. The values of the load amplitude $\hat{P}_i$ are reported

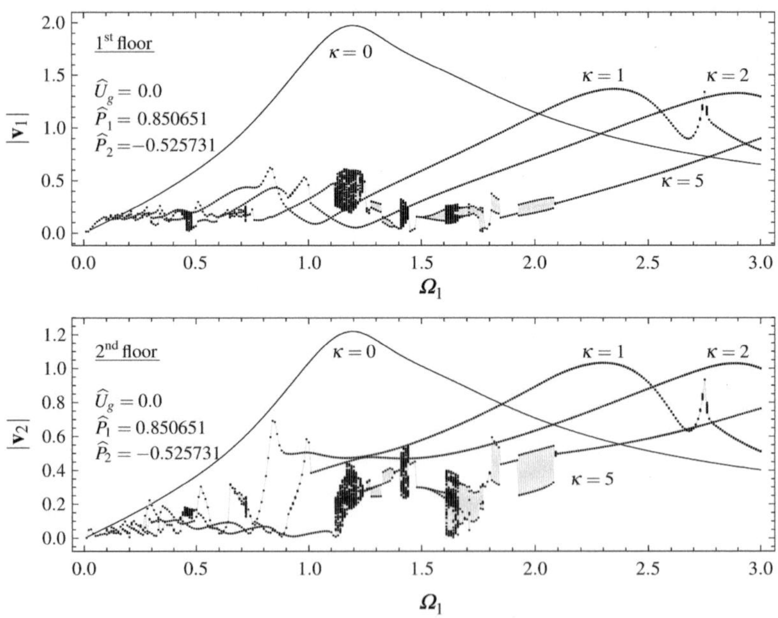

Fig. 21.11 The response of the 2-*DOF* system under loads applied along the height similarly to the second linear eigenmode, for different values of κ. The values of the load amplitude $\hat{P}_i$ are reported

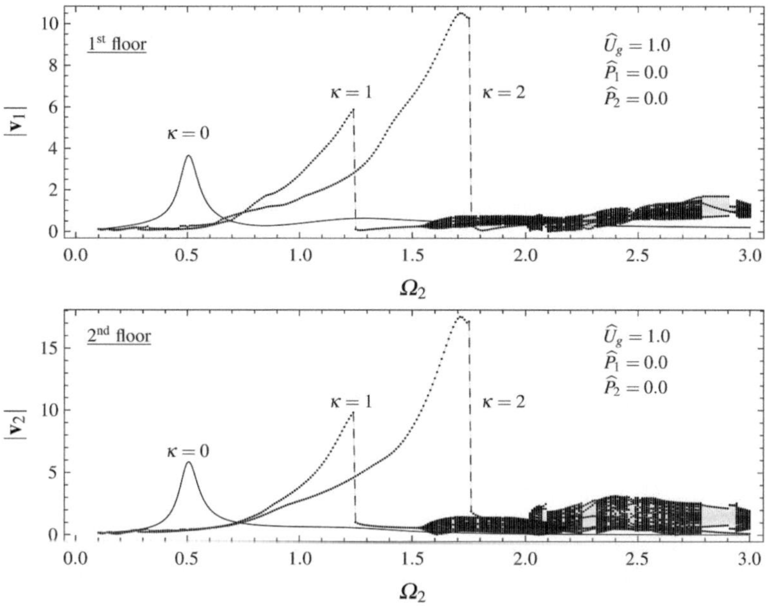

Fig. 21.12 The response of the 2-*DOF* system under ground motion, for different values of κ. The value of the ground motion amplitude $\hat{U}_g$ is reported

Table 21.3 Linear frequencies and damping ratios

$\omega_1 = 0.50462$	$\zeta_1 = \zeta^{a}$
$\omega_2 = 1.32112$	$\zeta_2 = 2.61803\ \zeta^{a}$

[a]Value of ζ reported in Table 21.2

While all the system components are linear, the behavior of the assemblage is nonlinear, being the restoring force of the bracing system given as a piecewise-defined function.

The behavior of the nonlinear system is compared with that of the linear one through a suitable concise descriptor. The obtained bifurcation diagrams are effective in detecting and identifying various different responses.

At the present stage of the research, it is possible to claim that endowing a frame, having linear elastic behavior, with the proposed devices has the advantage to reduce the stress experienced by the primary structure, near the linear resonance frequency and in wider ranges of the load frequency or amplitude, provided that the stiffness of the braces is tuned properly.

References

1. Babilio E (2012) A damping device based on bistable springs. In: NSC 2012 – IEEE 4th international conference on nonlinear science and complexity, Budapest, 6–11 August 2012, pp 57–62. doi:10.1109/NSC.2012.6304716

2. Butcher EA (1999) Clearance effects on bilinear normal mode frequencies. J Sound Vib 224:305–328
3. Cheng FY (2001) Matrix analysis of structural dynamics: applications and earthquake engineering. Marcel Dekker, New York, p 120
4. Choi HS, Lou JYK (1991) Nonlinear behaviour and chaotic motions of an SDOF system with piecewise-non-linear stiffness. Int J Nonlinear Mech 26:461–473
5. D'Aniello M, Della Corte G, Mazzolani FM (2006) Seismic upgrading of RC buildings by steel eccentric braces: experimental results vs numerical modeling. In: Mazzolani FM,Wada A (eds) Proceedings of the 5th international conference on behaviour of steel structures in seismic areas, STESSA, 2006, Yokohama,14–17 August 2006. Taylor & Francis, London, pp 809–814
6. D'Aniello M, Della Corte G, Mazzolani FM (2006) Seismic upgrading of RC buildings by buckling restrained braces: experimental results vs numerical modeling. In: Mazzolani FM,Wada A (eds) STESSA 2006: Proceedings of the 5th international conference on behaviour of steel structures in seismic areas, Yokohama, 14–17 August 2006. Taylor & Francis, London, pp 815–820
7. D'Aniello M, Portioli F, Landolfo R (2010) Modelling issues of steel braces under extreme cyclic actions. In: Mazzolani FM (ed) Urban habitat constructions under catastrophic events: Proceedings of the Cost C26 Action final conference. Naples, 16–18 September 2010. CRC Press/Balkema, Leiden, pp 335–341
8. Della Corte G, D'Aniello M, Mazzolani FM (2008) Inelastic response of shear links with axial restraints: numerical vs analytical results. In: Liew JYR, Choo YS (eds) Proceedings of the 5th international conference on advances in steel structures, ICASS 2007, vol III, General papers. Singapore, 5–7 December 2007. Research Publishing, Singapore, pp 651–656
9. Della Corte G, D'Aniello M, Landolfo R (2013) Analytical and numerical study of plastic overstrength of shear links. J Constr Steel Res 82:29–32
10. Ehrich F, Abramson HN (2002) Nonlinear vibration. In: Harris C, Piersol AG (eds) Harris' shock and vibration handbook, 5th edn. McGraw-Hill, New York, p 3
11. Gerber M, Engelbrecht L (1993) The bilinear oscillator: the response of an articulated mooring tower driven by irregular seas. Ocean Eng 20:113–133
12. Huang YM, Krousgrill CM, Bajaj AK (1989) Dynamic behaviour of offshore structures with bilinear stiffness. J Fluids Struct 3:405–422
13. Mazzolani FM, Della Corte G, D'Aniello M (2009) Experimental analysis of steel dissipative bracing systems for seismic upgrading. J Civ Eng Manag 15(1):7–19
14. Mendis P, Ngo T (2007) Vibration and shock problems of civil engineering structures. In: de Silva CW (ed) Vibration monitoring, testing, and instrumentation. CRC Press, Boca Raton, pp 32–33
15. Natsiavas S (1989) Periodic response and stability of oscillators with symmetric trilinear restoring force. J Sound Vib 134(2):315–331
16. Natsiavas S (1990) On the dynamics of oscillators with bilinear damping and stiffness. Int J Nonlinear Mech 25:535–554
17. Pall AS (1984) Response of friction damped buildings. In: Earthquake engineering research institute staff, Proceedings of 8th World conference on earthquake engineering, San Francisco, 21–28 July 1984, vol V. Prentice Hall, Upper Saddle River, pp 1007–1014
18. Pall AS, Marsh C (1982) Response of friction damped braced frames. J Struct Div ASCE 108(ST6):1313–1323
19. Scholl RE (1984) Brace dampers: an alternative structural system for improving the earthquake performance of buildings. In: Earthquake engineering research institute staff, Proceedings of 8th World conference on earthquake engineering, San Francisco, 21–28 July 1984, vol V. Prentice Hall, Upper Saddle River, pp 1015–1022
20. Thompson JMT, Bokaian AR, Ghaffari R (1983) Subharmonic resonances and chaotic motions of a bilinear oscillator. IMA J Appl Math 31:207–234
21. Wilson JF, Callis EG (2004) The dynamics of loosely jointed structures. Int J Nonlinear Mech 39:503–514

Chapter 22
In-Plane Free Vibration and Stability of High Speed Rotating Annular Disks and Rings

Hamid R. Hamidzadeh and Ehsan Sarfaraz

Abstract Analytical method is presented for the determination of free vibration characteristics of high speed viscoelastic rotating disks. In the development of this analytical solution, two-dimensional elastodynamic theory is employed and the viscoelastic material for the medium is allowed by assuming complex elastic moduli. The general governing equations of motion are derived and a solution for a single rotating disk with different boundary conditions is developed for a wide range of rotating speeds and any radius ratios, such as those for solid disks or thin rings. The proposed solution is used to investigate the influences of hysteretic material damping on dimensionless natural frequencies and modal loss factors for the rotating disks. Furthermore, the solution is expanded to consider the effect of adding disk segment with different material on the inner or outer sides of a disk on the natural frequencies and critical speeds of the equivalent single disk. The dimensionless results for these cases are presented for a wide range of rotational speeds.

Keywords In-plane free vibration • Plane stress • Annular disk • Rotating disks • Rotating rings • Natural frequency • Modal loss factor • Compound disks • Discontinuous medium • Critical speed

22.1 Introduction

Due to immense potential applications of the flexible thin rotating disks, the significance of their vibration characteristics has been emphasized in recent years. Rotating disks are the principal components in various rotating machinery. Their

---

H.R. Hamidzadeh (✉) • E. Sarfaraz
Department of Mechanical and Manufacturing Engineering,
Tennessee State University, Nashville, TN 37209, USA
e-mail: HHAMIDZADEH@Tnstate.edu; esarfara@my.tnstate.edu

J.A.T. Machado et al. (eds.), *Discontinuity and Complexity in Nonlinear Physical Systems*, Nonlinear Systems and Complexity 6, DOI 10.1007/978-3-319-01411-1_22,

applications can vary from space structures to torsional disk dampers and from turbine rotors to computer storage devices and brake systems. It is known that dynamic response and stability of rotating disk depends on its rotational speed. It should be noted that to design a rotating disk, the knowledge of modal vibrations and critical speeds are essential.

Vibration of rotating disk can occur as two types, in-plane and out-of-plane bending vibration. In-plane vibration occurs in the radial direction and can be coupled with the causing torsional vibration. Torsional vibration can occur in the disk surface angular displacement only which can vary with the radius. Out-of-plane bending vibration, the so-called transverse vibration, occurs on the direction occurring perpendicular to the plane of rotation.

Depending on the amplitude of vibrations, the established publications have used linear or nonlinear approaches. In the linear methods, the effect of higher-order terms in the strain–displacement relations is neglected. In the nonlinear theory of vibration, the effects of higher-order strain terms are taken into account and for most cases they have given approximate solutions. Based on both of these approaches, disk deflection will become unbounded at critical speeds corresponding to flutter or and divergence instabilities. In fact, in these unstable cases, the disk deflection is increased beyond the acceptable range of linear modal, and it is necessary to use nonlinear analysis for better predictions of the dynamics of spinning disks.

While the linear and nonlinear transverse vibrations of rotating disk have received higher attention; nevertheless, knowledge of the in-plane vibration of rotating disks is also essential for design of rotating disks. In practice, the problem of rotating disks is far more relevant to applications such as computer hard disks, turbine rotors, and circular saw blades. It should be noted that the vibration analysis of rotating disks has more complexities than that of a stationary disk subject to a rotating load. This complexity is due to the Coriolis and centripetal acceleration terms associated with the relative motion of the spinning disk.

The problems of in-plane vibration of rotating disks have been addressed by a few investigators. Bhuta and Jones [1] have presented a solution to the symmetric in-plane vibrations of a thin rotating circular disk for some specific modes. Burdess et al. [2] presented generalized formulation to consider asymmetric in-plane vibrations, while the effect of rotational speed on forward and backward traveling wave was discussed only for the mode with two nodal diameters. In their study, the equations of motion of a thin rotating disk were derived and a solution was achieved. Moreover, they studied free and forced vibrations and presented their results for the stability and resonant behavior of the disk. Before Chen and Jhu [3], in most of previous studies, the disk was assumed to be full. Chen and Jhu [3] determined the free in-plane vibration of a thin spinning annular disk and investigated the effects of clamping ratio on the natural frequencies and stability of disks. They extended their analysis to study the divergence instability of spinning annular disks clamped at the inner edge and free at the outer boundary. They also considered the effect of a radius ratio on the natural frequencies and critical speeds of the disk. Chen and Jhu [4] derived an analytical solution for the in-plane stress and displacement distributions in a spinning annular disk under stationary edge loads. Their numerical results showed that as the rotational speed of the disk approaches zero, the in-plane stresses and displacements are shown analytically to recover the solution derived through the Airy stress function in the classical theory of linear elasticity. Hamidzadeh and Dehghani [5] investigated the linear in-plane vibration of an elastic rotating disk and studied the effect of rotational speed and radius ratio on natural frequency and elastic stability of fixed–free vibration rotating disks. Hamidzadeh [6] also developed an analytical solution for in-plane vibration of spinning rings. Hamidzadeh's previous solution for the rotating disk was extended to investigate an analytical method for the determination of modal vibration of high speed double-segment compound rotating disks [7]. More specifically, a systematic approach for a compound rotating disk based on an established solution for linear in-plane vibration of each segment was developed by satisfying the displacements and stresses compatibilities. He also presented variation of the dimension natural frequencies for a number of modes versus non-dimensional speed of rotation for a fixed–free annular disk for the non-dimensional speeds ranging from 0 to 1.5 [8]. Deshpande and Mote [9] studied the stability of a spinning thin disk using a nonlinear strain in order to account changes in stiffness of the disk due to rotation. Their study suggested that the critical speeds were different using the linear strain assumption. Sarfaraz and Hamidzadeh [10] studied the effect of material hysteretic damping of the disk on the natural frequencies and mode shapes of a fixed–free rotating disk by considering constant complex elastic moduli.

This research report represents the linear in-plane free vibration of a thin viscoelastic annular rotating disk. In the development of the analytical solution, two-dimensional elastodynamic theory is employed and the viscoelastic material for the medium is allowed by assuming complex elastic moduli. The mathematical model is reduced to a wave propagation problem and time-dependent and time-independent modes are considered. The general governing equations of motion are derived by implementing plane stress theory. The natural frequencies and respective modal displacements and stresses are achieved by satisfying the inner and outer boundary conditions. The non-dimensional natural frequencies and modal loss factors for different boundary conditions are computed and presented for several modes, specific radius ratios, and material loss factors. Also, the critical speeds for rotating disks and rings are determined. Furthermore, the influences of embedded disk segments with a different material at one of the edges of the main disk on modal parameters are investigated.

22.2 Governing Equations

The material of the disk is assumed to be homogeneous, viscoelastic, and isotropic. The disk is rotating at a constant angular speed without any acceleration. The two-dimensional theory of elasticity is applied to derive the stress and strain in polar

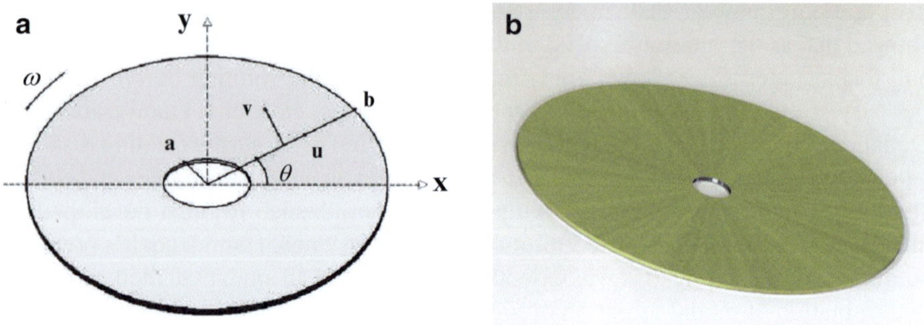

Fig. 22.1 (**a**) A typical rotating annular disk, (**b**) geometry of an annual disk in polar coordinate

coordinates. These relationships are then implemented into the dynamic equilibrium equations to derive the governing equations of motion. Figure 22.1 shows the radial and tangential displacements of a point in polar coordinates (r, θ). As it was presented by Hamidzadeh [8], equations of motion in terms of dilatation Δ and elastic rotation ψ or the freely rotating annular disk are given by:

$$\left.\begin{aligned} c_1^2\nabla^2\Delta - \ddot{\Delta} + \omega^2\Delta + 2\omega\dot{\psi} &= -2\omega^2 \\ c_2^2\nabla^2\psi - \ddot{\psi} + \omega^2\psi - 2\omega\dot{\Delta} &= 0 \end{aligned}\right\} \tag{22.1}$$

where

$$\left.\begin{aligned} \psi &= \frac{\partial v}{\partial r} - \frac{1}{r}\frac{\partial u}{\partial \theta} + \frac{v}{r}, \\ \Delta &= \frac{\partial u}{\partial r} + \frac{1}{r}\frac{\partial v}{\partial \theta} + \frac{u}{r}, \\ \nabla^2 &= \frac{\partial^2}{\partial r^2} + \frac{1}{r}\frac{\partial}{\partial r} + \frac{1}{r^2}\frac{\partial^2}{\partial \theta^2}, \\ E^* &= E\left(1 + i\eta\right), \\ G^* &= \frac{E^*}{2\left(1+\nu\right)}, \\ c_1^2 &= \frac{E^*}{\left(1-\nu^2\right)\rho}, \\ c_2^2 &= \frac{G^*}{\rho}. \end{aligned}\right\} \tag{22.2}$$

The functions u and v are radial and tangential displacements. υ, E^*, and G^* are Poisson ratio and complex elastic and shear moduli for the viscoelastic medium.

22.3 Solution to Governing Equation

The following solutions can be assumed for Eq. (22.1):

$$\left.\begin{aligned} \Delta &= \delta_0(r) + \Sigma\Delta_n(r)e^{i(n\theta+pt)}, \\ \psi &= \Sigma i\psi_n(r)e^{i(n\theta+pt)}. \end{aligned}\right\} \tag{22.3}$$

where Δ_n and ψ_n are time-dependent functions and $\delta_0(r)$ is a time-independent function. Also, Δ is time-dependent dilatation, ψ is time-dependent elastic rotation, n is any integer number, and p is the frequency of vibration. In the time-independent part, $\delta_0(r)$ is only a function of r so the solution of that can be given by

$$\left.\begin{aligned} \delta_0 &= A_0J_0\left(K_{1r}\right) + B_0Y_0\left(K_1r\right) - 2, \\ K_1 &= \frac{\omega}{c_1}. \end{aligned}\right\} \tag{22.4}$$

The governing equation for time dependence is

$$\left.\begin{aligned} c_1^2\nabla^2\Delta - \ddot{\Delta} + \omega^2\Delta + 2\omega\dot{\psi} = 0 \\ c_2^2\nabla^2\psi - \ddot{\psi} + \omega^2\psi - 2\omega\dot{\Delta} = 0 \end{aligned}\right\}. \tag{22.5}$$

The time-dependent equations have a significant role in determining the natural frequencies and mode shapes of the system. To continue with derivation of the final solution, it is convenient to introduce the following dimensionless variables:

$$\left.\begin{aligned} \Omega_1 &= \frac{\omega}{c_1}b, \Omega_2 = \frac{p}{c_1}b, \\ U_n^* &= \frac{U_n}{b}, V_n = \frac{V_n}{b}, \\ \sigma_{r_n}^* &= \frac{\sigma_r}{G}, \tau_{r\theta_n}^* = \frac{\tau_{r\theta}}{G}, \\ r* &= \frac{r}{b}, q = \frac{c_1}{c_2}. \end{aligned}\right\} \tag{22.6}$$

where Ω_1 and Ω_2 are non-dimensional speed and non-dimensional frequency. Substituting modal expression from Eq. (22.3) into governing Eq. (22.1), the two different ratios of modal elastic rotation to modal dilatation are expressed by the following equations:

$$\left.\begin{aligned} t_{1n} &= \frac{\psi_n(r)}{\Delta_n(r)} = \frac{2\Omega_1^*\Omega_2^*q^2}{-x_1^{*2} + q^2\left(\Omega_1^{*2} + \Omega_2^{*2}\right)}, \\ t_{2n} &= \frac{\psi_n(r)}{\Delta_n(r)} = \frac{2\Omega_1^*\Omega_2^*q^2}{-x_2^{*2} + q^2\left(\Omega_1^{*2} + \Omega_2^{*2}\right)}. \end{aligned}\right\} \tag{22.7}$$

where

$$x_1^*, x_2^* = \frac{-\left(1+q^2\right)\left(\Omega_1^2+\Omega_2^2\right) \pm \sqrt{\left[\left(1+q^2\right)\left(\Omega_1^2+\Omega_2^2\right)\right]^2 - 4\left[q^2\left(\Omega_1^2-\Omega_2^2\right)^2\right]}}{2}. \tag{22.8}$$

By using the Bessel function of the first and second kind, the solutions to the wave operators Δ_n and ψ_n are obtained:

$$\left.\begin{array}{l} \Delta_n(r) = B_n J_n\left(r^* x_1^*\right) + C_n Y_n\left(r^* x_1^*\right) + D_n J_n\left(r^* x_2^*\right) + E_n Y_n\left(r^* x_2^*\right), \\ \psi_n(r) = t_{1n} B_n J_n\left(r^* x_1^*\right) + t_{1n} C_n Y_n\left(r^* x_1^*\right) + t_{2n} D_n J_n\left(r^* x_2^*\right) \\ \qquad + t_{2n} E_n Y_n\left(r^* x_2^*\right). \end{array}\right\} \tag{22.9}$$

22.4 Modal Displacements and Stresses

The radial and tangential displacements in terms of time can be written by the following equations:

$$\left.\begin{array}{l} u\,(r,\theta,t) = U_n(r) e^{i(n\theta+pt)}, \\ v\,(r,\theta,t) = i V_n(r) e^{i(n\theta+pt)}. \end{array}\right\} \tag{22.10}$$

Substituting Eq. (22.9) and these displacements into equations of motions and rearranging, the result yields the modal solution for the non-dimensional radial and tangential displacements:

$$\left.\begin{array}{l} U_n^*(r) = m_1^* \left(\begin{array}{l} \left[x_1^* J_n'\left(r^* x_1^*\right) + \frac{n t_1 n}{q^2 r^*} J_n\left(r^* x_1^*\right)\right] B_n + \left[x_1^* Y_n'\left(r^* x_1^*\right) + \frac{n t_{1n}}{q^2 r^*} Y_n\left(r^* x_1^*\right)\right] C_n \\ + \left[x_2^* J_n'\left(r^* x_2^*\right) + \frac{n t_{2n}}{q^2 r^*} J_n\left(r^* x_2^*\right)\right] D_n + \left[x_2^* Y_n'\left(r^* x_2^*\right) + \frac{n t_{2n}}{q^2 r^*} Y_n\left(r^* x_2{}^*\right)\right] E_n \end{array}\right) + \\ m_2^* \left(\begin{array}{l} \left[\frac{n}{r^*} J_n\left(r^* x_1^*\right) + \frac{t_{1n}}{q^2} x_1^* J_n'\left(r^* x_1^*\right)\right] B_n + \left[\frac{n}{r^*} Y_n\left(r^* x_1^*\right) + \frac{t_{1n}}{q^2} x_1^* Y_n'\left(r^* x_1^*\right)\right] C_n \\ + \left[\frac{n}{r^*} J_n\left(r^* x_2^*\right) + \frac{t_{2n}}{q^2} x_2^* J_n'\left(r^* x_2^*\right)\right] D_n + \left[\frac{n}{r^*} Y_n\left(r^* x_2^*\right) + \frac{t_{2n}}{q^2} x_2^* Y_n'\left(r^* x_2^*\right)\right] E_n \end{array}\right) \\ \\ V_n^*(r) = m_2^* \left(\begin{array}{l} \left[x_1^* J_n'\left(r^* x_1^*\right) + \frac{n t_1 n}{q^2 r^*} J_n\left(r^* x_1^*\right)\right] B_n + \left[x_1^* Y_n'\left(r^* x_1^*\right) + \frac{n t_{1n}}{q^2 r^*} Y_n\left(r^* x_1^*\right)\right] C_n \\ + \left[x_2^* J_n'\left(r^* x_2^*\right) + \frac{n t_{2n}}{q^2 r^*} J_n\left(r^* x_2^*\right)\right] D_n + \left[x_2^* Y_n'\left(r^* x_2^*\right) + \frac{n t_{2n}}{q^2 r^*} Y_n\left(r^* x_2{}^*\right)\right] E_n \end{array}\right) + \\ m_1^* \left(\begin{array}{l} \left[\frac{n}{r^*} J_n\left(r^* x_1^*\right) + \frac{t_{1n}}{q^2} x_1^* J_n'\left(r^* x_1^*\right)\right] B_n + \left[\frac{n}{r^*} Y_n\left(r^* x_1^*\right) + \frac{t_{1n}}{q^2} x_1^* Y_n'\left(r^* x_1^*\right)\right] C_n \\ + \left[\frac{n}{r^*} J_n\left(r^* x_2^*\right) + \frac{t_{2n}}{q^2} x_2^* J_n'\left(r^* x_2^*\right)\right] D_n + \left[\frac{n}{r^*} Y_n\left(r^* x_2^*\right) + \frac{t_{2n}}{q^2} x_2^* Y_n'\left(r^* x_2^*\right)\right] E_n \end{array}\right) \end{array}\right\}, \tag{22.11}$$

where prime and double prime (′ and ″) represent first and second derivatives of the function and m_1^* and m_1^* can be presented by:

$$\left.\begin{aligned} m_1^* &= -\frac{\Omega_1^{*2}+\Omega_2^{*2}}{\left(\Omega_1^{*2}-\Omega_2^{*2}\right)^2}, \\ m_2^* &= -\frac{2\Omega_1^{*2}\Omega_2^{*2}}{\left(\Omega_1^{*2}-\Omega_2^{*2}\right)^2}. \end{aligned}\right\} \tag{22.12}$$

Similarly, the modal radial and shear stresses can be expressed by the following relations:

$$\left.\begin{aligned} \sigma_r\,(r,\theta,t) &= \sigma_{r_n}(r)e^{i(n\theta+pt)} \\ \tau_{r\theta}\,(r,\theta,t) &= i\,\tau_{r\theta_n}(r)e^{i(n\theta+pt)} \end{aligned}\right\} \tag{22.13}$$

The non-dimensional modal radial and shear stresses can be obtained by substituting from Eqs. (22.9) and (22.11) based on stress–strain relation, and after simplifications they are presented by

$$\left.\begin{aligned} &\sigma_{r_n}^* = \frac{\lambda}{G}\left[B_n J_n\left(r^*x_1^*\right)+C_n Y_n\left(r^*x_1^*\right)+D_n J_n\left(r^*x_2^*\right)+E_n Y_n\left(r^*x_2^*\right)\right]+ \\ &2\left[s_1^*x_1^{*2}J_n''\left(r^*x_1^*\right)+s_2^*\frac{n}{r^*}x_1^*J_n'\left(r^*x_1^*\right)-s_2^*\frac{n}{r^{*2}}J_n\left(r^*x_1^*\right)\right]B_n+2\left[s_1^*x_1^{*2}Y_n''\left(r^*x_1^*\right)\right. \\ &\left.+s_2^*\frac{n}{r^*}x_1^*Y_n'\left(r^*x_1^*\right)-s_2^*\frac{n}{r^{*2}}Y_n\left(r^*x_1^*\right)\right]C_n+2\left[s_3^*x_2^{*2}J_n''\left(r^*x_2^*\right)+s_4^*\frac{n}{r^*}x_2^*J_n'\left(r^*x_2^*\right)\right. \\ &\left.-s_4^*\frac{n}{r^{*2}}J_n\left(r^*x_2^*\right)\right]D_n+2\left[s_3^*x_2^{*2}Y_n''\left(r^*x_2^*\right)+s_4^*\frac{n}{r^*}x_2^*Y_n'\left(r^*x_2^*\right)-s_4^*\frac{n}{r^{*2}}Y_n\left(r^*x_2^*\right)\right]E_n \\ &\tau_{r\theta_n}^* = 2\left[s_2^*x_1^{*2}J_n''\left(r^*x_1^*\right)+s_1^*\frac{n}{r^*}x_1^*J_n'\left(r^*x_1^*\right)-s_1^*\frac{n}{r^{*2}}J_n\left(r^*x_1^*\right)\right]B_n+2\left[s_2^*x_1^{*2}Y_n\left(r^*x_1^*\right)\right. \\ &\left.+s_1^*\frac{n}{r^*}x_1^*Y_n'\left(r^*x_1^*\right)-s_1^*\frac{n}{r^{*2}}Y_n\left(r^*x_1^*\right)\right]C_n+2\left[s_4^*x_2^{*2}J_n''\left(r^*x_2^*\right)+s_3^*\frac{n}{r^*}x_2^*J_n'\left(r^*x_2^*\right)\right. \\ &\left.-s_3^*\frac{n}{r^{*2}}J_n\left(r^*x_2^*\right)\right]D_n+2\left[s_4^*x_2^{*2}Y_n''\left(r^*x_2^*\right)+s_3^*\frac{n}{r^*}x_2^*Y_n'\left(r^*x_2^*\right)-s_3^*\frac{n}{r^{*2}}Y_n\left(r^*x_2^*\right)\right]E_n \\ &-\left[t_{1n}B_n J_n\left(r^*x_1^*\right)+t_{1n}C_n Y_n\left(r^*x_1^*\right)+t_{2n}D_n J_n\left(r^*x_2^*\right)+t_{2n}E_n Y_n\left(r^*x_2^*\right)\right] \end{aligned}\right\}, \tag{22.14}$$

where

$$\left.\begin{aligned} s_1^* &= m_1^* + m_2^*\frac{t_{1n}}{q^2}, \\ s_2^* &= m_2^* + m_1^*\frac{t_{1n}}{q^2}, \\ s_3^* &= m_1^* + m_2^*\frac{t_{2n}}{q^2}, \\ s_4^* &= m_2^* + m_1^*\frac{t_{2n}}{q^2}. \end{aligned}\right\} \tag{22.15}$$

Using Eqs. (22.11) and (22.14), the modal displacements and stresses at any radius for each part of an annular disk can be expressed in the following form:

$$\begin{Bmatrix} U_n^*(r) \\ V_n^*(r) \\ \sigma_{rn}^*(r) \\ \tau_{r\theta n}^*(r) \end{Bmatrix} = \left[A_n(r)\right]\begin{Bmatrix} B_n \\ C_n \\ D_n \\ E_n \end{Bmatrix}, \tag{22.16}$$

where

$$[A_n(r)] = \begin{bmatrix} a_{11} & a_{12} & a_{13} & a_{14} \\ a_{21} & a_{22} & a_{23} & a_{24} \\ a_{31} & a_{32} & a_{33} & a_{34} \\ a_{41} & a_{42} & a_{43} & a_{44} \end{bmatrix}. \tag{22.17}$$

Elements of $A_n(r)$ are in terms of material properties and Bessel functions of first and second kinds. These elements are presented in the above-mentioned paper [8].

22.5 Natural Frequency Equation

To determine the modal parameters, the boundary conditions must be satisfied. For example, for the fixed–free boundary conditions, it is required that the modal displacements at the inner edge and the modal radial and shear stresses at the outer edge must be zero. By implementing the boundary conditions in Eq. (22.16) and combining them, displacements and stresses at the boundaries are related in the following form:

$$\begin{Bmatrix} U_n^*(b) \\ V_n^*(b) \\ 0 \\ 0 \end{Bmatrix} = [A_n(b)]\,[A_n(a)]^{-1} \begin{Bmatrix} U_n^*(a) \\ V_n^*(a) \\ 0 \\ 0 \end{Bmatrix}. \tag{22.18}$$

Considering that the matrix $[A_n(b)][A_n(a)]^{-1}$ is presented in the following form:

$$[A_n(b)]\,[A_n(a)]^{-1} = \begin{bmatrix} d_{11} & d_{21} & d_{31} & d_{41} \\ d_{21} & d_{22} & d_{23} & d_{24} \\ d_{31} & d_{32} & d_{33} & d_{34} \\ d_{41} & d_{42} & d_{43} & d_{44} \end{bmatrix}. \tag{22.19}$$

Then Eq. (22.18) can be reduced to the following expression in terms of the inner boundary stresses:

$$\begin{bmatrix} d_{31} & d_{32} \\ d_{41} & d_{41} \end{bmatrix} \begin{Bmatrix} U_n^*(a) \\ V_n^*(a) \end{Bmatrix} = \begin{Bmatrix} 0 \\ 0 \end{Bmatrix}. \tag{22.20}$$

In order to obtain a nonzero solution for the stresses, the determinant of the matrix in Eq. (22.20) must be zero. This results in the frequency equation for the system:

$$\begin{vmatrix} d_{31} & d_{32} \\ d_{41} & d_{41} \end{vmatrix} = 0 \tag{22.21}$$

The above equation is a function of circumferential wave number n, and other dimensionless parameters including Ω_1. For given values of $n, a/b, \upsilon, \Omega_1$, and material loss factor η_m there are infinite real values for Ω_2 that satisfy this equation. It should be noted that the dimensionless frequencies in the rotating coordinate system are given by the absolute values of Ω_2:

$$\Omega_R = |\Omega_2| \tag{22.22}$$

However, for viscoelastic disk, since modulus of elasticity for damping material is complex, then Ω_2 in Eq. (22.22) would be complex. In order to obtain the modal loss factor and the natural frequencies for the viscoelastic rotating disks, the following procedures are implemented:

$$\left.\begin{array}{l} \Omega_2^* = x + iy = \sqrt{x^2 + y^2} e^{i\alpha}, \\ \alpha = \tan^{-1}\frac{y}{x}, \\ \Omega_2^{*2} = \left(x^2 + y^2\right) e^{i2\alpha}, \\ \Omega_2^{*2} = \left(x^2 + y^2\right) (\cos 2\alpha + i \sin 2\alpha), \\ \Omega_2^{*2} = \left(x^2 + y^2\right) \cos 2\alpha \left[1 + i \tan 2\alpha\right], \\ \Omega_2^* = \Omega_2 (1 + i\eta_L). \end{array}\right\}, \tag{22.23}$$

where η_L is modal loss factor and is obtained in following form:

$$\eta_L = \tan 2\alpha \tag{22.24}$$

and the natural frequencies are given by:

$$\Omega_2 = \sqrt{\frac{x^2 + y^2}{\sqrt{1 + \eta_L^2}}}. \tag{22.25}$$

For mode shapes $n > 0$, if the direction of oscillating wave is the same as that of rotation of the disk ($p > 0$) in rotating coordinates, the wave is defined as forward wave in rotating coordinates. If the direction of oscillating wave is opposite to that of rotation of the disk ($p < 0$) in rotating coordinates, the wave is defined as backward wave in rotating coordinates. For mode shapes $n < 0$, if the direction of oscillating wave is the same as that of rotation of the disk ($p_F > 0$) in fixed coordinates, the wave is defined as forward wave in fixed coordinates. If the direction of oscillating wave is opposite to that of rotation of the disk ($p_F < 0$) in fixed coordinates, then the wave defined as backward wave in fixed coordinates. Thus, the relation between natural frequency in fixed coordinates (p_F) and rotating coordinates (p) and the relation between dimensionless natural frequencies in fixed and rotating coordinate system can be presented by the following equations:

$$\Omega_F = |\Omega_2 + n\Omega_1| \quad for \quad \Omega_2 > 0 \tag{22.26a}$$

$$\Omega_F = |\Omega_2 - n\Omega_1| \quad for \quad \Omega_2 < 0. \tag{22.26b}$$

Mode shapes for the in-plane free vibration of a rotating disk can be identified by the number of circular node numbers (m) and the number of nodal diameters (n). It should be noted that the lower modes ($m = 0, 1, 2, 3$ and $n = 0, 1, 2, 3$) have been found to be the dominant modes of vibration for vibration of the rotating disks.

22.6 Natural Frequencies and Critical Speeds

The lowest frequency at which the disk vibrates freely is called the fundamental mode. When the disk is excited at one of its resonance frequencies, respective nodal circle(s) and nodal diameter(s) appear. To determine the natural frequency of the system, the boundary conditions must be satisfied both at the inner radius of the disk ($r = a$) and the outer radius of disk ($r = b$). Considering that for the fixed–free rotating disks the modal displacements are zero at the inner radius and modal stresses are zero at the outer radius, then non-dimensional natural frequencies can be determined for any particular non-dimensional rotating speed and a given geometry by using Eq. (22.21).

The variations of the dimensionless natural frequencies of a thin annular disk with Poisson ratio of 0.3 and no material damping in the fixed coordinate for a number of modes are presented here. The boundary conditions considered are free–free, fixed–free, and free–fixed. It should be noted that critical speed for rotating disk is the speed of rotation at which the resonant frequency is zero. Needless to say that in general the annular disk has infinite number of natural frequencies with any combinations of positive integer values for n or m. Thus there are infinite possible numbers of critical speeds for any rotating disk. In this section, the results of the proposed solution are compared with the available data [2, 9]. The comparisons demonstrated excellent agreement among the present result and the available data. This comparison is depicted in Fig. 22.2.

Figure 22.3 presents the variation of dimensionless critical speeds for different modes of free–free boundary conditions versus radius ratios of the rotating disks. Illustrated results show that as the radius ratio increases, the critical speed decreases. In addition, for mode numbers of $n = 2$ and higher, the critical speed reduces to zero where radius ratio approaches to one. Figure 22.4 demonstrates the variation of dimensionless critical speed for fixed–free rotating disk versus radius ratio for different wave numbers of n. As depicted, since the disk is fixed at the inner and free at the outer radius, as the radius ratio increases, the critical speed increases, and for $n = 0$, as the radius ratio approaches zero, the critical speed approaches to zero.

Figure 22.5 shows the variations of dimensionless natural frequencies that are experienced in fixed coordinates for free–free conditions versus dimensionless speed for different modes (m, n) and a radius ratio $a/b = 0.1$. Figures 22.6 and 22.7 show the same results for disk with similar geometry for two different boundary conditions of fixed–free and free–fixed. The presented results are extended for a wide range of dimensionless rotational speed well beyond the speeds previously presented in the established publications. Please note that labels b and f refer to the backward and forward waves in the presented figures.

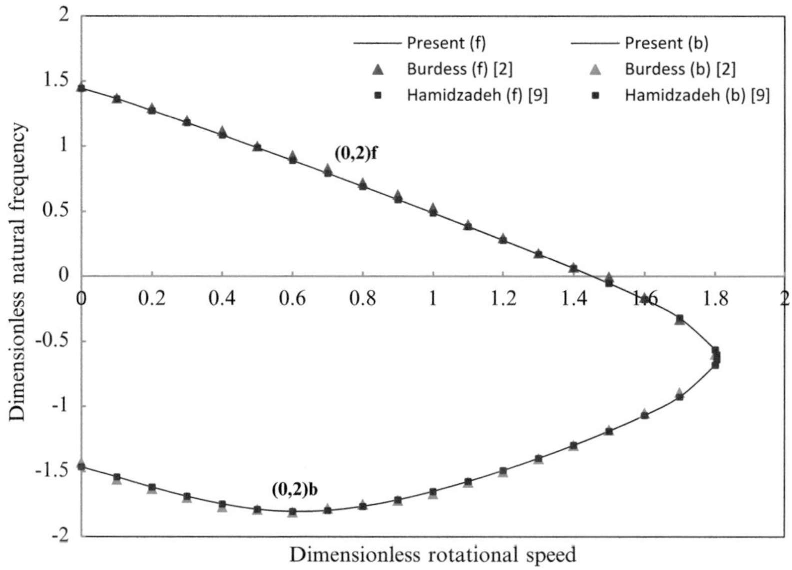

Fig. 22.2 Comparison of dimensionless natural frequencies for $m = 0$ and $n = 2$ with established results

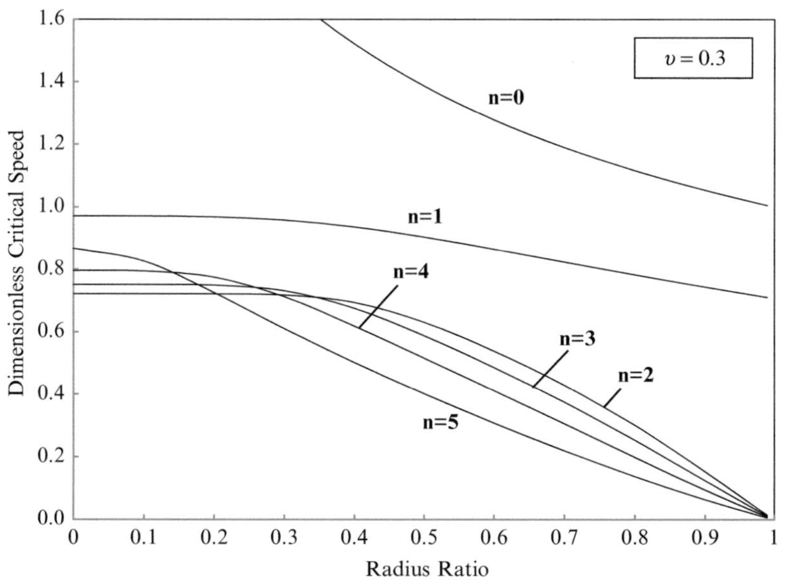

Fig. 22.3 Variation of dimensionless critical speed versus radius ratio for different modes of free–free rotating disks

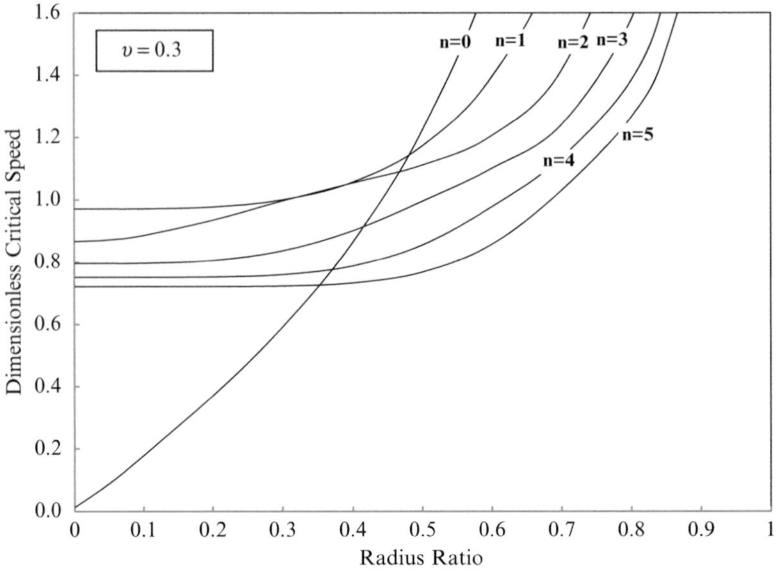

Fig. 22.4 Variation of dimensionless critical speed versus radius ratio for different modes of fixed–free rotating disks

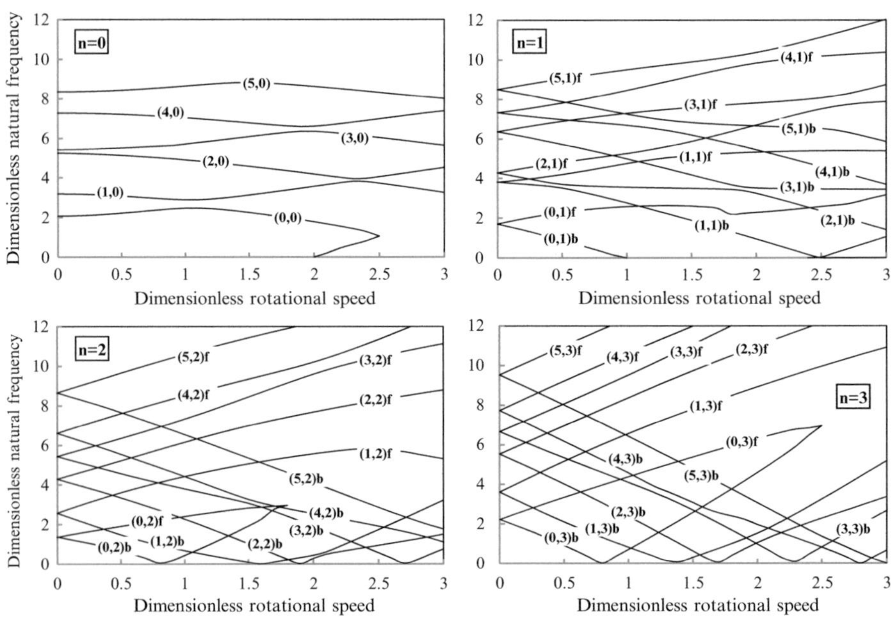

Fig. 22.5 Variation of dimensionless natural frequency versus dimensionless speed for different modes of a free–free disk with a radius ratio $a/b = 0.1$

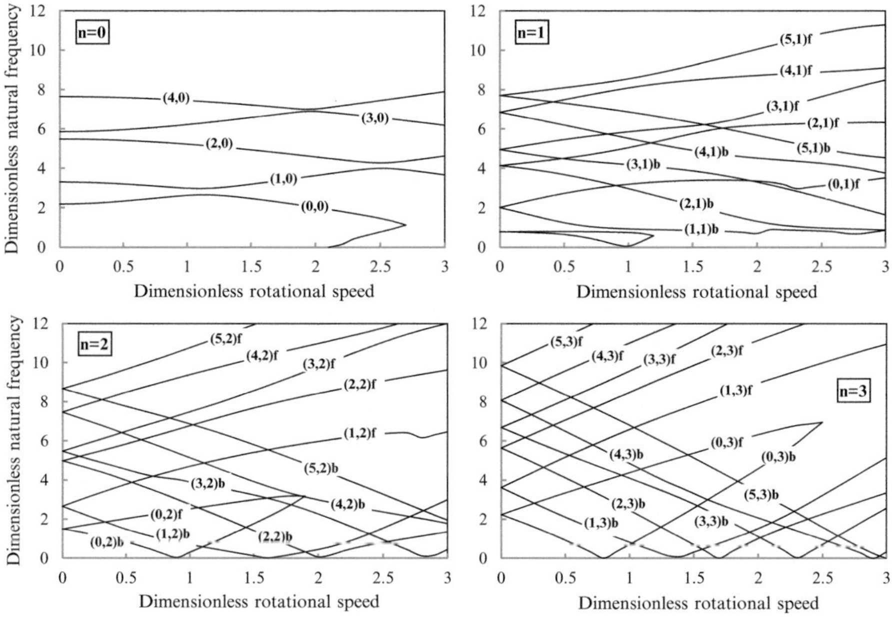

Fig. 22.6 Variation of dimensionless natural frequency versus dimensionless speed for different modes of a fixed–free disk with a radius ratio $a/b = 0.1$

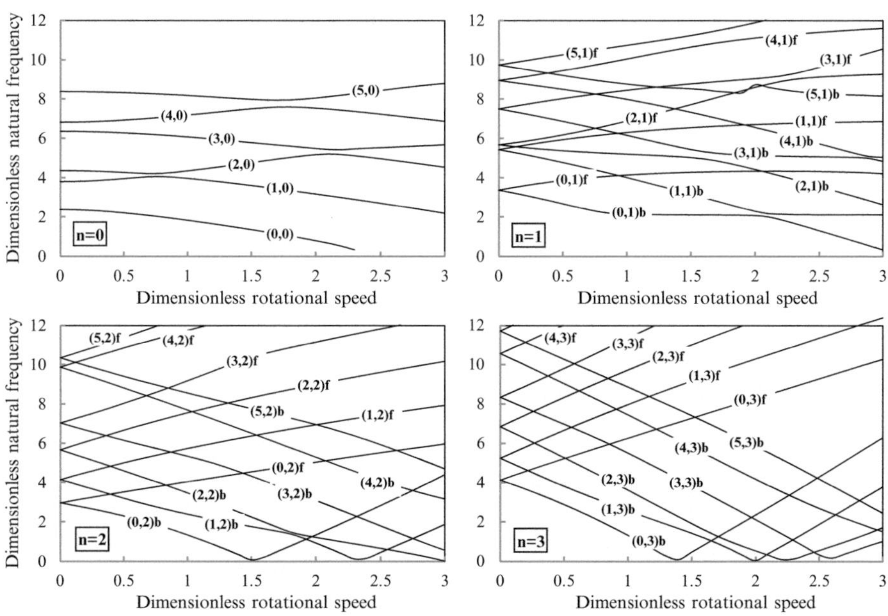

Fig. 22.7 Variation of dimensionless natural frequency versus dimensionless speed for different modes of a free–fixed disk with a radius ratio $a/b = 0.1$

22.7 The Modal Loss Factor of Viscoelastic Rotating Disk

The main objective of this section is to provide an accurate method for predicting the natural frequencies and modal loss factors for in-plane vibration of a rotating annular disk made of viscoelastic material for a specified boundary conditions. The material damping considered is based on typical hysteretic damping with complex elastic moduli. The viscoelastic material can provide the needed structure stiffness with possibility of dissipating vibration energy. To determine the influence of material loss factor on the non-dimensional natural frequencies and their corresponding modal loss factors, computed results for a certain radius ratio of *a*/*b*, Poisson's ratio of 0.3, and wide range of material loss factors are provided in this section.

Figures 22.8, 22.9, and 22.10 show variation of dimensionless modal loss factors versus dimensionless speed for a fixed–free viscoelastic rotating annular disk with a radius ratio 0.2 and different wave numbers. The presented results are for hysteretic damping with material loss factors of 0.05, 0.1, 0.3, 0.5, and 0.7. As shown, each curve presented in Fig. 22.11 depict the effect of different material loss factors on the non-dimensional natural frequencies for the mode associated with $m = 2$ and $n = 2$. It could be observed that by increasing wave number of n, modal loss factors are decreased.

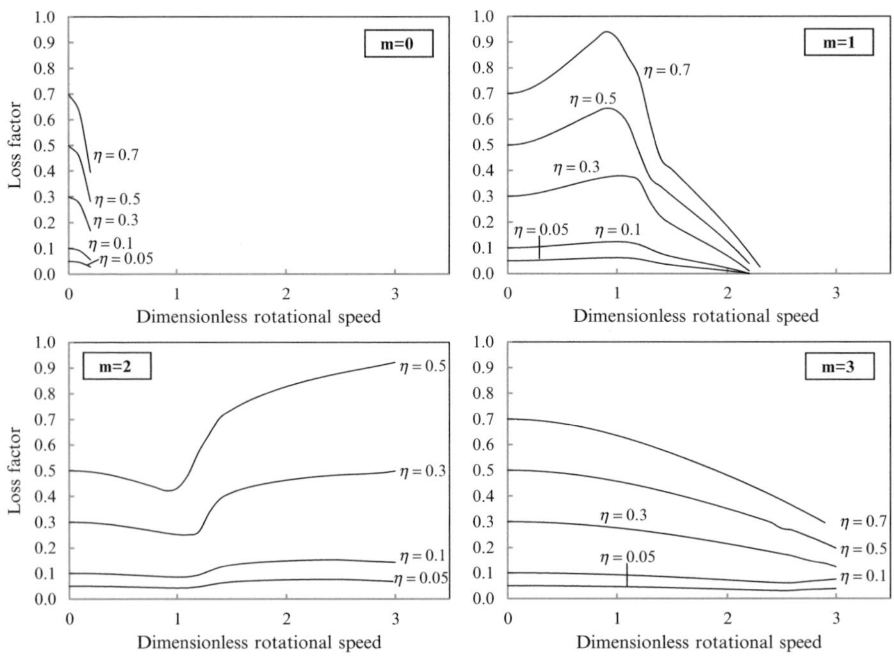

Fig. 22.8 Variation of dimensionless modal loss factor versus dimensionless speed for a fixed–free rotating disk with different material loss factor for radius ratio 0.2, $n = 0$, and $m = 0$ to 3

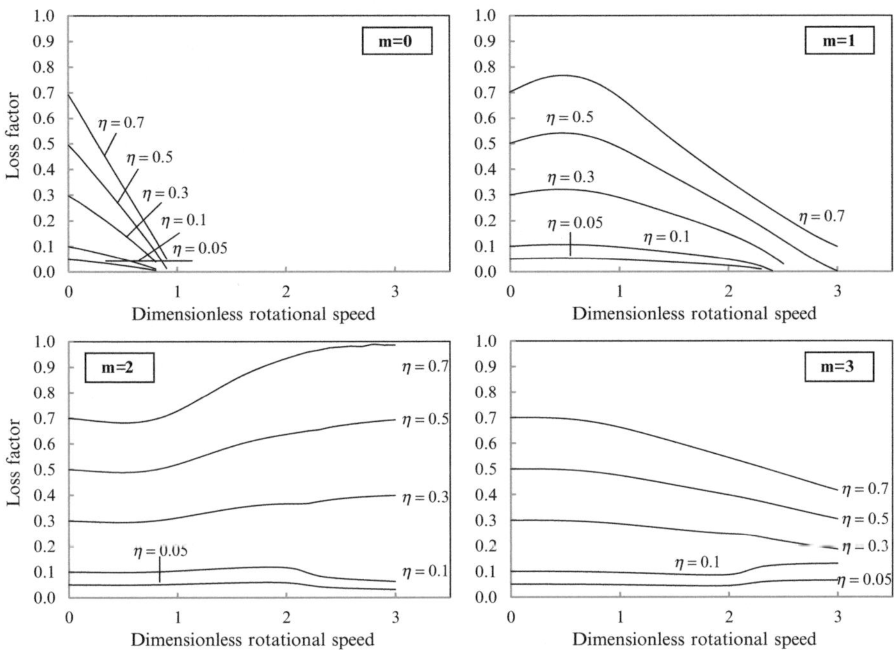

Fig. 22.9 Variation of dimensionless modal loss factor versus dimensionless speed for a fixed–free rotating disk with different material loss factor for radius ratio 0.2, $n = 1$, and $m = 0$ to 3

22.8 Natural Frequencies of Rotating Rings

Vibration abatement and structural stability in high speed rotating rings is one of the most prevalent problems in engineering practice. An important step in the study of these rotating structural components is the evaluation of the modal parameters such as mode shapes, natural frequencies, and critical speeds. This information has immense practical importance when designing for these components. It is known that in-plane motion of a point in the medium is combination of radial and circumferential displacements, and the natural frequencies depend on the rotational speed. The literature on dynamic response of rotating rings is mainly restricted to the application of shell or curved beam theories. The ring-like components is of great interest in mechanical systems. For the in-plane vibration of rings, they can be modeled by annular disks with radius ratios very close to one. Thus the general governing equation and natural frequency equation for the rotating annular disk are also valid to determine all the model parameters for ring when its boundary conditions are satisfied. Figures 22.12 and 22.13 show the variation of dimensionless natural frequencies in fixed coordinate versus dimensionless speed of a ring with radius ratio of 0.9 for two different boundary conditions of free–free and fixed–free and different wave numbers of $n = 0, 1, 2,$ and 3 and $m = 0$.

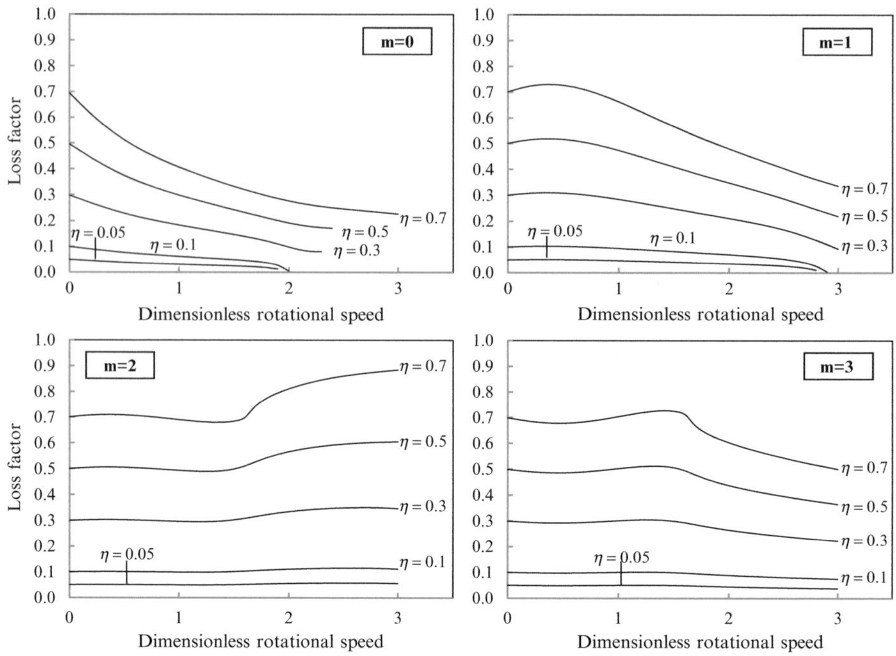

Fig. 22.10 Variation of dimensionless modal loss factor versus dimensionless speed for a fixed–free rotating disk with different material loss factor for radius ratio 0.2, $n = 2$, and $m = 0$ to 3

22.9 Effect of Embedded Different Material on Natural Frequencies

This section presents non-dimensional natural frequencies versus dimensionless rotating speeds for compound rotating annular disks with added disk segments with different materials at the inner or outer edge of the main disk. Figure 22.14 illustrates small embedded segments of higher stiffness and density at one of the edges of the rotating disk. Computation was performed to determine the effect of an added disk segment on the dimensionless natural frequency at different rotating speeds. This was done by considering the general solution for stresses and displacement at inner and outer edges of the main disk and the added disk segment using Eq. (22.14). The frequency equation for each mode can be determined by satisfying the compatibility of stresses and displacements at the interface between the main and the added disk segment as well as the boundary conditions of the compound disk. Analysis was conducted for three cases with the same inner to outer radius ratio of 0.2 and fixed–free boundary conditions. In case I, the disk is a single disk made of aluminum. Case II is for an aluminum main disk with added steel disk segment at the inner edge, and case III is an aluminum main disk with added steel disk segment at the outer edge. Non-dimensional frequencies in rotating coordinates for these three

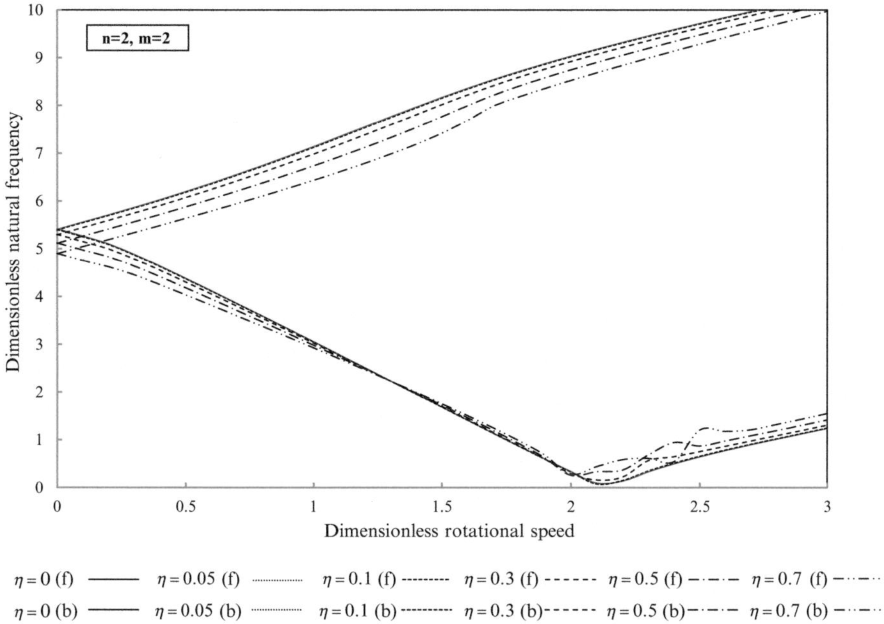

Fig. 22.11 Variation of dimensionless natural frequency versus dimensionless speed for different hysteretic material damping of a fixed–free disk with a radius ratio 0.2 and mode of $n = 2$, $m = 2$

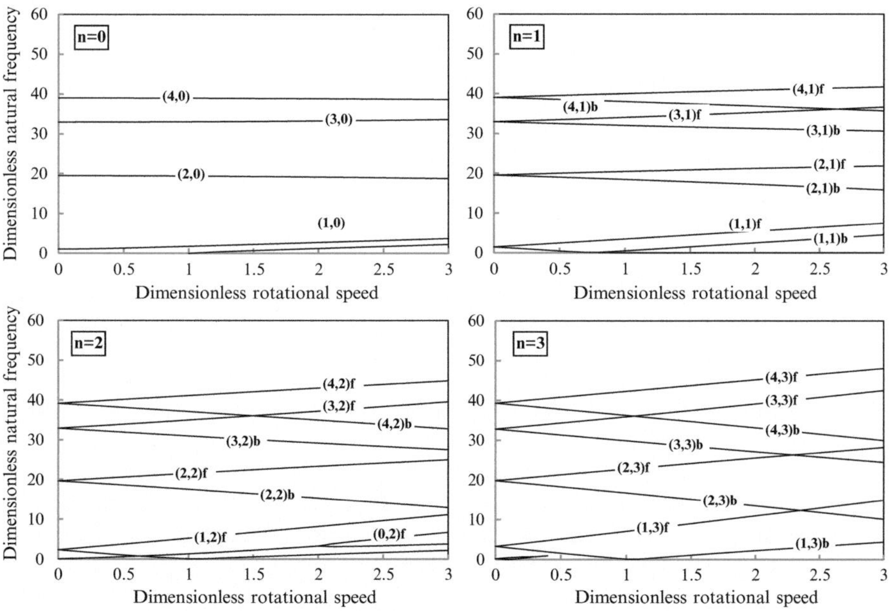

Fig. 22.12 Variation of dimensionless natural frequency versus dimensionless speed for different modes of a free–free ring with a radius ratio 0.9 for different modes

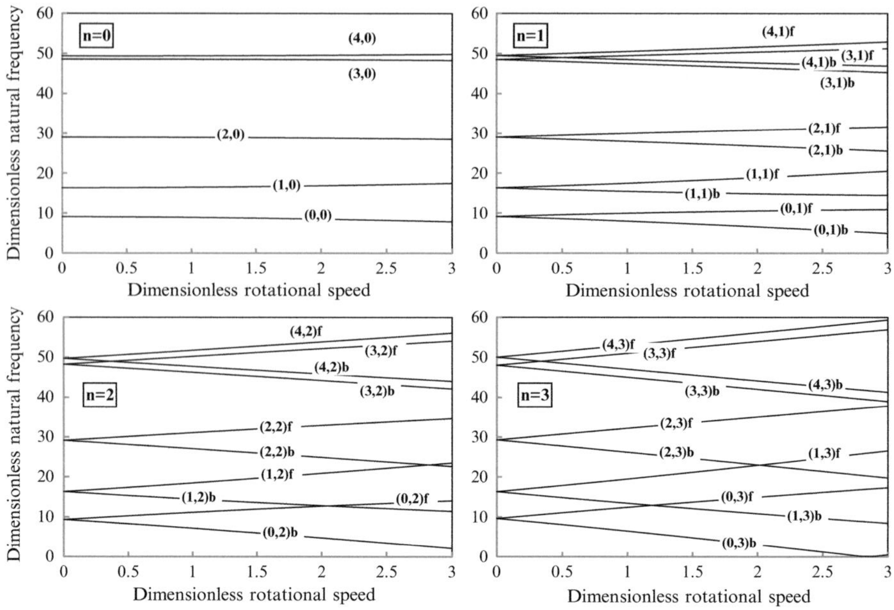

Fig. 22.13 Variation of dimensionless natural frequency versus dimensionless speed for different modes of a free–fixed ring with a radius ratio 0.9 for different modes

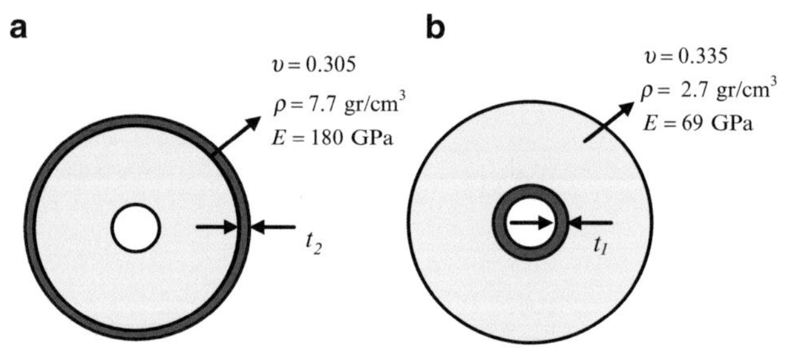

Fig. 22.14 Case (**a**): The disk has small segment of higher mass around of outer edge. Case (**b**): The disk has small segment of higher mass around of inner side

cases for a wide range of dimensionless rotating speeds and for $t_1 = (b - a)/c = 0.05$ and $t_2 = (c - b)/c = 0.05$ are illustrated in Fig. 22.15a.

Similar results for $t_1 = t_2 = 0.15$ are shown in Fig. 22.15b. The presented results are for $n = 0$, and $m = 0$, 1, 2, 3, *and* 4. The modulus of elasticity, mass density, and Poisson's ratio for aluminum disk and steel are assumed to be (180 GPa, 7,700 kg/m^3, and 0.305) and (69 GPa, 2,700 kg/m^3, and 0.335), respectively.

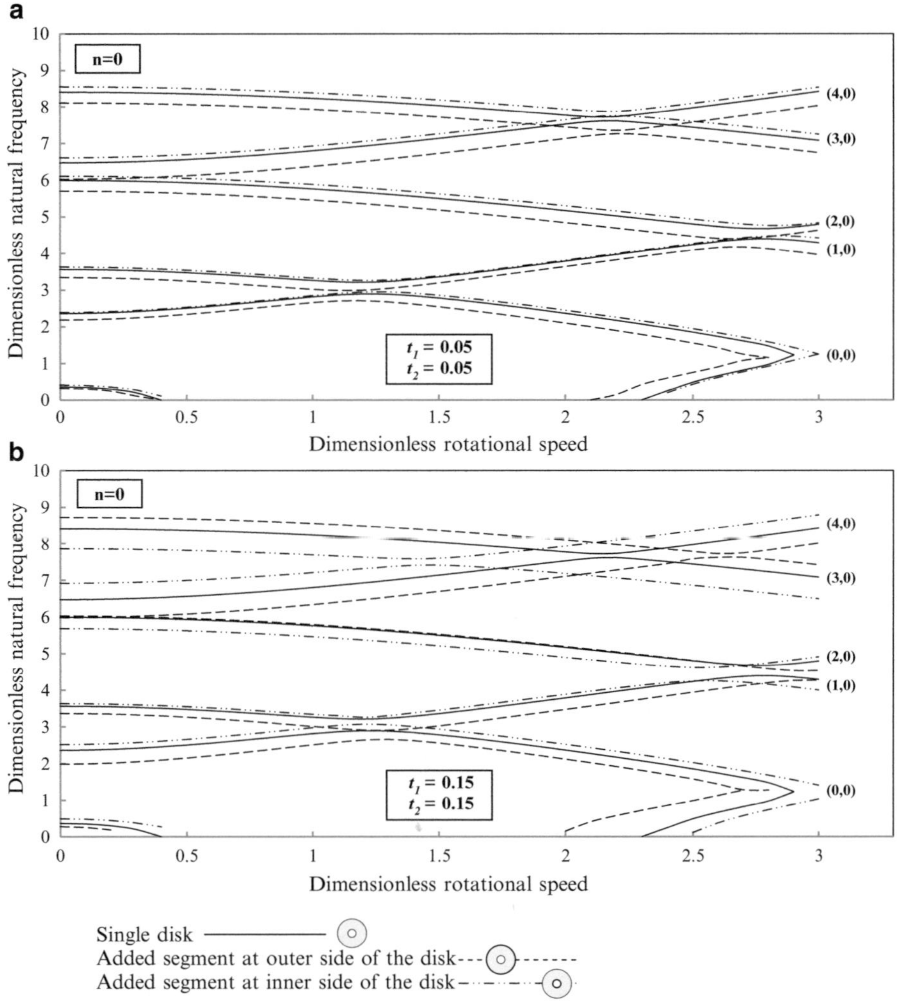

Fig. 22.15 Variation of dimensionless natural frequencies versus dimensionless speed of rotation for fixed–free disks with/without added segment for $t = 0.05$ and $t = 0.15$ for radius ratio of 0.2, $n = 0$ and $m = 0, 1, 2, 3, 4$

22.10 Conclusion

In this research report, an analytical method has been developed to determine the natural frequency and critical speed for in-plane vibration of a homogeneous, isotropic viscoelastic rotating disk for a wide range of rotational speeds. The modal vibration characteristics of in-plane vibration for annular rotating disks are studied for different types of boundary conditions, i.e., free–free, fixed–free, and free–

fixed. The proposed method of solution in this investigation can be effectively applied to determine the modal vibration characteristics of a high speed rotating annular disk. The provided method is capable of computing dimensionless natural frequencies for all modes at any rotating speeds. Furthermore, modal loss factor and stability of a rotating disk with hysteretic material damping ratio have been computed by considering complex natural frequencies. It was observed that the effect of rotational speed on natural frequency depended on the radius ratio, the mode of vibration, Poisson's ratio, stiffness, mass density of the material, and material damping. The presented solution is also capable of determining modal information for the in-plane vibration of rings by considering the radius ratio of the ring, which is slightly less than 1. Moreover, it was observed that a small segment of a material of higher density and elasticity modulus attached around the inner side of rotating annular disk induced higher natural frequencies and promotes a better dynamic stability for a disk. The presented results can provide a guideline to assist designers by choosing appropriate geometry and material properties to avoid critical speeds and possible resonances for obtaining desired operating speed.

References

1. Bhuta PG, Jones JP (1963) Symmetric planar vibrations of a rotating disk. J Acoust Soc Am 35(7):982–989
2. Burdess JS, Wren T, Fawcett JN (1987) Plane stress vibrations in rotating disks. Proc Inst Mech Eng 201:37–44
3. Chen JS, Jhu JL (1996) On the in-plane vibration and stability of a spinning annular disk. J Sound Vib 195(4):585–593
4. Chen JS, Jhu JL (1997) In-plane stress and displacement distributions in a spinning annular disk under stationary edge loads. J Appl Mech 64:897–904
5. Hamidzadeh HR, Dehghani M (1999) Linear in-plane free vibration of rotating disks. Proc of the 17th ASME Biennial Conference on Mechanical Vibration and Noise, Las Vegas, NV, September 12–16, DETC 99/VIB-8146
6. Hamidzadeh HR (2000) Free vibration of rotating ring – an analytical solution. ASME IMECE DE 108:9–16
7. Hamidzadeh HR, Karim RU (2001) In-plane free vibrations of the double – segment compound rotating disk. ASME IMECE DE 111:169–173
8. Hamidzadeh HR (2002) In-plane free vibration and stability of rotating annular disks. J Multibody Dyn 216:371–380
9. Deshpande M, Mote CD (2003) In-plane vibration of thin disks. ASME J Vib Acoust 125:68–72
10. Sarfaraz E, Hamidzadeh HR (2012) Influence of material damping on in-plane modal parameters for rotating disks. Proc of the ASME 2012 International Mechanical Engineering Congress & Exposition, Houston, TX, November 9–15, IMECE2012-86479, vol 4, pp. 89–97

Chapter 23
Patent Licensing: Stackelberg Versus Cournot Models

Oana Bode and Flávio Ferreira

Abstract In the present study we consider, on one hand, a differentiated Stackelberg model, and, on the other hand, a differentiated Cournot model, when one of the firms engages in an R&D process that gives an endogenous cost-reducing innovation. The aim of this study is two fold. The first is to study the licensing of the cost-reduction in the Stackelberg model. The second is to do a direct comparison between Stackelberg model and Cournot model. We analyse the implications of these types of licensing contracts over the R&D effort, the profits of the firms, the consumer surplus and the social welfare. By using comparative static analysis, we conclude that the degree of the differentiation of the goods assumes a great importance in the results.

Keywords Industrial organization • Optimization • Licensing • Differentiated Stackelberg model • Differentiated Cournot model

23.1 Introduction

The aim of the present chapter is to study the case of a patent licensing contract when the patentee is an insider and the innovation size is endogenous, in a differentiated-good duopoly. First, we do the analyses considering a Stackelberg model, then we do a direct comparison between the Stackelberg and Cournot patent licensing cases.

O. Bode (✉)
Faculty of Mathematics and Computer Science, Babeş-Bolyai University, Kogălniceanu Str., No 1, 400084 Cluj-Napoca, România
e-mail: oanabode@yahoo.com

F. Ferreira
ESEIG, Polytechnic Institute of Porto, R. D. Sancho I, 981, 4480-876 Vila do Conde, Portugal
e-mail: flavioferreira@eu.ipp.pt

J.A.T. Machado et al. (eds.), *Discontinuity and Complexity in Nonlinear Physical Systems*, Nonlinear Systems and Complexity 6, DOI 10.1007/978-3-319-01411-1_23,

We recall that the Stackelberg competition is a dynamic leadership model. It is a strategic game in economics in which the leader firm moves first and then the follower firms move sequentially, in a quantity competition. In contrast to the Cournot model, in which the firms choose simultaneously their quantities, in the Stackelberg model the decisions are made sequentially. So, the Cournot competition is an economic model used to describe an industry structure in which firms compete on the amount of output they will produce, which they decide independently of each other and at the same time. So, when competing in a Cournot model, firms do not cooperate, choose simultaneously the quantity of output it will produce in the market for a specific good, have market power (i.e. each firm's output decision affects the good's price) and are economically rational and act strategically, usually seeking to maximize profit given their competitors' decisions.

On the other hand, we recall that patent licensing covers a wide range of well-known situations. For example, a production firm might achieve the license for a proprietary production technology from another firm which owns it, in order to gain a competitive edge, rather than expending the time and money trying to develop its own technology.

The theoretical literature regarding patent licensing in the Cournot or Stackelberg model is vast and reveals three types of licensing contract: (per-unit) royalty licensing, fixed-fee licensing and two-part tariff licensing (fixed-fee plus royalty). Two types of licensors are revealed, namely, the outsider licensor (when it is an independent R&D organization and not a competitor of the licensee in the product market; for example, [8, 9, 12]) and the insider licensor (when competes with the licensee; for example, [4, 10, 15, 17–20]). There exists vast literature focusing on the decision of the optimal licensing contract by the patentee [1, 2, 5, 6, 11].

Nowadays, patent licensing is an important area of research which is becoming increasingly relevant because of the present trend of globalization and technology transfer between firms across countries. It takes place in many industries. It can be seen as a source of profit for the patentee (innovator) who earns rent from the licensee by transferring a new technology. In [21], the authors made an interesting and useful study concerning the intensity of licensing to affiliated and non-affiliated companies, its evolution, the characteristics, motivations and obstacles met by companies doing or willing to license, pointing out at the end the fact that patent licensing is widespread.

23.2 The Basic Framework

We consider a duopoly model where two firms, denoted by F_1 and F_2, produce a differentiated good. The inverse demand functions are given by $p_i = 1 - q_i - dq_j$, where:

- p_i represents the price of the firm F_i, $i = 1, 2$;
- q_i and q_j represent the outputs of firms F_i and F_j, $i, j = 1, 2, i \neq j$;
- d represents the degree of the differentiation of the goods, $d \in (0, 1)$.

The duopoly market is modeled either as a Stackelberg competition or as a Cournot competition. Initially, both firms have identical unit production cost $c_i = c$, with $i = 1, 2$ and $0 < c < 1$. We consider that one of the firms can engage in an R&D process in order to improve its technology. This allows a reduction of its production costs by an amount that we call *innovation size*. The cost-reducing innovation creates a new technology that reduces innovating firm's unit cost by the amount of k, while the amount invested in R&D is $k^2/2$. So, the innovation size is endogenous. There are many papers that use this approach to model process innovations, as [14, 16]. However, in other papers the innovation size is exogenous, as [5, 7].

In case that the duopoly market is modeled as a Stackelberg competition and there will be a technology transfer between the two firms, we consider the following five stages game. In the first stage, the innovator firm (the leader firm F_1) decides the value of the innovation size. In the second stage, the innovator firm F_1 decides whether to license the technology or not, because licensing reduces the marginal cost of the follower firm F_2. If decides to license it, then it charges a payment from the licensee (a per-unit royalty, a fixed-fee or a combination of both royalty and fixed-fee). In the third stage, the firm F_2 decides whether to accept or reject the offer made by the firm F_1. Then, both firms represent the players of a Stackelberg game. So, in the fourth stage the firm F_1 decides its output; and in the last stage, the firm F_2 being aware of the leader's output, chooses the output to produce.

The game will be solved by using backward induction. We also analyse, in each duopoly competition, the consumer surplus CS and the social welfare W, that are, respectively, defined by

$$CS = \frac{q_1^2 + 2dq_1q_2 + q_2^2}{2} \quad \text{and} \quad W = \pi_1 + \pi_2 + CS.$$

23.3 Stackelberg Competition

In the present section we analyse the benchmark case and the case of licensing by a two-part tariff in a differentiated-good Stackelberg duopoly,[1] when the leader firm (firm F_1) engages in an R&D process that gives an endogenous cost-reducing innovation. The results of this study are given in [3].

23.3.1 Benchmark Case: Pre-licensing

In the pre-licensing situation, if there exists no licensing between the two firms, firm F_1 owns a cost advantage on the market compared with firm F_2: $c_1 = c - k$ and

[1]Throughout the paper we use the notation superscript S to refer to the Stackelberg competition.

$c_2 = c$. Depending on the value of the differentiated parameter d, two cases can occur. Let d_1, $0 < d_1 < 1$, be such that $d_1^2 + 2d_1 - 2 = 0$.[2] Hence, we have the following:

(A) if $0 < d < d_1$, then firm F_2 competes with firm F_1 with its old technology and gets positive profit (non-drastic innovation);
(B) if $d_1 \le d < 1$, then firm F_2 find unprofitable to produce any positive output (drastic innovation). In this case, firm F_1 gains the monopoly.

The profit functions of firm F_1 and firm F_2 are, respectively, given by[3]

$$\pi^S_{1,nl} = (1 - q^S_{1,nl} - dq^S_{2,nl} - c + k^S_{nl})q^S_{1,nl} - (k^S_{nl})^2/2$$

and

$$\pi^S_{2,nl} = (1 - q^S_{2,nl} - dq^S_{1,nl} - c)q^S_{2,nl}.$$

(A) By using backward induction, standard computations yield that, in the case of non-drastic innovation, the optimal innovation size and firms' optimal outputs are, respectively, given by

$$k^S_{nl} = \frac{(2-d)(1-c)}{2(1-d^2)}, \quad q^S_{1,nl} = \frac{(2-d)(1-c)}{2(1-d^2)} \tag{23.1}$$

and

$$q^S_{2,nl} = \frac{(2-2d-d^2)(1-c)}{4(1-d^2)}. \tag{23.2}$$

Therefore, firms' profits, consumer surplus and social welfare are, respectively, given by

$$\pi^S_{1,nl} = \frac{(1-c)^2(2-d)^2}{8(1-d^2)}, \quad \pi^S_{2,nl} = \frac{(1-c)^2(2-2d-d^2)^2}{16(1-d^2)^2}, \tag{23.3}$$

$$CS^S_{nl} = \frac{(1-c)^2(5d^4+4d^3-20d^2-8d+20)}{32(1-d^2)^2} \tag{23.4}$$

and

$$W^S_{nl} = \frac{(1-c)^2(3d^4+28d^3-32d^2-40d+44)}{32(1-d^2)^2}. \tag{23.5}$$

[2] We note that $d_1 \simeq 0.732$.

[3] Throughout the paper we use the notation subscript nl to refer to the pre-licensing case.

From (23.2), we conclude that for $0 < d < d_1$ the innovation is non-drastic, and for $d \geq d_1$ the innovation is drastic, where d_1, $0 < d_1 < 1$, is such that $d_1^2 + 2d_1 - 2 = 0$.

(B) In the case of drastic innovation, firm F_1's monopoly arises.[4] Hence, we have firm F_2's output $\tilde{q}^S_{2,nl} = 0$, and so $\tilde{\pi}^S_{2,nl} = 0$. Furthermore, we obtain that

$$\tilde{k}^S_{nl} = \frac{(1-c)(4-2d-d^2)}{2d}, \quad \tilde{q}^S_{1,nl} = \frac{1-c}{d}, \tag{23.6}$$

$$\tilde{\pi}^S_{1,nl} = \frac{(-8+16d-4d^3-d^4)(1-c)^2}{8d^2}, \tag{23.7}$$

$$\tilde{CS}^S_{nl} = \frac{(1-c)^2}{2d^2} \quad \text{and} \quad \tilde{W}^S_{nl} = \frac{(1-c)^2(-d^4-4d^3+16d-4)}{8d^2}. \tag{23.8}$$

By evaluating the effects of the degree d of the differentiation of the goods over the amount that reduces the leader's unit cost, the profits of both firms (leader and follower), the consumer surplus and the social welfare, we state the following.

Theorem 1. *If there exists no technology transfer, then:*

(i) For $d \in (d_2, d_1)$ (resp., $d \in (0, d_2) \cup [d_1, 1)$), the optimal innovation size decreases (resp., increases) with the differentiation of the goods[5]*;*

(ii) For $d \in (0, 0.5) \cup (d_3, 1)$ (resp., $d \in (0.5, d_3)$), the profit of the innovator firm increases (resp., decreases) with the differentiation of the goods[6]*;*

(iii) For $d \in (0, d_4) \cup [d_1, 1)$ (resp., $d \in (d_4, d_1)$), the consumer surplus increases (resp., decreases) with the differentiation of the goods[7]*;*

(iv) For $d \in (0, d_5) \cup [d_1, 1)$ (resp., $d \in (d_5, d_1)$), the social welfare increases (resp., decreases) with the differentiation of the goods.[8]

Proof. From (23.1) and (23.6), it is easy to see that

$$\frac{\partial k^S_{nl}}{\partial d} < 0, \ \forall\, d \in (0, d_2), \quad \frac{\partial k^S_{nl}}{\partial d} > 0, \ \forall\, d \in (d_2, d_1),$$

$$\text{and} \quad \frac{\partial \tilde{k}^S_{nl}}{\partial d} < 0, \ \forall\, d \in [d_1, 1).$$

---

[4]Throughout the paper we will add a $\sim$ to identify the values we get in the drastic innovation case.

[5]We note that $0 < d_2 < 1$ is such that $d_2^2 - 4d_2 + 1 = 0$, i.e. $d_2 \simeq 0.268$.

[6]We note that $0 < d_3 < 1$ is such that $d_3^4 + 2d_3^3 + 8d_3 - 8 = 0$, i.e. $d_3 \simeq 0.812$.

[7]We note that $0 < d_4 < 1$ is such that $d_4^4 - 5d_4^3 - 3d_4^2 + 10d_4 - 2 = 0$, i.e. $d_4 \simeq 0.219$.

[8]We note that $0 < d_5 < 1$ is such that $7d_5^4 - 13d_5^3 - 9d_5^2 + 28d_5 - 10 = 0$, i.e. $d_5 \simeq 0.458$.

Furthermore, from (23.3) and (23.7), we get, respectively,

$$\frac{\partial \pi_{1,nl}^S}{\partial d} < 0,\ \forall\, d \in (0, 0.5), \quad \frac{\partial \pi_{1,nl}^S}{\partial d} > 0,\ \forall\, d \in (0.5, d_1),$$

$$\frac{\partial \tilde{\pi}_{1,nl}^S}{\partial d} > 0,\ \forall\, d \in [d_1, d_3), \quad \frac{\partial \tilde{\pi}_{1,nl}^S}{\partial d} < 0,\ \forall\, d \in (d_3, 1),$$

and

$$\frac{\partial \pi_{2,nl}^S}{\partial d} < 0,\ \forall\, d \in (0, d_1).$$

Based on (23.4), (23.5) and (23.8), we obtain that

$$\frac{\partial CS_{nl}^S}{\partial d} < 0,\ \forall\, d \in (0, d_4), \quad \frac{\partial CS_{nl}^S}{\partial d} > 0,\ \forall\, d \in (d_4, d_1),$$

$$\frac{\partial \tilde{CS}_{nl}^S}{\partial d} < 0,\ \forall\, d \in [d_1, 1),$$

and

$$\frac{\partial W_{nl}^S}{\partial d} < 0,\ \forall\, d \in (0, d_5), \quad \frac{\partial W_{nl}^S}{\partial d} > 0,\ \forall\, d \in (d_5, d_1), \quad \frac{\partial \tilde{W}_{nl}^S}{\partial d} < 0,\ \forall\, d \in [d_1, 1).$$

We note that if there exists no technology transfer and the innovation is non-drastic ($d \in (0, d_1)$), then the profit of the licensee firm increases with the differentiation of the goods.

23.3.2 Two-Part Tariff Licensing

Now we study in our differentiated Stackelberg duopoly model, the situation when there can be a technology transfer from the leader firm (the innovator) to the follower firm, based on a two-part tariff licensing contract, i.e. both fixed-fee and a royalty per-unit of output.[9]

Firm F_1's total profit in this case will be its own profit in the product market due to competition plus the fixed-fee it charges and the royalties it receives, i.e.

$$\pi_{1,l}^S = (1 - q_{1,l}^S - d q_{2,l}^S - c + k_l^S) q_{1,l}^S - (k_l^S)^2/2 + f_l^S + r_l^S q_{2,l}^S.$$

[9]Throughout the paper, we use the notation subscript l to refer to the two-part tariff licensing case.

Firm F_2's profit is given by

$$\pi_{2,l}^S = (1 - q_{2,l}^S - dq_{1,l}^S - c + k_l^S - r_l^S)q_{2,l}^S - f_l^S.$$

By using backward induction, standard computations yield that the profits of firms F_1 and F_2 are, respectively, given by

$$\pi_{1,l}^S = \frac{2r^2(1+2d-2d^2) - 4r(1-c)(2-d^2) - (1-c)^2(2-d)^2}{2(5d^2-4d-4)} + f_l^S \quad (23.9)$$

and

$$\pi_{2,l}^S = \frac{\left((1-c)(4-2d-d^2) - rd(3-2d)\right)^2}{(4+4d-5d^2)^2} - f_l^S.$$

Now, in order to determine the maximum fixed-fee that the leader firm can charge, we have to consider both non-drastic and drastic innovation cases.

For the case of non-drastic innovation ($d \in (0, d_1)$), the maximum fixed-fee that the leader firm can charge is such that the follower's profit equals its no-licensing profit, i.e $\pi_{2,l}^S = \pi_{2,nl}^S$. It results that the optimal royalty and the optimal cost reduction are, respectively, given by

$$r_l^S = \frac{(1-c)(3d^4-5d^3-4d+8)}{2(3d^4-3d^3-7d^2+6d+2)} \quad \text{and} \quad k_l^S = \frac{(1-c)(2d-3)}{3d^2-3d-1}. \quad (23.10)$$

Hence, the maximum fixed-fee is given by

$$f_l^S = \frac{(1-c)^2 g(d)}{16(d^2-1)^2(3d^4-3d^3-7d^2+6d+2)^2}, \quad (23.11)$$

where $g(d) = -9d^{12} - 18d^{11} + 205d^{10} - 234d^9 - 393d^8 + 684d^7 + 296d^6 - 848d^5 - 4d^4 + 592d^3 - 256d^2 - 64d + 48$.

Under the above circumstances, we get at the end that the optimal outputs and profits for the leader and follower firms, and the consumer surplus and social welfare, in the non-drastic innovation case, are, respectively, given by

$$q_{1,l}^S = \frac{(1-c)(2-d)(3d^2-d-4)}{2(2-d^2)(3d^2-3d-1)}, \quad q_{2,l}^S = \frac{(c-1)(5d^3-9d^2+4)}{2(2-d^2)(3d^2-3d-1)},$$

$$\pi_{1,l}^S = \frac{(1-c)^2 h(d)}{16(d+1)^2(d-1)^2(3d^4-3d^3-7d^2+6d+2)},$$

$$\pi_{2,l}^S = \frac{(1-c)^2(d^4+4d^3-8d+4)}{16(d^2-1)^2},$$

$$CS_l^S = \frac{(1-c)^2(15d^5 - 45d^4 + 17d^3 + 39d^2 - 8d - 20)}{4(d^2-2)(3d^2-3d-1)^2} \tag{23.12}$$

and

$$W_l^S = \frac{(1-c)^2(33d^5 - 75d^4 - 37d^3 + 163d^2 - 48d - 40)}{4(d^2-2)(3d^2-3d-1)^2}, \tag{23.13}$$

where $h(d) = -3d^8 + 15d^7 + 3d^6 - 82d^5 + 50d^4 + 132d^3 - 100d^2 - 88d + 72$.

Standard computations yield that the leader firm can license its technology based on a two-part tariff in the non-drastic innovation case, because its total profit (market profit + fixed-fee + per-unit royalty) exceeds the profit it makes with no-licensing, i.e. $\pi_{1,l}^S > \pi_{1,nl}^S$, $\forall\, d \in (0, d_1)$.

For the case of drastic innovation ($d \in [d_1, 1)$), the maximum fixed-fee that the leader firm can charge is such that the follower's profit equals its no-licensing profit, i.e. $\tilde{\pi}_{2,l}^S = \tilde{\pi}_{2,nl}^S$. We get that

$$\tilde{f}_l^S = \frac{(1-c)^2(5d^3 - 9d^2 + 4)^2}{4(3d^4 - 3d^3 - 7d^2 + 6d + 2)^2}. \tag{23.14}$$

Also, we obtain that the optimal royalty, optimal cost reduction and optimal leader's output are the same as in the non-drastic innovation case, i.e. $\tilde{r}_l^S = r_l^S$, $\tilde{k}_l^S = k_l^S$ and $\tilde{q}_{1,l}^S = q_{1,l}^S$, $\forall\, d \in [d_1, 1)$. Furthermore, we get that the leader's profit is

$$\tilde{\pi}_{1,l}^S = \frac{(1-c)^2(3d^3 - 2d^2 - 10d + 10)}{2(3d^4 - 3d^3 - 7d^2 + 6d + 2)}.$$

Obviously, $\tilde{q}_{2,l}^S = 0$ and $\tilde{\pi}_{2,l}^S = 0$. Therefore, we get that the consumer surplus and social welfare are, respectively, given by

$$\widetilde{CS}_l^S = \frac{(1-c)^2(2-d)^2(3d^2 - d - 4)^2}{8(d^2-2)^2(3d^2-3d-1)^2} \tag{23.15}$$

and

$$\tilde{W}_l^S = \frac{(1-c)^2 i(d)}{8(d^2-2)^2(3d^2-3d-1)^2}, \tag{23.16}$$

where $i(d) = 36d^7 - 51d^6 - 222d^5 + 405d^4 + 212d^3 - 644d^2 + 128d + 144$.

We note that in this case the leader firm can license its technology based on a two-part tariff, since its total profit (market profit + fixed-fee + royalties) exceeds the profit it makes with no-licensing, i.e. $\tilde{\pi}_{1,l} > \tilde{\pi}_{1,nl}$, $\forall\, d \in [d_1, 1)$. So, we have the following result.

Theorem 2. *A two-part tariff licensing strictly dominates no-licensing.*

Furthermore, by evaluating the effects of the degree d of the differentiation of the goods over the optimal innovation size, the optimal royalty rate, the maximum fixed-fee that can be charged by the leader firm, the consumer surplus and the social welfare, we conclude the followings.

Theorem 3. *If the innovation is non-drastic and the technology is licensed by a two-part tariff, then:*

(i) The optimal innovation size increases with the differentiation of the goods;
(ii) For $d \in (0, d_6)$ (resp., $d \in (d_6, d_1)$), the optimal royalty rate increases (resp., decreases) with the differentiation of the goods[10]*;*
(iii) The maximum fixed-fee that the innovator firm can charge increases with the differentiation of the goods;
(iv) The consumer surplus increases with the differentiation of the goods;
(v) The social welfare increases with the differentiation of the goods.

Proof. Based on (23.10), we obtain that $\frac{\partial k^S_{rf}}{\partial d} < 0,\ \forall\, d \in (0,1)$.

Furthermore, we get that

$$\frac{\partial r^S_2}{\partial d} < 0,\ \forall\, d \in (0, d_6), \quad \text{and} \quad \frac{\partial r^S_2}{\partial d} > 0,\ \forall\, d \in (d_6, d_1).$$

From (23.11), standard computations yield that $\frac{\partial f^S_2}{\partial d} < 0,\ \forall\, d \in (0, d_1)$.

Furthermore, based on (23.12) and (23.13), we get that

$$\frac{\partial CS^S_{rf}}{\partial d} < 0,\ \forall\, d \in (0, d_1), \quad \text{and} \quad \frac{\partial W^S_{rf}}{\partial d} < 0,\ \forall\, d \in (0, d_1).$$

Theorem 4. *If the innovation is drastic and the technology is licensed by a two-part tariff, then:*

(i) For $d \in [d_1, d_7)$ (resp., $d \in (d_7, 1)$), the optimal innovation size increases (resp., decreases) with the differentiation of the goods[11]*;*
(ii) The optimal royalty rate decreases with the differentiation of the goods;
(iii) The maximum fixed-fee that the innovator firm can charge increases with the differentiation of the goods;
(iv) The consumer surplus decreases with the differentiation of the goods;
(v) For $d \in [d_1, d_8)$ (resp., $d \in (d_8, 1)$), the social welfare increases (resp., decreases) with the differentiation of the goods.[12]

[10] We note that $0 < d_6 < 1$ is such that $6d_6^6 - 42d_6^5 + 125d_6^4 - 156d_6^3 + 14d_6^2 + 112d_6 - 56 = 0$, i.e. $d_6 \simeq 0.721$.

[11] We note that $0 < d_7 < 1$ is such that $6d_7^2 - 18d_7 + 11 = 0$, i.e. $d_7 \simeq 0.855$.

[12] We note that $0 < d_8 < 1$ is such that $54d_8^{10} - 99d_8^9 - 621d_8^8 + 1866d_8^7 - 42d_8^6 - 4446d_8^5 + 3146d_8^4 + 3020d_8^3 - 3276d_8^2 - 344d_8 + 736 = 0$, i.e. $d_8 \simeq 0.863$.

Proof. We easily get that

$$\frac{\partial \tilde{k}^S_{rf}}{\partial d} < 0, \ \forall \ d \in [d_1, d_7), \quad \text{and} \quad \frac{\partial \tilde{k}^S_{rf}}{\partial d} > 0, \ \forall \ d \in (d_7, 1).$$

Also, we note that $\frac{\partial \tilde{r}^S_{rf}}{\partial d} > 0$ and $\frac{\partial \tilde{f}^S_2}{\partial d} < 0, \ \forall \ d \in [d_1, 1)$.
From (23.15) and (23.16), we obtain that

$$\frac{\partial \widetilde{CS}^S_{rf}}{\partial d} > 0, \ \forall \ d \in [d_1, 1),$$

$$\frac{\partial \tilde{W}^S_{rf}}{\partial d} < 0, \ \forall \ d \in [d_1, d_8), \quad \text{and} \quad \frac{\partial \tilde{W}^S_{rf}}{\partial d} > 0, \ \forall \ d \in (d_8, 1).$$

23.4 Stackelberg Model Versus Cournot Model

In this section we do a direct comparison between our differentiated Stackelberg duopoly models and the ones of Cournot discussed by Li and Ji [13].

We recall that in the Stackelberg model, the innovation is non-drastic (resp., drastic) for $d \in (0, d_1)$ (resp., $d \in [d_1, 1)$), where $d_1 \simeq 0.732$. In the Cournot model studied by Li and Ji [13], the innovation is non-drastic (resp., drastic) for $d \in (0, d_9)$ (resp., $d \in (d_9, 1)$), where $d_9 \simeq 0.806$.

We begin by comparing the cost-reduction for those two models.[13]

Theorem 5. *(i) If there exists no technology licensing and the goods are sufficiently differentiated ($d \in (0, d_{10})$) (resp., sufficiently homogenous ($d \in (d_{10}, 1)$), then the innovator firm invests more (resp., less) in R&D under Stackelberg competition than under Cournot competition;*

(ii) If there exists a technology transfer based on a two-part tariff licensing contract, then the innovator firm invests less in R&D under Stackelberg competition than under Cournot competition.

We continue by investigating the profits of the innovator firms.[14]

Theorem 6. *(i) If there exists no technology licensing and the goods are sufficiently differentiated ($d \in (0, d_{11})$) (resp., sufficiently homogenous ($d \in (d_{11}, 1)$)), then the profit of the innovator firm is higher (resp., lower) under Stackelberg competition than under Cournot competition;*

(ii) If there exists a technology transfer based on a two-part tariff licensing contract, then the profit of the innovator firm is lower under Stackelberg competition than under Cournot competition.

[13] We note that $d_{10} \simeq 0.747$.

[14] We note that $d_{11} \simeq 0.828$.

Furthermore, we make a direct comparison of the consumer surplus for those two models.[15]

Theorem 7. *(i) If there exists no technology licensing and the goods are sufficiently differentiated ($d \in (0, d_9)$) (resp., sufficiently homogenous ($d \in (d_9, 1)$)), then the consumer surplus is higher under Stackelberg competition than under Cournot competition (resp., the same in both models);*

(ii) If there exists a technology transfer based on a two-part tariff licensing contract and the goods are sufficiently differentiated ($d \in (0, d_{12})$) (resp., sufficiently homogenous ($d \in (d_{12}, 1)$)), then the consumer surplus is lower (resp. higher) under Stackelberg competition than under Cournot competition.

Comparing now the social welfare for those two models, we get the following.[16]

Theorem 8. *(i) If there exists no technology licensing and the goods are sufficiently differentiated ($d \in (0, d_{10})$) (resp., sufficiently homogenous ($d \in (d_{10}, 1)$)), then the social welfare is higher (resp., lower) under Stackelberg competition than under Cournot competition;*

(ii) If there exists a technology transfer based on a two-part tariff licensing contract and the goods are sufficiently differentiated ($d \in (0, d_{13})$) (resp., sufficiently homogenous ($d \in (d_{13}, 1)$)), then the social welfare is lower (resp., higher) under Stackelberg competition than under Cournot competition.

23.5 Conclusions

The present chapter studied the licensing, one of the most used methods for technology transfer between firms. We analysed the benchmark case and the licensing case by a two-part tariff in a differentiated Stackelberg duopoly model when one of the firms engages in an R&D process that gives an endogenous cost-reducing innovation. We saw that in both cases, i.e. no-licensing or licensing by means of a two-part tariff, the innovation can be either non-drastic (both firms compete on the market using their own technologies or using the same technology, and get positive profit) or drastic (the non-innovator firm find it unprofitable to produce any output), depending on the degree of the differentiation of the goods.

We computed explicitly the main variables, i.e. the optimal innovation size; the optimal outputs the profits; the consumer surplus; and the social welfare, in both non-drastic and drastic innovation cases. Furthermore, we did a comparative static analysis and concluded that the degree of the differentiation of the goods represents a great importance in the results.

[15]We note that $d_{12} \simeq 0.928$.

[16]We note that $d_{13} \simeq 0.941$.

Furthermore, we compared our results obtained in the Stackelberg model and the results obtained by Li and Ji [13] in the Cournot model. We note that in each case we get different results, depending on if there exists no technology licensing; or if there exists technology licensing by means of a two-part tariff.

Acknowledgements The author Oana Bode thanks financial support from Babes-Bolyai University, Cluj-Napoca, Romania. The author F. Ferreira thanks financial support from ESEIG/IPP.

References

1. Chang MC, Lin CH, Hu JL (2009) The optimal licensing strategy of an outside patentee under an upstream supplier. Taiwan Economic Association. http://scholar.google.com/citations?view_op=view_citation&hl=en&user=1s4iaxkAAAAJ&citation_for_view=1s4iaxkAAAAJ:_FxGoFyzp5QC (25 Nov 2011)
2. Ferreira FA (2011) Licensing in an international competition with differentiated goods. In: Tenreiro Machado JA, Baleanu D, Luo A (eds) Nonlinear dynamics of complex systems: application in physical, biological and financial systems. Springer Science + Business Media Llc, New York, pp 295–305
3. Ferreira F, Bode OR (2013) Licensing endogenous cost-reduction in a differentiated Stackelberg model. Comm Nonlinear Sci Numer Simul 18(2):308–315. http://www.sciencedirect.com/science/article/pii/S1007570412002900
4. Ferreira F, Tuns (Bode) OR (2012) Per-unit royalty and fixed-fee licensing in a differentiated Stackelberg model. In: IEEE 4th international conference on nonlinear science and complexity proceedings, Budapest, pp 99–102
5. Filippini L (2005) Licensing contract in a Stackelberg model. Manchester School 73(5): 582–598
6. Fosfuri A, Roca E (2004) Optimal licensing strategy: royalty or fixed-fee? Int J Bus Econ 3(1):13–19
7. Kabiraj T (2005) Technology transfer in a Stackelberg structure: licensing contracts and welfare. Manchester School 73(1):1–28
8. Kamien MI (1992) Patent licensing. In: Handbook of game theory with economic applications, Amsterdam, Elsevier Science, vol 1, pp 331–354
9. Kamien MI, Tauman Y (1986) Fees versus royalties and the private value of a patent. Q J Econ 101(3):471–491
10. Kamien MI, Tauman Y (2002) Patent licensing: the inside story. Manchester School 70(1):7–15
11. Kamien MI, Oren SS, Tauman Y (1992) Optimal licensing of cost-reducing innovation. J Math Econ 21:483–508
12. Katz ML, Shapiro C (1986) How to license intangible property. Q J Econ 101(3):567–590
13. Li C, Ji X (2010) Innovation, licensing, and price vs. quantity competition. Econ Model 27:746–754
14. Lin P, Saggi K (2002) Product differentiation, process R&D, and the nature of market competition. Eur Econ Rev 46(1):201–211
15. Marjit S (1990) On a non-cooperative theory of technology transfer. Econ Lett 33(3):293–298
16. Qiu LD (1997) On the dynamic efficiency of Bertrand and Cournot equilibria. J Econ Theor 75(1):213–229
17. Rockett K (1990) The quality of licensed technology. Int J Ind Organ 8(4):559–574
18. Wang XH (1998) Fee versus royalty licensing in a Cournot duopoly model. Econ Lett 60:55–62
19. Wang XH (2002) Fee versus royalty licensing in differentiated Cournot oligopoly. J Econ Bus 54:253–266 (2002)
20. Wang XH, Yang BZ (1999) On licensing under Bertrand competition. Aust Econ Papers 38(2):106–119
21. Zuniga MP, Guellec D (2009) Who licenses out patents and why? Lessons from a business survey (2009). http://www.oecd.org/dataoecd/47/16/42477187.pdf. Accessed 25 Nov 2011

Chapter 24
Privatization and Government Preferences in a Mixed Duopoly: Stackelberg Versus Cournot

Fernanda A. Ferreira and Flávio Ferreira

Abstract We analyse the relationship between the privatization of a public firm and government preferences for tax revenue, by considering a (sequential) Stackelberg duopoly with the public firm as the leader. We assume that the government payoff is given by a weighted sum of tax revenue and the sum of consumer and producer surplus. We get that if the government puts a sufficiently larger weight on tax revenue than on the sum of both surpluses, it will not privatize the public firm. In contrast, if the government puts a moderately larger weight on tax revenue than on the sum of both surpluses, it will privatize the public firm. Furthermore, we compare our results with the ones previously published by an other author obtained in a (simultaneous) Cournot duopoly.

Keywords Stackelber duopoly • Cournot duopoly • Mixed duopoly • Privatization • Tax rate

24.1 Introduction

Tariff revenue may be an important source of government revenue for developing countries that do not have an efficient tax system. Brander and Spencer [1] have shown that a tariff has a profit-shifting effect in addition to its effect on tariff revenue. Larue and Gervais [9] studied the effect of maximum-revenue tariff in

F.A. Ferreira (✉) • F. Ferreira
ESEIG - Polytechnic Institute of Porto, Rua D. Sancho I, 981, 4480-876 Vila do Conde, Portugal
e-mail: fernandaamelia@eu.ipp.pt; flavioferreira@eu.ipp.pt

J.A.T. Machado et al. (eds.), *Discontinuity and Complexity in Nonlinear Physical Systems*, Nonlinear Systems and Complexity 6, DOI 10.1007/978-3-319-01411-1_24,

a Cournot duopoly. Ferreira and Ferreira [5] examined the maximum-revenue tariff under international Bertrand in competition with differentiated products when rivals' production costs are unknown. Clarke and Collie [3] studied a similar question, when there is no uncertainty on the production costs.

Furthermore, studies of mixed oligopoly models have been increasingly popular in recent years.[1] Some authors study Cournot models and others study Stackelberg models. In the first model the firms move simultaneously and in the second one the firms move sequentially, with at least one firm acting as a leader. We can say that the main concerns of these privatization studies are the welfare effect and the method of privatization. Chao and Yu [2] examined how either partial privatization or foreign competition affects the optimal tariff and found that foreign competition lowers the optimal tariff rate but partial privatization raises it. White [11] and Fjell and Heywood [7] introduced a subsidy into the mixed model. In these studies, the objective function of both the government and the public firm is the social welfare. Matsumura [10] considered an objective function that is a weighted average of a modified social welfare and the profit of the firm. The modified social welfare allows the government to prefer consumer surplus to the profits of the two firms.

Kato [8] considered a mixed (simultaneous) Cournot duopoly by assuming that the government puts a larger weight on tax revenue than on the sum of consumer and producer surplus, whereas the public firm only cares about the sum of consumer and producer surplus. In this context, he studied the relationship between the privatization of the public firm and the government preferences for tax revenue. He concluded that (1) the government sets a higher tax rate in a mixed duopoly than in a privatized duopoly; and (2) whether the government privatizes the public firm depends on the government preference for tax revenue.

In this paper, we consider the same objective functions for both the government and the public firm as in Kato's paper, but we analyse a mixed (sequential) Stackelberg market competition (see also Ferreira and Ferreira [6]). So, in our paper, the game runs as follows. First, the government chooses the tax rate t. Then, instead of a simultaneous decision on the quantities, the public firm chooses first the output level q_1 to produce, and after that and knowing this decision, the private firm chooses the output level q_2 to be produced. Furthermore, we compare the results in the two duopoly models: sequentially Stackelberg move model versus simultaneously Cournot move model.

The organization of this chapter is as follows. After this introductory section, we present and discuss the mixed model. In Sect. 24.3, we study the privatized model. Section 24.4 yields the results gained by a direct comparison between both the mixed and privatized models. In Sect. 24.5, we compare the main results concerned with both sequential and simultaneous move models. Conclusions are presented in Sect. 24.6.

---

[1] For a detailed survey, see De Fraja and Delbono [4].

24.2 The Mixed Duopoly

In this section, we consider a mixed Stackelberg duopoly where a public firm F_1 is the leader and a private firm F_2 is the follower that produce homogeneous goods. The inverse demand function is given by

$$p = a - Q,$$

where $a > 0$ is the demand parameter, p is the market price and $Q = q_1 + q_2$ is the total output, where q_1 and q_2 are the outputs of the public firm and the private firm, respectively. Both firms have the same production cost function $C(q_i) = q_i^2/2$, with $i = 1, 2$. The government imposes a specific tax rate t on both firms.

The model consists in the following three-stage game:

- In the first stage, the government chooses the tax rate t.
- In the second stage, the public firm F_1 chooses the output level q_1.
- In the third stage, the private firm F_2 chooses the output level q_2.

The payoff of the private firm is its profit:

$$\pi_2 = (a - q_1 - q_2)q_2 - \frac{q_2^2}{2} - tq_2;$$

the payoff of the public firm is the sum of consumer and producer surplus:

$$W = \frac{(q_1 + q_2)^2}{2} + (a - q_1 - q_2)(q_1 + q_2) - \frac{q_1^2 + q_2^2}{2} - T,$$

where $T = t(q_1 + q_2)$ is the tax revenue; the government's payoff is given by:

$$U = W + (1 + \alpha)T,$$

where the parameter α represents the weight of the government preference for the tax revenue. We will consider that the government puts a larger weight on T than on W, so we set $\alpha \geq 0$ (the other situation is inconsistent with reality). If $\alpha = 0$, the weight is the same on T and on W; For $\alpha > 0$, as α becomes larger, the more the government cares about T.

As usual in dynamic games, we solve our problem by backwards induction. Maximizing the private firm's profit π_2, we obtain

$$q_2 = \frac{a - q_1 - t}{3}.$$

Now, using this result and maximizing the objective function W of the public firm, we get

$$q_1 = \frac{5(a - t)}{14},$$

and, then,

$$q_2 = \frac{3(a-t)}{14}.$$

So, the government's payoff U can now be rewritten as follows:

$$U = \frac{(a-t)(9a + t(16\alpha + 7))}{28}.$$

Maximizing this objective function, the optimal tax rate in the mixed duopoly is given by[2]

$$t^M = \frac{a(8\alpha - 1)}{16\alpha + 7}.$$

Thus, if $\alpha > 1/8$, the optimal tax rate is positive; and if $0 \leq \alpha < 1/8$, the optimal tax rate is negative, so the government subsidizes the firms.

Proposition 1. *The tax rate imposed by the government increases with the weight of the government preference for the tax revenue.*

Proof. The result follows since $\partial t^M / \partial \alpha = 72a/(16\alpha + 7)^2 > 0$. □

Based on the expressions above, we get the following result.

Proposition 2. *In the mixed duopoly, the equilibrium outcomes are:*

$$q_1^M = \frac{20a(\alpha + 1)}{7(16\alpha + 7)},$$

$$q_2^M = \frac{12a(\alpha + 1)}{7(16\alpha + 7)},$$

$$Q^M = \frac{32a(\alpha + 1)}{7(16\alpha + 7)},$$

$$U^M = \frac{16a^2(\alpha + 1)^2}{7(16\alpha + 7)},$$

$$\pi_2^M = \frac{216a^2(\alpha + 1)^2}{49(16\alpha + 7)^2}.$$

We note that as α becomes larger, the quantity produced by each firm, and therefore the total quantity in the market, decreases. This is due to the fact that the optimal tax rate is positively correlated with respect to α.

---

[2]Throughout the paper, we use the notation superscript M to refer to the mixed duopoly.

24.3 The Privatized Duopoly

Now, let us consider the model where the public firm is privatized without cost. So, in this case, the objective function of firm F_1 is its profit:

$$\pi_1 = (a - q_1 - q_2)q_1 - \frac{q_1^2}{2} - tq_1.$$

Utilizing the same way of calculation as in the previous section, the output of the leader privatized firm F_1 is given by

$$q_1 = \frac{2(a-t)}{7}$$

and the output of the follower private firm F_2 is given by

$$q_2 = \frac{5(a-t)}{21}.$$

So, the government's payoff U can now be rewritten as follows:

$$U = \frac{(a-t)(20a + t(33\alpha + 13))}{63}.$$

Maximizing this objective function, we obtain that the optimal tax rate in the privatized duopoly is given by[3]

$$t^P = \frac{a(33\alpha - 7)}{2(33\alpha + 13)}.$$

Thus, if $\alpha > 7/33$, the optimal tax rate is positive; and if $0 \leq \alpha < 7/33$, the optimal tax rate is negative, so the government subsidizes the firms.

Proposition 3. *The tax rate imposed by the government increases with the weight of the government preference for the tax revenue.*

Proof. The result follows since $\partial t^P/\partial\alpha = 330a/(33\alpha + 13)^2 > 0$. □

Based on the expressions above, we get the following result.

Proposition 4. *In the privatized duopoly, the equilibrium outcomes are:*

$$q_1^P = \frac{33a(\alpha + 1)}{7(33\alpha + 13)},$$

$$q_2^P = \frac{55a(\alpha + 1)}{14(33\alpha + 13)},$$

---

[3]Throughout the paper, we use the notation superscript P to refer to the privatized duopoly.

$$Q^P = \frac{121a(\alpha+1)}{14(33\alpha+13)},$$

$$U^P = \frac{121a^2(\alpha+1)^2}{28(33\alpha+13)},$$

$$\pi_1^P = \frac{363a^2(\alpha+1)^2}{14(33\alpha+13)^2},$$

$$\pi_2^P = \frac{9075a^2(\alpha+1)^2}{392(33\alpha+13)^2}.$$

24.4 Effects of Privatization

We have derived preprivatization and postprivatization equilibria, and now we will examine the effects of privatization upon market equilibrium. The following proposition summarizes our results.

Proposition 5. *At equilibrium,*

(1) $t^M > t^P$;
(2.1) $Q^M > Q^P$, *for* $\alpha > 15/176$;
(2.2) $Q^M < Q^P$, *for* $\alpha < 15/176$;
(3.1) $U^M > U^P$, *for* $\alpha > 15/176$;
(3.2) $U^M < U^P$, *for* $\alpha < 15/176$;
(4) $\pi_2^M < \pi_2^P$.

We observe that the optimal tax rate in the mixed duopoly is always higher than that in the privatized duopoly. Furthermore, since

$$\frac{\partial(t^M - t^P)}{\partial\alpha} = -\frac{138a(44\alpha^2+88\alpha+29)}{(16\alpha+7)^2(33\alpha+13)^2} < 0,$$

the difference in the optimal tax rates between the mixed and privatized duopoly cases becomes smaller, when α becomes larger. We also note that when the aggregate output in the mixed market is larger than that in the privatized market, the government's payoff in the mixed duopoly is also larger than that in the privatized duopoly.

From the above proposition, we conclude that if the government puts a sufficiently larger weight on tax revenue than on the sum of both surpluses, i.e., if $\alpha > 15/176$, the government does not privatize the public firm. In contrast, if the government puts a moderately larger weight on tax revenue than on the sum of both surpluses, i.e., if $0 \leq \alpha < 15/176$, the government will privatize the public firm.

24.5 Stackelberg Model Versus Cournot Model

In this section, we do a direct comparison between our Stackelberg duopoly model and the ones of Cournot discussed by Kato [8].

We recall that the equilibrium outcomes in the mixed Cournot duopoly are the following[4] (see Kato [8]):

$$t_Q^M = \frac{a(15\alpha - 1)}{2(15\alpha + 7)},$$

$$q_{1,Q}^M = \frac{3a(\alpha + 1)}{15\alpha + 7},$$

$$q_{2,Q}^M = \frac{3a(\alpha + 1)}{2(15\alpha + 7)},$$

$$Q_Q^M = \frac{9a(\alpha + 1)}{2(15\alpha + 7)},$$

$$U_Q^M = \frac{9a^2(\alpha + 1)^2}{4(15\alpha + 7)};$$

and the equilibrium outcomes in the privatized Cournot duopoly are the following:

$$t_Q^P = \frac{a(4\alpha - 1)}{8\alpha + 3},$$

$$q_{1,Q}^P = \frac{a(\alpha + 1)}{8\alpha + 3},$$

$$q_{2,Q}^P = \frac{a(\alpha + 1)}{8\alpha + 3},$$

$$Q_Q^P = \frac{2a(\alpha + 1)}{8\alpha + 3},$$

$$U_Q^P = \frac{a^2(\alpha + 1)^2}{8\alpha + 3}.$$

We begin by comparing the tax rate imposed by the government for those two models.

Theorem 1. *1. In the mixed duopoly,*

(i) for $\alpha > 1/8$, the government imposes a higher tax rate if the firms act simultaneously than if they act sequentially;

[4] We use the notation subscript Q to refer to the Cournot duopoly.

(ii) *for* $1/15 < \alpha < 1/8$, *if the firms act sequentially, the government imposes a positive tax rate; and if the firms act simultaneously, the government subsidizes both firms;*
(iii) *for* $\alpha < 1/15$, *the government subsidizes both firms and the subside is higher if the firms act sequentially than if they act simultaneously.*

2. *In the privatized duopoly,*

(i) *for* $\alpha > 1/4$, *the government imposes a higher tax rate if the firms act sequentially than if they act simultaneously.*
(ii) *for* $7/33 < \alpha < 1/4$, *if the firms act simultaneously, the government imposes a positive tax rate; and if the firms act sequentially, the government subsidizes both firms;*
(iii) *for* $\alpha < 7/33$, *the government subsidizes both firms and the subside is higher if the firms act simultaneously than if they act sequentially.*

Proof. The results follow since

$$t^M = \frac{a(8\alpha - 1)}{16\alpha + 7} > 0 \Leftrightarrow \alpha > 1/8; \quad t_Q^M = \frac{a(15\alpha - 1)}{2(15\alpha + 7)} > 0 \Leftrightarrow \alpha > 1/15;$$

$$t^P = \frac{a(33\alpha - 7)}{2(33\alpha + 13)} > 0 \Leftrightarrow \alpha > 7/33; \quad t_Q^P = \frac{a(4\alpha - 1)}{8\alpha + 3} > 0 \Leftrightarrow \alpha > 1/4;$$

$$t^M - t_Q^M = -\frac{7a(\alpha + 1)}{2(15\alpha + 7)(16\alpha + 7)} < 0$$

and

$$t^P - t_Q^P = \frac{5a(\alpha + 1)}{2(8\alpha + 3)(33\alpha + 13)} > 0.$$

□

We continue by comparing the outputs produced by each firm and the aggregate quantity in the market.

Theorem 2. *(i) In the mixed duopoly, the public firm* F_1 *produces more if the firms act simultaneously than if they act sequentially (with the public firm as the leader).*
(ii) *In the privatized duopoly, the privatized firm* F_1 *produces less if the firms act simultaneously than if they act sequentially (with the privatized firm as the leader).*
(iii) *In the mixed duopoly, the private firm* F_2 *produces less if the firms act simultaneously than if they act sequentially (with the public firm as the leader).*
(iv) *In the privatized duopoly, the initial private firm* F_2 *produces more if the firms act simultaneously than if they act sequentially (with the privatized firm as the leader).*

(v) *In the mixed duopoly, the aggregate quantity in the market is higher (resp., lower) if the firms act simultaneously than if they act sequentially (with the public firm as the leader), for* $\alpha > 7/48$ *(resp.,* $\alpha < 7/48$*).*

(vi) *In the privatized duopoly, the aggregate quantity in the market is lower (resp., higher) if the firms act simultaneously than if they act sequentially (with the public firm as the leader), for* $\alpha > 1/44$ *(resp.,* $\alpha < 1/44$*).*

Proof. The results (i)–(iv) follow since

$$q_1^M - q_{1,Q}^M = -\frac{a(\alpha+1)(36\alpha+7)}{7(15\alpha+7)(16\alpha+7)} < 0,$$

$$q_1^P - q_{1,Q}^P = \frac{a(\alpha+1)(33\alpha+8)}{7(8\alpha+3)(33\alpha+13)} > 0,$$

$$q_2^M - q_{2,Q}^M = \frac{3a(\alpha+1)(8\alpha+7)}{14(15\alpha+7)(16\alpha+7)} > 0$$

and

$$q_2^P - q_{2,Q}^P = -\frac{a(\alpha+1)(33\alpha+8)}{14(8\alpha+3)(33\alpha+13)} < 0.$$

Result (v) follows since

$$Q^M - Q_Q^M = -\frac{a(\alpha+1)(48\alpha-7)}{14(15\alpha+7)(16\alpha+7)}$$

is negative if, and only if, $\alpha > 7/48$; and result (vi) follows since

$$Q^P - Q_Q^P = \frac{a(\alpha+1)(44\alpha-1)}{14(8\alpha+3)(33\alpha+13)}$$

is positive if, and only if, $\alpha > 1/44$. □

Furthermore, we make a direct comparison of the government's payoff for those two models.

Theorem 3. *(i) In the mixed duopoly, the government's payoff is higher (resp., lower) if the firms act simultaneously than if they act sequentially (with the public firm as the leader), for* $\alpha > 7/48$ *(resp.,* $\alpha < 7/48$*).*

(ii) In the privatized duopoly, the government's payoff is lower (resp., higher) if the firms act simultaneously than if they act sequentially (with the public firm as the leader), for $\alpha > 1/44$ *(resp.,* $\alpha < 1/44$*).*

Proof. Result (i) follows since

$$U^M - U_Q^M = -\frac{a^2(\alpha+1)^2(48\alpha-7)}{28(15\alpha+7)(16\alpha+7)}$$

is negative if, and only if, $\alpha > 7/48$; and result (ii) follows since

$$U^P - U_Q^P = \frac{a^2(\alpha+1)^2(44\alpha-1)}{28(8\alpha+3)(33\alpha+13)}$$

is positive if, and only if, $\alpha > 1/44$. □

24.6 Conclusions

We analysed the relationship between the privatization of a public firm and government preferences for tax revenue in a Stackelberg duopoly with the public firm as the leader. We concluded that if the government puts a sufficiently larger weight on tax revenue than on the sum of both surpluses, it will not privatize the public firm. In contrast, if the government puts a moderately larger weight on tax revenue than on the sum of both surpluses, it will privatize the public firm.

Furthermore, we compared our results obtained in the (sequential) Stackelberg model and the results obtained by Kato [8] in the (simultaneous) Cournot model.

Acknowledgements The authors would like to thank ESEIG and Polytechnic Institute of Porto for their financial support.

References

1. Brander JA, Spencer BJ (1984) Tariff protection and imperfect competition. In: Kierzkowski H (ed) Monopolistic competition and international trade: 194–207. Oxford University Press, Oxford
2. Chao C-C, Yu E (2006) Partial privatization, foreign competition, and optimum tariff. Rev Int Econ 14:87–92
3. Clarke R, Collie DR (2006) Optimum-welfare and maximum-revenue tariffs under Bertrand duopoly. Scot J Polit Econ 53:398–408
4. De Fraja G, Delbono F (1990) Game theoretic models of mixed oligopoly. J Econ Surv 4:1–17
5. Ferreira FA, Ferreira F (2009) Maximum-revenue tariff under Bertrand duopoly with unknown costs. Comm Nonlinear Sci Numer Simulat 14:3498–3502
6. Ferreira FA, Ferreira F (2012) Privatization and government preference in a public Stackelberg leader duopoly. In: IEEE 4th international conference on nonlinear science and complexity proceedings, Budapest, pp 87–89
7. Fjell K, Heywood J (2004) Mixed oligopoly, subsidization and the order of firms's moves: the relevance of privatization. Econ Lett 53:411–416
8. Kato H (2008) Privatization and government preference. Econ Bull 12:1–7
9. Larue B, Gervais J-P (2002) Welfare-maximizing and revenue-maximizing tariffs with few domestic firms. Can J Econ 35:786–804
10. Matsumura T (1998) Partial privatization in mixed Duopoly. J Publ Econ 70:473–483
11. White M (1996) Mixed oligopoly, privatization and subsidization. Econ Lett 53:189–195

Index

A

Adomian decomposition method, 37–46
Adomian polynomials, 38–40, 42–44, 58
Analgesia, 135–148
Annular disk, 389–408
Anomalous transport, 217, 226, 242–244
Approximate solutions, 37, 38, 46–50, 52, 53, 55, 57, 58, 164, 390
Assembly, 193–211

B

Basins of attractions, 315–321
Burgers, 5, 20, 23–33, 46, 48, 52, 348–357

C

Cahn Hillard equation, 157–162
Caputo derivative, 36, 37, 48
Chaos, 108, 144, 239–241, 255, 256, 305–313
Chaos detection, 109
Chaotic advection, 239–244
Chatter identification, 360
Chatter motion, 323, 324
Coarsening (interrupted coarsening), 153–167
Compound disks, 404
Conservation laws, 4, 5, 15–20, 23–33, 156–158
Constrained mass-damper-spring system, 323–342
Control algorithm, 120, 122, 126, 128, 132, 136, 144, 364
Copolymers, 157
Cournot duopoly, 422, 427
Critical speed, 390, 391, 398–401, 403, 407, 408

D

Decoupling FPGA implementation, 114, 115
Differentiated cournot model, 409
Differentiated Stackelberg model, 418
Discontinuity, 155, 171, 182, 183, 185–188, 338, 348, 351, 353–357
Discontinuous dynamical system(s), 256, 260, 271
Discontinuous medium, 389
Discrete local fractional Fourier transform, 87
Drilling system, 287–302
Dynamical model, 370
Dynamics of milling process, 360
Dynamics of phase transition, 154–159

E

Energy harvesting, 315, 316
Experimental modes decomposition, 361, 362

F

Fast local fractional Fourier transform, 87
FDE. *See* Fractional differential equations (FDE)
FOIMs. *See* Fractional-order impedance model (FOIMs)
Fractal space, 64, 67–70, 75–87
Fractional calculus, 35–58, 64, 92, 113–132, 144–145
Fractional derivative, 36, 37, 46, 58, 65–66, 92, 93, 96, 107–111, 144–147
Fractional differential equations (FDE), 41, 44, 45, 48, 49, 52, 55, 58, 92, 94
Fractional order controller, 92, 104, 114, 117–119, 124

J.A.T. Machado et al. (eds.), *Discontinuity and Complexity in Nonlinear Physical Systems*, Nonlinear Systems and Complexity 6, DOI 10.1007/978-3-319-01411-1,

Fractional-order impedance model (FOIMs), 135–148
Frequency response, 276, 362

G
The generalized local fractional Fourier transform, 87
G-Functions, 256, 261, 262, 268–270, 326–331
Ginzburg-Landau free energy, 155
Grazing motion, 337
Grünwald-Letnikov approximation, 96
Gyroscope system, 255–272

H
Hidden oscillation, 287–302
Hysteresis, 92, 94–97, 104, 107

I
Induction motor, 289, 293, 294, 298–302
Industrial organization, 409
In-plane free vibration, 389–408
Instability, 156, 159, 348, 350, 357, 390
Integral square error, 92, 94, 97–98, 101, 103, 104
Interfacial dynamics, 161, 167

L
Lateral loads, 370, 385
Lévy flights, 217, 225, 239–251
Licensing, 409–420
Lie symmetries, 24
Local fractional calculus, 64
Local fractional Fourier series, 63–87
Local fractional Fourier transform, 87

M
Micro-controller implementation, 114, 129–131
Mixed duopoly, 421–430
Modal loss factor, 391, 397, 402–404, 408
Model-based predictive control (MPC), 136, 137, 142–144, 147, 148
Modulated phase, 154, 162, 163
Multivariable fractional order controller, 114, 117–119
Multivariable processes DC motor speed control, 113

N
Natural frequency, 317, 391, 396–401, 403, 405–408
Non commutative tomogram, 217–225
Non-invasive pain sensor, 142–143
Nonlinear analysis, 169–190, 390
Nonlinear braces, 369–387
Nonlinear distortion, 275, 277–281
Nuclear fusion, 215

O
One-step numeric algorithms, 41
Optimization, 91–104, 114, 115, 117, 200, 275, 276
Ostwald ripening, 154, 159, 161

P
Pain relief level, 136, 143
Pattern formation, 154
Phase-locked loop, 169–190
Piecewise linearity, 313
Piecewise linear system, 305–313
Plane stress, 390, 391
Privatization, 421–430
Projective synchronization, 255–272

R
Recurrence plots, 359–366
Reflectometry, 217, 225–231, 236, 237, 246, 251
Respiratory system, 281, 282
Response surface methodology, 91–104
Riemann-Liouville derivative, 36
Riemann-Liouville fractional derivative, 36, 92, 93, 96
Robustness, 114–116, 120, 124, 132, 245, 250, 321
Rotating disks, 389–391, 397–400, 402–404, 407, 408
Rotating rings, 403–404

S
Saddle-node bifurcation, 347–357
Segregation, 154, 157, 159–161
Self-adjointness, 3–20, 23–33
Shear-type buildings, 371
Shock wave, 347–357
Signal analysis, 216, 223

Simulation, 37, 110–111, 121, 162, 170, 179, 240, 250, 263, 267–272, 288, 299, 317–320, 333, 334, 336, 338–342, 380, 385
Soliton lattice, 159
Spinodal decomposition, 158, 160, 163
Stackelber duopoly, 421–430
Static complexity, 194
Stuck motion, 323, 324, 328, 329, 332, 333, 338, 342
Supply chain, 193–211

T
Tax rate, 422–428
0–1 test, 108–111
Time delay experimental results, 113
Topological classes, 194, 196, 210

V
Van der Pol system, 107–111
Variational iteration method (VIM), 46–58
Viscoelasticity, 113, 144, 275

W
Weak self-adjointness, 5, 10–11, 13–14, 23–33
Weierstrass theorem, 306

MIX
Papier aus verantwortungsvollen Quellen
Paper from responsible sources
FSC® C105338
FSC
www.fsc.org

If you have any concerns about our products,
you can contact us on
ProductSafety@springernature.com

In case Publisher is established outside the EU,
the EU authorized representative is:
Springer Nature Customer Service Center GmbH
Europaplatz 3, 69115 Heidelberg, Germany

Printed by Libri Plureos GmbH
in Hamburg, Germany